WITHDRAWN 134778

```
R        The History of
737      medical education.
.H53

              134778
```

UNIVERSITY LIBRARY
Governors State University
Park Forest South, Il. 60466

not weeded 1/2008

UCLA FORUM IN MEDICAL SCIENCES

Victor E. Hall, *Editor*
Martha Bascopé-Espada, *Assistant Editor*

EDITORIAL BOARD

Forrest H. Adams William P. Longmire, Jr.
Mary A. B. Brazier H. W. Magoun
Carmine D. Clemente C. D. O'Malley
Louise M. Darling Sidney Roberts
Morton I. Grossman Emil L. Smith
Reidar F. Sognnaes

UNIVERSITY OF CALIFORNIA, LOS ANGELES

THE HISTORY OF MEDICAL EDUCATION

UCLA FORUM IN MEDICAL SCIENCES
NUMBER 12

THE HISTORY OF MEDICAL EDUCATION

An International Symposium held February 5-9, 1968
Sponsored by the UCLA Department of Medical History, School of Medicine
Supported by the Josiah Macy, Jr. Foundation

EDITOR

C. D. O'MALLEY

UNIVERSITY OF CALIFORNIA PRESS
BERKELEY   LOS ANGELES   LONDON
1970

In Homage to

Dr. and Mrs. John A. Benjamin

CITATION FORM
O'Malley, C. D. (Ed.), *The History of Medical Education*. UCLA Forum Med. Sci.
No. 12, Univ. of California Press, Los Angeles, 1970

University of California Press
Berkeley and Los Angeles, California

University of California Press, Ltd.
London, England

© 1970 by The Regents of the University of California
Library of Congress Catalog Number: 72-85449
Standard Book Number: 520-01578-9
Printed in the United States of America

PARTICIPANTS IN THE CONFERENCE

C. D. O'MALLEY, *Chairman* and *Editor*
Department of Medical History, UCLA School of Medicine
Los Angeles, California

L. R. C. AGNEW
Department of Medical History, UCLA School of Medicine
Los Angeles, California

LUIGI BELLONI
Istituto di Storia della Medicina, Università degli Studi
Milan, Italy

JOHN Z. BOWERS
Josiah Macy, Jr. Foundation
New York, New York

CHARLES COURY
Histoire de la Médecine et de la Chirurgie, Faculté de Médecine
Université de Paris
Paris, France

JOHN FIELD
Department of Medical History, UCLA School of Medicine
Los Angeles, California

MIRKO GRMEK
Centre National de la Recherche Scientifique
Paris, France

FRANCISCO GUERRA
Wellcome Institute of the History of Medicine
London, England

SAMI HAMARNEH
Division of Medical Sciences, Smithsonian Institution
Washington, D.C.

PIERRE HUARD
Laboratoire d'Anatomie, Faculté de Médicine
Université de Paris
Paris, France

NANDKUMAR H. KESWANI
Department of Anatomy, All-India Institute of Medical Sciences
New Delhi, India

WOLFRAM KOCK
Medicinhistoriska Museet
Stockholm, Sweden

FRIDOLF KUDLIEN
Institut für Geschichte der Medizin und Pharmazie
Christian-Albrechts-Universität
Kiel, Germany

ERNA LESKY
Institut für Geschichte der Medizin der Universität Wien
Vienna, Austria

G. A. LINDEBOOM
School of Medicine, Free University
Amsterdam, The Netherlands

WM. FREDERICK NORWOOD
Department of Legal and Cultural Medicine, Loma Linda University
Loma Linda, California

F. N. L. POYNTER
Wellcome Institute of the History of Medicine
London, England

HANS H. SIMMER
Departments of Obstetrics and Medical History, UCLA School of Medicine
Los Angeles, California

CHARLES TALBOT
Wellcome Institute of the History of Medicine
London, England

FOREWORD

There is at present a great interest in the problems of medical education, and many medical schools have recently revised or are in the course of revising their curricula in an effort to solve those problems presented by medicine's perplexing and ever changing demands. It was in recognition of this strong current interest that the Department of Medical History, with the support of the Josiah Macy, Jr. Foundation, organized and held an international symposium on the history of medical education in the Bio-Medical Library, 5-9 February 1968. The group of distinguished medical historians who participated in the symposium had been invited as authorities on the subject as it related to national areas or chronological periods, but the history of medical education is so extensive that despite nineteen presentations some *lacunae* were inevitable. Indeed, some of the participants were faced with the problem of treating the subject in a single nation either broadly but without detail, or of selecting certain aspects for consideration in depth. Nevertheless the total body of information presented here represents a very real contribution to knowledge of medical education through the centuries and contains much material not readily available or in fact not available at all elsewhere. One need only examine Theodor Puschmann's *Geschichte des medicinischen Unterrichts von den ältesten Zeiten bis zur Gegenwart* (1889), the only preceding general work on the subject, to realize the achievements of the symposium. Not its least important lesson, incidentally, is the fact that the present concern with medical education, the scrutiny and revision of training courses and methods of teaching, do not represent a solely modern phenomenon. The problem has existed from the time of medicine's origin.

    The symposium would not have been possible except for the generous support of the Josiah Macy, Jr. Foundation to which sincere thanks are herewith expressed, as well as to all those in the School of Medicine, and especially the Historical Division of the Bio-Medical Library, who helped to make the occasion a pleasant and successful one.

C. D. O'M.

CONTENTS

The Earlier Period in the West

Medical Education in Classical Antiquity
Fridolf Kudlien — 3

Medical Education and Practice in Medieval Islam
Sami Hamarneh — 39

Medical Education in the Middle Ages
C. H. Talbot — 73

Medical Education During the Renaissance
C. D. O'Malley — 89

The Modern Period

Italian Medical Education After 1600
Luigi Belloni — 105

The Teaching of Medicine in France from the Beginning of the Seventeenth Century
Charles Coury — 121

Principles and Problems of Medical Undergraduate Education in Germany During the Nineteenth and Early Twentieth Centuries
Hans H. Simmer — 173

Medical Education in the Netherlands 1575-1750
G. A. Lindeboom — 201

The Development of Bedside Teaching at the Vienna Medical School from Scholastic Times to Special Clinics
Erna Lesky — 217

Medical Education in England Since 1600
F. N. L. Poynter — 235

Scottish Medical Education
L. R. C. Agnew — 251

Medical Education in Scandinavia Since 1600
Wolfram Kock — 263

Eastern Europe and the Far East

The History of Medical Education in Russia
M. D. Grmek — 303

Medical Education in India Since Ancient Times
N. H. Keswani — 329

Medical Education in South-East Asia (Excluding Japan)
Pierre Huard — 367

The History of Medical Education in Japan: The Rise of Western Medical Education
    John Z. Bowers   391

Western Hemisphere

Medical Education in Iberoamerica
    Francisco Guerra   419

Medical Education in the United States Before 1900
    Wm. Frederick Norwood   463

Medical Education in the United States: Late Nineteenth and Twentieth Centuries
    John Field   501

Index   531

THE EARLIER PERIOD IN THE WEST

MEDICAL EDUCATION IN CLASSICAL ANTIQUITY

FRIDOLF KUDLIEN
Christian-Albrechts-Universität
Kiel, Germany

It is with some hesitation that I undertake to deal with the present theme in view of the excellent studies of the late I. M. Drabkin (1), but I shall, I hope, present something more than a mere borrowing. In every case, the picture to be drawn must, I think, cover the whole of classical antiquity, and therefore I shall begin not with Hippocrates, as Drabkin did, but with Homer, and end with Galen. Furthermore, my account must have clearly recognizable lines; wherefore I intend, progressing in chronological sequence, to put the following four questions again and again: *Who* taught medicine? *How* was medical knowledge acquired and transmitted? *What* was the content of medical knowledge? *To whom* was medical knowledge given?

I take it to have been of great and stimulating importance for the early Greek that, as for his cultural predecessors, all knowledge, and therewith medical knowledge, came from the gods (2). There was a gulf between gods and men that in some way had to be transgressed or passed by. But in early Greek medicine it was not the priests who acted as intermediaries between gods and men (3); early Greek physicians were not, or usually not, semi-religious shamans. And so the question is: Where or whence one could learn medicine in the Homeric times of ancient Greece? How did such a rational figure as the surgeon-physician of the *Ilias* acquire his special skill, as depicted, for example, in *Ilias*, XI, 515?

The early Greek had two solutions for this problem: If medicine was a manual work, a craft, the physician was like a *"tekton" (cf. "techne")* and could even be so named (4). The source of his learning was therefore merely an empirical one (5). But in so far as his knowledge, be it a technical or a theoretical one, tended to be or to become a science, there was something irrational involved, for the gods alone possessed an infallible science, that is, a "wisdom" or "sophia." A man gifted with such wisdom, as for instance a poet, must have had a "revelation" (6). This seems to be one of the reasons why medicine could have a supernatural or mythical origin for the early Greeks. Apollo "donated the remedies to the Asclepiads, that is, to the physicians," or Chiron, son of Kronos, "instructed" young Asclepius in medicine (7). In the same sense early Greek physicians were "Asclepiads," descendants and disciples of a hero or even a god (8).

To sum up, in early Greece medical knowledge was acquired either empirically and transmitted from man to man—usually, perhaps, from father to son, or to an elect who received it from a supernatural being and eventually allowed a disciple, as an elect, to participate. In every case medical teaching was oral, at least until the sixth century B. C. Under such circumstances, two different types of physicians were developed: the resident or wandering craftsman, representative of a rational *"techne iatrike,"* like the famous physician Democedes of Crotona, son of a physician; or, on the other hand, the "sacred" physician or *"iatromantis,"* eventually a kind of shaman such as Empedocles was in some degree.

Thus we have found a first answer to our questions number 1 and 2, concerning the *who* and *how* of early Greek medical teaching. Our question number 4 can be answered here only vaguely; we may guess that in early Greece the higher social classes exclusively were allowed to learn medicine. Podaleiros and Machaon, the exemplary physicians of the *Ilias,* were noblemen. A man of lower rank, and especially a slave, was certainly prohibited from becoming a physician, and in this respect cultural tradition may have been decisive.

Our question number 3, however, concerning the *what* of medical knowledge, can help to reveal a very distinct situation. In his dialogue *Politeia* (405 C, 8 ff.) the conservative-minded philosopher Plato, who had his own opinion about the social role of medicine, stated that early Greek medicine was nothing other than traumatology and epidemiology and that Asclepius intentionally "demonstrated" to his disciples only such a narrow knowledge (*op. cit.*, 406 C, 1 ff.). Dietetics, that is, the knowledge of non-traumatic or internal diseases, was, as Plato declared, involved with old-fashioned traumatology in later times. His statement seems to be in accord with the historical situation as far as we can recognize it. Medicine in Homer was indeed nothing other than traumatology or, at best, contained a bit of epidemiology. It was this alone that could be perceived rationally and therefore could be taught and learned by men alone. On the other hand, internal diseases could not yet be understood or, in consequence, treated in a rational form— they belonged to the divine jurisdiction.

This is precisely the lacuna early Greek "physiologists," from the sixth century B. C. onwards, were trying to fill. We have in mind the well-known pre-Socratic *naturphilosophen,* the most important of whom for our purpose was Alcmeon of Crotona (*c.* 525 B.C.). This man, who is said to have been in his youth a personal attendant to Pythagoras, thought and spoke not only about "physiology" but "mostly about medical matters" (9). It was he who gave the first definition we know of internal diseases (fragment B 4, ed. Diels-Kranz), and he did not fail to recognize the audacity of his thinking, as is demonstrated by his own words (fragment B 1): "The gods alone have certainty . . . about unseen things. For men there remains nothing but guess [*tekmairesthai*]." This statement was fully characteristic of the limits of early Greek thinking relative to recognition of the inner processes of the human body and the nature of internal diseases.

It is not known if Alcmeon, though so strongly interested in medicine, was himself a physician, but it is beyond question that he was a philosopher. Moreover, he was the first to write a "scientific" book in Greek prose concerning medical or natural themes. This immediately brings us to two problems of importance for medicine of the fifth century B.C. At first, it was absolutely necessary for this amplified medicine—now comprising internal medicine as well as surgery—to possess not only a technical but also a theoretical foundation. Under the circumstances the physician could learn the theories from no one except a philosopher—unless he was himself a philosopher. That is to say, the philosopher had followed the god as a teacher of theoretical medicine, giving therewith a new answer to our question number 1. Not before Hippocrates who "separated medicine from philosophy," as Celsus stated in a more or less symbolic way (*De medicina*, "prooemium"), did the physicians take the practice *and* the theories into their own hands (10).

The other problem arose with the new possibility for teaching medicine from books. Here is a second answer to our question number 2. There now existed the books that constitute the oldest part of the Corpus Hippocraticum, stemming from the two concurrent medical schools of Cos and Cnidos. This birth of a Greek medical literature, books which were not concealed in the secret archives of priests or in libraries of esoteric schools, was highly influential upon the development to come, provoking many further problems and consequences. Antiquity was fully aware of this, as is evident from the following statement of Galen (Kühn, II, pp. 280-282): In ancient times it would have been superfluous, as he declares, to write medical books (*hypomnemata*), since medicine was in the hands of families of physicians and one had to learn it during childhood, in the course of daily life. But in later times strangers came into the medical profession, and then the good old family teaching, begun in childhood, disappeared (11), and one needed books for teaching as well as for retaining medical knowledge (*theoria*).

True, this statement of Galen sounds somewhat romantic and looks with more than one eye upon the dilemma of contemporary medicine. But despite this, it undoubtedly touched upon two further cardinal problems of the new medicine of about 400 B.C.: How must this medicine think about the categories "esoteric" and "exoteric," a problem we shall discuss later. And what was, in fact, the value of the rapidly increasing number of medical books (*cf.* Xenophon, *Memorabilia*, IV 2, 10) as instruments for teaching and as a kind of bridge between theory and practice? This last point was a very complex one, as Aristotle, himself the son of a physician, tells us (*Ethica Nicomachea*, 1181b, 2 ff.): "Clearly you do not become a physician by books (*syngrammata*). Nevertheless the writers of books try to describe not only the remedies but the general and special methods of therapeutics, too, with respect to the individual case. That would be useful for the skilled man, but the untrained one gains no use from it" (12).

The greatest difficulty lay in the fact that the new medicine did not yet have a *communis opinio* about the *how* and the *what*. That is to say, in the fifth century B.C. the old-fashioned surgery and the "modern" dietetics fought a major battle over the question of the better medical basis. The old-Hippocratic surgeon and author of *De fracturis*, for instance, wished to prove (ch. 31), *mutatis mutandis*, that all diseases could be considered as wounds or "helkea"; as he says, "in many respects the things here are like brother and sister." And so for him a single, and especially traumatological, observation could be a very useful dogma (*mathema*) of *general* importance. In contrast, the old-Hippocratic author of *De diaeta in acutis* sometimes brought up certain doctrines (*didakteria*) inspired by internal medicine which he declared to be "like brother and sister," because being clinical matters they were of general importance within the whole of medicine. In consequence the question was to be put: Is medicine, methodically considered, more surgical or more dietetical? The words "*mathema*" and "*didakterion*" and their use in this medical literature (13) seem to prove that here, as for example in the two books quoted—which were certainly also, and in some degree even primarily, addressed to medical apprentices—the very basis of medicine was under discussion. Hence it was, in fact, a didactical discussion among the physicians themselves, and not among or for their disciples. Thus Aristotle was evidently right, and we can add to his statement a further remark: The medical textbook as such was not only a problem, but it reflected at the same time the methodical uncertainty of the whole of medicine about 400 B.C. It is in no way surprising that there were even physicians who questioned if medicine could be taught at all. And under the circumstances it seems absolutely natural that there was sometimes resignation or even nihilism among them concerning this question (14).

Here I should like to remind the reader that in discussing early Greek medicine and the struggle for the best basic method, one must not think of any term like "specialization." There were in Greece no medical specialists as there were in Egyptian medicine (*cf.* Herodotus, II, 84). Older Greek physicians were trained with a view to a *uniform* medicine, whatever might be its constitution. Aristophanes who caused a character in his comedy *Ecclesiazusai* to ask (v. 363 f.): "who brings to me a physician, and which? Who is skilled about the region of the anus?" was certainly not a wholly credible witness for the existence of Greek proctologists in the fifth century B.C. (15).

This century was primarily under the influence of the sophists, filled with a strong urge for teaching, and all was believed to be teachable, even the social virtues. With such a claim the sophists, wandering all over Greece, required payment for their teaching. The physicians may have handled things in a similar manner. In every case one could, if Plato was correct (*Protagoras* 311 B 1-C 1), learn medicine from Hippocrates himself for payment. At least it seems to be clear that from the end of the fifth century B.C. onwards nobody other than a physician taught the new "*techne iatrike.*" We cannot say if the relations between medical teacher and medical apprentice were generally fixed in the form of a contract (16). But if so, this happened certainly not in so peculiar a way as described in the pseudo-Hippocratic "Oath" (*cf.* below and note 17 for Edelstein's interpretation). The teaching went on either within an ambulant praxis or in an "iatreion," located in a city and belonging to the city-magistrate or to a physician. Here the apprentices lived (*cf.* Aeschines, *Timarchus,* 40) and were instructed in practice by oral instruction and from books. We may recall that such an "iatreion" was originally a *surgical* practice-room, as some of the Hippocratic writings prove, such as *In the surgery* and the much later *On the physician.* Although medicine in the fifth century had also become internal medicine, nevertheless the "iatreia" perhaps continued to practice primarily surgery. Internal diseases may have been treated in the homes of the patients themselves. At best such remedies as emetics or purgatives were given in the "iatreion" (*cf.* Plato, *The Laws,* 646 C 2f.).

Given the circumstances of Greek medicine about 400 B.C., the impetus for teaching must have provoked not only methodical but social problems. What Galen deplored as the breakdown of the old medical family was, in other words, a developing "exoteric" educational procedure. For now it was not exclusively the sons of physicians who studied medicine, and because of this so radically altered situation we suspect that some conservative-minded physicians would not have been disinclined to see medicine as exclusive, that is, as an esoteric field. Indeed, we can recognize about 400 B.C. two extremely opposite positions: the pseudo-Hippocratic "Oath" has the almost overemphasized character of a document speaking for an esoteric family or club-circle, a character in which the "Oath" stands alone within the Corpus Hippocraticum (17). On the other hand, the treatise *De affectibus,* addressed deliberately to laymen, even propagandizes a kind of popular medicine (18). Setting aside such extravagant positions, one must admit that there must have been a very urgent need to gain a clear opinion relative to our question number 4: *Whom* could a Greek physician, about 400 B.C. accept as his disciple? This problem was the more important because of the lack of any governmental control over physicians. True, we may guess from a passage of Xenophon (*Memorabilia,* IV 2,5) that a man who wished to become a "public physician" had to name his teacher and account for his medical training, but in no way were all physicians "public" (19). Be that as it may, it was in every case of primary interest for the regular medical profession to put itself at as great a distance as possible from charlatans or other non-serious "medical" persons.

What personal and social qualifications, then, were required for a man who intended to become a physician? Galen, speaking about the classical epoch of Greek medicine, vaguely refers to the *"arete"* of such persons (see above, p. 5). Later Hellenistic writings of the Corpus Hippocraticum such as *On the Physician* and *The Law* discuss this question more substantially. Regarding the most general claims of such books to have been old ones, we can say that it was the famous Greek value-concept of *"kalokagathia,"* that is, a high quality of body and mind as well as of social status, which the future physician had to possess.

Considering the socio-cultural aspects of the selection of future Greek physicians, it must be emphasized that in Greece about 400 B.C. it was impossible for two groups of persons to become physicians: women and slaves. Women could be concerned only with midwifery, although in this field they were highly esteemed because of their special skill (*cf.* Hippocrates, *De carnibus*, ch. 19, and *Gynaikeia*, I 68). As midwives, a status in which they might know gynecology as well, women were especially designated as *"maiai"* or *"maieutriai"* and had their own tradition and profession. But the books they must have needed for instruction seem to have been written usually by male physicians and not by the midwives themselves (20).

At that time slaves were distinctly prohibited from becoming physicians (for slaves as lower assistants of physicians, see below). Moreover, there exists even an explicit literary testimony that slaves at Athens were forbidden to study medicine. The judgment of modern experts in this field seems unanimous (21), but in spite of that, there appears to have been an obstacle to such a unanimous judgment for the fourth century. One may recall that Plato, himself a contemporary, alleges the existence of two absolutely different classes of physicians, that is to say, one class of freeborn physicians which had to treat free patients, and another class of slave-physicians for slave-patients, with wholly different therapeutics for the two classes of patients (*cf.* Plato, *The Laws*, 720 A-E, 857 C-D, and *Politeia*, 405-409). However, examining this "description" carefully one comes to the conclusion that Plato neither gives here "a picture of conditions as they existed in Athens in the 4th century B.C." nor "a recommendation or an ideal" (22). On the contrary, his picture turns out to be nothing but an allegory which in no case is to be taken literally nor is it a historical testimony in the sense of the word and can only be understood in its Platonic context. Having discussed this problem in greater detail elsewhere, I can only draw attention now to the misconceptions modern medical historians fall into by following Plato too carelessly on this point (23). It must be stated once more that there cannot have been in Greece of the fifth and fourth centuries B.C. any slaves who practiced medicine as physicians.

It was quite another thing in regard to the lower assistants, that is, the sometimes so-called *"hyperetai"*. Here Plato was surely correct in saying that these *"hyperetai"* could have been freeborn men as well as slaves (24).

But Plato would have done better not to have said that the *"hyperetai"* were "very likely" called physicians, too (*Laws* 720 A 8). On the contrary, in ancient Greece the *"hyperetai"* were fully distinguished from physicians. Within surgery, for instance, they could play no part other than that of assistant, working "in silence and attentively" under the direction of the "superior" physician (*cf.* Hippocrates, *De officina medici*, chs. 5-6). They were not allowed or able to act in any way voluntarily. In the Corpus Hippocraticum such subordinates figure only within the field of surgery (25). For later times, Galen (26) suggested a larger role for them. Then their profession was described as *"rhizotomoi"* (herbalists) and "unguent-makers" as well, and the *"hyperetai"* had to undertake compresses, clysters, phlebotomy and scarification. In earlier times all these activities, or at least some of them, were occasionally performed by the physician himself—but if a *"hyperetes"* performed them, he was in every case under the supervision of a physician. The *"hyperetai"* as such were not allowed to transgress the boundaries of their status. On the other hand, it seems very unlikely that the medical apprentice, that is, the future physician, was in any way identified with such a *"hyperetes"*. There was, it appears, no reason or justification for it.

As has been said before, the new *"techne iatrike"* had to see to the careful selection of medical apprentices, since under the circumstances there was danger of "profanation" of the medical profession. A sharp distinction was drawn between physicians, be it regular or future, and non-physicians as *"hyperetai"* or, on the other hand, "non-serious" persons, not only within their own medical area (27) but also outside the medical profession. These last were persons who practiced without being trained in any form of the medical art as such. From early times onwards there were "magicians, religious absolvers, beggar-priests and charlatans" (*cf.* Hippocrates, *Sacred disease*, ch. 1) who treated diseases in an entirely irrational manner but, nevertheless, with noteworthy success, at least in the view of the run of the patients. Furthermore, there were some less mysterious healers such as the so-called *"pharmakopolai"* who both sold remedies and treated diseases. They could be ill-reputed as charlatans without being handicapped in their activities (28). Since this was so, there had grown up a kind of vulgar and non-serious "para-medicine" the effects of which, objectively viewed, were not always positive apart from the fact that the physicians as "rivals" in every case tried to promote the bad reputation of such persons. Perhaps, too, matters were not always as correct as they should have been among the *"iatreia"*. Indeed, we have some testimonies concerning the somewhat non-serious character of such *"iatreia"* in the fourth and third centuries B.C., a circumstance which would make such an *"iatreion"* or its practitioner not so different in degree from a medieval barber-surgeon (29), or, on the other hand, a *"pharmakopoles"* since the latter seems to have had some traits not very dissimilar from those of the barber-surgeon.

Such were, to sum up, some of the tendencies and dangers the regular and serious medicine of the time had to avoid, the more so since about 400 B.C. there was, as has been said, neither a governmental control nor a genuine and professional institutionalization for physicians (30). At best, there were medical schools, but one should not imagine for the early period a fixed corporation like that, for instance, of the later Herophileans. Originally such a "school" was a more or less loose congregation of physicians all practicing at the same place, as was the case of the "Crotonian and Cyrenaic physicians" of whom Herodotus speaks (III 131). The most famous of all were, of course, the schools of Cos and Cnidos. Here we can see, at a relatively early stage, a "head" of the school (Hippocrates in Cos and Euryphon in Cnidos), and we can recognize certain established doctrines which the students had to know and to follow. There is no question that disciples were taught in these schools, and we know some of the disciples of Hippocrates himself by name: Dexippos, Apollonides. Here attention ought to be drawn to an error which continues to exist: there was never a *school* of "dogmatists" in the precise sense. Nor was Hippocrates the real *"archegetes"* of a school of dogmatists, nor does it make any sense to call the post-Hippocratic physicians of the fourth century B.C. by this name (31).

Of course, we should like to know something substantial about medical education within the great schools, especially that of Cos. But the source which alone could inform us in detail on this matter—the inscriptions of Cos —remain unpublished (32), so that just at this decisive point we must confess to a restriction in our delineation.

As it was an actual question, about 400 B.C., who should learn medicine and who not, so at the same time there began to be another question that called for attention. Who would be, in every respect, the best type of physician? (33). Or, in other words, what was the basic theoretical knowledge a physician must possess to be highly esteemed. Considering the Greek discussion on this question, we must not forget that theoretical medicine had been for a time more or less in the hands of philosophers. In general, medicine as such had fluctuated between science and craft (34). Under these circumstances discussions and quarrels were not to be avoided, and it is clear enough that a physician who maintained the standpoint of the Hippocratic author of *On ancient medicine* must contradict the opinion (ch. 20) that a physician *a priori* ought to be trained in "natural science" in respect to the human body, in order to be able to treat the body effectively. The author of *On ancient medicine* believed such knowledge to be out of reach for the moment and, in any case, not necessary *in principle* for treating patients. On the other hand, it was natural that philosophers strongly interested in medicine would propose a type of physician who was a good scientist and philosopher as well, and, perhaps, primarily the latter. In the *Laws* (720 A-E, 857 C-D) Plato sharply contrasted the type of medical craftsman—crude and uncultivated, as he declares—to the philosophically inspired medical scientist (35). Plato's disciple Aristotle likewise advocated the "educated" (*charieis*) physician, opposing him to the *"hoi polloi"* (36). In consequence there were imagined, in a somewhat schematic way, three types of men with a knowledge of medicine: the simple craftsman (*demiurgos*); the physician educated as a scientist (*architektonikos iatros*); and the educated layman who had some medical knowledge (*pepaideumenos peri ten technen iatriken*) (37). The question, then, was which of these types was to gain the field and was, herewith, to provide a model for medical education in the future.

Although there never was a definite decision on this question—indeed, the same man could be inconsistent on this point, as Plato demonstrates (*cf.* note 35)—one can state nevertheless that, at least for the immediate future, the "scientific" type was the more attractive. This may have been influential in some degree as Cohn-Haft points out (*op. cit.*, p. 24 f.), in regard to a certain lack of physicians in the Hellenistic medical schools—at least in the earlier ones—since the conditions and requirements for becoming a "*charieis*" physician were undoubtedly much more difficult than in Roman times (*cf.* below) (38). And so, although Hellenistic medical schools were certainly not rare, as Cohn-Haft believes, the number of "educated" physicians may indeed not have been great. But if true, it could not have provoked such substantial and urgent problems as they were to be found later in Roman society where there seems to have existed, at least for a time, a true lack of physicians.

In the long run, the standard of the "*charieis*" physician must have produced a new need for medicine and medical education. There was need for anatomy as a genuine way to the correct and exact knowledge of the human body. It is true that in classical antiquity there were obstacles to a thorough awareness that only systematic dissection of human bodies could lead away from the *naturphilosophische* speculations and towards an objective knowledge of the normal and pathologic organism, founded on phenomena alone. To be sure, we are told that old Greek philosophers like Alcmeon and Democritus dissected, but this was dissection of animals, the results of which these men believed to be, by analogy, valid for human bodies. On the other hand, it is well known that the anatomical knowledge of the old Hippocratic physicians was slight. Physicians as well as philosophers failed to recognize clearly the scientific value of knowledge of human anatomy.

The stimulus to overcome this failure arose, at the same time, among both physicians and philosophers. The old Hippocratic surgeon and author of *De articulis* claimed emphatically that "You must *first* know the anatomical structure (*physis*) of the spine, how it is constituted—that may be important in regard to many diseases" (ch. 45). Here, as elsewhere in the same treatise (chs. 11, 30), the physician was required to gain a proper anatomical knowledge which could satisfy the demands of *clinical* medicine (which means, obviously, an anatomical knowledge limited in some degree, but conceived otherwise strictly as *human* anatomy). Where, however, could a Greek physician gain such knowledge, since it was impossible in classical

Greece to dissect a human body, and since there was no special literature on human anatomy? Under the circumstances it was a matter of mere chance if one were to meet with related anatomical phenomena.

Hence it must be said that the first Greek anatomical monograph was a great step forward in discussion and even the teaching of anatomy. This was the work of the famous physician Diocles of Carystus, called a "second Hippocrates" (*cf.* Galen, Kükn ed. II, 282). For us, and especially for our theme, it would be very interesting to know if Diocles, writing such a book, was inspired by philosophers. We must recall in this connection a thesis of Werner Jaeger according to which Diocles was a personal attendant of Aristotle and, therefore, himself a Peripatetic philosopher. Beyond any question, the Peripatos was highly interested in anatomy—were the Peripatetics then, in consequence, the real inaugurators of *medical* inquiries in this field? Had the philosopher now become the "anatomical teacher" of the physicians?

As far as the situation can be judged this must be denied, and it must now be said that Jaeger's thesis has thoroughly failed to find a firm and acceptable basis (39). The time of Diocles must be located around 360 B.C., and neither he nor any other physician of the fourth century appears likely to have been a disciple of a philosopher (40)—that there were interdependencies and contacts between physicians and philosophers is, on the other hand, a fact as undoubted as it is irrelevant to this particular question. We must assume, therefore, that Diocles' anatomical inquiries were stimulated by the needs of medicine as well as, and perhaps even more, by general philosophical or scientific tendencies of contemporary thought. To that degree they were integrated firmly within the framework of medicine itself.

At the same time it must be said that even Diocles' anatomy was based upon the dissection of animals and followed that false analogy previously described. By what factors, then, was anatomy forced to use human subjects? Here we must note the strong interests in systematic and comparative anatomy among the Peripatos as inaugurated by Aristotle, son of a physician. About 300 B.C., furthermore, there was widespread tendency towards what one may call strict concentration upon exact observation of phenomena. Last but not least was the foundation of a new metropolis, Alexandria, the melting-pot of peoples, behaviors, and ideas so altogether different, a non-Greek city without any fixed traditions or prejudices.

It is known that Clearchus of Soloi, formerly a personal attendant of Aristotle, wrote about 300 B.C. a treatise *peri skeletōn* which contained a description of human bones and muscles and their nomenclature (41). At that time the word "skeleton" meant at best a mummy. However, with a lack of mummies as well as of dead bodies to be inspected and dissected in Greece itself, where could Clearchus—or his source—have found the material for such a book except in Egypt, that is to say, in Alexandria where Ptolemy I was devoted so deeply to Greek, and especially Aristotelian, science and had created, at the end of his life, the world-famous "Mouseion." A contemporary of Clearchus was Herophilus of Chalcedon who inaugurated in Alexandria the first human anatomy in the true sense, based on systematically performed dissections of human bodies—perhaps living as well as dead. Seen in this way, medicine and philosophy stood side by side—each of them independent of the other—at the birthplace of human anatomy.

Therewith, no doubt, anatomy had immediately to become a genuine part of medicine and medical knowledge. Moreover, since the "Mouseion" was a place not only of research but also of teaching (42), anatomy now became an accepted part of medical education.

Here again is a new answer for our standard question number 3, concerning the *what* of medical education. But for the moment, before discussing the further vicissitudes of Greek anatomy as a part of medical teaching, we must consider briefly our questions number 2 and 4 in the light of the Hellenistic epoch. The structure of the "Mouseion" as a place of research, teaching, and way of life as well for many scholars, was certainly influential in producing a type of teaching which can be called the lecture-type. Two centuries later Apollonios of Cition (*c.* 50 B.C.) tells us that in Alexandria one "sat before" a teaching physician, hearing and looking upon him (43). Later we shall have to discuss the fact that and also how, in Alexandria, such a lecture, which seems originally to have been a rather lively give and take between teacher and disciples, came to degenerate gradually into a desiccated, formal "reading" by an "authority" or, at best, a rhetorical display. In every case, from the third century onwards Alexandria was, on the whole, the school for the more ambitious physicians as well as for other scholars and professions. In antiquity one could say that "the Alexandrines are educating all the Greek and Barbarian people," and such a remark could be made just at the moment when, in the last decades of the second century B.C., the Alexandrine scholars—and among them the physicians—were expelled by Ptolemy VIII. For these people now filled "islands and cities [in Greece], and being urged on by fear for material existence, they taught all their knowledge and produced many educated people" (44).

Now we must try to find an answer to the question whether or not in Hellenistic times there were any slaves as physicians. This was believed by not a few scholars, because of a statement in the so-called *Suda* (Suidas) and a single inscription. In the *Suda*, Aristogenes of Cnidos is said to have been the slave of the stoic philosopher Chrysippus as well as the personal physician of King Antigonos Gonatas (45). But modern experts such as Forbes (46) strictly deny the existence of slave-physicians in the pre-Roman Greek world, calling attention to the fact that the testimony of the *Suda* is obviously confused in itself, and therefore, not a credible witness, and that the only related inscription, a Delphic manumission of 155 B.C., cannot be interpreted as a testimony for a slave-physician (47). Lacking further related testimonies or references, we must assume that it continued to be impossible in Hellenistic times for slaves to learn medicine and to become physicians.

Let us now return to the anatomy which had become a new part of medical knowledge in the Hellenistic period. By its aid the physician could become genuinely "educated," that is to say, a man who really knew the things of nature, human and general, according to the standard Plato and Aristotle had proposed. In view of this, it is very remarkable that immediately after the time of Herophilus anatomy, performed as a systematic dissection of human bodies, as well as its value for medicine, was strongly opposed by a group of physicians, the so-called "Empiricists", a school which was in fact inaugurated by a former disciple of Herophilus, Philinus of Cos (48).

According to the oldest report about Empirical doctrines we possess, that of Celsus (49), these physicians declared nature to be unintelligible (*incomprehensibilis*) and, in consequence, any research concerning the "hidden natural processes" to be in vain. It is clear that, following such presumptions, the Empiricists not only must have driven anatomical and physiological research out of the field of medicine, but also abandoned the Platonic-Aristotelian ideal of the educated physician. Instead of this, as they said, medicine was to be learned most correctly and infallibly through treating patients and so gaining experience and skill; it could continue to be learned and practiced only in the same way. In accordance with this view, the Empiricists developed their three main doctrines, the first two of which were "*empeiria*" and "*historia.*" For want eventually of a direct course to "*empeiria*" or autopsy—as it might be only during the period of training—one must have "*historia*", that is, the observations collected by others (50). In this connection the Empiricists declared the following in detail: Only "anatomy by chance", such as the relation of observations from the body of a wounded man was permissible or even necessary for the physician. But researches performed *ad hoc* were not allowed for two reasons: If the object of such anatomical inquiry were a living human body, the inquiry itself was cruel and, moreover, gave a distorted picture not corresponding to the real situation. If, on the other hand, the object were a cadaver, anatomical inquiry was disgusting, and the picture to be gained was likewise distorted since there was a fundamental difference between a dead and a living body.

One may readily imagine how influential such a standpoint could have been and, indeed, was for further development. It provoked a veritable splintering of medicine into opposed doctrines. For there were, of course, Herophileans and Erasistrateans who had, following their masters, a totally different opinion of the value and range of anatomy and, generally speaking, of theoretical medicine. Lacking any "official" norm of studies, all the opposed schools produced differently educated types of physicians. We must confess that this was a great pity because uniformity of medical education was under the circumstances highly desirable. Take, for example, the question of whether or not anatomy ("pure anatomy") belonged to medical education. There must now have been physicians who, because of the education program of their school, intentionally remained unconcerned about any then existing scientific anatomy and theoretical medicine within the natural borders of ancient science, as such, a serious possibility to gain such knowledge. In every case, anatomy should have been much discussed within medicine in the two pre-Christian centuries, as Apollonius of Citium stated (51).

But the Empiricists had a further decisive influence. Since the *"historia"*, that is, the use of observations made by others, was necessary for their doctrine and training program, they found such observations mostly in the Hippocratic writings. The Empiricist physicians tended to avoid strenuously any original and independent literary production. To them, Hippocrates seemed to be nearly infallible, the "most divine" man (52). By such presumptions and by developing the first—and for the Empiricists, first-ranking Hippocratic exegesis—they undoubtedly fostered a tendency towards increasing degeneration of Greco-Roman medicine into a sterile and repetitive medical commentary. To such a medicine, the teaching from books was, of course, no problem at all, as, in contrast, it had been to the older physicians. It seemed even possible, without any scruple, to instruct laymen in complicated medical matters. Thus, for instance, Apollonius of Citium could at request write a treatise about dislocation and setting of joints, addressed to one of the Ptolemy's, a book that pretended rather self-consciously to demonstrate these things "in a written form with pictures" (*cf.* Apollonius of Citium, *op. cit.*, p. 14 1.9), although in fact it was almost nothing other but a collection of Hippocratic sentences.

To sum up, the traditional, uneasy position of medicine which had to stand by itself, with some misgiving between science and craft, or, in other words, between theory and practice, had not yet found an effective solution. Instead, it seems paradoxical that only the Empiricists, with their love of commenting, had taken a great step away from a desirable balance of powers. What a young physician needed to know in order to become at least an efficient practitioner must have been less clear than at any time before. In view of this confusion, there is obvious significance in a Greek papyrus of the first century A.D. (53), of which the text seems to be based upon Archibios' statements concerning medical education. The physician Archibios, most likely living in the first century B.C. (54), speaks here of the training of surgeons and complains that the beginning was wasted in fruitless discussions over the origin (age) and position of surgery in comparison to internal medicine (55). He proposed that instead the students ought to be instructed in the names of diseases and then their education be limited *a priori* to the "necessary" things alone, sweeping aside all "discussion about theoretical problems [*problematikos logos*]."

Surgery is, and must be, the most craft-like part of medicine, and so it was, as we have seen, from older times onwards in some contrast to internal medicine. Since surgery was already handled in an overly theoretical way— and this, moreover, during the training period—the practical side of medicine as such may then have been handicapped and not without danger (56), even admitting that Archibios had in some degree exaggerated.

Under the circumstances it seems evident that in later Hellenistic medicine the never totally extinguished connections between medicine and philosophy became rather close again. Plato's and Aristotle's claim for the "scientific" physician could be interpreted in a twofold way, as the further development of Hellenistic medicine indicates: Either the physicians ought to be concerned for a genuine natural science, that is, a medicine including anatomy and physiology, as were Herophilus and Erasistratus; or the physicians ought to "philosophize" in a more or less speculative sense. The Hellenistic, pseudo-Hippocratic treatise *De decenti habitu* (ch. 5) proposed "bringing philosophy into medicine and medicine into philosophy, since the physician, being a philosopher, is godlike—for there is almost no difference between both," that is, philosophy and medicine. A Greek medical treatise following such a concept can only be understood as influenced by specifically Hellenistic philosophy. In fact, Stoa and Epicureanism declared the philosopher to be godlike (57). Therefore the treatise *De decenti habitu* is, for us, a witness of a certain part of Hellenistic medicine which one may call the "philosophic" in contrast to the "scientific" of Herophilean-Erasistratean and the craft-like part of Empiricist provenance (58).

As it appears, the physician as a philosopher meant a true unity of person and not a combination of two professions or jurisdictions different in themselves, as it was defined by Galen. In contrast, if the things were visualized not so "mysteriously" but in a more realistic and practicable manner, a physician who wanted to know something about philosophy could go to a philosopher and become his disciple. But in such case the philosopher in no way taught medicine—as occurred in Byzantine times (59). Instead, being himself interested in science and medicine, he could give ideas and mental stimuli to an interested physician. In such manner the famous Stoic philosopher Posidonius (first century B.C.), who himself had participated in medical lectures in Alexandria (60), had a connection as "teacher" with the physician Athenaios of Attaleia (61). The medical sect of the so-called Pneumatists was inaugurated under the influence of middle Stoicism and, especially, of Posidonius—an influence we can observe in not a few details (62).

Concerning later Hellenistic medicine and our question number 1, it may be said that the philosopher had become again a "teacher" of the physician in the limited sense that has been outlined. Consequently, in respect to our question number 3, a physician who wished to become really educated, now had to study philosophy, at least if he belonged to a school which had a related program. But, on the other hand, this was in no way obligatory for medicine. In contrast, we can observe that during the turn of the first century B.C. into the first century A.D. the situation must have been the same as it had been in early Hellenism: medicine and medical education were heavily impaired by a new, and perhaps even stronger, splintering into absolutely different meanings and sects. The old gap between science and craft became deeper than ever before. And again anatomy, discussed in this or that direction, was a good indicator of the disunion.

While Herophileans and Erasistrateans praised anatomy highly, and the Pneumatists did more or less the same (63), in total contrast to them, and also in some degree to the Empiricists, there now arose a new medical sect under the influence of Asclepiades of Bithynia—the so-called Methodists. The fact that this sect, roughly speaking, tended to an exaggerated antiscientific simplification of medicine as well as to shameless publicity, may be interpreted by the specific behavior and needs of Roman society and culture which this school wished especially to satisfy. One may recall the hostility with which Roman culture had reacted originally to Greek medicine. This hostility caused the Greek physicians in Rome to seek sympathy and publicity; thus, for instance, they claimed to treat patients in the simplest but at the same time most effective and most pleasant way (*cito tuto jucunde*). Furthermore, one may recall the often cited "practical sense" of the Roman mind which was not so willing to become concerned with philosophy and science. Roman culture itself must have been hostile to things like anatomy, the dissection of dead human bodies. After all, it cannot be surprising that the Methodists, unlike the Empiricists, denied in principle the concept of medicine as science, including each form of anatomy—be it anatomy *per se* or by chance—and physiology (64).

The medical system of the Methodists (65) was based upon a few simple and schematic presuppositions. People had to be treated as masses—that was what mattered, and such was the need of the Roman mass population. Therefore in the first century A.D. the chief Methodist, Thessalus, was greatly interested in attracting as many people as possible to the study of medicine through the promise that they could reach the status of physician within six months (66). True, Galen, who reports this, was a polemicist and not thoroughly trustworthy witness here, as well as in some other matters. But we may dare to say that such principles, existing as they undoubtedly did in some fashion, were at any rate not healthy for medicine as a whole, advancing the decadence of medical education as well as reputation—as Galen is correct in criticizing (67)—and providing an entrance into medicine for not a few unfit persons (68).

Under the circumstances, Rome had a peculiar influence upon the *how* and *what* of medical education. Let us now examine in more detail the question how such influence was effective for our question number 4, concerning the *who*, that is, the object of medical education. As we have seen in older Greek medicine, women were allowed to practice medicine only within very restricted limits. But in Hellenistic times there appeared at once the term *"iatrine"* (= woman physician). Evidently that meant originally, and even in Galen's time, not exactly a woman with equal rights in every respect in medical practice. At first an *"iatrine"* continued to be, as she was in earlier times, a midwife with the role of gynecologist. But it is remarkable that the special term now changed from *"maia"* to *"iatromaia"* and *"iatrine"*. Evidently it seems to have been intended therewith as a kind of social and professional sublimation. In consequence, such an *"iatrine"* could now be given a greater and greater jurisdiction (69). For example, a male physician could write a book for the *"iatrine"*, so entitled, and could address her as a colleague and give her advice. This, as it seems, was done in the first century B.C. by the Empiricist Heracleides of Taras writing to the woman physician Antiochis (70). No doubt within the course of time the *"iatrinae"* (*medicae*) became more and more nearly equal partners of their male colleagues, be it within medical education or professional rights (71).

Now let us come to the other group of persons who, in earlier times, had been excluded from the medical profession—the slaves. Here we must emphasize that in Rome for the first time slaves were allowed to learn and to practice medicine. Several factors may have been contributory to this novelty. The physicians who at first began to practice in Rome were mostly *peregrini*, that is, they had come from Greece. Such Greek physicians may have been imported, at least in part, as prisoners of war and thereby as enslaved persons. On the other hand, there was a specifically Roman *medicina domestica* practiced on the great farms as well as in the urban households by *medici domestici et familiares* or *servi medici*, that is by slaves who belonged to the household (72). Although offering a kind of model for future developments, the *medicina domestica* does not concern the question of admission of slaves into the medical profession (the word "admission" is used in view of the corporation of physicians, but not of any governmental licence), in so far as it seems very unlikely that the *servi medici* learned medicine in another way than as "household-medicine", that is, as roughly empirical treatment with remedies approved and transmitted by old wives or similar persons.

But in this connection a further and very decisive factor must be recalled; from the beginning of the Roman Empire onwards the enormous and, so to speak, explosive increase of population in Rome must have heavily augmented the need for physicians. Already, on the occasion of a famine, the Emperor Augustus had striven to diminish the number of Roman people by expelling all strangers and slaves—except such of them as were physicians (73). The Methodists, as we have seen, tried to overrride the increasing lack of physicians by admitting and training as many persons as possible by instructing them as rapidly as possible. Thessalus, as we are informed explicitly, admitted and trained slaves (74). In consequence, physicians could have many disciples at the same time; and these disciples going in a body with their teacher to the beds of patients could by the character of such a mass as well as by the manners of the individuals produce an unpleasant effect (75).

There is no question but that some negative reactions to such degeneration of the medical profession were to be expected—a degeneration born from a genuine need but declined to a deplorable state of affairs. Since such reactions came from without they caused a remarkable edict of the Emperor

Domitian known to us from an inscription (76). Domitian believed the distress of the medical profession to have been caused by the admission of so many slaves, despite the fact that relatively few persons were able *by nature* to act as physicians. For this the Emperor believed the physicians themselves responsible, since they were covetous of ever increasing teaching fees. Therefore he forbade them to accept slaves as disciples for payment, that is to say, to educate slaves as physicians in the usual manner. This edict, and it must be confessed that it was a rather superficial one, had evidently been intended as a means of purifying the medical profession. Nevertheless, the edict may have been of some help in inducing the physicians to be more careful in the selection of their disciples. However, in every case the *servi medici* continued to exist in many households.

It seems remarkable that just at this time when these evils were at their height the physicians were considering and discussing the problems of professional ethics. One will scarcely fail to interpret this as an inner medical reaction to existing evils as well as to criticisms from without. For example, in the first century A.D. the Roman physician Scribonius Largus outlined a lofty standard of ethics for physicians, following the Hippocratic model (77). Furthermore, it was in this century that the so-called Hippocratic Oath was quoted by Scribonius Largus and Soranus for the first time, as it seems, in medical literature, and from then onwards it became more and more important for medical ethics and education (78). In the first two centuries A.D. there appeared at least three "Introductions" into medicine, concerned with the personal qualities of the future physician: the lost treatise, *The physician* by the Pneumatist Herodotus (79); the pseudo-Galenic *Introductio sive medicus;* and the so-called *Quaestiones medicinales* that appeared under the name of Soranus and included a short prefatory introduction into medicine. It must also be remembered that one or another of the deontological treatises of the Corpus Hippocraticum, *The Law* (80), *The Physician* (81), *Peri euschemosynes,* and *Parangeliai,* stemming from later Hellenistic times, even possibly from the first century A.D., belong to the same sort of Introductions (82).

The physicians and treatises mentioned here dealt with the personal aspects of the medical profession, that is, with the personal qualities a future physician ought to possess. The future physician, it was declared, ought to be as young as possible at the beginning of his training (83). Here perhaps was an attempt to revive the old ideal of learning medicine from childhood (*cf.* above, p. 5). In regard, to his moral character, the young student ought to be in the status of "*kalokagathia*", meaning, as a standard term of Greek values, a "gentleman" in body, mind and social position. Consequently the slaves, never able to be attributed "*kalokagathia*", a priori and implicitly ought to have been excluded from medicine. Furthermore, the special talents for becoming a physician were discussed somewhat vaguely in *The Law:* there must be a "natural talent" (*physis*) as a kind of soil into which the teacher could plant the seed (par. 2,3). In regard to the preparatory knowledge believed to be necessary for the study of medicine, it is pleasant to hear something more specific from the pseudo-Soranus: the future medical student must have, he says, a certain basic knowledge in natural sciences as well as in grammar, rhetoric, geometry or arithmetic, and astronomy (*cf.* p. 244, 1.23 ff. ed. Rose). He should, therefore, be a cultivated man who follows the standards of the higher social classes.

As stated above, I believe that with such a claim the physicians wanted to secure themselves and their profession against any blame resulting from the existing decline, and in this connection attention must be called to a further proposal that the physicians advanced about the turn from the first century B.C. to A.D. I mean the proposal that medicine ought to be treated as part of the general cultural education. This may have been a demand of society, at least of the more educated part of it, as well as of the physicians. It may have been demanded for a variety of reasons of which one, perhaps, was recognition of the old need for popularizing medicine as it had been proposed by Hippocratic physicians (see above p. 7). But now there was an additional point of view to which, in the first century B.C., the physician Athenaios of Attaleia calls our attention (84): It may have been, as he states, not only useful but even necessary for laymen to "hear" medicine within the frame-work of their general education, for then they could, if there were later need for it, help themselves.

From such a statement, sounding surprisingly somewhat "modern", we may draw the two following conclusions: Athenaios' own wording ("necessary") could be a reference to the above-mentioned special problem of the lack of physicians in the Roman empire, a problem which might be, if not solved, perhaps eased in such an honest way, since the physicians would be less burdened through the eventual effects of such a "do it yourself" appeal. And, furthermore, the remark about the "hearing" of medical "*logoi*" may show that there were, in Roman times, public lectures on medicine in addition to the special training lectures given by physicians (85) or by "Iatrosophists" (to be discussed in more detail below). Related to Athenaios' proposals are such medical treatises as that of Rufus of Ephesus, *For laymen* (86), or the *Euporista* (= remedies easily acquired) of Dioscorides and Galen, tending to be aids and advice for laymen in sudden need of medical treatment, as for example, on a journey.

Now let us stop for a moment at this point and direct a brief glance back over the terrain thus far traversed. It cannot be denied that the varying colors of the picture drawn in the foregoing pages scarcely mask the insufficiency which was, to speak freely, a characteristic trait of medical education in classical antiquity. Indeed, we must confess that all attempts to elevate the medical profession and to regulate the supply of personnel must have been frustrated in principle by two reasons: there was, within Greco-Roman antiquity, neither any effective control nor any effective institutionalization of the medical profession (as V. Bullough, for example, has rightly pointed out). The above quoted demands and proposals were given out officially, that is to say, by any corporation as such, but by physicians acting as private persons. Their colleagues could accept the proposals or they could ignore them. The "literary" character, moreover, of such proposals may have been superior to their practical effectiveness. Otherwise, medical education could not be regulated in any obligatory form within and by the physicians themselves so long as there was a lack of any *communis opinio* of the most useful curriculum and of the elementary subjects to be learned by a future physician.

Once again anatomy and its image can be an indicator for the varying standpoints of physicians. True, the radically anti-scientific position of the Methodists was finally in no way victorious and might quickly have ceased to be of any influence at all despite Galen's long and somewhat tiresome tirades against it. On the other hand, there were the Empiricists whose negative standpoint concerning the *"anatome kat' epitedeusin"*, that is, systematically performed dissections, certainly remained unchanged. That this standpoint, by the way, was essentially inconsequential Galen makes clear in a very instructive manner (87): even the Empiricists readily admitted, as he states, that a "clinical" anatomy was indispensable for physicians. But they maintained that one can (and must!) gain such an anatomical knowledge "by chance". However, this claim was for Galen that of "a man who is sitting elevated on a throne and speaking to his audience". But in simple reality, a good knowledge of all important parts of the human body never could be acquired in such a superficial and casual way.

While reading Galen's statement we must ask: How and where at that time could one acquire a good anatomical knowledge, be it "clinical" or scientific? Human bodies had not been dissected for a long time, not even in Alexandria where one could, at best, inspect human skeletons. Rufus of Ephesus who believed a good anatomical knowledge to be as necessary for medical education as the knowledge of letters for learning to write, complained bitterly of having at hand for anatomical demonstration of the inner parts of the body to his students, only the bodies of dead animals (88). Even Galen, so greatly interested in anatomy, had never dissected systematically a human body. As another possibility for learning anatomy, there was the study of related monographs. Under these conditions, we can well understand that there was a veritable flood of anatomical literature in the first two centuries A.D. One may recall, as a voluminous standard work, the "Anatomy" of Marinus (epitomized by Galen) and, furthermore, some more small or introductory treatises such as those of Rufus and Galen concerned with osteology or the anatomy of the vessels and greater organs—e.g., uterus —and destined especially for medical students (89). Turning from the principal and general lack of anatomical knowledge in antiquity, we may even dare to say finally that this was not such a deplorable state of affairs, since *mutatis mutandis,* it corresponded to a modern one (*cf.* for example, the situation in a European country like Norway, where the medical students dissect only apes).

But, as one must continue to remember, the situation was not uniform. Moreover, the disagreement of ancient physicians concerning problems of principal importance must have been one of the factors which provoked a general uncertainty which, naturally, also had its effect upon medical education. Under the circumstances, our questions concerning the *how* and *what* of medical education continue to find new and varying answers. Anatomy now being, in whatsoever form, a much contested part of medical education, there was also another discipline that was discussed as a part of medical studies—mathematics (geometry and arithmetic). As we have seen already, in the pseudo-Soranus, geometry played its part within a useful preparatory knowledge for medical studies. In the same sense, geometry was discussed in the pseudo-Hippocratic letter, number 22 (like most of these "letters," a late forgery). Here "Hippocrates" is giving advice to his son Thessalus, regarding medical studies. One point alone is under discussion: the "son", serving evidently as a kind of symbol for the medical student, is asked to study especially mathematics and geometry as being multipurpose and useful for medical thinking. Furthermore, Galen in his autobiography (90) calls attention to his own methodical knowledge of mathematics and geometry, which he had acquired during childhood from his father, an architect. To Galen, his own education in this field seems to have been a model of great importance.

Mathematics was believed to be useful not only for medical studies but also for medical practice. The above-cited pseudo-Hippocratic letter calls attention to this point in some detail: geometry can, for instance, be an aid for the surgery of bones as it relates to certain mechanical problems. Arithmetical knowledge, on the other hand, can be helpful for understanding fever periods and the processes of internal diseases. Moreover Galen, who was very sensitive to the uncertainty of contemporary medicine, freely confessed mathematics and geometry to have been a veritable safeguard which had rescued him from the morass of Pyrrhonic scepticism. In consequence, in his younger years he used the "certain" geometry as a methodical model for medicine, understanding clinical problems in a way which I may dare to formulate as *more geometrico* (91).

Mathematics as a remedy against the uncertainty of the medical art and thinking, brings us to that medical sect called "Iatromathematicians" (92). As must be emphasized, however, this sect had a concept and character which was by far more primitive and narrow than the level to which Galen aspired. It was simply a medical astrology or a prognostical art for health and diseases, using planets and constellations as its main instrument. It was originally practiced in Egypt in this form and thereafter during the Roman Empire came to the Greeks and Romans. Here we must note the fact that even this non-serious "mathematical" medicine was recommended by some persons as a panacea against the mental distress of medicine. The famous astronomer Ptolemaeus, living in Galen's time, reproached the physicians generally for their almost nihilistic scepsis and uncertainty, which was, in his opinion, totally without basis, and recommended imitation of the methods of the "iatromathematicians" (93).

In terms of medical education, all this could mean was that the student of medicine ought to have not only a preparatory knowledge in mathematics and astronomy (astrology), but also he ought to learn and deal regularly with such sciences within medicine. However, we must recognize the variations that the notion of "mathematics" could also have in this respect. It seems to have been nearly the same situation with philosophy. A physician bringing medicine into relationship with philosophy (or likewise, with mathematics) could at times have in mind something different in principle, that is to say, either a scientifically founded and clearly shaped philosophy or something else quite vague and even irrational in itself. The wedding between medicine and philosophy, as discussed above (p. 16), continued to be an alluring goal for the physicians within the Roman Empire. But now, this purpose was to have some curious consequences.

By giving to one of his treatises the title *That the best physician is also a philosopher*, Galen proposed an ideal which he himself was personifying exactly in possibilities as well as in limitations. Galen had in mind the concept of the "*charieis iatros*" as proposed by Aristotle and Plato, that is, a physician knowing the "*physis*" as such and using his philosophical knowledge in the interest of medicine. However, Galen did not have in mind a type of "philosopher-physician". He recognized very well, as we learn from his treatises, the boundaries between medicine and philosophy (94). Other physicians, however, failed to do this, the more so if philosophy was conceived of as an ambiguous conglomerate, including rhetoric and sophistics as well as mystical tendencies and behavior. A physician who was devoted to such a "philosophy" never could be named a genuine "*charieis iatros*" but, at best, a "*logiatros*" who did not possess any practical experience in medicine or in medical research but "spoke" to his hearer only on medical themes.

During these centuries there was, no doubt, a serious danger of decline for medicine, in some measure at least into "lecture-medicine". Under the circumstances it seems apparent that at this time the study of medicine could be called "*iatrologia*" (95). That seems to recognize that the "*logos*", that is, the use of sterile theories, was threatening to become overwhelming. According to Galen (96), the ancients had already named as "*logiatros*", a sophist who discussed medical problems without practical experience. Now, this type was reborn and even tended to become a medical teacher. Again we must quote Galen (97) who in this connection used the term "Alexandrian prophets"—what a degeneration of scientific Alexandrian medicine! Such men were now concerned with "scholastic problems" (98). Galen sharply compared the *"tribones"*, that is, the medical practitioners, with them. It seems to me that there was a more or less straight pathway from the *"logiatros"* and the "Alexandrian prophets" to the well-known *"iatrosophists"* of late antiquity and, finally, to the *philosophi et medici* of medieval times. They had in common the fact of speaking and writing on medicine without being medical practitioners, or even having studied medicine at all. As one can see, in every case the roots of this phenomenon lay in a rather early epoch, although some scholars are prone, as it seems, to neglect this. Concerning the *"iatrosophists"*, one should not overemphasize the fact that the term as such was first employed in the fourth century A.D. (99). As a type, the *"iatrosophist"* existed in earlier epochs. Coming from philosophy, rhetoric, sophistics or natural sciences, he tended towards medicine. Or, he had been originally a physician and came, thereafter, to philosophy. In this respect special attention ought to be called to Timocrates of Heracleia who lived in the first century A.D. and was, beyond any doubt, a true *"iatrosophist"* (100).

In consequence we feel caused to look upon another tendency which was peculiar to that time. In the light of our theme, this tendency may be called a kind of "conversion" of physicians. Specifically, I mean the fact that some physicians as, for example, Timocrates, were leaving their original profession and seeking a mental sphere which had to be more "elevated" in a manner either philosophical or religious or mystical (101). As it seems to me, the best way to interpret this phenomenon is to think of it as a genuine reaction to a deeply rooted mental dissatisfaction or even despair. In the foregoing pages we had to speak in some greater detail about the mental uncertainty of Greco-Roman medicine, a situation from which escape had been sought in the recommendation of the mental assistance of some related disciplines. There remained, on the other hand, the way towards philosophy in a large sense. But now, in view of this way the situation had grown into a somewhat different one; the philosopher could now be addressed to help the physician by carrying him *away from medicine* towards a mental "verity" which was not to be recognized by study and research but believed by faith.

The physician deserting medicine was a phenomenon well adapted to a more common trait of that time. I mean the widespread belief that all sciences were useless and vain (102). An often quoted and, indeed, very typical testimony for such "conversion" of a former follower of science is the letter of a physician named Thessalus (103) who was certainly not identical with the famous founder of the school of Methodists. The letter writer had studied medicine, as he says, in Alexandria under "dialectic" physicians—here one may recall Galen's "Alexandrian prophets"—but was not yet satisfied in his search for a true revelation of medical knowledge. Finally, he succeeded in finding such a revelation by the aid of a priest who procured for Thessalus a vision of Asclepius. In this way the god privately revealed medical knowledge to Thessalus who wrote it down immediately from the god's dictation. Asclepius explicitly called attention to the fact that a self-acquired medical knowledge was worthless in comparison to a medical revelation given by a god to an elected person.

That seems to be, in my opinion, the ultimate and most radical consequence to be drawn from the mental distress of medicine and medical education. At the end of classical antiquity one could, therefore, return to the most archaic answer given to our questions number 1 and 2, that is to say, one could again believe a god—and a god of the first rank—to be the genuine "teacher" of medicine. Medical knowledge, seen as a mystery, that is, a formula we find explicitly shaped in the pseudo-Hippocratic *Law* (par. 5). Superficially considered, one could assume this to be a mere metaphor or a highly stylized expression. But recalling Thessalus, one will admit that it could, indeed, have been believed to be a genuine pathway for gaining knowledge, and that medical education as such, that is, as the learning and teaching of medicine in whatsoever *rational* form, ought properly to be abolished. For this, as Aristotle pointed out so appropriately (104), is peculiar to a mystery, that the initiated persons need not *learn* their knowledge.

Having now reached the most extravagant point of development, we must return to the normal level of conditions. Something was said in the foregoing pages about the mental disharmony and uncertainty of Greco-Roman medicine and of the effects this uncertainty had upon medical education. It remains to speak briefly of the practical and professional splintering of medicine, that is, of specialization as such. In the third century A.D. (105), the sophist Philostratus declared that nobody was able to master medicine as a whole, since one physician was a specialist for wounds, another for fevers, a third for ophthalmology, and so on, and therewith depicted a situation of which many physicians must have been conscious. Indeed, there had been medical specialists in Greco-Roman medicine for a long time, perhaps even in pre-Roman Hellenistic medicine, as there had been since very early times, in Egyptian medicine. But whatever the situation may have been, in view of a possible Egyptian influence upon Hellenistic medicine, it is clear that in the first and second centuries A.D. the mere *feeling* of "I can master only a small part", taken together with the common uncertainty, must have played an effective role in specialization. Within the borders of our present theme, I wish to restrict myself to the following questions: Was there already a tendency towards specialization within the training-period? and, furthermore, in what estimation did the specialists stand either within or outside medicine?

Let us take, as an example, the surgeons. From Hellenistic times onwards beyond any doubt they held a kind of special status in comparison to the other physicians, though the situation is in no way simple and clear. Celsus (*cf.* preface to book VII of *De medicina*) considered the surgeons to be a peculiar sort of physicians, having a special and separate education from childhood onwards. Archibios (see above p. 15) seems to have been of the same opinion. Galen, too, speaks in a similar way of the "so-called" surgeons to whom the physicians, that is, internists or *medici clinici* (106) had to cede certain cases. Plutarch considered surgeons and physicians as being clearly two different categories (107). Surely we know the difference between surgeon and internist from old Greek medicine onwards (see above p. 6). Both of them, however, had in earlier times to constitute a unity in regard to the *"techne iatrike"* as a whole. But in Hellenistic times the situation changed considerably. If Celsus and Archibios were correct, surgeons underwent a special education from the beginning of their studies. If so, what about other medical specialists such as ophthalmologists or dentists? Did they come to their special areas after having ended a regular medical curriculum, or was their case the same as that of the surgeons? Or did they become specialists after having been general practitioners because they lacked (as Philostratus said) any possibility of mastering medicine as a whole? We do not know. The above-mentioned "Introductions" discuss the special areas within medicine seen as a unity. Moreover, Galen emphasized with great earnestness that all special areas were to be seen only as *parts* of a uniform medicine (108). But for us, just such an emphasis could be very indicative in view of the reality, by suggesting additionally the impression that there was surely *not* a unity of medicine. Therefore, it seems to be a reasonable conclusion that physicians of the mental level of Galen were strictly opposed to specialization and specialists. And it may be well understood that a man standing outside medicine could say that "you can *perhaps* call physicians those men who treat only a certain part of the body or a certain pain" (109). Such an attitude towards specialists—surgeons, first of all—may have weighed heavily upon medicine and medical education as an additional problem. Herein medieval medicine did nothing but pursue an inherited situation.

Medicine in classical antiquity ended with Galen. This undoubtedly great physician, trying to comprehend all the past and present of medicine in a thoroughly personal way and through his personality impressing contours upon medicine to come, had likewise inherited the problems of medical education. Taking his influence within this field to be as important as in others, we must ask: What did Galen think and do concerning medical education? This theme, which is a demanding one and may require a special study, will be considered here, at the end of our survey, at least in a few words.

Immediately two negative points come to mind which cannot be slurred over: first, despite the urgent importance of the theme, Galen did not write a special work on medical education (110). Second, the example given by Galen himself as a teacher to all the physicians he met seems to have been ineffective in a surprising measure during his life, since we don't know of any school founded by him, or even of a physician who may be called his disciple. The reason for this cannot have been any inability of Galen as a teacher or any weakness in his personality—he was, as is clear from his commentaries alone, a strenuous as well as a very able teacher. Instead, we must confess that Galen was an exceptional and almost anachronistic personality, a personality which must have been, because of his encyclopedic knowledge and greatness, isolated among contemporary physicians. In this regard, Galen's own medical curriculum seems to be very indicative, which, as has been rightly stated (111), cannot be taken as a typical one of this time.

Throughout twelve years Galen endeavored to become a "cultivated" physician according to the standards of Plato and Aristotle, which had become the flesh and bones of Galen's belief, already personified in Hippocrates. Galen proposed a type of physician who was fully balanced between theory and practice and a thorough master of medicine as a whole. You will admit, I believe, that such a standard is the most perfect to be proposed by any rational thought. A most perfect standard, to be sure, holds good for a well-conditioned or at least normal situation, but, on the other hand, is scarcely apt for compromise if there is a need for it. And a practicable compromise would have been a much more desirable thing for the mass of physicians at a time when medicine was in such general distress. Galen, as a man of exceptional standards, failed to provide a "normal" standard for his colleagues, and this, perhaps, was the main reason for his failure to have disciples and to write the necessary, exemplary book on medical education.

As we have seen in our survey, the right balance between theory and practice was, in Greek and Greco-Roman medicine, often sought for but, at the end, never achieved. Moreover, it seemed to be a symptomatic paradox that the very medical school which claimed experience to have the first, if not the unique, place in medicine, had used for its teaching the *"historia"*, that is, the books written by the old authorities. Nothing, I dare say, can be more indicative of the dilemma that confronted even Galen than his own ambiguous attitude towards books and their value for medical knowledge and teaching. Galen, though stating that the need for books was symptomatic of a kind of medical degeneration (see above p. 5) or demanding that an anatomist should "trust his own eyes but not the books" (Kühn, ed., III, p. 98), himself wrote so many books with a hand so unrestrained by any scruples that it evidently enjoyed its activity (112). In contrast, you may remember what a difficult problem it had been for some of the old Hippocratic physicians to write books making a medical problem clear to a man who had no experience in medicine (see above p. 6). Of such scruples Galen, indeed, was free. Instead, he really enjoyed presenting a medical problem in a manner which is understandable even for the simplest mind.

With this facility to write so many understandable books on so many medical themes and problems, Galen fostered a danger he wanted to avoid, the victory of books, as the most practicable and most fascinating instrument of learning for medical students. A book can too easily have the flair of "authority", and one can learn so comfortably from it, committing its contents to memory without any criticism. Galen's treatise *On medical examina-*

tion (113) seems to reveal this danger in a very significant way. As he says, the future physician must prove that he has "learned" fully not only practice but also theory. In this connection, as Galen emphasized with a peculiar *élan*, the young physician must know and understand *in the first place* the books of "the ancients" (114). True, Galen, as is said, could hold both together in balance—books as well as practice. But the physicians to come after him evidently misunderstood his intentions, and herein Galen is, in our judgment, not free of sin, the more so since even for him the books of the ancients had been largely *infallible*—in so far, at least, as they were believed to stem from Hippocrates. Hence what need should a post-Galenic physician have felt to use his own experience together with the "pure doctrine" if not, at best, as a means for verifying—but not for correcting—the infallible book-science (115). In so far, Galen indeed helped to make the Hippocratic writings—as well as his own—a kind of bible for the medicine to come, especially so in Alexandria (116).

The great and indefatigable teacher Galen did not solve the urgent problems and demands of medical education. And how he could have done so effectively in the face of a situation which was so complicated and insoluble in itself, since there continued to be a lack of any control over the medical profession or any approbation of physicians and the related institutions (117). With this melancholic statement we have reached the end of our survey. For an unprejudiced observer, medical education in classical antiquity must be a phenomenon heterogeneous as well as unsatisfactory. Nevertheless the picture cannot be considered hopelessly chaotic. Surely Cohn-Haft is right in some sense, saying that it is "remarkable how little is known about the training of the Greek physician" (118). And he would have been right if he had extended his judgment in the direction of the Roman physician. But, on the other hand, taking "medical education" in a sense which is large enough—to have done so is, I think, the greatest merit of Drabkin—you will find a fair number of lines and colors which together may give a sufficiently clear and impressive picture. Drawing this picture, we have tried to broaden the scope even more than Drabkin did. Herein our four standard questions lead us, in view of the development of Greco-Roman medical education, to the following general answers.

(*a*) The medical teacher as a personality did not continue from the beginning to the end to be a physician, but could be sometimes a philosopher or even a supernatural being. (*b*) Corresponding to this, the form of medical teaching and learning could vary, at first being oral instruction alone, be it in a rational sense or as a "revelation"; after that it became a genuine learning with books, lectures and practical instruction in both anatomy and physiology and in bedside instruction; it ended in some more or less degenerated forms—pure book-study, "hearings" or even kinds of "medical mysteries". (*c*) In connection with the differences in form, the content and extent of medical studies changed, as well as the leading models of physician-types. At first, rational medicine was nothing but traumatology. After that, it increased to an internal medicine which, together with surgery, in the fifth century B.C. formed a new "*techne iatrike*". In further development, anatomy, under controversial discussion became a part of medicine and medical education. Next, philosophy came again into medicine and medical education. Finally, mathematics (geometry and astronomy) were recommended as a fundamental part of medical education and even of medicine itself. Corresponding to this, the length of medical studies proposed varied very greatly—from six months to eleven years. (*d*) The personality of the medical student was subject to standards of selection corresponding to the changing cultural or social patterns. In pre-Roman Greece slaves were prohibited from becoming physicians, and women could be nothing but midwives. In Roman times slaves as well as women were allowed to become physicians. At the end, Galen seems to have believed the Jews to be a people that were generally too "primitive" to learn medicine in the correct sense, an opinion which was to become highly influential and effective in the time to follow.

Perhaps you are willing to admit—as I am—that Greco-Roman medical education, within its own borders and the borders of Greco-Roman culture in general, discovered and discussed some, if not many, of the main topics of "European" medical education. True, Greco-Roman medical education did not succeed in composing a satisfactory whole from all these topics and motives. But it did offer some models that were to be influential for a long time to come. And this alone may justify, at least, the fact that now and then physicians and scholars, interested in historical facts and developments, turn their attention to medical education in classical antiquity.

NOTES

1. I. M. DRABKIN, On medical education in Greece and Rome. *Bull. Hist. Med.*, 1944, 15: 333-351; Medical education in ancient Greece and Rome, *J. Med. Educ.*, 1957, 32: 286-295. There is a popular but interesting essay by W. B. WARTMAN, *Medical teaching in western civilization*, Chicago, 1961, pp. 13-31. To be used with some caution is V. BULLOUGH, *The development of medicine as a profession*, Basel, 1966, pp. 13-29. For the early Greek situation see the first two chapters of my book, *Der Beginn des medizinischen Denkens bei den Griechen von Homer bis Hippokrates*, Zürich, 1967. Puschmann's classical *Geschichte des medicinischen Unterrichts*, Leipzig, 1889, pp. 29-112, can scarcely be called a contribution to our theme, being in fact a kind of general history of Greco-Roman medicine.
2. Cf. in general B. SNELL, Menschliches und göttliches Wissen, in *Die Entdeckung des Geistes*, Hamburg, 1955, pp. 184-202.
3. The false construction of Hippocratic medicine as being derived from the archives of the temples of Asclepius had already emerged in later antiquity. The medicine of the Romantic Era had a special liking for this fiction, cf. K. SPRENGEL, *Versuch einer pragmatischen Geschichte der Arzneikunde*, 2nd ed. Part I. Halle, 1800, p. 212; as a romantic W. LEIB-

BRAND continues to call Greek physicians "Priesterärzte," cf. J. SCHU-MACHER, *Die Anfänge abendländischer Medizin in der Antike*, Stuttgart, 1965, p. 7. This fiction may be caused by a misinterpretation, which likewise is ancient, of the old term "Asclepiads", i.e., physicians, falsely taken as "priests of Asclepius", as may be seen, e.g., in the so-called Suda (Suidas) *sub voce* "Democedes," where the father of this famous physician is said to have been a *"hiereus Asklepiou"*, what in one of the sources of the Suda may have been simply "Asklepiades", cf. in general E. T. WITHINGTON, The Asclepiadae and the priests of Asclepius, in: *Studies in the history and method of science*, ed. CH. SINGER, II, Oxford, 1921, p. 192 ff., and L. EDELSTEIN, *Asclepius*, Baltimore, 1945, p. 55 ff.
4. Cf. *Odyssey*, XVII 382 ff., and PINDAR, *Pythian Odes*, III 6.
5. In the seventh century B.C. the poet ALCMAN (Frag. 109, ed. DIEHL) says "experience is the beginning of the acquisition of knowledge" *(peira toi mathesios archa)*.
6. He can, then, be a "divine man" who "knows", for example, that and why one is allowed to urinate only at certain times of day, at certain places, and under certain conditions (cf. HESIOD, *Works and days*, v. 727 ff., especially 731 f.). In the extreme, he feels himself to be like an "immortal god, being no longer a mortal man" and behaves as such, as did EMPEDOCLES (Frag. B 112, 4, and 146, ed. DIELS-KRANZ). Herein the physician Empedocles was imitated in the fourth century B.C. by the physician Menecrates Zeus, a follower of the ancient idea of *Gottmensch* and, besides, a veritable psychopath, as pointed out in a brilliant study by O. WEINREICH.
7. HESIOD's lost poem "Chiron's Advice" *(Chironos hypothekai,* cf. Frag. 170-173, ed. Rzach) may have contained, among other matters, some "medical" remarks, as did pseudo-EPICHARM's poem "Chiron" (Frag. 58-62, ed. DIELS-KRANZ). It should be noted that the passage "whatever houses I may visit, I shall come for the benefit of the sick, remaining free of all intentional injustice" of the pseudo-Hippocratic "Oath" looks, at least to some degree, like a special transformation of HESIOD's "Chiron's Advice", "In the moment of entering a house you must do things which are pious and pleasant for the eternal gods" (cf. HESIOD, Frag. 170, ed. RZACH).
8. "Apollo donated", etc., cf. EURIPIDES *Alcestis*, v. 970, where additionally there is mention of remedies that "the Orphic voice wrote on Thracian tablets"—revealed medicine, as is apparent, in a genuine and mystical sense. On Chiron as teacher of Asclepius, cf. PINDAR, *Pythian odes* III. The Asclepiads, as representatives of a *"techne"* being inaugurated by a hero, provide an exact analogy to, for example, the Daedalids, cf. EDELSTEIN, *Asclepius, op. cit.*, II, p. 57.
9. See DIOGENES LAERTIUS, VIII, 83. In general one may use the collection of pre-Socratic fragments by DIELS-KRANZ from which I am quoting.
10. The physician has to master his art *"kata gnomen"* and *"kata cheirourgien"*, as is stated by the author of *On diseases* (*Oeuvres complètes d'Hippocrate*, ed. LITTRÉ, VI, Paris, 1849, p. 150).
11. In an epoch when this was already very unusual, PLATO demanded that, for any *"techne"* the education should be from childhood, within and by the games the child plays. See the *Laws*, 643 B 4 ff.

12. For his special field, the old-Hippocratic author of *On joints*, ch. 33, freely confesses the same difficulties: "Indeed it is not easy to demonstrate surgical procedures in a book, for the reader has to gain from the written description a clear impression of them." In general one may compare Greco-Roman textbooks and for their development M. FUHRMANN's study, *Das systematische Lehrbuch*. Göttingen, 1960.
13. Cf. G. H. KNUTZEN, Technologie in den hippokratischen Schriften peri diaites oxeon, peri agmon, peri arthron emboles, *Abh. Akad. Wiss. Mainz*, 1963, no. 14, p. 29 ff., and especially for "being like brother and sister", p. 15. Knutzen considers the writings discussed by him to have been written by the same author, i.e., Hippocrates.
14. Cf. HIPPOCRATES, *On the sites in man*, ch. 41, "It is impossible to learn medicine, since there cannot be developed in it a firm and constant technical method." For this passage see my book quoted in note 1, p. 46 n.1. In a way that is not thoroughly pessimistic but, nevertheless, quite sceptical, the author of *On diseases*, I, 9, discusses the "principle" *(arche)* of "techne iatrike" as a whole.
15. In this regard S. REINACH's remarks in *Dictionnaire des antiquités grecques et romaines*, ed. DAREMBERG-SAGLIO, Paris, 1877-1919, III, pt. 2, p. 1677, are to be used with caution.
16. To be sure we have considerable number of contracts for apprentices, written on papyrus, which, although belonging to the time of Roman Egypt, reflect an older Greek model. But among them there are no medical ones, cf. L. COHN-HAFT, *The public physicians of ancient Greece*, Northampton, Mass., 1956, p. 15, n. 27, with further references.
17. Cf. the important interpretation of L. EDELSTEIN in his *Hippocratic Oath*, Baltimore, 1943. In my judgment Edelstein's arguments regarding the esoteric and isolated character of the Oath are thoroughly convincing, although this must not inevitably lead us to the Pythagorean origin that Edelstein believed he had found in the Oath.
18. For this cf. J. ILBERG, Die Ärzteschule von Knidos, in: *Verh. sächs. Akad.*, etc., 1924, 76 (3): 7 f.
19. COHN-HAFT, *op. cit.*, p. 42 and n. 49, has pointed out that PLATO, *Gorgias*, 514 D ff., and *Statesman* 259 A, cannot be taken as a witness for the real existence of "public" and "private" physicians, as does, e.g., the *Greek-English Lexicon* of LIDDEL-SCOTT, *sub voce* "Idioteuein". In simple terms, there were general physicians and city-physicians, and the difference between them cannot have been very remarkable.
20. From SORANUS, cf. *Corpus medicorum graecorum*, IV, p. 186, *sub voce* "Herophilos" we know that there was, e.g., a "Maiotikon" written by Herophilus.
21. Cf. C. A. FORBES, The education and training of slaves in antiquity, *Trans. Am. phil. Ass.*, 1955, 86: 343.
22. We refer to the judgment of H. E. SIGERIST, *History of medicine*, II, New York, 1961, p. 310, and L. COHN-HAFT, *op. cit.*, p. 15.
23. See, furthermore, P. LAIN ENTRALGO, *Arch. Gesch. Med.*, 1962, 46: 194 f.; T. A. SINCLAIR, *Bull. Hist. Med.*, 1951, 25: 386 f.
24. Cf. PLATO, *Laws*, 720 B 2, and HIPPOCRATES, *Oeuvres*, ed. LITTRÉ, V, p. 26 f.

25. Cf. LITTRÉ's index to *Oeuvres d'Hippocrate*, X, sub voce "aides".
26. In his commentary to the Epidemics of Hippocrates, *Corpus medicorum graecorum*, V, 10, 2, 2, p. 257, ll. 4 ff.
27. The older surgical writings of the Corpus Hippocraticum are especially polemic against some questionable and showy therapeutic methods used not by charlatans but by colleagues, cf. for example HIPPOCRATES, *On joints*, chs. 42, 78.
28. Concerning the *"Pharmakopolai"* cf. the careful article by W. MOREL in PAULY-WISSOWA, *Realencyclopädie der classischen Altertumswissenschaft*, Vol. XIX, 2 (Stuttgart, 1938), p. 1840 f. The cynic philosopher TELES (p. 18, ll. 9-12, ed. Hense), third century B.C., declared that it was a paradox to prefer a *"pharmakopoles"* to a physician; this statement undoubtedly corresponded to a general attitude. However, it could be observed otherwise that occasionally a *"pharmakopoles"* was superior in his skill to a physician, cf. the testimony of Diodorus that I discussed in *Clio Med.*, 1966, 1: 319-324.
29. Cf. PLAUTUS, *Amphitruo*, v. 1013; HORACE, *Satires*, I. 7, v. 2; AELIAN, *Varia historia*, III 7, where *"iatreia"* as well as barber-rooms are called places of idle talk and scandal.
30. For the professional organization of Greco-Roman physicians, cf. STÖCKLE in PAULY-WISSOWA, *op. cit.*, Supplement IV (1924), pp. 176 f., with further literature. Before the 3rd century B.C. such organizations seem to have been very unlikely.
31. See my article Dogmatische Ärzte, in PAULY-WISSOWA, *op. cit.*, Supplement X (1965), p. 179 f.
32. RUDOLF HERZOG, who alone knew all the Coan inscriptions in the original, continued to publish many, sometimes very problematic, statements, conclusions and theories concerning Coan medicine which were taken from these inscriptions; however he did not succeed in publishing the documents themselves; for that see COHN-HAFT, *op. cit.*, pp. 62 ff. It is to be hoped that the Deutsche Akademie der Wissenschaften, where Herzog's manuscripts were deposited after his death, will undertake an edition of them.
33. *"tis orthōs iatreuken"*, as ARISTOTLE (1281b 40) formulated the question.
34. Cf. in general O. TEMKIN, Greek medicine as science and craft, *Isis*, 1953, 44: 213-225.
35. In his *Statesman*, 293 B, PLATO evaluated the situation in a different way; for him the physician whether educated or not, poor or wealthy, seems to be good if his treatments are well performed and successful.
36. Cf. ARISTOTLE 463a 5, and 1102a 21; K. DEICHGRÄBER is preparing a paper on the *"charieis iatros"*.
37. Cf. ARISTOTLE, 1282a 3 f. For further references to the discrepancy between medical craftsmen and medical scientists, see A. J. FESTUGIÈRE, *Hippocrate: l'ancienne médecine*, Paris, 1948, p. 28.
38. In classical times for example, a medical apprentice could not become a physician by mere book study (see above, p. 10), as he could in some degree, in Roman times (see below, p. 28). When the pseudo-Hippo-

cratic letter no. 25, an alleged honorary *degree* for Hippocrates himself, states that Hippocrates had written so many books for creating as many physicians as possible; such a statement seems apt for the Roman but not for the older Greek situation, cf. COHN-HAFT, *op. cit.*, p. 25, n. 77.
39. Concerning Jaeger's thesis, see my paper, Probleme um Diokles von Karystos, *Arch. Gesch. Med.*, 1963, 47: 445-464.
40. Hippocrates as a "disciple" of Democritus—who was, indeed, interested in medicine and seems to have dissected animals—or of the rhetor Gorgias (cf. SORANUS, *Corpus medicorum graecorum*, IV, p. 175, ll. 7-9) is a fiction of the ancient Hippocrates-romance and cannot be taken for historical fact.
41. For the fragments of "peri skeleton" see F. WEHRLI, *Die Schule des Aristoteles*, pt. III, Basel, 1948, p. 39 (Frags. 106-110).
42. In Alexandria, perhaps in connection with the Mouseion, were to be found, we may assume, the first professional organizations of Greek physicians, as later within the Mouseia of, for instance, Pergamon. They were called *"synhedreia"* (Latin, *"collegia"*). Cf. STÖCKLE, *op. cit.*, above, note 30.
43. The Greek term is *"parhedreuein,"* cf. APPOLLONIUS of Citium, *Corpus medicorum graecorum*, Vol. XI 1, 1 p. 12, ll. 2-5.
44. See ATHENAEUS, *Deipnosophists*, book IV, 184 B-C, taken from the historian Andron of Alexandria.
45. See S. REINACH, *op. cit.*, Vol. III, pt. 2, p. 1671.
46. FORBES, *op. cit.*, p. 344.
47. Apparently this slave had been a *"hyperetes."* Cf. COHN-HAFT, *op. cit.*, p. 14 f., with n. 19, and p. 21 with n. 58.
48. See in general K. DEICHGRÄBER's standard monograph, *Die griechische Empirikerschule*, Zürich-Berlin, 1965.
49. CELSUS, *De medicina*, "Prooemium" par. 27 ff.; K. DEICHGRÄBER, *op. cit.*, p. 92-94.
50. For the *"historia"* and its role within the teaching of the Empiricists, see Galen's delineation, quoted by DEICHGRÄBER, *op. cit.*, pp. 126 f. The third main principle of Empiricist doctrine was the conclusion by analogy, cf. the testimonies collected by DEICHGRÄBER, *op. cit.*, pp. 128 ff.
51. *"polythryletos anatome,"* as APOLLONIUS formulates it in his polemics against the Herophileans. See *Corpus medicorum graecorum*, Vol. XI 1, 1 p. 78, l. 24.
52. See DEICHGRÄBER, *op. cit.*, p. 148, ll. 9 ff., and pp. 317 f., and APOLLONIUS of Citium, *op. cit.*, p. 10, line 5: *"Hippokrates ho theiotatos"*.
53. Ed. with commentary by H. SCHOENE, in *Berliner Klassikertexte*, Vol. III, Berlin, 1905, pp. 22-26.
54. The papyrus is, as has been said, apparently from the first century A.D. Ignoring it, M. Wellmann placed Archibios in the first century A.D., cf. PAULY-WISSOWA, *op. cit.*, Vol. II, 1, p. 466, without a cogent reason. SCHOENE, *op. cit.*, p. 26n, is fully right judging more cautiously about the date of Archibios.
55. Such discussions are known to us from pseudo-GALENIC *Eisagoge*. Besides, the papyrus itself is a fragment of an *Eisagoge*.

56. Sometime it may have happened that a *"pharmakopoles"* was more skilled or at least had more therapeutic ability than a physician. See above note 28 and my paper referred to there.
57. See E. R. DODDS, *The Greeks and the irrational*, Berkeley-Los Angeles, 1964, pp. 238, 255 and notes 4-6.
58. For the date and the historical place of the pseudo-HIPPOCRATIC writing "Peri euschemosynes" see U. FLEISCHER, *Untersuchungen zu den ps. hippokratischen Schriften Parangeliai, Peri ietrou und Peri euschemosynes*, Berlin, 1939.
59. I call attention to LEO, the author of a *"synopsis tes iatrikes"*, (ed. F. Z. ERMERINS in *Anecdota medica graeca*, pp. 79 ff.) who seems to have been identical with the philosopher and mathematician Leo, a professor in Constantinople, cf. SARTON, *op. cit.*, Vol. 1, pp. 554 f., and G. E. VON GRÜNEBAUM, *Der Islam in Mittelalter*, Zürich-Stuttgart, 1963, pp. 74 f., 468 n. 93.
60. See APOLLONIUS of Citium, *Corpus medicorum graecorum*, Vol. XI, 1, 1 p. 12, l. 5.
61. Cf. my paper in *Hermes*, 1962, 90: 419 f.
62. Cf. *ibid.*, and my remarks in Untersuchungen zu Aretaios von Kappadokien, *Abh. Akad. Wiss. Lit., Mainz*, 1963, pp. 36 ff., and in *Clio Med.*, 1966, 1: 319 ff.
63. But the Pneumatists favored, as it seems, a conception of anatomy which gave more importance to clinical than to morphological facts, cf. ARETAIOS in *Corpus medicorum graecorum*, Vol. II, p. 53, ll. 27 ff.
64. Cf. *Opera Galeni*, ed. KÜHN, Vol. X, pp. 5, 9, 17, 349.
65. Cf. in general for that TH. MEYER-STEINEG, Das medizinische System der Methodiker, *Jenaer med.-histor. Beiträge*, Jena, 1916, pts. 7-8.
66. Cf. H. DILLER's article on Thessalos, in PAULY-WISSOWA, *op. cit.*, Vol. VI A 1 (1936), p. 169, ll. 38 ff., with references.
67. See *Opera Galeni*, ed. KÜHN, Vol. X, pp. 352, 381, 390, 909.
68. Undoubtedly it is incorrect to say that the Methodists, and especially Thessalos, had great merit within medical education because of having introduced bedside teaching, cf. H. DILLER, *op. cit.*, p. 169, ll. 52 ff. Clinical teaching in this sense had been practiced by other schools for a long time, for instance, by the Empiricists. The pseudo-HIPPOCRATIC "Peri euschemosynes," ch. 17, demanded that the medical apprentice stay *alone* at the bedside of the patient for the latter's supervision.
69. See also P. DIEPGEN, *Geschichte der Medizin*, Vol. I, Berlin, 1949, p. 152, and for the terms *"iatromaia"* and *"iatrine"*, the *Greek-English Lexicon* of LIDDELL and SCOTT.
70. This is said on the premise that the woman doctor Antiochis, sometimes referred to by GALEN, is identical with the addressee of HERACLEIDES' work "Pros Antiochida," cf. DEICHGRÄBER, *op. cit.*, pp. 259 f., and frags. 203 ff. We cannot say if the woman doctor Antiochis of Tlos in Lycia, daughter of Diodotos, having been honored by her city with a statue, cf. *Tituli Asiae Minoris*, II, 595, is identical with the Antiochis quoted by Galen and Heracleides. For Antiochis and further women doctors known by inscriptions, see J. OEHLER in *Janus*, 1909, 14: 12 f.

71. This is proved by a constitution of the Emperor Justinian from the year 530 A.D., where it is explicitly stated *"sin autem sit . . . medicus sive masculus sive femina,"* cf. K.-H. BELOW, *Der Arzt im römischen Recht*, Munich, 1953, p. 9. It may be remarked that in some of the testimonies the women doctors are praised not only with admiration but also with amazement as equal to male colleagues, cf. Below, *op. cit.*, pp. 9 f., note 36. The woman doctor seems to have been a phenomenon not so usual in itself.
72. See K.-H. BELOW, *op. cit.*, pp. 7 f., with further references; and for the old Roman *"medicina domestica"*, *ibid.*, pp. 2 f.
73. Cf. SUETONIUS, *Divus Augustus*, 42, 3, and OROSIUS, *Adversus paganos*, VII, 3, 6. V. BULLOUGH, *op. cit.*, p. 24, in this connection confounds Caesar with Augustus. We must confess with regret that this is not the only carelessness of Bullough's work.
74. *Opera Galeni*, Vol. X, pp. 4 f., and Diller's interpretation in PAULY-WISSOWA, *op. cit.*, Vol. VI A 1 (1936), p. 170.
75. See, for example, Martial's often cited epigram V 9.
76. Cf. BELOW, *op. cit.*, pp. 8 f., who prints the text, which has many *lacunae*, following Herzog's quite acceptable reconstruction.
77. For this generally see K. DEICHGRÄBER, Professio medici—Zum Vorwort des Scribonius Largus, *Abh. Akad. Wiss. Lit.*, Mainz, 1950, no. 9.
78. See L. EDELSTEIN in *Bull. Hist. Med.*, 1956, 30: 402, note 22.
79. See GALEN in *Corpus medicorum graecorum*, Vol. V 10 2, 2 p. 117, ll. 24 f.
80. Concerning the date of the "Law," the last word has not yet been spoken. See my remarks in *Hermes*, 1966, 94: 56, note 2.
81. For this treatise see my paper referred to in the foregoing note.
82. For the last-quoted three writings, see U. FLEISCHER's monograph referred to in note 58.
83. See *Law* 2 ("paidomathes"); "Parangeliai" 13 (no "opsimathia"); CELSUS *De medicina* VII, "prooemium" ("*adulescens*," concerning the surgeon especially); pseudo-Soranus, in: *Anecdota graeca et graecolatina*, ed. ROSE, Vol. II, p. 244, ll. 8 ff. ("*aetate illa, ex qua maxime a pueris homines transeunt ad juvenem, qui est in annis XV*").
84. See ORIBASIUS in *Corpus medicorum graecorum*, Vol. VI 2, 2 p. 139, ll. 19 ff.
85. For public lectures ("*akroaseis*") given by physicians and destined for public enlightenment, there are inscriptional testimonies from about 100 B.C., that is from the time of Athenaeus, cf. COHN-HAFT, *op. cit.*, p. 23 and note 72.
86. Regrettably lost, cf. J. ILBERG, Rufus von Ephesos, *Abh. sächs Akad. Wiss.*, 1930, no. 41, 1, pp. 41 f.
87. *Opera Galeni*, ed. KÜHN, Vol. II, pp. 288 f.; for the anatomy "*kat' epiteudeusin*" and "*kata periptosin*" see pseudo-Galen, Vol. XIX, p. 357 K.
88. RUFUS of EPHESUS in the preface to his work "About the names of the parts of the human body", ed. Daremberg-Ruelle, pp. 133 f.
89. Cf. GALEN, *On the dissection of muscles; On the bones; On the dissection of nerves; On the dissection of the uterus.*
90. For Galen's autobiography, see G. MISCH, *Geschichte der Autobiographie*, 3 ed., Vol. I, Bern, 1949, pp. 344-347.

91. For Galen's autobiographical statements see MISCH, *op. cit.* For Galen's estimation of mathematics relative to the medical art see K. DEICHGRÄBER, *Die griechische Empirikerschule,* p. 89, note to line 25. Galen's little treatise on pleurisy (*Opera,* ed. KÜHN, Vol. I, pp. 274, l. 2-279, l. 5) which is written in a nearly mannered way, *more geometrico,* i.e. in the form of a mathematical proof, may originally have been an independent treatise as, indeed, it is delivered to us. Cf. H. DIELS, *Die Handschriften der antiken Ärzte,* Vol. I, p. 129—photocopies of the two Greek manuscripts are in my hands. Subsequently Galen could have integrated it into the larger treatise *On the constitution of the medical art,* which stems from his youth.
92. See E. BOER's article in *Der kleine Pauly,* part 13, 1967, pp. 1326 f.
93. PTOLEMAEUS, *Apotelesm.* I 3 (Vol. III 1, p. 14, ll. 16 ff. and p. 16, ed. BOLL-BOER).
94. See O. TEMKIN, *op. cit.,* note 34 above, pp. 224 f.
95. Cf. PHILO of ALEXANDRIA, ed. COHN-WENDLAND, Vol. I, p. 302. On *"akouein"* as a term for medical teaching, see J. OEHLER, *Janus* 1909, 14: 14. Sometimes the medical teacher was called *"kathegetes",* see e.g., J. OEHLER, *op. cit.,* a term that becomes more frequent in Roman times and sounds, in some degree, more "theoretical" than the usual *"didaskalos."* See for references, the *Greek-English Lexicon* of LIDDELL-SCOTT *sub voce.*
96. Cf. *Corpus medicorum graecorum,* Vol. V 9, 1 p. 81, ll. 23 f.
97. *Ibid.,* p. 88, ll. 1 ff.
98. In the text referred to in the foregoing note I would prefer the word *"scholastikois"*—which is either explicitly rendered or, at least, to be supposed in some manuscripts—instead of *"stochastikois,"* which, in view of the context, does not provide the proper sense. Furthermore, there are arguments to be drawn from textual criticism. The crux of the matter seems to lie in the word *"chalastikois",* which is more self-evidently a mistake for *"scholastikois"* than for *"stochastikois,"* cf. the *apparatus criticus ad locum.*
99. Cf. for example, O. TEMKIN in *Kyklos,* 1932, 4: 41 f.; and H. DILLER in PAULY-WISSOWA, *op. cit.,* Vol. XVIII 3 (1949), p. 211, ll. 45 f.
100. On him see W. KROLL in PAULY-WISSOWA, *op. cit.,* Vol. VI A 1 (1936), pp. 1270 f. The name-like physician quoted in ORIBASIUS, *Corpus medicorum graecorum,* Vol. VI, 1, 1 p. 266, ll. 13 ff.—Kroll failed to mention this reference—may be identical with our iatrosophist.
101. See for this my paper, Der Arzt des Körpers und der Arzt der Seele, printed in *Clio Med.,* 1968, 3: 1-20, where I have discussed some further medical "conversions".
102. See F. BOLL in *Zeitschr. f. neutestamentl. Wiss.,* 1916, 17: 139 ff.
103. For this cf. M. P. NILSSON, *Geschichte der griechischen Religion,* 2 ed., Vol. II, Munich, 1961, pp. 532 f., with further references, and A. D. NOCK, *Conversion,* Oxford, 1965, pp. 108 f.
104. ARISTOTLE, ed. BEKKER, p. 1483a 19, frag. 45.
105. PHILOSTRATUS, *Gymnastikos,* ch. 15. On the phenomenon of specialization within the medicine of the Roman Empire, see K. H. BELOW, *op. cit.,* pp. 4-6, with further references.

106. *Ibid.*, p. 4, note 15 for the *"medicus clinicus."*
107. *Opera Galeni*, ed. Kühn, Vol. X, p. 450; Plutarch, *De fraterno amore*, ch. 15.
108. For example, cf. *Opera Galeni*, ed. Kühn, Vol. V, pp. 846 ff.
109. Cf. the fragment from the "Digestae" quoted by K. H. Below, *op. cit.*, p. 4, which stems either from the famous jurist Ulpian himself or from his time and environment.
110. The treatise *Peri aristes didaskalias* (special edition by A. Brinkmann, Bonn, 1914) is nothing but a philosophical polemic against Favorinus, composed in the form of a public disputation. Another treatise delivered under Galen's name, *Peri iatron didaskalon kai matheton* or *Epistula Galeni de instructione medici*, cf. H. Diels, *op. cit.*, Vol. I, p. 123, has not yet, so far as I know, received any attention. It may be a pseudepigraphon or a forgery. For the treatise *De examinando medico* see below.
111. Cf. G. Sarton, *Galen of Pergamon*, Lawrence, Kansas, 1954, p. 19.
112. In the first two centuries A.D. the mass of medical books was increasing enormously. Even prices were given for physician-writers, cf. J. Oehler, *Janus*, 1909, **14**: 19, with references.
113. Delivered in an Arabic version. See the specimen of translation by A. Iskandar, *Bull. Hist. Med.*, 1962, **36**: 362-365. An edition of this important Galenic writing is highly desirable.
114. On this point, Galen seems to prepare a new answer for our question no. 4 for he believes Jewish physicians to be principally unable to understand the books of the ancients, cf. *Corpus medicorum graecorum*, Vol. V, 10, 2, 2 p. 413, ll. 37 f. For further ancient judgments on an alleged Jewish "illiteracy" and "stupidity", see I. Heinemann in Pauly-Wissowa, *op. cit.*, Supplement V, 1931, p. 22, ll. 1-21. This "inborn illiteracy" was for times to come one of the readily used subterfuges for excluding the Jews from medical studies and practice.
115. How Late Antiquity understood the Hippocratic writings in this sense, and especially the famous first aphorism, may be learned from a "Preface to Hippocrates" discussed by H. Flashar in *Hermes*, 1962, **90**: 402 ff., especially p. 405.
116. One may recall the late commentaries on Galen's writings and the so-called "Summaria Alexandrina".
117. Galen's *De examinando medico* gives its author's private opinion and is in no way "official".
118. Cohn-Haft, *op. cit.*, p. 15.

MEDICAL EDUCATION AND PRACTICE IN MEDIEVAL ISLAM

SAMI HAMARNEH
Smithsonian Institution
Washington, D.C.

An accurate, concise description of early Arabic attainment was given by Ibn Ṣā<id (1029-1070), Judge of Toledo, in his *Des categories des nations* (1). He remarked that:

> The acquisition of knowledge the Arabs attained, and of which they are proud, is the precise use and eloquence of their language, the composing of poems and the improvising of speeches. . . . They have also knowledge of narrations, biographies, and the geography of lands which they enter for trade and thus endeavor to become acquainted with the history of its inhabitants. The Arabs are, moreover, a people of well-trained memory and of ready wit, as appropriate and idiomatic words come easily to them . . . as a result of long and repeated observation, they have acquired knowledge of the rising and setting of the stars and of meteorology not by deducing data from experimentation or by following scientific reasoning, but in as much as such disciplines have bearings on their everyday living. . . . Then in the early days of Islam, the Arabs cared for none of the sciences or arts save the rules and syntax of their mother tongue, with the exception of the healing art, a profession that has been practiced and highly regarded by some individuals and held in high prestige by the majority of the people because all have need of it (2).

Soon after the triumph of Islam, caliphs and rulers gave further support and impetus for promoting the health professions. It was in the ninth century (3rd A. H.), however, that medical education and practice in Islam developed into a well-defined and established profession, with solid foundations and a scholarly and secular outlook (3). One might list several reasons for this progress. Nonetheless, there were, in my estimation, three primary factors responsible for such development.

The first was the transmission of intellectual legacies of earlier great civilizations. Under the Abbaside caliphs, competent scholars in Iraq and neighboring countries embarked on the translation of the best available writings from Syriac, Persian, Sanskrit, Nabataean, Coptic and Greek into Arabic. The richest, finest and most influential in medicine, pharmacy and the allied sciences were translations from the Greek legacy (4).

The second factor was the acceleration and abundance in locally manufactured good-quality paper which facilitated and enhanced literary production and the art of writing and copying in Islam. After learning the process from Chinese artisans in Trans-Oxiana, the Arabs established the first paper factory in Baghdad in 794. Very shortly, this industry together with that of making writing tools and inks reached an unprecedented degree of significance unknown before in the history of man (5). From Iraq, this knowledge passed to Syria where high quality paper was made, and thence to Egypt, North Africa and Spain (al-Maghrib). Thus the copying of manuscripts, cataloging, binding, selling and collecting of books led to new crafts, industries and trades which brought added wealth and high measures of prestige to those involved in them. In turn, the abundance of manuscripts—some of them elegantly and artistically inscribed—generated a demand for more reliable and precisely executed copies and manuals in the various fields of human knowledge, including those on the healing arts. It is interesting that several authors during that period warned against relying on copyists who used cheaply manufactured inks and poor-quality writing materials (6).

The third factor was the rise of educational and medical institutions—public (state-sponsored) and private—such as libraries, hospitals and medical schools which greatly contributed to the sound foundation and steady advancement of medical education. The first hospital, in the modern sense of the word, opened in Baghdad during the reign of al-Rashīd (786-809). It continued to receive adequate attention and financial and moral support from succeeding caliphs. In this and other hospitals established in the Abbaside capital and other cities, physicians of great renown such as Yūḥannā b. (for ibn) Māsawayh (d. 855), al-Rāzī (d. 925) and Sinān b. Thābit (d. 941) practiced and also trained others (7). As word spread of the accomplishments at Jundī-Shapūr (the ruins of this town in southwestern Iran are near the present village of Shāh Ābād less than sixty miles north of Ahwāz) and Baghdad, hospital construction spread into other big cities which included Marw, Rayy, Mayyāfāriqīn, Damascus, Antioch, Makkah, al-Madīnah, Cairo, Al-Qayrawān, Marrākush and Granada. The <Aḍudī, the most renowned hospital in Baghdad, was founded in 978-979 by the illustrious Buwayhid King <Aḍud al-Dawlah. It was generously endowed and provided with a full medical staff of twenty-four physicians, surgeons, oculists and pharmacists, in addition to administrative personnel for the building and the kitchen and hospital management. With its rich medical library, the hospital served as a focal point and a convenient center for teaching medicine and for the training and practicing of young doctors (8). It was al-Rāzī who enthusiastically recommended that a physician should acquire theoretical as well as practical training in hospitals at the patient's bedside. The libraries in each hospital served as lecture and reference rooms as well as for reading and consultation of medical and pharmaceutical texts.

In less than a century, the <Aḍudī hospital was surpassed by the construction of the Nūrī hospital, built in Damascus by King Nūr al-Dīn Zinkī (1118-1174). Ibn Jubayr (1145-1217), nearly half a century later in his *Riḥlat*, de-

scribed it as "one of the great glories of Islam". The hospital with its library and fully equipped pharmacy shop served as one of the greatest medical-educational centers in Syria for over four centuries despite the periods of decline and stagnation in Arabic medicine. Ibn Uṣaybiʿah devoted a good section of his <Uyūn al-Anbā> to biographies of eminent Syrian doctors who practiced and taught at the Nūrī hospital. One of the physicians, Muhadhdhab al-Dīn ʿAbd al-Raḥīm b. ʿAlī known as al-Dukhwār (d. 1231), continued to treat his patients at the Nūrī hospital while teaching medicine to a great number of students in his private school that functioned even after the founder's death (9). Among his students was Uṣaybiʿah himself. His uncle the oculist Rashīd al-Dīn ʿAlī b. Khalīfah, practiced at the Nūrī hospital in Damascus as well. Uṣaybiʿah also speaks of his classmate Badr al-Dīn b. al-Qāḍī ʿAbd al-Raḥmān of Baalbak who enlarged and renovated several halls in the hospital by buying properties adjacent to it and then donating them to the hospital (10).

A counterpart hospital, constructed over a century later, was the Manṣūrī of Cairo founded by King al-Manṣūr Qalāwūn about 1285. It was operated for the first two years under the competent leadership of the famous physician Ibn al-Nafīs, a graduate of the Damascus Nūrī hospital who later taught medicine in Cairo. Before his death in 1288, he donated his library and other property to the hospital which he had administered and served so well. The hospital resisted the ravages of time for centuries and it was my good fortune to visit and see the remaining parts of the original building during my stay in Cairo in 1964.

The good reputation these and other hospitals enjoyed, especially during the medieval period, attracted eminent physicians as well as brilliant students to complete their training and study, and to work in them. By and large, hospitals in Islam were well financed and patients enjoyed good medical care, service and healthy environment. Here, no matter what class of society they belonged to, poor people found a helping hand, and chronic cases met with constant attention. Lepers and others with a variety of contagious diseases were admitted into special halls and quarters provided for that purpose and were separated from patients with uncontagious fevers, eye infections and cases requiring minor surgery. Nevertheless the well-to-do and the ruling classes relied on private medical treatment by family doctors in the comfort of their homes.

The history of many of these hospitals—the Nūrī and the Manṣūrī previously mentioned are good examples—show that they were well furnished with sufficient supplies of drugs and food and equipped with adequate libraries and lecture halls. Thus, they could operate without delay when construction was completed. As long as the administrative body functioned efficiently, the hospital was able to obtain additional properties, philanthropic endowments and gifts to make it self sustaining (11).

Libraries in Islam were both state and privately owned, and were particularly numerous in state capitals and the larger cities (12). They were established and financed by rulers as well as learned men and could be found adjacent to palaces, schools, religious centers and hospitals. Caliph Al-Ma᾽mūn enlarged and generously endowed the House of Wisdom (*Bayt al-Ḥikmah*) in Baghdad. Al-Ḥakam II founded an imperial, rich library in Cordova. Ibn Sīnā (d. 1037), in his autobiography, tells of the extensive library of Bukbārā where he made good use of the manuscripts in his early youth (13). We are also told that ᶜAbd al-Raḥmān b. Muh. b. Futays, one of the most learned men in Cordova and a contemporary of Ibn Sīnā, had collected, in his time, more books than any private citizen in al-Andalus. In addition to purchases, he hired six men to copy still more books to enrich his own library (14). Similarly, history records that the judge and bibliophile-historian Jamāl al-Dīn al-Qifṭī (d. 1248) (15) had accumulated a large number of books and donated them to the city of Aleppo shortly before his death. ᶜAlī b. Yaḥyā al-Munajjim (d. 888) had a rich library in his palace which he called the Treasure of Wisdom and it was open to readers of every type. Also the Maqrīzī in his *Khiṭaṭ* (1:445, 458-9) describes the House of Knowledge Library founded in Cairo by the Fatimids.

That legacies of many cultures were transmitted from other languages into Arabic through translations, and that paper and writing material were abundant to facilitate literary productivity, contributed to great interest in books and in learning in a way that can hardly be exaggerated. The translation activity that flourished in Baghdad (after 800) and the availability of manuscripts and libraries in hospitals or owned by eminent physicians helped to establish medical education and practice on solid foundation and provided for continued progress.

Translations and the Development of Medical Education

Teaching the healing arts in hospitals and private medical schools under tutorship of competent doctors and providing adequate and high-caliber texts and references to students and practitioners were not only possible in the ninth century, but became firmly and professionally established as well. These educational centers and their libraries were either privately administered or were in connection with hospitals. Books, translated or original, soon became plentiful, and students flocked to prepare for a highly respected profession. The literary productivity was accompanied by a wide public interest and ever growing demand for manuscripts. Because of this, Arabic became—and for three centuries remained—the *lingua franca* for all branches of health and other sciences. The texts, although based mainly on the Greek, provided the pattern and source for information and teaching in Islam for several centuries to come and were important in the revival of medical and pharmaceutical teaching and practice in the West in Latin versions.

The name that deserves to start the list of contributors to such development is that of the Bakhtīshūᶜ (meaning servant of Jesus). This family of physicians that faithfully guarded the Hippocratic tradition rendered a great service to the rise of medical education in Islam and helped in the establishment of hospitals in the Abbasid capital. Jūrjīs b. Bakhtīshūʿ of Jundī-Shāpūr was summoned to Baghdad in 766 for the treatment of Caliph al-Manṣūr (16). His son Bakhtīshūʿ and grandson Jibrāʾ il and their descendants served the caliphs as well as in hospitals for several generations (17).

The star that shone brightly in the first half of the ninth century was Yūḥannā b. Māsawayh (c. 777-857), a Nestorian, the son of an apothecary, who served several caliphs as court physician from al-Rashīd to al-Mutawakkil. He seems to have been an intelligent and shrewd physician who not only made a fortune for himself but taught and promoted the healing arts enthusiastically as well. His private medical school was recognized as one of the best in Iraq. He made several translations from the Greek and composed about twenty medical books, according to the *Fihrist* which was completed in 987 (18). Important among them was *Miḥnat al-Ṭabīb* in which, following the example of Galen, he warned against quacks and charlatans who deceive the public. He urged his readers to seek the advice of learned and highly trained physicians. Less than a century later, al-Rāzī wrote on the same theme also to help turn the tide against the increasing number of quacks intruding on the practice of medicine and pharmacy (19).

In 1964, I found at the Rabāt National Library a manuscript dated 9th century A. H. (c. 1459), No. J 404, of special interest. It contains treatise one and part of treatise two of *al Taṣrīf* by al Zahrāwi (d. c. 1013) bound in one volume with two treatises and a medical text ascribed to Ibn Māsawayh. The first manuscript, entitled *al-Taqsīm*, on the classification of medicine, ascribed to Māsawayh, seems spurious since no such title is mentioned by either al-Nadīm, Qifṭī nor Uṣaybiʿah and is not in Māsawayh's style. It appears that this short treatise is a compilation taken from contemporary and later works, such as those of Ḥunayn to be discussed shortly. It divides medicine into theory and practice, and defines the four elements, their qualities, temperaments, humors, faculties and the three spirits. Then it speaks of diet, work, rest, respirations, hygiene and health, causes of diagnosis, and apparent and hidden symptoms of diseases. Finally, it discusses medical treatment, methods, and the ten areas to be considered in treatment including condition of patient, his temperament, age, climate and season.

The genuine Māsawayh's contributions are no doubt the second and third parts in this volume. The second, the *Mushajjar*, on diseases, was praised by Qifṭī as an important text (20). It discusses skin diseases, headaches, amnesia, lethargy, melancholia, epilepsy, hydrophobia (21), pleurisy, colic, dropsy, diabetes (mellitus), gout and sciatica, and the diseases of the eye, heart, stomach and liver, with definitions, causes, symptoms, methods and recommended medications for treatment of diseases in 65 chapters. In most cases, the Syriac and Greek synonyms for diseases are given. We find later on that al-Rāzī wrote on similar topics in his book entitled *al-Tashjīr wāl-Taqsīm* of which there is a copy in the British Museum (Add. No. 5932).

The third part is Māsawayh's book on fevers, *al-Ḥummayāt*, the first known, independent treatise in Arabic, which influenced later authors on the subject (22). As in the earlier manual, he intended the text to be in the form of tabulated charts (*mushajjar*). He defines fever as: "An external heat [rise in temperature] not originated from man's nature, but spread out from the heart through veins [blood vessels] to all parts of the body thus hurting the natural activities [functions]. It is of three types: [1] fevers in the psychic, animal or natural spirits recurring daily; [2] hectic or putrid fevers in the solid organs or the humors; [3] septic fever in the chyme." He then gives detailed description of causes, symptoms and methods of treatment including several recipes.

A copy of Māsawayh's *al-Nawādir al-Ṭibbīyah* is to be found in a bound volume at El-Escorial in Spain (No. 5240). The author seems to have written it for his student, Ḥunayn b. Isḥāq. It stresses the limitless boundaries of the healing art and shows the need for studying what the ancients wrote and for using the drugs that they had found most useful by experimentation. He sounded the Hippocratic call for a return to nature, equilibrium in the body's humors, use of contrasting symptoms and remedies (the *contraria contrariis*—old therapeutic theory), and emphasized a preference for using simple remedies whenever and wherever possible rather than the compounded drugs. Māsawayh said: "The physician should imitate nature in his approach to the treatment of diseases even if such an arrangement took a longer time and greater effort than other seemingly useful methods contrary to nature. These [unnatural] concepts for treatment are deceitful as are the claims of alchemy." This indicates that alchemy was flourishing in the early ninth century, contrary to the claims that put it about a century later. Then Māsawayh gives advice on having a competent family doctor whom one trusts, and who has considerable knowledge in relation to meteorology and general hygiene. In his concluding remarks, he praises Ḥunayn as being the first to translate into Arabic one of the most worthy books of Galen—that on the treatment of diseases entitled *Ḥīlat al-Burū* (23).

Māsawayh's treatise on barley water (Māʾ al-Shaʿīr) also was an original contribution. According to Arabic medical history, this short work was highly recommended throughout this medieval period, and it influenced later authors (24). The subject of his treatise on preventive therapy against the harm caused by certain diet was taken over and expanded by al-Rāzī and was taught to students of the health professions for several generations (25). He also wrote on bathing (*al-Ḥammām*) and on headaches (*al-Ṣudāʿ*), two favorite topics for later practitioners. Through Greek influence, his two other treatises on phlebotomy and cupping (*al-Faṣd wal-Ḥijāmah*) and on dentifrices, pioneered these practices in Arabic medicine.

The *Paradise of Wisdom* (*Firdaws al-Hikmah*), by Māsawayh's contemporary ᶜAlī b. Sahl Rabbān al-Ṭabarī (d. *ca* 855 (26), although a magnificent contribution to Arabic medicine, was in less circulation than later medical texts. In it, al-Ṭabarī seems to have relied on Greek, as well as eastern sources for medical information. Several copies of the book are extant, and it was edited by Siddiqi in 1928 (27). The majority of the medical treatises by al-Kindī, his companion at the Muᶜtaṣim court, however, are lost (28).

The man who contributed the most to the development of medical education and practice and promoted professional ethics and standards was Abū Zayd Ḥunayn b. Isḥāq al-ᶜIbādī (Johannitius, 809-873), a one-time student of Māsawayh and a contemporary of the Philosopher of the Arabs, al-Kindī. His books and translations constituted the basis of medical and pharmaceutical teaching in Islam. The *Fihrist* lists about 30 works by Ḥunayn on natural sciences, particularly the healing arts. For instance, he wrote on milk, diet, bathing (as Māsawayh did earlier), drugs, stomachaches and what today is termed embryology (29).

His writings on ophthalmology, especially the ten treatises on the eye, which were completed by his nephew Abū al-Ḥasan Ḥubaysh b. al-Aᶜsam, surpassed in importance the *Daghal al-ᶜAyn* of Māsawayh (30). Ḥunayn's sources of information, besides personal experience and observations as an oculist—like those of Māsawayh—were based on earlier Greek writings, most of which he translated into Syriac and Arabic. With this manual, he established on a firm basis the study of the anatomy and physiology of the eye and the treatment of its diseases in Arabic (31). Authors of leading texts on ophthalmology written later relied heavily on Ḥunayn's work. One of the best known of these texts, the *Thesaurus for the Oculists (Tadhkirat al-Kaḥḥālīn)* by ᶜAlī b. ᶜIsā, completed at the close of the tenth century, is a good example. It was one of the most comprehensive and widely read texts by both the students and the oculists, and the one most often consulted by the *muḥtasib* when examining eligible oculists for official licenses. In the introduction, the author explicitly gives credit to both Ḥunayn's and Galen's writings on the subject (32). *Kashf al-Rayn* on the anatomy, physiology and treatment of the eye by Muh. b. Ibrāhīm al-Anṣārī, known as Ibn al-Akfānī (d. 1348 in Cairo), also was influenced by these earlier texts. Because of Ḥunayn and succeeding oculists, Arabic ophthalmologic texts reached a high standard unparalleled in medieval medical literature, and influenced authors writing in Latin.

Another important treatise by Ḥunayn is his *Fī Ḥifẓ al-Asnān wāl- Laththah*, on medically protecting the teeth and gums. Unlike the work on dentifrices by Māsawayh, this independent treatise was the first to give special emphasis to oral hygiene and the function of the teeth in relation to the entire digestive operation of the body. Thus it is of particular interest to the history of dentistry. His treatise confirms the fact that bad eating habits, indulgence and indigestion seriously affect the teeth and vice versa. To prevent tooth decay, he warns against chewing gummy substances and biting hard matter. He also recommends that for effective mouth hygiene, one should avoid drinking sour milk and very cold or snow water. The treatise urges regular cleaning of teeth after meals and gives several recipes for dentifrices, some of which contain *nūshādir* (sal ammoniac) and alum. It warns, however, against use of henbane *(banj)* "which has side effects (causes trouble) and which, unfortunately, many doctors use" (33).

Ḥunayn's best known work on the healing art was his *Masā> il fī al-Ṭibb lil-Muta <allimīn*, generally known as *Masā> il Ḥunayn*. Written primarily for the medical students, it became the most widely circulated text, referred to, and consulted by students and practitioners of the health professions and often used by examiners such as the *muḥtasib*. It was completed, with additions, by Ḥunayn's nephew and very close associate, Ḥubaysh b. al-Hasan al-A <sam (34). This work did more to define and fashion Arabic medical thought and teaching throughout the medieval period—its doctrines, scientific and technical approach, procedures and concepts—than any other Arabic text of its size. Despite what now seems a cumbersome, catechismal form of arrangement, the question and answer manner in which this work was written had been, at the time, a simple tool for teaching and facilitating reading, reciting and understanding of the subject matter. It assisted students and practitioners in their studies. It also guided teachers in quizzing students and aided medical examiners, such as the *muḥtasib*, in examining newly graduated physicians and oculists previous to issuing licenses for the legal practice of their profession. In the classification, the author divides the healing art into two parts: the theoretical and the practical (*<ilm wa-<amal*), divisions that were followed faithfully by later authors and which influenced such classification in Latin medical texts in the late Middle Ages (35). This system was fully utilized by <Ali b. <Abbas al-Majūsī (Haly Abbas d. 994) in his *Kāmil al-Ṣinā <ah*, known as al-Malakī, for this whole medical encyclopedia is divided into two parts—each consisting of ten treatises—the theoretical and the practical.

In his *Masā> il* Ḥunayn also discusses what he termed the "seven natural matters"; namely the elements, the body's temperaments, humors, major and minor organs, natural powers (faculties), actions and reactions (af <āl) and spirits, which he divides into natural, animal and psychic. Obviously, he based his interpretations on Greek writings which he studied and translated from the works of Hippocrates, Galen and others. These ideas thus, were founded upon certain modified definitions and, with ramifications, on the Greek humoral theory. This theory still continues to dominate medical thought, and has left its impact on language and literature as well. Of particular medical and historical interest is the space Ḥunayn devoted to a discussion of the "six common and natural (health essential) principles (*al-asbāb al-sittah*)" which ultimately govern sickness and health. No doubt, he derived them from the Greek legacy, mainly from the writings of Galen. Strangely enough, these six principles were known as the "non-naturals" in later Latin texts. He explains the ways and means in which they are applied to maintain equilibrium, the time, quality and quantity. He lists the six principles (essentials) as: the air we breathe; the regular intake of food and drink; work (including physical exercises) and rest; wakefulness and slumber; vomiting and the use of enemas; and whatever affects us emotionally such as worry, fear, anger and joy. Together with these principles, he applied psychic therapy. This praiseworthy system, which is derived primarily from Greek sources, was widely adopted by later Muslim authors and became an integral part in the structure of Arabic medical teaching (36). Later, Ibn Sīnā (980-1037) incorporated this idea into his medical encyclopedia *al-Qānūn fī al-Tibb* without giving any credit to Ḥunayn or the Greeks. Therefore, some practitioners believed that the system was genuinely Ibn Sīnā's. Before the middle of the eleventh century, al-Mukhtār b. Butlān provided his valuable, tabulated text, *Taqwīm al-Ṣiḥḥah* (Latin, *Tacuini Sanitatis*, 1531), most of which is based on the six principles as presented by Ḥunayn. The interpretation and additions by Ibn Butlān deserve further attention, and I hope to devote a brief discussion to it in a later publication. Before the close of the twelfth century, Ḥubaysh b. Ibr. al-Tiflīsī wrote *Taḥṣīl al-Ṣiḥḥah* which promotes the same idea—that good health can be achieved by careful observation of the "six principles" and by maintaining an equilibrium.

The *Masā>il* also contains several passages of interest in pharmaceutical history which were incorporated into the teaching curriculum. The author gives instructions and recommendations regarding simple and compounded drugs, techniques and methods of preparation, rules and means of testing those drugs for the purpose of determining their physical properties, and the pharmacological effects they have on man. Ḥunayn was the first Arab physician to justify the use of compounded drugs and presented certain practical cases where such therapy was needed, such as a patient having more than one ailment at the same time. This concept was dogmatically opposed by physicians who never practiced what they preached and who used compounded drugs, electuaries and theriacs in what is known, too often, as polypharmacy. Fortunately, there were others who were more sensible and honest who admitted that such need existed and repeated in their texts most of the reasons given earlier by Ḥunayn. We can mention in particular, the very interesting work *al-Irshād* by Hibat Allāh Ismā <īl ibn Jumay< (d. 1198), a Cairo physician to Saladin (Ṣalāḥ al-Dīn) (37).

In his *Masā>il* Ḥunayn discusses other questions regarding the preservation of health, the senses, symptoms of chronic and acute diseases and fevers, and the diagnosis and methods of medical treatment.

There were some Muslim writers who argued against the concepts and teaching of Ḥunayn such as <Alī b. Riḍwān al-Miṣrī (d. 1068), who seems to have been prejudiced and tried to alter certain details in the *Masā>il*. However, he was refuted by later authors (38). Despite refutations (rudūd) and counter refutations, Ḥunayn's contributions and influence, as well as that of some of his associates, such as Ḥubaysh, on the development and solidification of medical education, ethical standards and practice in Arabic medicine can hardly be exaggerated. Through their translations and books, they established the foundations and drew the outlines for the important doctrines and concepts of the health profession which were generally followed by later generations of practitioners in medieval Islam and through translations of these works in the West (39).

The *Masā> il* of Ḥunayn was further commented upon, abridged and imitated by later authors. For example, about the middle of the eleventh century, Abū al-Ḥasa Sa<īd al-Nīlī made an abridgement entitled *Talkhīṣ (Ikhtiṣār) Masā> il Ḥunayn* in three treatises (40). Later on, the Christian physician Abū Naṣr Sa<īd b. Abī al-Khayr (d. 1193) wrote a medical catechism, *al-Iqtiḍāb <alā Ṭarīq al-Mas> alah wāl-Jawāb* in which he not only copied many of the doctrines by Ḥunayn, but also followed the same style and method of approach (41).

On medical botany, Ḥunayn helped in reviewing and correcting the Arabic translation of the five treatises of Dioscorides' *Materia Medica* by Iṣtifān b. Basīl. Although the translation was not entirely complete, the text and terminology helped to promote the dissemination of further knowledge of drugs and therapy. The book was studied by students of medicine and pharmacy and was often consulted by later authors. Muslim herbalists and physicians built upon this firm foundation and added personal observations on crude drugs in their natural origins. Thus, by the thirteenth century, Arabic manuals on *materia medica* and the therapeutic and pharmacological effects of drugs, as in the works of Ibn Juljul, al-Zahrāwī, al-Ghāfiqī, al-Sharīf al-Idrīsī, Ibn al-Bayṭār and Ibn al-Ṣūrī, reached their highest expression (42).

It has been previously explained that Ḥunayn and his associates translated into Arabic the best of the Greek medical legacy. This included a large part of the Hippocratic corpus, the writings of Galen (especially those known as *Jawāmi Jālīnūs,* compiled and organized in sixteen texts and annotated by the Alexandrian physicians), and the works and compilations of Oribasios, Aretaeus of Cappadocia and the seven books of Paul of Aegina (43). In addition, Ḥunayn enriched other areas of Arabic culture. He was probably the first to translate the New Testament (probably from the Greek and Syriac), a translation that must have been consulted by such historians as al-Dīnawarī (d. *ca.* 889), Muh. b. Jarīr al-Ṭabarī (833-923), and al-Ya< qūbī (d. 897) (44). In addition, Ḥunayn's treatise on understanding and grasping the truth (*Idrāk Ḥaqīqat al-Diyānah*) is a masterpiece in the history and philosophy of religion (45). Unfortunately, this work is almost unknown in learned circles. The short epistle, *al-R. al-Shāfiyah,* on drugs used to cure amnesia and other medical recipes to promote good health is not by Ḥunayn. It is bound together with a copy of the medical formulary of the *Aqrābādhīn* of Ibn al-Tilmīdh, which dates 902 A. H. (1496), and obviously belonged to his son Isḥāq (d. 911) (46). This is a rare and important manuscript.

By late ninth century, students as well as practitioners found adequate material to continue their studies, teaching and work, and an opportunity to promote professional standards. Copyists, some with a high degree of education, were kept busy meeting the ever-increasing demand for books or to enrich their own libraries or those of their patrons. The highly organized book market is a reminder of similar development of stationaries in university neighborhoods in Europe in the late Middle Ages (47).

MATURING OF ARABIC MEDICAL EDUCATION AND PRACTICE

During the late ninth and the tenth centuries, Arabic medicine and pharmacy, teaching and practice, reached its maturity. At the top of a myriad of eminent physicians, teachers, and practitioners who helped to bring about this great achievement stands the figure of Abū Bakr Muh. b. Zakariyā al-Rāzī (d. 925). He was the first physician known to have refused to surrender blindly to the authority of Galen's doctrines and speculations as evident in his book *al-Shukūk ᶜalā Jālīnūs* (Doubts on Galen's writings). He also challenged al-Kindī's refutation of the alchemists' claim that they could transmute base metal into silver and gold. Al-Rāzī affirmed, however, that the probabilities for alchemy and transmutation were greater than the argument for denial. He wrote about fifteen books and three epistles on alchemy and the philosopher's stone by which he established chemical teaching and techniques which pioneered what we now term chemotherapy. His treatise *On Smallpox and Measles,* has been translated into several languages. Thāhit b. Qurrah (826-901) wrote on the same topic earlier, but his epistle is lost. He also preceded al-Rāzī in his treatise *On the Pains of the Joints,* including gout (also, unfortunately, lost). Above all, al-Rāzī was a great medical educator and his fame attracted students from far and near to attend his lectures. Al-Nadīm gives us an excellent but brief description of the setting and the methods in which al-Rāzī conducted his class and lists his numerous books and epistles (48). Many of these works, which emphasized clinical medicine and teaching at the patient's bedside, were utilized as references and medical texts and were consulted, taught and copied by educators, practitioners and students. From the late twelfth century when they were translated into Latin, and for over three centuries, they influenced medical education and practice in Europe as well (49). Because of the importance of these writings a few of them will be considered briefly.

The largest and most famous among al-Rāzī's medical texts is his *al-Ḥāwī fī al-Ṭibb* (*Liber Continens*), a comprehensive compilation constituting personal data collected for teaching and as reference material. It was generally divided into twelve parts (apparently there are more than that) on the treatment of disease from head to foot including fevers, skin diseases, hygiene, treatment of fractures and bone setting, drugs, diet, pharmacy in medicine, weights and measures, general anatomy, physiology, an introduction to the healing art and its study, terminology and definitions (50). Due to the nature of the contents, style and scope, it is safe to assume that the author kept adding and altering parts of it until his death. An earlier and widely circulated work was his *al-Mansūrī* (*Liber de medicina ad Almansorem*) in ten treatises (51). The ninth treatise, in particular, played an important role in Western medical teaching (52).

As a true follower of Hippocrates, al-Rāzī was deeply influenced by the parts which were translations of the Hippocratic corpus, such as *Prognostic, Regimen in health and in acute diseases, Fractures and joints, Nature of man* and the *Aphorisms*. This last work was highly regarded by medical educators and practitioners in Islam. Al-Rāzī felt that the text needed, and indeed it lacked, organization. As a result, he revised it with annotations and personal comments in his *al-Murshid,* or *al-Fuṣūl fī al-Tibb* (53).

Medical educators and physicians in Islam, moreover, following the Hippocratic tradition, emphasized the importance of adequate diet for maintaining good health. Al-Rāzī pioneered writings on diets for the sick and how to prevent harm caused by inadequate, inappropriate, or spoiled food or diet to which people are "allergic". Several authors followed his lead. A good example was the Heret physician, Najīb al-Dīn al-Samarqandī (d. 1222), who wrote a treatise on diets and drinks for the healthy as well as for the sick (54). Al-Rāzī also gave special attention to pharmacy as an important part of the medical curriculum. He wrote on available, easy-to-get and inexpensive drugs, the price of which even poor families could afford and thus be cured. Furthermore, despite controversies concerning the authenticity of *al-Fākhir,* I have reasons to believe that it was al-Rāzī's and that several copies of it are still extant (55). In twenty-six chapters, it discusses recommendations on drug and diet recipes by eminent physicians.

Al-Rāzī also fought against fraud and quackery in medicine and in his writings deplored the ignorant practitioners who deceived the public. He worked hard to uphold professional standards. He insisted on theoretical as well as practical training at the patient's bedside and in the hospital. He wrote on his clinical observations at the hospital and in most of his books, including the *Continens,* he directed his students to report their observations. He took a further step and wrote on medications to be used in minor illnesses when there was no physician available (56). Several Muslim physicians later composed books on the same topic. These texts in Arabic were in circulation and often well written, a forerunner of later home remedies, people's common-sense medical advisers, and modern household dictionaries for diagnosis and treatment in Europe and America.

Of special interest to medical education was the attention al-Rāzī paid to the physician-patient relationship and the maintenance of high professional ethics. He urged the physician to ask his patient in a friendly way questions concerning his condition, then allow him to relate his case history. The physician, he insisted, should show that he is interested in his patient's answers and in his explanations in expressing his feeling. The physician, meanwhile should keep careful watch over the patient's general condition, keeping in mind the following seven points, which al-Rāzī mentions in his *al-Murshid* and asks that his students and readers follow:

1. To define the disease from the symptoms and clinical examination.
2. To find out the reasons for this particular type of illness.
3. To deduce from causes as well as symptoms whether the particular case involves one or more diseases or types and to define them.
4. To distinguish adequately one type from another.
5. To recommend treatment by diet, drug or both.
6. To gain the patient's confidence and his readiness to respond willingly to his physicians advice and also to build up the patient's general attitude and morale.
7. To forecast what was going to take place and thus warn the patient of what might happen before it does, as suggested in the Hippocratic *Prognostics*.

In the *Fākhir*, moreover, al-Rāzī advised physicians against indulgence in bad habits and the lusts of the flesh. He also explained that, "since man uses drugs to combat diseases, he therefore should devote further attention to seeking what is good and noble that will bring joy to the soul and drive sorrows away."

There was an abundance of available books which led learned men to introduce criteria for what constituted the best and the most useful reading. Al-Rāzī's countryman, al-Majūsī (d. 994), in the introduction to his *al-Malakī* outlined eight rules for judging the worth of any book, of which every author ought to be reminded (57).

These are:

1. To make clear the author's objectives and motives for writing it.
2. To explain the benefits that can be derived from reading it.
3. To have a title relevant to the subject matter.
4. To spell out methods, concepts and doctrines adopted by the author.
5. To name the author.
6. To give the author's competence.
7. To explain contents and validity of his writings.
8. To have proper organization.

Of further interest to medical education and ethics were the recommendations made in *al-Malakī's* first treatise, chapters one and two, by al-Majūsī to physicians to adhere to the Hippocratic oath (which Ḥunayn translated into Arabic). He urged the physician to be honest, skilled, resourceful, kind and compassionate, to shun evil and cling to what is good. During his visits, the physician's only aim should be to relieve the patient's suffering and to care for and promote his health and general condition without divulging the patient's secrets. The physician should abstain from over drinking, refrain from vices, and care for the sick by regular hospital visits. In so doing, he would gain a good reputation and the respect of all concerned.

Earlier, in the late ninth century, Isḥāq al-Ruhāwī wrote a complete text on the ethics of the physician. He lamented how untutored practitioners, who were unable to comprehend the intricate principles involved in the health field and owing to lack of skill, turned to quackery and perversion and disgraced the profession. To guard against such intruders, he was challenged to write on deontology and explain the duties, responsibilities and honors associated with the practice of the healing arts. He devoted chapter twelve of his book to an explanation of the high calling of the physician and the importance of his profession. Then he spoke of the honors (chapter thirteen) and trust bestowed on highly qualified physicians. In it he tells the story of caliph al-Mu<taṣim who gave more respect and honor to his physician Salmawayh b. Bīnān than he gave to the supreme Justice (Qāḍī al-Quḍāt) (58).

Another text with a closer connection to deontology as it relates to medical education and practice, is a manuscript at the Ẓāhirīyah Library (no. 4883, pt. 2) bound with a copy of Uṣaybi<ah's <Uyūn. This thirteenth century, five-chapter treatise might very well be the book Uṣaybi<ah wrote earlier in life and referred to in his <Uyūn. In chapter one, he speaks of what a physician should be—moderate, virtuous, religious, prudent, honest and faithful, willing to do good to others, diligent in his studies of the writings of the ancient sages, friendly, compassionate, especially mindful of the poor, and one who follows and upholds the Hippocratic oath. If need be, he should travel to far-off places to further his knowledge, and he should be ready to listen to the truth and follow it. He explains how some entrusted the care of their patients to charlatans with the result that they worsened or died instead of recuperating. The author also criticizes those who consulted the physician only when they were sick and needed medication and urged that each individual should have periodic check-ups even when healthy so that by following his physician's advice he could preserve his health.

Chapter two is on the knowledge and academic training needed by anyone preparing for the health profession. That person should study the essentials of logic, anatomy, geometry and mathematics, metrology, astrology, climatology, geography, meteorology, optics and music. He should, in addition, know the various medical schools, philosophies and what each stood for. The Greco-Roman medical sects never flourished as such in Islam but their methods and concepts influenced authors and practitioners. He also added that the physician should be acquainted with other natural sciences. For each of the above-mentioned disciplines, he lists the utilities derived from each by the physician.

Chapter three concerns what a physician should acquire after meeting academic requirements. It thus explains the need for regular attendance at hospitals and visiting and examining of patients with various diseases.

Through hospital training at the bedside, the physician examines common and rare cases which previously he only read about in books. After observing these cases, he should document their causes, symptoms and treatment. Then, in entering the hospital, he should observe certain regulations: to maintain silence, be dignified, have a kindly attitude to his patients and their attendants, listen patiently to them and explain exactly what he wants each to do. He should not spare advice or service to his patients and should always be ready to answer their call, the rich as well as the poor. He should remain at the patient's bedside, facing him, and not prescribe the needed medication until he has questioned, examined and observed the symptoms of the disease for which he gives the prescription to be filled.

Contrary to what was mentioned in chapter one, chapter four concerns the things physicians should avoid; hastiness, impetuousness, cowardice, inconsiderateness, greediness, anger, ridicule of others, impiety, conceit and pride. The physician should never tell a patient whether he will live or die for he may be proved wrong.

Chapter five is concerned with the testing of the competence of a physician to beware of the charlatan and honor the competent and the virtuous. Earlier, Sinān b. Thābit convinced al-Muqtadir (908-932) to issue and enforce an edict that all physicians in Baghdad should take and pass an examination by Sinān before receiving a license to practice the profession.

Insight into practical aspects of medical education can be deduced from biographies of the eminent physicians that history records. For example, there is the biography of Sulaymān b. Juljul who was born about 332 A. H. (c. 943) in Cordova. He studied Muslim tradition in 952 at the mosques of the Muslim capital in Spain and continued his studies of the Arabic language and syntax, logic and theology until 968. When he was fourteen years of age, he studied medicine, and began practicing when twenty-four. He wrote a commentary on Dioscorides in 982 and completed his best known book on the categories of physicians in 987. He died at the end of the century (59).

In the first treatise (chapter three) of *al-Malakī*, al-Majūsī wrote: "If one wishes to reach the highest level of preparation for practicing medicine he should study the books of logic and the four principal branches of knowledge, [the quadrivium]: arithmetic, geometry, astronomy and music (60). Inasmuch as logic balances our reasoning and straightens our dialectic approach and utterances and is useful in every field of knowledge, so are the four principal sciences." His junior contemporary, al-Zahrāwī (d. c. 1013) in the second treatise of *al-Taṣrīf*, gave interesting remarks concerning the psychology of child education. He urges fathers to take special interest and care in bringing up their children. He advises that children should be trained in doing what is good and should be directed toward virtue and good behavior. If so brought up, they would be more apt to follow these ways when older. When of age, each child should be sent to a *kuttāb* (elementary school) under the direction of a good, gentle and compassionate tutor for Qurʾān and religious teaching and the study of Arabic syntax, reading, writing and grammar. Thereafter, the child should be sent to another tutor for the learning of arithmetic and geometry, then astronomy and music. After doing well in these courses and passing the examinations, the child should study logic and philosophy. Thereafter, he could specialize in one of the other sciences such as the healing art. Concerning the tutor, al-Zahrāwī adds "if the tutor notices special talent in a student in a particular field of study, he should encourage and develop such talent until he excels in such line of specialization" (61). This shows how seriously interested al-Zahrāwī was in adequate education, who gave us a delightful, valuable, yet critical approach to general education with emphasis on the force of habits in reshaping future conduct. It agrees with Hitti's statement that: "A child's education begins at home ... when six years old the child is held responsible for the ritual prayer and his formal education begins." The *kuttābs* were mainly adjunct to mosques or in them, and the curriculum centered at the beginning upon the readings from the Qur ᵓān and recitations of selected religious passages (62).

The division between elementary and higher education, however, was sharp, and while tutors of elementary classes were regarded only as religious educators, teachers of advanced studies were highly honored. Thus, much of the importance of a scholar depended on the esteem enjoyed by his teacher (shaykh), and their licenses (ijāzah) were honored in the same way that a degree was regarded when obtained by graduates from European universities in the late Middle Ages and thereafter (63). The system of education in Islam, as stated by K. Bakhsh was based largely on voluntary efforts with considerable freedom in teaching and the selection of topics. Debates, the fear of criticism and frequent ridicule compelled teachers and competent physicians to prepare their lectures well. The establishment of large libraries was a matter of pride to some and a devotion to learning to others. Unlike universities in Europe, teachers in Islam were not restricted to a rigid syllabus. Physicians frequently gave lectures in logic and astronomy in addition to those related directly to the healing art. Fees were collected from students, and often patrons and rulers gave living allowances to famous and competent teachers (64).

In Arabic Spain, the teaching of medicine was characterized by discussions, seminars and frequent free debates between students and teacher, a precedent to thesis defense practiced later on in European universities (65).

Of another nature in consequence and application, were the debates among educators themselves which, in certain cases, enriched Arabic medical literature with worthwhile additions. A case in point was the debate and correspondence between Ibn Sīnā and Abū al-Rayḥān al-Bīrūnī (66). Still more publicized and producing wider consequences was the controversy between al-Mukhtār b. Buṭlān of Bagdad (d. *c.* 1068 near Antioch, Syria), and his contemporary Ibn Riḍwān of Cairo. The letters exchanged between them contributed greatly to the evaluation of Greek learning among the

Arabs and its appeal in learned circles (67). Moreover, rulers and patrons, such as the Fatimid al-Ḥākim in the early eleventh century, held regular scientific symposia to which famous teachers in mathematics, logic and medicine were invited for discussion and debate. Calif al-Mustanṣir (d. 1242) appointed a skillful doctor in al-Mustanṣirīyah school in Baghdad to lecture on medicine, and for this service he paid him well.

Earlier, Niẓām al-Mulk (d. 1092) founded in Baghdad the Nizamiyah College, which was completed in 1067. It was the first of its kind in Islam, and was followed in 1168 by one in Damascus by Nūr al-Dīn, the founder of the famous Nūrī hospital. The latter college included lecture, conference and teachers' rooms, in addition to a mosque, dormitory for male students, dining room, kitchen and storage, all of which was financed by the state. Teachers who organized private schools for advanced and specialized subjects such as medicine were well paid and enjoyed prestige. Students traveled even from far off places and flocked to hear the famous scholars. In these schools they were allowed to change from one subject to another depending on the field of study they preferred or in which they proved to be more competent.

Women were taught separately by private tutors, often in their own homes or palaces. Many were trained as nurses, mid-wives and gynecologists (68).

The Leveling-Off Period

With the writings and enthusiastic and prudent efforts of Ḥunayn, al-Rāzi and al-Majūsī under the Abbasid caliphate in the East, Isḥāq b. Sulaymān and Ahmad ibn al-Jazzār (d. 1009) in North Africa, ʿArib b. Saʿīd al-Qurṭubī, Sulaymān b. Juljul and Khalaf al-Zahrāwī (Abulcasis d. c. 1013) in Spain, Arabic medicine reached its peak and unmistakably exhibited its independent features and characteristics. Both teaching and practice became firmly established at this time for supplies of high quality texts were available in abundance. It has already been noted that there were three types of schools or methods of instruction in which medical education was conducted and can be listed as follows:

1. Schools connected with hospitals such as those that flourished at the ʿAḍudī hospital in Baghdad, the Nūrī in Damascus and the Manṣūrī in Cairo. In them there were lecture rooms, libraries, pharmacy shops, and storage and manufacturing rooms for drug preparations—electuaries, syrups, ointments and decoctions—in addition to the storage of medicinal herbs. These schools were ideal for teaching theoretical courses by attending physicians, and the students obtained practical training by visiting patients in regular rounds with their teachers and by sitting at the bedside. The direct influence in the development of such an institution came in a minor way from the school of Alexandria, but in a more major way from the famous Christian medical school and hospital of Jundī-Shāpūr in southwestern Persia. This last served as a pattern for early hospitals in Islam and physicians

who were first trained there came later to Baghdad to establish or to supervise such institutions. In connection with hospital organization came the use of medical titles such as chief physician, instructor, attendant, and what are known today as the position of a resident physician or intern. Sinān ibn Thābit, mentioned earlier, headed the Muqtadirī hospital in Baghdad and was appointed chief physician by al-Muqtadir (908-932). In this capacity, he attempted to promote public health in cities and rural areas and the use of hygienic measures for prisoners (69). Hibat Allāh ibn al-Tilmīdh, whose student body numbered fifty, was the chief physician at the <Aḍudī hospital (from the Syriac sā<ūr) until his death in 1165 (70). The title of chief physician is found more frequently in Syria and Egypt after the eleventh century. In Egypt, it was given to <Abd Allāh al-Shaykh al-Sadīd (d. 1196), Ibn al-Nafīs (d. 1288), and to Yūsuf ibn al-Maghribī (d. 1374) (71).

2. Private medical schools which were run by eminent physicians whose fame attracted students from near and far to attend their lectures. Uṣaybi<ah quotes Yūsuf b. Ibrāhīm as saying: "I found that the lectures (classroom or seminar circle) of Yūḥannā b. Māsawayn were the most popular in the whole of Baghdad [which indicates that there was more than one medical school] for a physician, logician or philosopher." (72) Al-Nadīm also reports that al-Rāzī's lecture room was crowded with advanced and beginning medical students who flocked to hear him (73). Many such physicians wrote texts with the view toward their use as manuals by their students and followers. These works continued to be used as reference books by later generations. Al-Zahrāwī in the introduction to his *al-Taṣrīf* speaks of the students as his children. During the course of their study, one or more of the advanced students would come across ambiguous passages, indecisive, dubious or undefined statements which needed further explanation. Therefore, they would ask their teacher to write a commentary or an abridgment of the text under discussion to make its contents more readily comprehensible. The teacher not infrequently agreed. The outcome would be a new commentary; often voluminous, but seldom practical enough to meet the need for which the task was undertaken in the first place. Arabic (and for this matter, Persian and Turkish) libraries were amply supplied with commentaries on such books as the *Aphorisms* of Hippocrates, the *Masā>il* of Ḥunyan and the *Qānūn* of Ib Sīnā. In the long run, one faces the situation of having commentaries or elucidations on commentaries, on commentaries.

In Arabic Spain many prominent physicians during the first half of the eleventh century had studied under the famous educator-mathematician-physician Maslamah al-Majrīṭī (74). In Damascus, the chief physician <Abd al-Raḥīm al-Dakhwār established a medical school named for him, al-Dakhwārīyah, and many students were attracted to his lectures. This school continued to operate after his death (75).

3. Private medical tutoring by which a student became an apprentice to a well-known doctor whom he would call his master (shaykh), or to his father

or older members of the family as in the Bakhtīshū<, Qurrah and Zuhr families. The tutor would teach his apprentice the venerable, traditional texts in sequence of subject: the *Masā>il* of Ḥunayn as an introduction to medicine, and his translations on anatomy and surgery, medical sects, regimen, prognostic and aphorisms; the *Manṣūrī* of al-Rāzī; and the *Malakī* of al-Majūsī. In addition, the master would instruct the student in the ethics, objectives and social aspects of the profession. As the apprentice advanced, he would be allowed to join the master on his visits to patients and when he performed surgical operations. When fully trained in the theory and practice of medicine, the apprentice was allowed to practice medicine. In later centuries, a certificate (*ijāzah*) would often be issued and signed by the master testifying to the qualifications and competence of the new "graduate".

In his *al-Malakī*, on the one hand, al-Majūsī repeatedly referred to the fact that as a student he was trained under the tutorship of Abū Māhir Mūsā b. Sayyār of Jurjan whom he seemed to revere. On the other hand as a medical tutor, Uṣaybi<ah spoke of his brilliant student Abū al-Faraj ibn al-Quff (d. 1286) and how he taught him the healing art. At the time of writing al-Quff's biography, he seemed proud of the fact that this one-time student of his had become a famous physician-surgeon during the teacher's life-time (76).

In the tenth century, the two sons of Yūnis al-Ḥarrānī, <Izz and Aḥmad traveled from Spain to Iraq to obtain medical training and to attend the lectures of such renowned physicians as Thābit b. Sinān and Ibn Wāṣif. After ten years of such medical education they returned to practice their profession in the then great capital of the Western Caliphate, Córdova. Caliph al-Ḥakam II (961-976) allowed Aḥmad al-Ḥarrānī to establish the first-known, government-sponsored pharmacy shop at the palace and to dispense medications to patients (77).

The excellent compensation and prestige that the health professions attained in Islam left the door open for competition. In order to safeguard and promote medical services and public health, some rulers issued—on the advice of their most trusted physician—edicts to establish minimum requirements for the licensing of qualified practitioners, especially in larger cities. This arrangement, however, never attained complete enforcement of the law in most Muslim cities. The *muḥtasib* and his aides inspected the drugstores, examined practitioners, oculists and even surgeons. He issued licenses, enforced laws against adulteration of drugs or any other commodity, and inflicted punishment for cheating in weights and balances (78). However, these attempts to protect the interest, safety and well-being of citizens did not stop peddlers from selling their commodities or prevent untutored practitioners and old women from dabbling in drugs or using bloodletting and cauterization for cures.

In his epistle on Egypt, Umayyah b. Abī al-Ṣalt (1068-1134), lamented the decline in medical education and practice and commented on the increase in the number of quacks and charlatans among practitioners. He pointed out that they were only interested in deceiving people for monetary gain. All they knew were the names of a few drugs which they prescribed for practically every disease. He wrote that the outstanding physicians were primarily Christians and Jews, and added "when I first came to Egypt I was extremely interested in the writings of Galen and Hippocrates, and I searched hard to find a good teacher to benefit from his knowledge and found none . . . only a group with which mere contact and friendship causes the brain to rust" (79).

Two other dangers threatened the continued progress of medical education and practice in Islam. The first generated from religious prejudice and taboos and the blind interpretations of the faith by narrow-minded theologians. The introduction of religious restrictions dimmed the secular image of the healing art that was vividly exhibited by early Arabic, medical educators. Practitioners were less free to think and interpret their observations; hence, initiative and originality were curtailed.

The second danger was the awesome power of authority, whereby ancient writings were revered and accepted as gospel truth. Anyone suggesting changes was considered to be either odd, a heretic, or lacking in educational qualifications. In Iran and Iraq, from the eleventh century on, a new and important Muslim figure in medicine and philosophy dominated the scene, the chief physician (al-Shaykh), Abū ᶜAlī al-Ḥusayn Ibn Sīnā (Avicenna 980-1037). His bio-bibliography is available in several languages (80). He was born a prodigy who in his youth showed intellectual genius to the extent of surpassing his own teachers; however, he took the study of medicine very lightly. He gave little time to it and throughout his adult life never devoted full time to its practice. He was engaged in a multiplicity of intellectual, social and political affairs. Thus, when he embarked on the writing of his medical books, he relied heavily on compilation, reorganization and regrouping of data from books by earlier authors, most of which sources he never acknowledged. Nonetheless, he contributed important and well organized interpretations and personal speculations. His style, except for verbosity, is elegant and admirable. Through his intelligent, persuasive approach, and the compiling and organizing of contemporary medical knowledge, the star of Ibn Sīnā shone brightly. There is no doubt that his commendable contribution and competence in fields other than medicine and pharmacy—philosophy, logic, metaphysics and natural science—helped to widen his fame. His reputation as a medical author soon surpassed that of most of his predecessors and spread tremendously for centuries after his death not only in Islam but also in Christendom. Unfortunately, and in Islamic countries particularly, this prestige held a sway over the minds of medical educators and practitioners who followed in his footsteps. As a result, progress in Arabic medicine and pharmacy has been thwarted and impeded up to the modern period. In many universities in Europe, the chair of Avicenna was recognized even after the Renaissance.

His voluminous, five-book medical encyclopedia, *al-Qānūn fī al-Ṭibb*, is verbose and full of unnecessary classifications despite the author's repeated claim of being brief and to the point (81). The same can be said of his other medical works: the *Urjūzah* (*Cantica Avicennae*, a poem of mediocre quality); the treatise *On Cordial Remedies*, which provides neither an adequate physiologic or pharmacologic discussion nor any contribution to *materia medica*; and the *Sakanjabīn* epistle (on honey, vinegar and water mixture) (82). Ibn Sīnā apparently transmitted this practice to his admirers, an attitude almost foreign to the early scholarly activities led by Ḥunayn in Iraq, al-Rāzī in Iran and al-Zahrāwī in al-Andalus. A great many Muslim physicians after Ibn Sīnā, especially in the Eastern part of the Islamic domain were followers and blind imitators. This fact caused an immediate leveling off of medical education and development in those places where his prestige was held high, and hindered new progressive ideas. Under the spell of his elegant and persuasive style, his type of reasoning, and his great fame as a philosopher and natural scientist, no one dared to violate, disapprove or think ill of his teaching. His books were read and recited as authoritative references by medical students and practitioners alike. It was fashionable to consider that the closer a physician followed in the steps of the "master of physicians," and comprehended his doctrine, the more reputable he became. His medical books, especially the *Qānūn* and the *Urjūzah* were diligently studied, copied and commented upon by authors and compilers. Numerous copies of texts and commentaries are still extant in many libraries as an evidence of their great popularity. Furthermore, the *Unānī* medicine which has thrived through the centuries up to the present in India, Pakistan, and neighboring countries was mainly the offshoot of Ibn Sīnā's medical teaching and that of his time.

Several highly qualified physicians and pharmacists in al-Andalus, Egypt, and Syria, however, awakened to this danger which was threatening creativity, initiative and originality in medical education. They refused to accept blindly such dogmas unchallenged. Moreover, they criticized Ibn Sīnā's medical books and labeled them as worthless and repetitious of earlier works. From an anonymous commentary, probably on Ibn Jumayʿ's *Taṣrīḥ*, a copy of which is in the Ẓāhirīyah Library, there seems to have been a bitter attack and even ridicule of Ibn Sīnā's work and fame. Involved in this controversy were such important medical educators of the twelfth century as Ibn al-Tidmīdh of Baghdad and Ibn Jumayʿ of Cairo. In his *al-Taṣrīḥ*, Ibn Jumayʿ tells of a certain merchant who brought a handsomely decorated copy of the *Qānūn* from Iraq to al-Andalus and presented it as a gift to Abū al-ʿAlāʾ ibn Zuhr (Avenzoar, d. 1131 in Córdova). After reading the work, the latter found fault with it and refused to enter it among his library holdings. Instead, he used the paper for the writing of his prescriptions (83). Ibn Rushd (Averroes, d. 1199) had the same opinion regarding the *Qānūn*, but not of the *Urjūzah* about which he wrote a commentary for his patron. There is also documentation that ᶜAbd al-Laṭīf al-Baghdādī (d. 1231) was at first an admirer of Ibn Sīnā and enthusiastically read his books, as he did with most available material on alchemy. After being exposed to other authors and gaining more experience and deeper knowledge, he wrote, "The more I read the writings of the ancient [Greeks] the more I admire them. But the more I read Ibn Sīnā the more I despise him, as I have also discovered the falsehood and futility of alchemy. . . . I thus was saved from two great and disastrous errors, a fact for which I double my thanks to God who deserves my praise, for more people were destroyed by the books of Ibn Sīnā than by alchemy" (84).

TEACHING AND PRACTICE OF HEALTH PROFESSIONS OTHER THAN MEDICINE

Although previously the health professions have been discussed in general, medicine was the more emphasized field. For this reason, certain qualities of these closely related professions will be presented here. It is difficult, however, to draw a line of separation. An overlapping in medical practice was the rule rather than the exception as in the case of Abulcasis al-Zahrāwī who was simultaneously a pharmacist, a physician, a surgeon and an oculist. Nevertheless, specialization was taught and honored in Islam and in certain cases adhered to strictly.

In the apothecary's art, pharmacy became—for the first time in history—an independent, well-defined profession in the early ninth century in Islam. Privately owned drugstores accessible to the public, as well as pharmacy shops connected with hospitals and palaces, were firmly established and often run by qualified pharmacists in central areas (85). Pharmacy thus became a twin profession helping with medicine to promote health. Al-Bīrūnī (d. 1048) defined the pharmacist as a professional "who collects the best and purest of drugs—simples and compounded—and prepares them as best described by eminent physicians." He then added that "pharmacy became independent from medicine as language and syntax are separate from composition, the knowledge of prosody from poetry, and logic from philosophy, for it [pharmacy] is an aid [to medicine] rather than a servant." He also wrote of training brilliant students for this profession so that they could differentiate good drugs from bad. He criticized charlatans, untutored perfumers, and apothecaries and gave them a different title (ᶜaṭṭār) from that he gave to the highly qualified pharmacist (ṣaydalī naṭāsī). He also differentiated between pharmacy and pharmacology: "The knowledge of the faculties and characteristics of the *materia medica*" (86). Members of this profession in cooperation with competent physicians helped to produce some of the finest compendia and manuals of *materia medica* written in the Middle Ages.

The first text on pharmacy was written by Sābūr (d. 869) and was used in private as well as hospital pharmacies (87). Although such a text was not rigidly followed or enforced, pharmacists used and regarded it as their guiding formulary. Less than a century later in Spain, a similar text was written by Ibn ʿAbd Rabbih, a copy of which exists at the Ẓāhirīyah (*Catalogue*, 236-41). These two works were surpassed by the compendium of Ibn al-Tilmīdh (d. 1165), a text that was the pharmacist's guide for a century. In 1260, al-Kohen b. al-ʿAṭṭār of Cairo completed his detailed text on the art of the apothecary, which continued in use for centuries (88).

Several eminent herbalists, physicians and medical botanists wrote manuals on *materia medica*. They were influenced by the translated work of Dioscorides that was almost surpassed in many areas by the works of al-Zahrāwī, al-Ghāfiqī (d. 1165) and Ibn al-Bayṭār (d. 1248), each of which contained original contributions.

The development of ophthalmology followed a similar pattern to that of medicine, for many physicians were oculists as well. The training was mainly in the hospital or by apprenticeship. Oculists devised several instruments, some of which showed ingenuity and fine taste for surgical treatment of the eye. Treatises on anatomy and physiology of the eye, its diseases and treatment, were written as early as the first half of the ninth century. Manuals for oculists and students were plentiful, and show the high caliber of Arabic ophthalmology.

Dentistry continued to be a part of the medical profession. Physicians occasionally devoted sections in their medical texts to discussions of oral hygiene and the instruments used in dental treatment. The best known are the illustrated chapters in *al-Taṣrīf*, by al-Zahrāwī, which include tooth extraction, plugging, crown and bridge appliance, filling and prosthetics.

Physiology and surgery in Islam, as in Christendom during the Middle Ages, had very few competent promoters. This was mainly due to the taboo on human dissection. Fewer books were written on surgery than on any other theme related to the healing arts. This gives added reason to admire those few who dared to explore and tackle such a difficult and often unpopular subject among practitioners.

It must be noted, however, that educated surgeons during this period were never considered less important than the well-qualified physicians. But only a few were devoted to practicing the art of surgery in view of the dangers, difficulties and uncertainties involved. To meet the ever increasing demand for surgeons the door was left wide open for bone setters, leechers, barbers, bloodletters and charlatans. More and more practitioners turned to internal medicine using diet and drug therapy and avoiding surgery. Notwithstanding, leading physicians from the ninth through the thirteenth centuries taught anatomy and physiology and emphasized their importance in the appreciation and practice of the healing art.

The first essential step in developing this field was to translate earlier texts on anatomy and surgery, especially from the Greek, a task undertaken by Ḥunayn and his associates in the ninth century. Indian surgery had also some bearing on Arabic medicine and few doctors were reported to have come to practice in the Abbasid capital (89). At the same time, Ibn Masawayh emphasized and encouraged special attention to the subject in medical education. Al-Majūsī also wrote on the importance of applied surgery, especially minor manipulations and on fractures and treatment of wounds.

Foremost among the surgeons, however, was Abū al-Qāsim (Abulcasis) al-Zahrāwī (d. c. 1013). In his *al-Taṣrīf*, he devoted a complete treatise to surgery which included cauterization, bloodletting, cupping, bone setting, extraction of arrows, bladder and kidney stones, midwifery and use of instruments, delivery as in parturition, treatment of fractures and wounds, use of various types and makes of threads for binding in operations, and methods of checking bleeding. Moreover, he emphasized the need and value for teaching anatomy and for training in surgery. In reading the text, one cannot escape the feeling that the author himself performed operations such as tonsillectomy and treatment of "thyroid sacs". He seems to have been a careful observer, a diligent instructor and a competent practitioner. His drawings of the various instruments he used, devised, and recommended for instruction and practical application, are an important contribution to illustrative surgical texts and are the earliest extant illustrations known. These were copied and imitated by later authors and scribes. His regret was that attention to surgery was declining in his time. It deteriorated further after his death. *Al-Taṣrīf* received scant attention in Arabic educational circles in Spain and in the East. In the West, however, it received better treatment and met with more enthusiasm, especially after the surgical treatise was translated into Latin by Gerard of Cremona (d. 1187) (90).

In Islam, it was not until the thirteenth century that competent anatomists and surgeons appeared on the scene. Ibn al-Quff of Jordan (d. in Damascus, 1286) attempted to revive the teaching and practice of anatomy and surgery. He also wrote the most comprehensive Arabic manual on the subject, *al-ʿUmdah*, in twenty treatises (edited in Hyderabad, 1937, 2 vols.). In it, he interpreted the connection between arteries and veins and discussed the capillaries as tiny pores unseen to the naked eye. He was the first to explain and to point out the physiology of the cardiac valves and the direction in which they open and close. He also appealed for pan-Arab unification of standards of weight and measure in medicine and pharmacy. His pleas, like those of his predecessor al-Zahrāwī, were scarcely heeded.

Ibn al-Quff's classmate at the Nūrī hospital, ʿAlāʾ al-Dīn ibn al-Nafīs (d. in Cairo, 1288), although praised in contemporary Arabic records for his great knowledge of drugs and medical therapy, made his best contribution in the field of anatomy and physiology. In his commentary on *Tashrīh al-Qānūn* of Ibn Sīnā, he discovered the pulmonary circulation of the blood and gave it the first simple yet clear explanation. He refuted Galen's theory of the permeability of the interventricular septum of the heart (91).

By the end of the thirteenth century, surgery had declined into the hands of charlatans and bone-setters and, with the exception of a few treatises compiled from earlier works, the decline continued until the nineteenth century.

Duration of the Arabic Medieval Period

Arabic medical education exhibited its highest expression and vigor from the ninth to the eleventh centuries, but survived between weakness and strength until the early fourteenth century. Shortly thereafter and at the time Europe was experiencing the Renaissance and entering into the modern period, Arabic medical education and practice declined to its lowest ebb. The exceptions are few; for example, Dāwūd al-Anṭākī (d. 1599) in his *Tadhkirah* lamented the degraded state which the healing art had reached and tried with slight success to breathe a new life into it (92). A still more promising attempt was made by Ṣāliḥ ibn Sallūm (d. 1670) of Aleppo. He was acquainted with Latin, and possibly the German language, and transmitted some European writings (especially those related to Paracelsus and his teachings) into Arabic. The choice was not always the best, but it was different and sensational. The value of his work, however, was that it opened a window into newer and more vigorous cultures and attempted to get some enlightenment from them. About the middle of the eighteenth century, Muh. s. al-Rūmī presented original ideas on the origin of the universe as well as on bodily digestion. He referred to digestive "juices" in the mouth, stomach, and intestines (93). These men, no doubt, were generations ahead of their time.

The effective realization of modern medical education, however, came into Egypt with the founding of the school of medicine at Abū Zaʿbal, near Cairo, in 1827. The credit goes to the French physician-surgeon and educator, Antoine B. Clot (d. 1868) summoned to Egypt by Muh. ʿAlī. This was the first effective modern encounter in the Arab world of the new European learning. In this first stage of founding a modern medical school, French, German and Italian teachers helped to establish and maintain the educational curriculum. In 1837, the school was moved to the Qaṣr al-ʿAynī in the Egyptian capital to become the nucleus of the present college of medicine at Cairo University. The buildings, clinics, laboratories and hospitals expanded greatly, but it still occupies the old mansion of Qaṣr al-ʿAynī in which it began over 130 years ago (94). In 1866, the American University of Beirut was founded as the Protestant Syrian College. It inaugurated a school for the teaching of medicine and resulted in a revival of medical education in Lebanon and neighboring countries.

In the hundred years that followed, other colleges opened that were devoted to medicine and the other branches of the health professions: pharmacy, dentistry and nursing. In rapid succession colleges opened in Beirut, Damascus, Baghdad, Cairo, Alexandria, Asyut, Tunis, Algiers, Rabat and Tripoly. They are all modeled after European or American systems of medical education or a combination of the two, but they have kept their indigenous and authentic features. There was hardly any inclination to revive or associate curricula with the medieval Arabic legacy. It was a new start based and closely integrated with modern Western advances in the health sciences. With the exception of the Syrian University, Western languages were used as the teaching media. At present, native teachers, many of whom are graduates of European and American universities, are now carrying the torch of medical and pharmaceutical education. Since 1827, therefore, the Arabic people borrowed and learned a great deal from Europe in developing medical education and practice in their respective countries, to which they had transmited much in medieval times.

NOTES

1. ABŪ AL-QĀSIM ṢA<ID B. [for ibn] AḤMAD B. ṢA<ID AL-ANDALUSĪ AL-QURṬUBĪ, *Ṭabaqāt al-Umam*, ed. L. Cheikho, Beirut, 1912, pp. 44-47, and Cairo edition p. 74. The *Ṭabaqat* was translated into French (Paris, 1935), by R. Balchère.
2. See the detailed description given also by <ABD AL-MALIK AL-AṢMA<Ī (d. 832) in <*Antarah*, Cairo, 1961, pp. 4-8; AL-JĀḤIẒ (d. 869), *Rasā>il*, ed. J. Finkel, Cairo, pp. 13-16; and AḤMAD AL-BALĀDHURĪ (d. 892), *Futūḥ al-Buldān*, Leiden, 1866, pp. 2-90, 181-183.
3. S. HAMARNEH, *Index of Arabic manuscripts on medicine and pharmacy at the National Library of Cairo*, Cairo, 1967, pp. 19-24, 68.
4. This Greek influence was clearly and repeatedly emphasized by ABD AL-RAḤMĀN B. KHALDŪN, *Muqaddimah* (Prolegomena), Cairo, n. d., pp. 38, 493-494; GEORGE SARTON, *Introduction to the history of Science*, Vol. 1, Baltimore, 1927, pp. 542-548, 583-588, 621-624; ALDO MIELI, *La science arabe et son rôle dans l'evolution scientifique mondiale*, ed. A. Mazahéri, Leiden, 1966, pp. 57-59, 72, 80, 86-106, 119-127 and 293-294; and DE LACY O'LEARY, *How Greek science passed to the Arabs*, London, 1949.
5. Papyri and several kinds of parchment were used in ancient Egyptian and Greco-Roman civilizations. Parchment continued in wide use in Europe up to the fourteenth century; four centuries after it was replaced by paper in Arab lands.
6. See MUḤAMMAD B. AL-ḤĀJJ, *al-Madkhal*, Vol. 4, Cairo, 1929, p. 84; IBN KHALDŪN, Muqaddimah, 421-423; and MARTIN LEVEY, *Medieval Arabic book making and its relation to early chemistry and pharmacology*, Philadelphia, 1962, and my review, in *Bull. Hist. Med.*, 1963, 37: 384-385.
7. JAMĀL AL-DĪN<ALĪ B. YŪSUF AL-QIFṬĪ (d. 1248), in his *Tārīkh al-Ḥukamā>*, known also as *Ikhbār al-<Ulamā>*, Cairo, 1908, pp. 250-251, mentions that al-Rashīd appointed Ibn Māsawayh to head the hospital he had founded in Baghdad. Māsawayh apparently directed its affairs with the

help of Ibn Bakhtīshū<, thus becoming the first known physician-director of a hospital in Islam. See LUCIEN LECLERC, *Histoire de la médecine arabe*, Vol. 1, Paris, 1876, pp. 558-572.

8. S. HAMARNEH, "Development of hospitals in Islam," *Jour. Hist. Med.*, 1962, **17**: 366-384; and AḤMAD B. ABĪ UṢAYBI<AH, *<Uyūn al-Anbā>*, Vol. 1, Cairo, 1882, pp. 233, 244, 253-254. Here, Uṣaybi<ah tells when and why the hospital at Mayyāfāriqin (capital of Diyār Bakr, known as the city of the Martyrs, Martyropolis) was founded by Naṣīr al-Dawlah under the direction of the Christian physician, Zāhid al-<Ulamā>. He also hints that <Aḍud al-Dawlah merely renovated and enlarged the older hospital (possibly al-Sayyidah's or al-Muqtadirī's) in which al-Rāzī and early tenth-century physcians practiced. See also QIFṬĪ, *Tārīkh*, pp. 250-251, 264; and J. SAUVAGET, *Les monuments historiques de Damas*, Beirut, 1932, pp. 49-54.

9. UṢAYBI<AH, *<Uyūn*, Vol. 1, pp. 239-245; and IBN JUBAYR, *Riḥlat*, Cairo, 1949, pp. 283-284, 325-326, 383-384.

10. UṢAYBI<AH, *<Uyūn*, Vol. 2, pp. 274-260; CYRIL ELGOOD, *A Medical History of Persia*, Cambridge, 1951, p. 162; and SARTON, *Introduction*, Vol. 1, pp. 631, 641.

11. AHMAD AL-MAQRĪZĪ, *al-Khiṭaṭ*, Vol. 1, Cairo, 1853, pp. 407-420; Vol. 2, pp. 55, 379-380, 405-408; IBN BAṬṬŪTAH, *Riḥlat*, Cairo, 1871, p. 134, and Cairo, 1908, Vol. 1, pp. 10-11, 58.

12. See the fine articles: "Background of the history of Moslem libraries", "Arabic books and libraries in the Umaiyad period", and "Moslem libraries and sectarian propaganda" in *Am. J. Semit. Langs. Lit.*, 1935, **51**: 83-113, 114-125; 1936, **52**: 22-33, 104-110, 245-253; 1937, **53**: 239-250; 1937, **54**: 41-61; and 1939, **56**: 149-157; and "Four great libraries of medieval Baghdad," in *Libr. Q.*, 132, **2**: 279-299, by RUTH STELLBORN MACKENSEN; and J. ZAYDAN, *T. al-Tamaddun al-Islāmī*, Vol. 3, Cairo, 1931, pp. 206-212.

13. ZAHĪR AL-DĪN AL-BAYHAQĪ (1106-70), *Tārīkh Hukamā> al-Islām*, ed. M. Kurd <Alī, Damascus, 1946, pp. 52-64; and QIFṬĪ, *Tārīkh*, pp. 269-271.

14. FRANCISCO PONS BOIGUES, *Ensayo bio-bibliografico*, Madrid, 1898, pp. 101-102; and *The Encyclopaedia of Islam*, 1936, Vol. 3, pp. 352-353.

15. <ALĪ AL-QIFṬĪ (see Yāqut's *Dictionary of Learned Men*, ed. Margoliouth, London-Cairo, 1929, Vol. 5, pp. 477-494) is named after his native town of Qifṭ (ancient Coptos) in upper Egypt. The foundation deposits of King Thotmes (III Dynasty) were excavated there early this century, and several of the specimens unearthed are related to pharmacy and drug manufacturing. I have examined these at the National Museum of Cairo, and they include ointment and pigment jars (some of them still contain remains of the pigments), mortars and pestles, spatulas for mixing of ointments, surgical instruments, sieves and stone and metal weights.

16. NADĪM, *Fihrist*, pp. 426-427; Qifṭī, *Tārīkh*, 109-111.

17. Besides the Bakhtīshū<'s, other families which served to promote professional standards were the Tayfūrī's, the Qurrah's and the Zuhr's to name a few, UṢAYBI<AH, *<Uyūn*, Vol. 1, pp. 158, 215-226 and Vol. 2, pp. 64-70.

18. Muh b. Isḥāq b. al-Nadīm, *al-Fihrist*, Cairo, 1929, pp. 425-426, 454; and Uṣaybi＜ah, ＜*Uyūn*, Vol. 1, pp. 175-176.
19. See A. Z. Iskandar, "al-Rāzī wa-Miḥnat al-Ṭabīb," *al-Machriq*, 1960, 54: 471-522; and S. Hamarneh, "Origin and functions of the hisbah system," *Arch. Gesch. Med.*, 1964, 48: 161-173.
20. Qifṭī, *Tārīkh*, pp. 248-290.
21. In this manuscript (written by the practitioner Nasir al-Din Muh. b. Khidr for the oculist ＜Abd al-Qādir b. M. al-Jazīrī), it seems that Māsawayh recognized cases of rabies as well as those with symptoms resembling ascending paralysis (hydrophobia).
22. Later books and treatises on fevers were written by Isḥāq b. Sulaymān al-Isrā＞īlī, al-Rāzī and Abulcasis al-Zahrāwī, al-Tidmīdh and others. See S. Hamarneh, *Index of Arabic manuscripts, op. cit.*, pt. 2, 1967, pp. 64 (Arabic 29-30).
23. This Escorial manuscript 5240 in Maghribī script, dated 1424, deserves further study.
24. Ibn Māsawayh, *Le livre sur l'eau d'orge*, ed. Paul Sbath, *Bull. Instit. d'Egypte*, 1938-1939, 21: 13-24, with French translation. For Māsawayh's other edited works, see my *Bibliography on medicine and pharmacy in medieval Islam*, Stuttgart, 1964, pp. 73-74.
25. Uṣaybi＜ah, ＜*Uyūn*, Vol. 1, pp. 181-182, reports that Māsawayh recommended that one should not drink sour milk (laban) when eating fish, for the combination causes sickness. This idea, which most probably is a myth, is still a Middle Eastern tradition.
26. Qifṭī, *Tārīkh*, p. 155; and Carl Brockelmann, *Geschichte der arabischen Litteratur*, Vol. 1, Leiden, 1943, p. 265 and Supplement, Vol. 1, pp. 414-415.
27. See my *Bibliography* (1964), pp. 106-107, and the *Index of medical manuscripts at the Zahiriyah Library*, Damascus, 1968, pp. 77-82.
28. S. Hamarneh, al-Kindi, a ninth century physician, philosopher and scholar, *Med. Hist.*, 1965, 9: 328-341.
29. Ibn al-Nadīm, *Fihrist*, 1929, pp. 423-424. See also Ibn Juljul, *Tabaqāt al-Aṭibbā＞*, F. Sayyid, Cairo, 1955, pp. 68-72; Qifṭī, *Tārīkh*, pp. 117-122; Uṣaybi＜ah, ＜*Uyūn*, Vol. 1, pp. 184-200; and Moritz Steinschneider, *Die arabischen Übersetzungen aus dem Grechischen*, Graz, 1960, pp. 13-29. Casimir Petraitis in *The Arabic version of Aristotle's Meteorology*, Beirut, 1967, pp. 12-35 speaks of important translations from the pre-Ḥunayn period.
30. This treatise on ophthalmology, the earliest known in Islam, was extracted, annotated by C. Prufer and Max Meyerhof under the title, Die Augenheilkunde des J. b. Masawaih, *Der Islam*, 1915, 6: 217-268.
31. See M. Meyerhof's edition, *The book of the ten treatises on the eye ascribed to Ḥunayn ibn Ishaq*, Cairo, 1928.
32. S. Hamarneh, *Index of the National Library of Cairo*, 1967, pp. 33, 50-51, 53, and Albert Z. Iskandar, *A catalogue of Arabic manuscripts on Medicine and Science*, London, 1967, p. 194.
33. This survey is based on my study of the only known manuscript housed at the Ẓāhirīyah Library, *Catalogue*, Damascus, 1968, pp. 227-230 (in press).

34. Nadīm, *Fihrist*, p. 424; and Qifṭī, *Tārīkh*, p. 122.
35. See Heinrisch Schipperges, "Die arabische Medizin als Praxis und als Theorie," *Arch. Gesch. Med.*, 1959, 43: 317-328.
36. I examined several copies of the *Masā>il*, and reference can be made to the fine manuscript in Dr. Ḥaddād's collection in Beirut, dated 11 Jumādā II, 787 A. H. (1385), and the one preserved at the Medical Library, University of Cairo (no. 20936) which dates 8 Rajab 526 A. H. (1131), the earliest known of this work.
37. Uṣaybi<ah, <*Uyūn*, Vol. 2, pp. 112-115; Brockelmann, *op. cit.*, Vol. 1, p. 643, and *Supplement*, Vol. 1, p. 892; and Hamarneh, *Ẓāhirīyah Catalogue*, 1968, pp. 306-308 (in press).
38. *Ibid.*, pp. 185-191; Uṣaybi<ah, <*Uyūn*, Vol. 2, pp. 99-105; Brockelmann, *op. cit.*, Vol. 1, pp. 637-638, and *Supplement*, Vol. 1, p. 886.
39. Aḥmad B. Khallikān (1211-82), in his *Wafayāt al-A<yān*, Vol. 1, Cairo, 1885, pp. 209-210, states that "without these translations [of the Greek legacy by Ḥunayn] no one could ever have benefited from the books of the ancient physicians and sages." See also Leclerc, *Histoire*, Vol. 1, pp. 133-152. His works were also praised by al-Majūsī (d. 994) in the introduction to his *al-Malakī*.
40. Uṣaybi<ah, <*Uyūn*, Vol. 1, pp. 253-254. The abridged copy of the *Masā<il*, Ṭibb 1206, in the National Library of Cairo was consulted.
41. I examined several manuscripts of *al-Iqtiḍāb* including Ẓāhirīyah no. 4715 (T. 53), *Catalogue*, 1968, pp. 316-318; and Ṭibb M. Fāḍil no. 5 at the National Library of Cairo (in 69 fols., 16 lines, 16 × 20 cm. in size and dates c. 1639). See Brockelmann, *Supplement*, Vol. 1, pp. 892-893.
42. For over a decade following 951, work continued on the translation of Dioscorides into Arabic by translators in the Moorish capital in Spain. See Ibn Juljul, *Tabaqāt*, Cairo, 1955, introduction by F. Sayyid; Uṣaybi<ah, <*Uyūn*, Vol. 2, pp. 46-53, 133; and Meyerhof, Arabian pharmacology, *Ciba Symp.*, 1944, 6: 1847-1875.
43. Nadīm, *Fihrist*, pp. 416-421; Qifṭī, *Tārīkh*, pp. 85-92, Uṣaybi<ah, <*Uyūn*, Vol. 1, pp. 108-126 and Vol. 2, p. 136; J. Zaydān, *T. al-Tamaddun*, 1931, 3: 152-163, 182-190; and G. Bergstraesser, Neue Materialen zu Hunain ibn Isḥāq's Galen-Bibliographie, *Abh. Kunde Morgenlandes, D.M.G.*, 1932, 19: 7-80.
44. See al-Ṭabarī, *T. al-Rusul wāl-Mulūk*, Vol. 1, pp. 417-432; and Aḥmad B. Abī Ya<qūb, *T. al-Ya<qūbī*, Vol. 1, al-Najaf, Iraq, 1964, pp. 2-67. Unfortunately, such acquaintance with the Biblical accounts diminished in Islamic records and in the nineteenth century scarcely influenced Arabic cultural life. Even today, the Bible is almost a closed book in many areas and segments of Arabic culture. This accounts for what I believe caused a lack of great Arabic literary contributions in these areas to world literature. In the British Museum, I found a Ms. Or. 8612 comprising parts of the New Testament dating from about 900. From its style and phraseology, it appears to be based on Ḥunayn's translation, and thus becomes the first-known version in Arabic.
45. Ḥunayn's treatise was edited by Paul Sbath in *Vingt traités philosophiques et apologetiques d'auteurs arabes chrétiens*, Cairo, 1929, pp. 181-200.

46. The manuscript at the National Library of Medicine in Washington, D.C. was consulted. See USAYBI<AH, <Uyūn, Vol. 1, pp. 200-201; and BROCKELMANN, op. cit., Vol. 1, p. 227, and Supplement, Vol. 1, p. 369.
47. For Arabic influence on European higher education, licensing, academic costumes, book markets and lectures, see HEINRICH SCHIPPERGES, "Einflüsse arabischer Wissenschaft auf die Entstehung der Universität", Nova Acta Leopold. 1963, 27: 201-212; and Arabische Medizin und Pharmazie auf europäischen universitäten, Pharmac. Ztg. 1963, 108: 1197-1202; CHARLES H. HASKINS, The rise of universities, Ithaca, N.Y., 1957, pp. 4-5, 34; DAVID RIESMAN, The story of medicine in the middle ages, New York, 1936, pp. 49-66; and PEARL KIBRE, The Nations in the Medieval Universities, Cambridge, Mass., 1948, pp. 99-100. For further comparison see The Encyclopaedia of Islam, Vol. 3, Leiden, 1936, p. 365, where there is mention of some sort of non-official guild of practitioners in Islam.
48. IBN AL-NADĪM, al-Fihrist, pp. 429-434. See also QIFTĪ, Tārīkh, pp. 178-182; and M. CASIRI, Bibliothèque Arab.-Hispan. Escur., Vol. 1, Madrid, 1776, pp. 262-266.
49. FERDINAND WÜSTENFELD, Geschichte der arabischen Ärzte und Naturforscher Göttingen, 1840, pp. 40-49; A. R. HALL, The scientific revolution (1500-1800), London, 1954, pp. 1-3; and The legacy of the middle ages, ed. C. G. Crump and E. F. Jacob, Oxford, 1962, pp. 62, 255-278.
50. D. V. SUBBA REDDY, Al-Rāzī's al-Hāwī, Bull. Dept. Hist. Med., Hyderabad, 1963, 1: 163-187; 1964, 2: 23-32; 1965, 4: 220-225, describes the contents of the Arabic edition of the Rhazes' Continens and gives a biobibliography. A valuable study was made by A. Z. ISKANDAR, Catalogue, op. cit., pp. 1-32. Personally, I consulted the manuscripts at Escorial, Yale University (containing parts 1-8, dated 1674), the two parts at McGill's Osler Library nos. 449-450; the Cairo and Vienna copies, and that at the National Library of Medicine (A 17). An abridgement of the Continens was made by DĀWŪD B. ABĪ AL-BAYĀN, a copy of which is listed by Arthur Arberry, A handlist of the Arabic MSS. in the Chester Beaty Library, Vol. 1, Dublin, 1955, p. 11, no. 3029. IBN JULJUL, Tabaqāt, 1955, pp. 77-80 mentions how Ibn al-<Amīd bought the original copy of al-Hāwī, comprising 70 treatises, from al-Rāzī's sister and then asked his students to put it in order.
51. In the introduction, AL-RĀZĪ declared, "I am writing to my patron, the son of my Lord, Prince Abū Sāliḥ Mansūr ibn Isḥāq . . . a book in ten treatises." They deal with the anatomy of bodily organs, temperaments, natural faculties, humors, drugs, diet, hygiene; medical treatment, cosmetology and skin diseases; antidotes, surgery, pharmacy and the compounding of drugs; diseases from head to foot and their treatment; and fevers. I consulted the copy in the National Library of Medicine (A 28) and others.
52. Liber nonus ad Almansorem cum expositione Sillani de Nigris, by PETRUS DE CURIALTI TUSSIGNANO (d. 1410), and Pandectae medicinae, by MATTHAEUS SYLVATICUS (d. 1347), printed by Matthaeus Moretus at Brescia 1486, and in Venice, 1542; and SCHIPPERGES, "Bemerkungen zu Rhazes und seinem Liber Nonus," Arch. Gesch. Med., 1963, 47: 373-377. Judging from a still

extant copy, Andreas Vesalius (1514-16) seems to have read the text carefully and commented on it.
53. I consulted the copy in the Garrett collection at Princeton University, no. 1076 (243 B, in 88 fols. dated 1282), and the Ṭibb Ṭalʿat manuscript no. 594 (fols. 446-456 A) at the National Library of Cairo. The *Murshid* was edited with annotations by A. Z. Iskandar, Cairo, 1961.
54. He wrote *Fī Aghdiyat al-Marḍā*, and *Dafʿ Maḍārr al-Aghdhiyah*. Also a Samarqandī (d. 1222) manuscript, Ṭibb no. 1 at the National Library in Cairo was consulted. See also G. S. A. Ranking, *The life and works of Rhazes*, London, 1914, pp. 237-268.
55. Ḥājjī Khalīfah, *Kashf al-Ẓunūn*, Cairo, Vol. 2, p. 162; and Wahlwardt, *Die Arab. Handsch.-Verzeichnisse der Koen. Bibliothek zu Berlin*, Vol. 5 (1893), pp. 516-517. I also consulted the Rāzī's manuscript on drugs, Yale University, no. 6 in the Cushing collection; the manuscript number 9521 at the Ẓāhirīyah, and the Ḥaddād copies of the *Fakhir*.
56. Muh. al-Sinjārī al-Akfānī (d. of the plague in Cairo 1348), wrote *Ghunyat al-labīb ʿind Ghaybat al-Ṭabīb*. I examined a copy of it at the American University Library of Beirut (see Brockelmann, *op. cit.*, Vol. 2, p. 171), in which the author emphasizes the six health principles introduced earlier by Ḥunayn. See also Ṭibb Manuscript no. 1118 at the National Library of Cairo.
57. A manuscript on ophthalmology at the American University of Beirut, the Maʿlūf collection no. 415 (fol. lr) mentions that these eight points were first outlined by Ḥunayn. See also al-Majūsī's *Kāmil al-Ṣināʿah*, Būlāq, Vol. 1, p. 9.
58. For Ruhāwī's work, see Martin Levey, *Medical ethics of medieval Islam*, Philadelphia, 1967, pp. 18-19, 70-72; and the fine article Die Bildung des Arztes, *Arch. Gesch. Med.*, 1966, **50**: 337-360 by Christoph Bürgel. For Salmawayh, see Uṣaybiʿah, *ʿUyūn*, Vol. 1, pp. 164-170.
59. Muh. bin al-Abbār (1199-1260), "Tecmila" de Ebn al-Abbar, ed. González Palencia, Madrid, 1915, pp. 297-298. See also Juljul's *Ṭabaqāt*, ed. F. Sayyid, Cairo, 1955, Introduction.
60. This reminds us of a similar arrangement in the medieval educational curriculum in Europe, the seven liberal arts: the trivium and the quadrivium. See Nathan Schachner, *The medieval universities*, London, 1938, pp. 14, 42, 51-53, 95-98, 132, 248, 327. It is of interest that Arab physicians were possibly the first to recognize the therapeutic value of music and to write in detail concerning its commendable psychological effect on patients.
61. The Leningrad Manuscript of al-Zahrāwī's *al-Taṣrīf*, fols. 164r-165v, and Taymūr Ṭibb 137, fols. 97-98.
62. Philip Hitti, *History of the Arabs*, London, 1956, pp. 408-409; see the *Encyclopaedia of Islam*, Vol. 3 (1936), p. 350.
63. Arthur S. Tritton, *Materials on Muslim education in the Middle Ages*, London, 1957, pp. 1, 27-41.
64. Khuda Bakhsh, "The educational system of the Muslims in the Middle Ages." *Islamic Culture*, 1927, **1**: 442-443. See also H. F. Wüstenfeld, *Die Academien der Araber und ihre Leher*, Göttingen, 1837, pp. 8-9.

65. Zakī <Alī, R. al-Ṭibb al-<Arabī, Cairo, 1931, p. 40.
66. Uṣaybi<ah, <Uyūn, Vol. 1, p. 239, Vol. 2, pp. 20-21.
67. The epistles were edited with introduction, translation and annotation by Joseph Schacht and Max Meyerhof, *The medico-philosophical controversy between Ibn Butlan and Ibn Ridwan*, Cairo, 1937.
68. Aḥmad Shalaby, *History of Muslim education*, Beirut, 1954, pp. 1-3, 21-23, 41-57, 63-73, 139, 172-174, 198-200; and al-Maqrīzī, *Khiṭaṭ*, Vol. 1, pp. 409-413, Vol. 2, p. 341.
69. Uṣaybi<ah, <Uyūn, Vol. 1, pp. 220-224.
70. *Ibid.*, Vol. 1, pp. 259-264; Yāqūt, *Dictionary of Learned Men*, Vol. 7, London-Cairo, 1925, pp. 243-246. See also Maqrīzī, *Khiṭaṭ*, Vol. 2, p. 406.
71. Jalāl al-Dīn al-Suyūṭī, *Ḥusn al-Muḥāḍarah*, Cairo, 1904, pp. 259-262.
72. Uṣaybi<ah, <Uyūn, Vol. 1, pp. 175-176.
73. Ibn al-Nadīm, *Fihrist*, pp. 429-430.
74. Ibn Ṣā<id, *Ṭabaqāt al-Umam*, pp. 107-108.
75. Uṣaybi<ah, <Uyūn, 2:239-246; and Muh. A. Ghunaymah, *T. al-Jāmi< āt al-Islamiyah*, Tetuan, 1953, p. 127.
76. Uṣaybi<ah, <Uyūn, 2: 273-274; and S. Hamarneh, Thirteenth century physician interprets connection between arteries and veins, *Arch. Gesch. Med.*, 46: 17-26.
77. Ibn Juljul, *Ṭabaqāt*, pp. 97-98; and Ibn Ṣā<id, *Ṭabaqāt*, p. 124.
78. S. Hamarneh, Origin and functions of the hisbah system, *Arch. Gesch. Med.*, 1964, 48: 161-173.
79. See Umayyah, b. <Abd al-<Azīz b. Abī al-Ṣalt, *al-R. al-Miṣrīyyah*, in F. *Nawādir al-Makhṭūṭāt*, by <Abd al-S. Hārūn, Cairo, 1951, pp. 30-34; and Uṣaybi<ah, <Uyūn, Vol. 2, pp. 52-55.
80. Ẓahīr al-Dīn al-Bayhaqī, *T. Ḥukama> al-Islām*, ed. M. Kurd <Alī, Damascus, 1946, pp. 52-72; Qifṭī, *Tārīkh*, pp. 268-278; Uṣaybi<ah, <Uyūn, Vol. 2, pp. 2-19; and Brockelmann, *op. cit.*, Vol. 1, pp. 589-599, *Supplement*, 812-828. Vol. 1, pp. 812-828.
81. The Rome and Cairo editions of the *Qānūn* were consulted in addition to several other manuscripts. The Ẓāhirīyah copy no. 9729, for example, is written in elegant Naskh, with two decorated title pages (1 and 14) and colophon. It dates from Rajab, 988 A. H. (1580), and was copied by the scribe Muh. Sharif, a nephew to the physician Maḥmūd b. Mas< ūd. The major part of the *Qānūn*, Book I, was translated with introduction and annotation by M. H. Shah under the title, *The general principles of Avicenna's Canon of Medicine*, Karachi, 1966.
82. Several extant manuscripts of these two works exist in oriental libraries. See Iskandar, *Wellcome Catalog op. cit.*, pp. 27-34, 156-165; and my *Index of the National Library of Cairo*, pp. 29-31, 52.
83. Uṣaybi<ah, <Uyūn, Vol. 2, pp. 64-65.
84. *Ibid.*, Vol. 2, pp. 201-212; and Brockelmann, *op. cit.*, Vol. 1, pp. 632-633, and *Supplement*, Vol. 1, pp. 880-881.
85. S. Hamarneh, The rise of professional pharmacy in Islam, *Med. Hist.*, 1962, 6: 59-66.
86. I consulted the Baghdad copy of *al-Ṣaydanah fī al-Tibb*, no 1911, and M. Meyerhof's edition in *Quell. Stud. Gesch. Naturw. Med.*, 1933, 3: 1-52,

and the Arabic text. Al-Bīrūnī confirms that the word *saydanānī* or *saydalānī* (Arabic for pharmacist) comes from the Indian "chandal" for sandalwood, used first as spice and perfume (<*itr*) and later, as a drug. See the *Ẓāhirīyah Catalogue*, pp. 104-120.

87. S. Hamarneh, "Sābūr's Abridged Formulary," *Arch. Gesch. Med.*, 1961, **45**: 247-260.
88. *Minhaj al-Dukkan* by al-Kohen b. al-<Aṭṭār al-Isrā> īlī exists in numerous manuscripts and several editions printed in Cairo and elsewhere.
89. Ibn al-Nadīm, *Fihrist*, pp. 356, 392, 435-439, 448-453, 498-500; and N. H. Keswani, Evolution of Surgery, *Medicine and Surgery*, 1961, **1**: (8-9): 1-3; and Ancient Hindu orthopaedic surgery, *Indian J. Orthop.*, 1967, **1**: 80-93.
90. H. Schipperges, Schmerzbekämpfung in der arabischen Chirurgie durch Schmerzhafte Applikationen, *Ther. Ber.*, 1963, **35**: 89-94; and S. Hamarneh, Drawings and pharmacy in al-Zahrawi's 10th-century surgical treatise, Paper 22 in *Contributions from the Museum of History and Technology*, Smithsonian Institution, *Bulletin 228*, Washington, D.C. 1961, pp. 83-94.
91. Paul Ghaliounguі, *Ibn al-Nafīs*, Cairo, 1966, pp. 109-129; and Iskandar, *Catalogue op. cit.*, pp. 38-42.
92. The medical thesaurus or *Tadhkirah* of Dāwūd al-Anṭākī was edited more than once in Cairo, and was also abridged and commented upon by his students.
93. This primitive concept of the digestive system was thus recognized over half a century before the monumental *Experiments and observations on the gastric juices and the physiology of digestion* by William Beaumont. See S. Hamarneh, *Index of the National Library*, pp. 65-66; and Wilhelm Pertsch, *Die Arab. Handschrift. Bibl. zu Gotha*, Vol. 3 (1881), p. 481, describing the Arabic manuscript on the iatrochemistry of Paracelsus.
94. Antoine B. Clot, *Aperçu générale sur l'Egypte*, Paris, 1840, 2 vols.; and *Compte rendu des Travaux de l'École de Médecine d'Abou-Zabel (Egypte) et de l'examen général des élèves*, Paris, 1833; Aḥmad H. Wahbs, The history of medicine in the U.A.R. *J. Minist. Health*, Cairo, 1960, **1**: 75-77; and A. I. <Abd al-Karīm, *T. al-Ta< līm f. <A. Muh. <Alī*, Cairo, 1938, pp. 251-294.

MEDICAL EDUCATION IN THE MIDDLE AGES

CHARLES TALBOT
Wellcome Institute of the History of Medicine
London, England

The medieval period in the history of medicine, particularly that section of it usually called "Monastic Medicine" has been described by many scholars as a bookish one. By this they mean to imply that medical texts were copied into manuscripts in a more or less haphazard fashion, that the copyists did not understand what they were writing, and that there was no rational purpose behind it. They were, to use Singer's phrase, the result of "scribal accident" or "monastic stupidity" (1). Furthermore, these medical texts (so they say), even when copied down, were seldom if ever studied; they were merely collected and stored away and rarely put to practical use.

Anyone who pauses for a moment to consider these two statements will realise immediately how misleading they are. In the first place, book production was not simply a matter of scribbling something down on parchment. There was an economic problem involved, for sheep and cattle had to be present to provide the skins. There was a highly technical process involved in transforming the skins into parchment. There was a difficulty, owing to the distances between monasteries or intellectual centres, in procuring texts to copy from. And finally there was the physical labour demanded by the mere act of transcription. No one, not even a monk, could do all this in a fit of absent-mindedness, by scribal accident or sheer stupidity.

In the second place, an examination of these texts proves that they were not simply meant to lie on library shelves unread, but intended for the use of pupils and teachers. The dialogue form in which many of them are couched, a form employed in the teaching of the *trivium* and *quadrivium* by such celebrate masters as Donatus, Victorinus, Augustine and Alcuin, is a clear indication of this. Texts beginning with such questions as "What is medicine", "What is an aphorism", "What is phlebotomy", "What is a midwife", obviously connote a teacher-pupil relationship (2). They were, if you like, medical catechisms, to be learned by heart by students under the supervision of a teacher. From the ninth century onwards a spate of medical texts of this kind appeared in monastic libraries, dealing with pulse-lore, phlebotomy, dietetics, pathology, obstetrics, surgery and *materia medica*, and it is from these that we learn how medical instruction was imparted.

The ground covered by these question-and-answer treatises is best revealed by the *Medical responsions* of Caelius Aurelianus and the *Questions* of pseudo-Soranus (3). They deal both with the theory and practice of medicine. On the theoretical side they deal with the definition of medicine, whether it is an art or not and, if so, to what disciplines it is subject; they deal with the principles of medicine and its parts according to the different sects (Hippocratic, Methodist and Empiric), with what is involved in pathology, physic and therapeutics, the division of therapeutics into diet, remedies and surgery, and so on. The question-and-answer texts also show that the students were taught the Galenic principles of reasoning by inductive analogy, that is, of proceeding by way of the more obvious causes and effects to the hidden causes of disease (4). They show too that the students were taught the processes of definition and resolution, and what is more, the need for confirmation of theory by experiment or experience (5). This should cause no surprise, since all these topics had been discussed in Galen's *De sectis* (6) and *Ars medica* (7), both of which had been translated into Latin and commented on at Ravenna as early as the sixth century.

On the practical side there were instructions on how to take the pulse (8), how to diagnose disease from urine (9), and how to recognise illness from pathological symptoms. For instance: "From what signs do you diagnose melancholia?" Answer: "From a distaste for life and dislike of other people's company: from the sadness of the countenance, from the silence, suspicion and irrational weeping of the patient: from the inflammation of the precordia, from the coldness of the limbs accompanied by slight perspiration, from the thinness of the body and the general debility of the subject," and so on (10). This is as good a description as you will find in late classical times.

Complementing these question-and-answer treatises were medical glossaries which explained the Greek terms employed in the description of diseases and physical processes. For instance: "What is orexis? What is cataposis? What is pepsis? What is anadosis?" (11). The elucidation of these terms shows conclusively that medical texts were closely scrutinized by both teacher and pupil in order to ensure a complete comprehension of the material. Allied to these were the illustrations, which were intended to impart instruction by visual means. Many complex problems in the physical sciences, which would have needed a lengthy text to disentangle, could often be resolved by a diagram or series of diagrams. The whole interplay of the elements, the humours, the temperaments, the relation of man to earth and of earth to the macrocosm with much else could be explained in this way (12). Even in the late eleventh century, as is shown by the *Liber floridus* of Lambert of St. Omer, a whole chapter on some aspect of natural science was often comprised in a single drawing; and modern scholars have only just begun to realise that lack of an accompanying text does not signify that the manuscript is incomplete, but that a close study of the diagram makes verbal explanation unnecessary. In medical teaching there had existed an ancient tradition of this kind perpetuated in the so-called *Funfbilderserie* (13), from which, almost at a glance, the student could discover the disposition of the bones, muscles, veins and main organs of the body, a visual method that was later exploited in the teaching of anatomy by Henri de Mondeville, Vigevano and others.

Other types of evidence that must not be neglected in our assessment of medical education in these early centuries are the deontological texts, which not only evince a grave concern for the ethical standards of those who are to practise medicine, but also draw up in some detail the preliminary studies which a candidate was assumed to have completed before embarking on the study of medicine (14). When we consider that the would-be doctor was expected to possess a competent knowledge of all the liberal arts by the age of fifteen, we can, perhaps, better appreciate why his texts were written in a catechetic style.

Already, then, in the ninth and tenth centuries there was, apart from the longer books of Caelius Aurelianus and the so-called Aurelius-Aesculapius complex, a corpus of medical works drawn up solely for teaching purposes. They covered both the practical and theoretical aspects of medicine. Furthermore, they dealt with surgery. It is usually taken for granted that early teaching on surgery was confined to a handful of texts on instruments and surgical practice, all of them so garbled as to be practically unintelligible (15). But the pseudo-Soranus treatise in question-and-answer form is admirably clear; it not only provides precise instruction for the pupil on many surgical procedures, but even gives descriptions of complicated operations, such as that for hydrocephaly, which would not have disgraced writers of later centuries (16). Whether or not such surgical operations were performed is another matter; we are speaking here only of medical education.

What these question-and-answer treatises bring out most cogently is the teacher-pupil relationship. The very brevity of the texts and the complicated configurations of the drawings meant that the authority of the teacher was paramount. The best teacher was he who could expand and elucidate these texts from his long experience and wider acquaintance with other writings, the best pupil was he who could absorb and remember what the teacher said. So learning was a matter of listening, not necessarily of reading. Hence the insistence, not merely in these early centuries but throughout the whole of the Middle Ages, on the biblical phrase *"fides ex auditu"*. Hence also the reiterated quotations from Seneca and Horace, found in numerous prologues to medical books, namely, *Non satis dicitur, quod non satis discitur* and *Saepius repetita placebunt*. It followed logically from this that memory was the chief faculty to be cultivated. In book after book on medical matters a tender sollicitude is shown for care of the memory. Not only was there a special regimen prescribed for the improvement of the memory (combing of the hair included), but in manuals of student behaviour all kinds of pitfalls, such as drunkenness, overeating, *consortium mulierum,* were signalized as destructive of a good memory (17). Needless to say, methods for training the memory, handed down from classical times, were constantly used, and as time progressed these methods became more and more elaborate (18). In his *Practica,* written in the middle of the fifteenth century, Michael Savonarola devotes a long chapter to showing medical students how to recall difficult details and doctrines by assigning them to different parts of a building. Builders of colleges in the late fourteenth and fifteenth centuries even went so far as to stipulate that the ceilings in students' rooms should not be too low, lest they adversely affect *"facultatem memorialem"* (19).

Medical education, however, did not consist solely in learning texts off by heart or in assimilating the words of the teacher. There were discussions between master and pupil, or between the pupils themselves, to ensure that the student thoroughly understood what had been learned. So there developed a different kind of *questio,* not one that required a stereotyped answer, but one that demanded intellectual analysis based on the use of dialectic. This form of *questio* had become common in the schools of theology by the end of the eleventh century. It had precipitated the Eucharistic controversies associated with the name of Berengar of Tours and was responsible for St. Anselm's ontological argument for the existence of God. During the twelfth century, when dialectic became an all-consuming passion, this technique even invaded the schools of grammar, and FitzStephen describes how boys from various schools in London met on festival days to dispute against each other with every kind of syllogism, enthymeme and sophistry (20). In France the same phenomenon could be observed; the *questio* passed from the theological school of Laon to Chartres and from Chartres to Paris. Stephen of Tournai, who had studied at both the latter schools complained bitterly of this new development, calling it *"verborum strepitus et disputationum anfractus"*. The purpose of such "questioning" was not necessarily to get at the real truth of the matter, but rather to show off one's dialectical skill. The result was that in many cases the longer and more complicated the argument, the further away from the original problem the disputants receded.

This passion for employing dialectic coincided, more or less, with the introduction to the West of Constantine's translations of Arabic medical works. The Arabic writers, excluding Rhazes and Avenzoar, were addicted to philosophical argument and subtle analysis of all kinds. They loved classifying, systematizing, dividing and subdividing subjects down to their smallest detail. They had a passion for constructing encyclopedic works in such a logical fashion that it was impossible to separate one part from another without disturbing the whole complex.

Previous to the introduction of these Arabic medical writings into southern Italy, Salernitan medicine, exemplified in the *Passionarius* of Gariopontus, differed little, if at all, from the medical teaching diffused throughout the rest of Europe. This is not wholly unexpected, since the same basic sources, which we have already outlined, were available to teachers and students everywhere. In the prologues and treatises that emanated from the medical school of Salerno we can discern not only the same essential doctrines that had been taught in the ninth and tenth centuries, but also the same traditional methods of teaching by question and answer. At one point, for instance, in the anatomic work of Copho, the teacher is described as holding up the nerves of a pig for the students to examine, and remarking that in the previous year one of the pupils had disagreed with the teacher's explanation (21). Even the seven methods of testing theory by experiment, adumbrated in the earlier ninth century texts (and usually attributed to Petrus Hispanus and Jean de St. Amand of the thirteenth century) are, if we can trust the texts, found fully described in the writings of Bartholomew of Salerno.

Gradually, however, the dialectical methods of the Arts schools and the Arabic medical writers infiltrated into Salernitan medical teaching, and by the end of the twelfth century the philosophical approach to medicine became the established norm. This facet of educational method finds its fullest illustration in the writings of Urso of Calabria.

Meanwhile a corpus of medical teaching, known as the *Articella*, was gradually gaining acceptance as the established curriculum for students of medicine, though whether its origin is to be sought at Salerno or Chartres cannot for the present be decided.

Now, in what did the *Articella* consist? This is an important question, for we want to know exactly what the students were taught, and in what it differed from the older curriculum of medical education. The *Articella* consisted in a set of short texts: the *Isagoge* of Johannitius, the *De pulsibus* of Philaretus, the *De urinis* of Theophilus, three texts of Hippocrates (*Aphorisms, Prognostics* and *Regimen in acute diseases*), and the *Ars parva* of Galen. The *Isagoge* was a mere outline of Galen's system of medicine, a collection of universal concepts on elements, humours, temperaments etc., easy to grasp and more or less self evident. The *De pulsibus* was a compilation, gathered from twenty-six books of Galen, extremely brief, and formulated on a system of question and answer. The *De urinis* was an abridgement of material taken here and there from Hippocratic and Galenic writings. The *Aphorisms*, sententious in form and dealing with both theoretical and practical medicine, were obviously chosen as suitable for being committed to memory, whilst the *Prognostics* and *Regimen* and Galen's *Ars parva* were also comparatively short. All in all the course of medicine outlined in the *Articella* was rather limited and could be read through by an ordinary reader in a few hours. Whether in coverage or content it appreciably improved on the texts available in earlier centuries is debatable. But there was a definite change in the direction of ideas, for whereas through the influence of Caelius Aurelianus, pseudo-Soranos and the work connected with and derived from them, the Methodist school of thought had previously prevailed, the new texts directed students towards Galenic doctrine.

It is quite obvious that these short texts would need extensive commentary if they were to provide students with a complete medical education, so that when the *Articella* with certain modifications and additions, was adopted by the Universities at the end of the twelfth century, the method of teaching was dominated by this necessity. This method of commentary had already been employed in the study of theology, canon law and the liberal arts. As in theology the words of Scripture were interpreted by reference to other parts of the Bible; as in law every word of a decretal was explained by its meaning in some other papal or legal context, so also the medical texts were subjected to this minute scrutinising of every word and sentence. And as the scriptural, legal or literary texts on which a commentary was made were treated as sacrosanct, so the medical texts were set up as monumental authorities; they were to be interpreted and explained, but not doubted or questioned. In this way medical education became a process of learning a set body of doctrine, and then of amplifying it and manipulating it by subtle analysis and dialectic argument.

The best illustration we can give of this development is found in the work of two Englishmen, one from Montpellier, the other from Salerno and Paris; the first belongs to the end of the twelfth century, the other to the first quarter of the thirteenth.

Ricardus Anglicus wrote his *Micrologus*, at the request of the Dean of Beauvais, for young students desirous of obtaining a complete medical education in a short time. Like Gilles de Corbeil, the Paris teacher, Ricardus did not approve of short cuts to eminence, travelling, as he expressed it, from Gad to Ganges in one day. But bowing to necessity, he agreed to provide a succint epitome of medical knowledge in five books. His work, written in elegant and clear Latin deals with the whole of medicine, surgery included. It contains no theorising, no argument, no subtle distinctions, no philosophical analysis. It is purely factual, simple and straightforward.

His compatriot, Gilbertus Anglicus, though a pupil of Salerno, had been influenced by the dialectical methods of the Parisian school and by the writings of the Moslems. The result of this amalgamation, especially of his training in scholastic philosophy, becomes obvious when he begins to speak of efficient, formal, material and final causes of disease, when he distinguishes between substance and accident, genus and species, essence and being, and brings to his analysis of medical cases all kinds of syllogistic reasoning. On the slightest provocation he glides from discussion of symptoms to philosophical argument. If he is discussing fever, he cannot forbear to argue about the essence of heat. If he is treating of tinnitus of the ear, he immediately trails off into an analysis of the essence of sound. If he is talking about the eye, he goes off at a tangent and begins to argue about Aristotle's theory of images. In short, though he is writing about medical topics, he oscillates between two extremes, the purely practical and the purely theoretical or philosophical. Moreover, he fully exploits the method of commentary we have described above. He takes a sentence word by word, meticulously explains each one, puts them together again and finally gives the complete meaning. This was to become the fixed approach of teachers throughout the whole of the Middle Ages, particularly of those who concentrated their attention on the theoretical parts of medical instruction.

The *Articella*, then, formed the basis of all medical instruction; the method of teaching it was by extended commentary.

In what order the various treatises contained in the *Articella* were studied we have no precise evidence, but it is quite obvious that students must have begun with the more general and simple ideas and then progressed to the more specific and complicated parts of medicine. It is only later in the fourteenth century, when the scope of medical education had been widened by the incorporation into the curriculum of additional texts from Galen and the Arabic writers, that we are afforded a glimpse of the organisation of these studies. If we take our example from Bologna, as illustrating what was happening at the same time in other European Universities, we shall see exactly how the medical student was taught during a four years' course (22).

There were usually four lectures a day, two in the morning devoted to theory, and two in the afternoon devoted to practice. In the first year all of the preliminary book of Avicenna (with the exception of anatomy) was dealt with. The first morning lecture (*lectura de mane*, or *prima*) was based on the chapters dealing with the necessity of death, diseases, children's ailments, dietetics and regimen; the second morning lecture explained the book on fevers, and, when this was completed, Galen's *De causis diversis malitie* and *De simplici medicina.* In the afternoon (*lectura de nonis*) attention was given to part of the fourth Fen of Avicenna about the prognostication of crises, the sixth book *De regimine sanitatis*, the second book of Galen's *De diebus criticis* and the last book of Hippocrates' *Aphorisms*.

In the second year the first morning lecture explained the *Tegni* of Galen, then Hippocrates' *Prognostics* and part of his *De victus ratione in morbis acutis*, followed by part of the *De viribus cordis* of Avicenna. The second lecture was concerned with Galen's *De morbo et accidenti*, then *De crisi*, the third book of *De diebus criticis* and finally the *Ars medendi ad Glauconem*. The afternoon lectures in this second year were not devoted to any new texts at all; they merely repeated with extended comment what had been taught during the first year, so that the student should have further opportunity of recalling, in the light of his new knowledge, what he had previously learned.

The two morning lectures of the third year were occupied in explaining Hippocrates' *Aphorisms* (except the seventh book, which had already been studied in the first year), chapters seven to twelve of the *Therapeutics*, some chapters from Averroes' *Colliget*, the first and third book of the *De virtutibus naturalibus*, and the second book of Galen's *De diebus criticis*. The afternoon lectures were taken up by repeating the material learned in the morning lectures of the second year.

In the fourth year the whole of the lecture programme seems to have been taken up with a recapitulation of the material that had already been studied. The first morning lecture went over the subject of the first afternoon lecture of the second year. The second morning lecture explained those parts of Avicenna that has been dealt with during the second afternoon lecture of the first year, whilst the afternoon lectures repeated parts of the *Aphorisms,* much of the *Colliget,* and part of the *De virtutibus naturalibus,* which had been already commented on in the afternoon lectures of the third year.

As can be seen, there was a continual overlapping of instruction, so that the material which had not been fully grasped by the student in the first instance, was carefully repeated at a later stage, in order to ensure that he had a full comprehension of all the subjects dealt with. Here we see put into practice the two classical passages, which writers on medical matters so sedulously quoted: *Non satis dicitur, quod non satis discitur,* and *Saepius repetita placebunt.*

Besides these cathedratic lectures (and we are now speaking of the fourteenth century which, naturally, showed certain signs of progress over the earlier medical course of instruction), there were practical lessons on anatomy, surgery, operations and dissections.

This brings us to an examination of the instruction provided for students of surgery. Until the end of the twelfth century, no distinction was made between the education of the physician and the education of the surgeon. Both had been comprised under the heading "medicine." But after the decree of the Lateran Council of 1215, when clerics in major Orders were forbidden to practise surgery or cautery, it became clear that physicians and surgeons could not be housed (at least in ecclesiastically dominated universities) under one and the same academical roof. Surgeons had to fend for themselves. At Montpellier, for instance, we know that William de Congenis gave lessons on surgery in his own house. Students from the university attended his lectures, and even the university lecturers on medicine came to him, out of term time, for instruction. His method of teaching was to comment on the text of Roger Frugardi of Salerno written sometime before 1170, "but instead of merely repeating the text, he followed the order of the chapters and showed how he himself operated, sometimes approving of what Roger said, at other times changing it for something better, and sometimes even discarding it as contrary to ordinary surgical practice." Instruction was not merely theoretical, it was practical also. The students were taken to the hospital of the Holy Spirit where they could watch actual operations carried out, "because only in this way can a student attain full comprehension of surgical procedures. And besides, this gives him courage, for an essential element in the practice of surgery is boldness." They were shown how to operate on the intestines and other internal organs and on the head; "and once when the Master was trephining a wounded man, one of the students fainted as soon as he saw the brain pulsating. So my advice is, that no one should undertake an operation until he has first seen it performed" (23). Surgery, therefore was considered an art gained by experience and not necessarily dependent on book knowledge, and consequently, though ordinary physicians might disdain the activities of itinerant practitioners, established surgeons often admired their skill. William's pupil remarked: "But take note that these surgeons, who move about from place to place, are often more skillful than famous surgeons because of their constant practice."

In the north Italian cities where a more secular outlook prevailed, the teaching of surgery at the Universities appears to have been allowed. Though we have no evidence which would connect Hugh of Lucca, Theodoric of Cervia (his son) and Bruno Longoburgo with university teaching, we have the express statement of William of Saliceto, a native of Piacenza and a contemporary of Theodoric, that he had taught surgery at Bologna for four years. He was a university educated man, who had practised his art on the battlefield, in hospitals and prisons and had effected remarkable cures in Cremona, Piacenza, Bergamo, Pavia and elsewhere. He ended his career at Verona, where he was surgeon to the city, and where in 1275 he wrote his book, dedicated to Bono del Garbo, son-in-law of the celebrated medical theorist, Taddeo Alderotti. He is a resounding contradiction of the long-accepted assertion that after the decree of 1215 the craft of surgery fell into the hands of illiterate barbers, for besides his excellent work on surgery he also wrote for the benefit of his son, Leonardino, a *Summa conservationis et curationis,* a book on internal medicine. William's ideas were carried into France by one of his pupils, Lanfrank of Milan, who was, perhaps, one of the earliest to teach surgery at Paris, for he was invited by the dean of the Faculty of Medicine there, Jean de Passavant, to give lectures on the subject to the members of the medical faculty. The teaching of surgery, however, was not allowed as part of the University medical course at Paris and even so eminent a surgeon as Henri de Mondeville, who had lectured at Montpellier, had studied in Italy and was obviously a university-trained man in medicine was compelled to instruct his students outside the university framework. The same tradition prevailed in England, where neither Oxford nor Cambridge made any provision for the education of surgeons. In these circumstances surgical knowledge was imparted through a system of apprenticeship, so that skill and experience were passed down from master to pupil. In some ways this method of education was more fruitful than the methods of teaching medicine in the Universities and certainly led to progress not only in the devising of better and better instruments for performing surgical operations, but also in the concocting of medicaments for healing wounds. Often this knowledge and experience would be confined to a particular family, so that crafts and skills would be handed down from father to son throughout several generations; but more frequently an eminent surgeon would have two or three apprentices to whom he would communicate his expertise and to whom, after his death, he would bequeath his books, his instruments, his drug shops and his clientele (24).

By the fourteenth century surgical teaching had become an established element in Italian medical instruction; it was classed as part of *Practica*, to be studied during the afternoon lectures, and was based, generally, on the text of William of Saliceto. But outside the universities, as elsewhere in Europe, Italy had its colleges of surgeons imposing their own standards of competence, which were subject, as elsewhere, to the ultimate control of University physicians.

During the four years devoted to medical instruction in the Universities, students were called upon to display in disputation with other students their grasp of the material which they had studied. Besides the disputations concerned exclusively with their course, there were other disputations *De quolibet* based on any kind of topic which may have occurred in the texts they were reading. These dialectical exercises were intended to sharpen the intellect, to see how deeply the student had understood certain debatable points and to assess his ability in arguing with others. At these disputations the master would be present and would be able to act as arbiter in judging who showed the greatest promise.

At the end of the four-years' course the candidate for the degree of Bachelor of Medicine would have to defend a thesis on some medical subject in public, when all kinds of objections, subtly phrased, would be put to him by chosen members of his audience. On his display in this kind of intellectual tournament would depend, to a certain extent, his receiving the coveted degree or not. Disputations such as these obviously were not concerned with practical problems of diagnosis, prognosis and treatment, but merely theoretical questions. The tendency therefore was for a greater stress to be laid upon the dialectical skill of a student than upon his clinical experience. Indeed, he could have no clinical experience whatever until he had attained his degree, for no one below the status of a bachelor of Medicine was allowed to visit patients even in the presence of a fully qualified doctor.

Once the student had gained his degree of Bachelor, he passed into a kind of intermediate stage in which he was allowed, still under the supervision of his own teacher, to give elementary lectures on texts that he had already studied. At the same time, he pursued a higher course himself for at least two years, at the end of which he could apply for the Licentiate. The granting of the Licentiate was subject to the same conditions as that of the Baccalaureate, namely oral examination by masters of the medical faculty in the presence of the masters of the higher faculties, theology, law and arts. This degree was the most important of all, for it conferred the right to teach, that is, the right to interpret texts on a personal basis without any restrictions whatever. The Doctorate was the culminating point of all these efforts presupposing as it did not merely the completion of courses of study, but actual experience as a teacher in the schools. This also was conferred after the submission of a thesis and interminable discussions on it.

The surgeons, in default of a university curriculum with its examinations, disputations, degrees and so on, which ensured a certain standard of professionalism, banded themselves together into guilds, which, apart from safeguarding the ethics of their members, imposed certain conditions and levels of proficiency before a candidate could be admitted. In England and France these guilds have a decisive influence on the education of surgeons, and the fact that manual dexterity was not the sole criterion of a man's professional qualifications is proved by the rigorous examinations he had to pass in all branches of medicine at the hands of the university-trained physician.

It was more or less at this period, i.e., in the late fourteenth century, that a latent antagonism between two sections of the medical faculty rose to the surface. Already at the end of the twelfth century a distinction had been drawn between those who studied medicine from a purely theoretical point of view and those who put it into practice. This distinction had first been made by the Moslems, notably by Avenzoar. But it had been taken over by Western teachers when they came into contact with Arabic texts and had been emphasized by such people as Gilbertus Anglicus and Alfred Sareshal. Hence the two terms physicus and medicus, the first being applied to a man who approached medicine from a philosophical or theoretical angle, the second being merely "a mercenary healer of bodies". At some Universities, like Montpellier, Oxford and Cambridge, the practical aspect of medicine was more closely pursued, but in others, particularly in those of Italy, a heavier stress was laid on theory. This theoretical trend had been fostered by the works of Taddeo Alderotti, a brilliant dialectician, who brought to his commentaries on the *Aphorisms* and the *Isagoge* every kind of scholastic subtlety. It was encouraged and promoted by other teachers like Peter of Abano and Ugo Benzi (appointed professor of medicine mainly because of his reputation as a brilliant logician, after only six months study), by Jacopo di Forlì, Marsilio de Sancta Sofia, and others. It was further pushed to the forefront by the attack on the medical profession by lawyers, who poohpoohed the idea that medicine was a real science and asserted that it was a mere empirical and mechanical art (25). As a result of the controversy that ensued greater attention was paid in the schools to pure speculation, on the assumption that the dignity of medicine could be assured only by a preoccupation with universal ideas which would involve its prerogative of absolute certainty. In order to support this attitude teachers were compelled to distinguish between medical science and medical practice. It is true that both theory and practice had been taught in the schools from the beginning, theory in the morning lectures and practice in the afternoon. But it is obvious from the texts circulating in the fourteenth and fifteenth centuries that more prestige was attached to dialectical prowess and the subtle arguments on medical principles than was given to the practical application of these principles. So true is it, that writers and teachers of *Practica* were often reduced to invoking quite unworthy motives in order to lure students away from the pursuit of speculative medicine and attract them to practice (26).

Practical medicine enjoyed, in spite of this, a certain degree of importance. This was ensured by two factors: a greater interest in public hygiene, and a fervid enthusiasm among students for dissection. There seems little doubt that the recurrent outbreaks of plague and pestilence had awakened men's attention to the essential part played in community health by clean water, wholesome food, sewage disposal, segregation of the infected, and many other elements. Though all these topics had been dealt with in earlier texts concerning *Regimen*, and particularly in the Plague tracts issued subsequent to the Black Death, the medical teachers had not, as a whole, insisted enough upon them. The emphasis laid on them by various municipalities, the regulations laid down by kings and princes, and the general pressure of public opinion meant that physicians could no longer afford to dismiss them summarily as an insignificant part of the study of medicine. Hence the appearance at this time of more detailed commentaries in the schools on the texts of *Regimen*, the popularity of illustrated copies of the *Tacuinum sanitatis*, and the extensive treatment of the subject in the *Consilia* written by eminent physicians. The gradual disappearance of "leprosy" in the fifteenth century may credibly be attributed to this insistence on the elementary rules of hygiene.

As for dissection, it was no longer pursued exclusively as a means of teaching human anatomy. An awareness had developed that the cause of disease could be more easily ascertained by an examination of the internal organs after death. As a result, surgeons were particularly interested in this source of extra knowledge, and it became quite normal for them to ask parents and relatives for permission to open up the bodies of the dead. This was invariably done on the plea of public utility. The frequency with which these autopsies were carried out (the results being communicated to both physicians and surgeons) is apparent from even a cursory reading of the case book of Benivieni. Though the explanations of morbid conditions were often as faulty as the descriptions of organs advanced by Mondino di Luzzi, and though theoretical considerations usually distorted the interpretations of observed facts, a certain amount of practical experience was gained which could not fail, in the end, to have a beneficial effect. Ordinary people were quick to realize that surgeons had a more immediate grasp of the practical implications of disease and were better qualified in an emergency to deal with sickness than were the professors of theory propounding, with Olympian aloofness, their opinions *ex cathedra*. And since, in the people's view, practising physicians were not far removed from the surgeons, the popularity of the surgeons brushed off on to those medical men whose concern was more with patients than with books. Whilst the pronouncements of the theorists may have been widely acclaimed by the intellectuals, the ministrations of the practitioners were more prized by the populace.

Under the influence, then, of popular opinion the tide gradually began to turn, and by the early part of the fifteenth century Michael Savonarola was deploring the fact that the theoretical methods of Benzi, Jacopo di Forlì, and Marsilio de Sancta Sophia had engulfed all the schools up to his time. He strove, and not in vain, to turn students' attention to the actual curing of the sick. From this time onwards we find a greater preoccupation with practice, manifested in the writings (that is, the lectures given to students and afterwards consigned to manuscript) of Giovanni Arculano of Verona, Cristoforo Barziza of Padua, Gianmatteo Ferrari of Pavia, Marco Gattinaria of Milan, and Baverio de Baveriis of Bologna.

The close relationships between the universities of Italy and the northern countries, some of which had been founded from Italy and depended on it in some degree for their teachers, could not fail to involve the transmission of fashion from one centre to another. And so the insistence on the teaching of *Practica* in contradistinction to the formalised analysis of theoretical concepts became more widely accepted.

This does not mean to say that the curricula of the medieval universities suffered any change or that the competence of candidates supplicating for degrees was judged more on practical than theoretical knowledge. As far as the universities were concerned the basic medical course remained the same, namely the texts of Hippocrates, Galen, Avicenna, Theophilus and Philaretus. The teaching methods by commentary and disputation underwent no radical reconstruction, the examinations for degrees by public discussion on an aphorism of Hippocrates and a text of Galen continued uninterrupted. Long after the advent of the Renaissance the whole paraphernalia connected with the teaching of medicine remained undisturbed. Here and there some rebel raised his objections to this sterile process and from time to time some courageous and talented individual would break out from this charmed circle to discover something new. But on the whole the education and training of candidates for the medical profession lay untouched, like some antiquated fossil buried beneath layers of tradition and an inert mass of indifference.

NOTES

1. J. H. G. GRATTAN and C. SINGER, *Anglo-Saxon magic and medicine*. Oxford, 1952, pp. 16, 24.
2. A. BECCARIA, *I codici di medicina del periodo presalernitano*. Rome, 1956, pp. 114, 116, 127, 168, 215, 300; V. ROSE, *Anecdota graeca et graecolatina*. Vol. 2, Amsterdam, 1963, p. 248.

3. Rose, *op. cit.*, p. 183 f., p. 243 f. This whole question is exhaustively dealt with by B. Lawn, *The Salernitan questions.* Oxford, 1963, pp. 1-15.
4. Rose, *op. cit.*, pp. 253-254.
5. *Ibid.*, pp. 246-247.
6. O. Temkin, Alexandrian studies on Galen's De sectis. *Bull. Inst. Hist. Med.*, 1935, 3: 405-430.
7. Beccaria, *op. cit.*, pp. 289-290.
8. Rose, *op. cit.*, pp. 275-280.
9. Beccaria, *op. cit.*, pp. 161, 201, 257, 258, 295, 309, 348, 362, 376.
10. Rose, *op. cit.*, p. 221.
11. *Ibid.*, p. 255.
12. See the illustrations in Grattan and Singer, *op. cit.; Bryhtferth's manual,* ed. S. J. Crawford. London, 1929.
13. *Arch. Gesch. Med.,* 1910, 3: 165-187, 353-368; 1913, 7: 363-366 and plates.
14. L. McKinney, Medical ethics and etiquette in the early middle ages: the persistence of Hippocratic ideas. *Bull. Inst. Hist. Med.,* 1952, 26: 1-31.
15. See the texts printed by D. de Moulin, *De heelkunde in de Vroege Middeleuwen.* Leyden, 1964.
16. Rose, *op. cit.*, p. 251.
17. Annaliese Maier, *Ausgehendes Mittelalter,* Vol. 2, Rome, 1967. *Un manuale per gli studenti di diritto in Bologna del secolo xiii-xiv,* p. 100.
18. Frances Yates, *The art of memory.* London, 1966, pp. 50-129.
19. Maier, *op. cit.*, p. 101.
20. *Materials for the history of Thomas Becket,* Ed. J. Craigie Robertson. Vol. 3, London, 1877, pp. 3-4; the passage is translated in Stow's *Survey of London,* ed. C. L. Kingsford. Vol. I, Oxford, 1908, pp. 71-72. Stow recalls that this tradition endured to his own day, *ibid.*, p. 73.
21. S. de Renzi, *Collectio salernitana.* Vol. 2, Naples, 1853, pp. 397-398.
22. I refer the reader to the full bibliography on this point given by Vern Bullough, *The Development of medicine as a profession.* New York, 1966, pp. 115-116.
23. K. Sudhoff, *Beiträge zur Geschichte der Chirurgie im Mittelalter.* Vol. 2, Leipzig, 1918, p. 301.
24. Vern Bullough, Training of the non-university-educated medical practitioners in the later middle ages. *J. Hist. Med.,* 1959, 14: 446-458, and the bibliography cited there.
25. Coluccio Salutati, *De nobilitate legum et medicine,* ed. E. Garin. Florence, 1947; further texts in E. Garin, *La disputa delle arti nel quattrocento.* Florence, 1947; the whole question resumed by L. Thorndike, *Science and thought in the fifteenth century.* New York, 1929, pp. 24-58 and more briefly by R. de Rosa, Die Stellung der Medizin in der Frührenaissance, in *Aktuelle Probleme aus der Geschichte der Medizin.* Basel-New York, 1966, pp. 259-266.
26. Let one example suffice: Joannes Arculanus, *Practica,* Venice 1493, fol. 2r: "*Si enim pro questionibus Jacob* [Forliviensis] *aut Ugonis Senensis tantum ocii temporisque terimus summa cum advertentia animique applicatione quarum notitiam parum decoris aut lucri nobis allaturam*

speramus, quanto magis pro huius libri (Practicae) scientia capescenda invigilare debemus, ex quo tantorum bonorum assecuturos nos speramus. Credite mihi, experto credite, quisquis honores, opes, gloriam, apud homines gratiam consequi desiderat, huic libra intendat. Facile enim brevi temporis decursu hec omnia assequatur." This is perhaps more high-minded than the prospects held out in the *Disputa delle arti*, p. 90: "*Sed si stercus et urina sint medici fercula prima, sunt etiam puellarum ac iuvencularum interdum amoenae indoles, nollesque manuum ac brachiorum pulsus, et tenerae mamillae dulcia fercula, atque tibiarum et pectinis suaves aspectus.*"

MEDICAL EDUCATION DURING THE RENAISSANCE

C. D. O'MALLEY
UCLA School of Medicine
Los Angeles, California

Medical education of the later Middle Ages continued pretty much unchanged into the early sixteenth century. In consequence, the present paper is concerned with approximately the hundred years from 1500, during which time medical education underwent alteration, expansion in some directions and restriction in others. It is further limited by exclusion of any consideration of medicine in Spain that will be dealt with in another presentation.

Throughout this closely defined renaissance period the schools of Italy remained the most powerful magnet for medical students from all parts of Europe, although in France the tradition of the medicine of Montpellier continued attractive to French, German, Swiss and, indeed, Scottish students, and the medical faculty of Paris was growing in significance. The university movement that had got underway in the Empire by the mid-fourteenth century became more pronounced in the fifteenth and continued to develop into the sixteenth. Nevertheless, with some few exceptions the medical schools of these German universities were not of particular significance, at least during the first half of the period now under review. Nor in England could the medical faculties of Oxford and Cambridge, despite their venerable years, claim true distinction. During the sixteenth century the most famous medical school in Europe was that of the University of Padua, Shakespeare's "nursery of arts", and it was particularly attractive to German and English students whose native schools had, as was said, much less to offer.

The sixteenth century witnessed some degree of governmental recognition of the significance of medicine to the state and therewith some small assistance to medical faculties. The most notable example of this, although apparent even in the previous century, was the Venetian government's intelligent support of Padua, contributing in no small way to the distinction of that university and, for our purpose, of its medical faculty.

Hence it was that in the course of the sixteenth century medical schools began to assume a position of some greater importance relative to the other faculties of the universities. It is true that in consequence of earlier and consistent governmental support and, as it seems, a related excellence, both

Padua and Bologna had already gained such status. The medical faculty of Paris, on the other hand, was still dependent on students' fees, and through lack of substantial resources could only more slowly develop what may be called a "fulltime" faculty and the necessary physical facilities. It is for this reason that an examination of the Parisian Commentaries reveals such careful recording of fees received from students. Despite the fact that Montpellier possessed the oldest existing medical faculty of any university, it was not until the close of the fifteenth century that it gained any royal support. In 1498 four chairs of medicine were established there by Charles VIII, but it was not until almost the passage of a century, in 1593, that a fifth chair of anatomy and botany was added. Elsewhere, as in England, the medical schools of the two universities were heavily overshadowed by the other faculties, even though Henry VIII did establish a Regius Professorship of medicine at Cambridge in 1540, and at Oxford in 1546; in the Empire, with a few exceptions, schools of medicine were far from significant, and, indeed, in a few cases virtually somnolent appendages of their universities.

Nevertheless the study of medicine had everywhere certain attractions. In addition to whatever few students recognized medicine as their vocation, its practice could be lucrative, and in consequence it was a profession to which at least some poor students aspired. Their numbers seem to have been sufficient to cause Jacobus Sylvius, a member of the Parisian faculty and himself once a poor student, to write a small treatise of advice for penurious aspirants to medicine. Medicine was also a means of lifelihood for some who in fact had a primary interest in other subjects such as the physical sciences. Thus the Netherlander Gemma Frisius practised medicine to support his mathematical studies, and as a more notable instance, the practice of medicine permitted Copernicus, one-time student of Padua, to be an astronomer.

The absolute number of medical students was not large, at least in northern Europe. At no time during the sixteenth century did the annual supplicants for the degree of Bachelor of Medicine in Paris surpass eighteen. At Montpellier where the medical course was shorter, living cheaper, and more requirements honored in the breach, the annual number of matriculations varied from about twenty to forty, and in the year 1546 reached a peak of seventy-one. At Padua and Bologna the numbers were somewhat greater.

Throughout the sixteenth century medicine, as a professional, graduate study, everywhere required completion of the arts course as a necessary prerequisite to entrance. This meant essentially a grounding in Latin, philosophy and argumentation, usually acquired in four years and recognized by the degree of Master of Arts. At Oxford the requirement of both the Bachelor of Arts and the Master of Arts degrees could extend this preliminary work over a much longer period of time, and this greater amount of preparatory study was an inducement to its students to transfer to one of the less demanding continental medical schools. For the most part, however, there seems to have been a growing uniformity in premedical requirements during the sixteenth century.

The course leading to the degree of Bachelor of Medicine began in the autumn, often, officially at least, on St. Luke's day, and was normally comprised of about four nine-month years of study. This represented some reduction in the length of the curriculum. Indeed, the medical course at Montpellier had been reduced to three years of study and seems to have been further compressible, since Thomas Platter, despite some extended absences from the university, acquired the degree of Bachelor of Medicine at Montpellier between September 1595 and April 1597. Forty years earlier his half-brother Felix, despite conscientious attendance of lectures took about twice as long for completion of the same course of study.

Standardization of the length of the medical curriculum appears to have been the result of a growing pragmatic attitude *vis-à-vis* the medieval, scholastic approach to medicine that was being replaced by that of the new Galenism. The latter, the new Galenism, emphasized a more precisely defined body of knowledge, as contrasted to medieval, less contained verbosity, as did likewise the new spirit of science that was introduced through the revolution in anatomical studies. The two attitudes were well illustrated by an episode as late as 1544, when Andreas Vesalius, passing through Bologna and invited to present an anatomical demonstration, dissected out the venous system of a human body. The consequent discussion of the spectators, physicians and philosophers, waxed furiously far into the night, not over the decisive evidence of the body but rather over the contrasting doctrines of Galen and Aristotle. Vesalius, a representative of the new, pragmatic school, made a point of early departure from Bologna on the following morning in order to avoid the continuance of the debate which left him uninterested and completely bored with the whole incident.

Temporal uniformity and precision as to the length of the course of study, standard reductions in time for those who had studied partly at another university, and generally precision of detail in matters of curriculum, were some of the prominent characteristics of the period that help to distinguish its medical education from that of earlier centuries. Such things may be illustrated from conditions at Padua, where the classes were organized on a strict, hourly routine, announced by a bell in the tower of the bishop's palace, and where the teacher who failed to heed the routine was subject to penalty.

The important degree was that of Bachelor of Medicine. This had for long been acquired after the student had sat through courses of lectures on such works as the medieval translations of the *Aphorisms, Prognostics,* and *Regimen* of Hippocrates, the *Ars parva* of Galen, the *Viaticum,* then attributed to Isaac Judaeus, as well as that author's works *On fevers* and *On diet,* the books *On urines* and *On pulses* of Theophilus Protospatharius, the *Antidotarium* of Nicholas, the *Isagoge* of Hunain, and, of course, the writings of Rhazes, Avicenna and other major Moslem physicians. These works were either medieval in origin or medieval translations such as those of Constantine the African and Gerard of Cremona; they continued to comprise the literary content of medical instruction into the early years of the sixteenth century. Indeed, as late as 1516 the Commentaries of the medical faculty of Paris recorded an admonition to three Bachelors of Medicine who had read other books than those assigned and in consequence were reprimanded for opposing their misbegotten views to those of their teachers. In addition to these texts the student had to learn the names and declared medicinal values of a variety of herbs, to attend the annual dissection of a human body—if that event took place—and gain some knowledge of Mondino's *Anathomia* that had been composed in 1316. Ultimately the student took a series of examinations in the subjects related to these books and activities, sustained a number of theses of little value except as demonstrations of ability in argumentation, provided a banquet and various gifts for the medical faculty, and so received his degree and the right to learn the practice of medicine upon such patients as were at his mercy. The degree of Licentiate and of Doctor of Medicine normally required no great further outlay of time but chiefly the successful defence of a series of theses and the presentation of lectures on set themes.

The opening of the sixteenth century roughly coincided with two events of importance: the production of books in Venice, Basel, Lyons, and Paris at prices that made them available to students; and new translations of classic, Greek medicine, notably the works of Galen, directly from Greek into Latin. Obviously it was of advantage to medical students to possess printed copies of medical texts, and the new cheap books were also the vehicle for the propagation of the concurrent wave of Galenism that was to oust much of the hitherto dominant influence of medieval Moslem authorities and their medieval commentators.

It became the belief of sixteenth century physicians that direct access to Hippocratic and Galenic writings would be of inestimable value to the medical art, and for a time there was some truth in this belief that progress lay with the Greeks. With a few exceptions, such as Niccolò da Reggio's early fourteenth century translation of Galen's *Use of parts* directly from the Greek, medieval medical doctrine had diluted such Greek medical thought as was known in the West; terminology had been sadly confused in the course of translation; spurious writings, notably those ascribed to Galen, falsified the Greek position, and, generally speaking, the rational, scientific spirit of the Greeks had been sacrificed. All this is more clearly apparent if one examines manuscripts of medieval texts rather than such printed editions of them as were published considerably emendated by their later editors. Moreover, in some instances only fragments of works had been known through medieval Latin translations, hence out of context, and still other Greek writings remained completedly unknown except by title, until they were studied for the first time during the sixteenth century in the original Greek and translated directly from that language into Latin.

The physician-translators, or medical humanists, such as Niccolò Leoniceno and Thomas Linacre, and later John Caius, were impressed and even dazzled by their ancient Greek predecessors, so much so that they willingly subscribed to the newly recovered classical medical doctrines and, presuming an ancient omniscience, placed themselves voluntarily under bondage not only to the truths of Greek medicine but also to its errors. As this attitude was expressed by John Caius, "Except for certain trivial matters, nothing was overlooked by [Galen], and everything that recent authors consider important could have been learned solely from Galen."

Obviously this new attitude of medical humanism would and did bring about changes in the content of the curriculum, chiefly through discard of medieval writings and medieval translations of classical medicine and their replacement by classical medical texts newly translated directly from Greek into Latin. There are a few signposts along the way indicating the growth of this new influence, such as the controversy that arose about 1517 in Paris over the procedure to be followed in venesection as the result of the discovery that classical Greek methods differed from the medieval-Moslem practice hitherto employed. In 1526 the faculty of medicine of Paris demonstrated its allegiance to the new movement by purchase of the Greek *opera omnia* of Galen published by Aldus in the preceding year, and shortly thereafter by purchase of the Aldine edition of the works of Hippocrates. The *Articella*, that collection of most of the shorter works required by the medieval curriculum, last appeared in 1534, since further editions were nullified by lack of demand.

Meanwhile, the new translations of Galen were appearing in Italy, France, and Switzerland in numerous, cheaply priced editions to meet student demand. In addition to the already extant translation of the *Use of parts* by Niccolò da Reggio, mention may be made of such sixteenth century versions as Linacre's *Natural faculties, Method of treatment,* and *On hygiene,* and Leoniceno's translation of the *Art of treatment* and the *Differences of diseases.* With the exception of the *Use of parts,* which had implications for anatomy and, for example, was employed for that subject in Paris by Jacobus Sylvius, these earliest new translations may for the most part be looked upon as related to clinical medicine. Within a decade, however, they were to be followed by those relating to anatomy, the basic science of the time, such as *Dissection of nerves, Dissection of veins and arteries, On the bones,* and most important of all, in 1531, Guinter of Andernach's Latin version of the *Anatomical procedures.* The Galenic triumph was quickly followed by a like Hippocratic conquest.

The speed with which the newly recovered classical Greek medicine conquered the medical profession and made its appearance in the medical curriculum varied from one school to another, obviously depending upon the progressive or conservative nature of the medical faculty or sometimes of some strong personality in it. The Italian schools, as might be expected, quickly adopted the new Galenic and Hippocratic medicine, although, curiously or not, at Padua a chair for lectures on Avicenna was retained until the final quarter of the sixteenth century. In this regard it should be noted that the statutes of the different schools did not always reveal the true state of affairs. Often the statutes had been drawn up in the medieval period and remained essentially unchanged in later, more modern times. In such case they were often quietly, although in fact illegally, ignored. Thus the Paduan statutes, despite some slight revisions remained essentially medieval throughout the sixteenth century. In consequence the Paduan lecturer on Avicenna may perhaps have presented the medicine of that Moslem physician to such audience as he was able to assemble, or he may well have turned his attention to other more modern topics in the same way that lecturers on surgery and anatomy tended to deemphasize surgery, which was recognized in medieval statutes as the more important subject, in favor of anatomy that hitherto had been looked upon as of less significance.

The Parisian faculty gave its allegiance in particular to Galen during the third decade of the sixteenth century, and since some of the most prominent translators of Greek medicine were associated with that faculty, such as Jean Ruel, Nicholas Cop, and Guinter of Andernach, it was inevitable that the influence of classical medicine would be especially strong in Paris. Indeed, the faculty of Paris retained a highly orthodox Galenic position well and ludicrously far into the seventeenth century. The Parisian statutes, too, are misleading, since they remained unrevisedly medieval until almost the close of the sixteenth century, and in consequence some at least of the required medieval texts mentioned in them were simply ignored by the new Galenists.

In contrast, even though Rabelais had lectured at Montpellier on the Greek texts of Hippocrates and Galen in 1531, he seems to have made little immediate impression there, and it was not until about twenty years later that the required medieval texts were ousted by the Greeks. At Louvain the situation was revealed in the words of one of the professors who declared that he would rather be wrong with Avicenna than right with any Greek physician. It was only through a student revolt that such medievalism was brought to an end in 1544. In England, too, despite the pioneering efforts of Linacre, the new Galen and Hippocrates were relatively late arrivals.

The German medical faculties may also for the most part be described as conservative during the first half of the sixteenth century. At Tübingen reform of the medical curriculum in favor of the new Greek medicine was undertaken through the efforts of Leonhart Fuchs who came to the university in 1535, and it was at about the same time that there was a similar introduction of classical medicine into the curriculum of Vienna. The statutes of the medical faculty of Heidelberg were modernized in favor of the Greeks in 1558, and about that time the faculties of Ingolstadt and Freiburg swung over to the new, classical doctrines only to be compelled, through the interference of government, some twenty years later, to return to what was called the "old, more solid medicine", that is, a continued recognition of works of Galen and Hippocrates but with a strong admixture of those of Rhazes and Avicenna.

As the revival of classical medicine affected the literary content of the curriculum so, too, it brought about changes in the courses offered within the curriculum and in methods of instruction. Hence it was, under the influence of the revival and recovery of Greek medicine, and especially of Hippocratic medicine, that clinical, bedside teaching gained a new importance and a definite place in the curriculum.

It is true that throughout the Middle Ages there had been recognition of the fact that the medical student ought to have some practical experience in treatment of the sick, and as early as the thirteenth century the laws of the Emperor Frederick II specified that the student must gain such experience with a qualified physician before undertaking his own, independent practice of medicine. This guild principle continued to be recognized at least verbally, so that in the fourteenth century, according to the statutes of the medical faculty of Vienna, and in succeeding times in the statutes of other medical schools, a period of such practical training was required for the degree of Doctor of Medicine. There is no satisfactory information, however, as to the thoroughness or consistency with which the requirement was enforced at any of the universities. If we turn again to the memoirs of Thomas Platter, we find that after he had gained the degree of Bachelor of Medicine at Montpellier—where there was a requirement of this sort—he immediately began practice in a small French town, even undertaking such drastic procedures as amputations, without any guidance or direction except his own independent judgment based upon less than two years of instruction. It should be noted, moreover, that the requirement was to be met, according to the statutes of the various universities, not within the medical school, but outside, through a relationship between an individual student and an individual physician.

The introduction of clinical teaching within the medical school so that the students as a group accompanied their teacher to the bedsides of the ill, both to observe their physician-teacher's methods and to practise those methods under his observation and direction, was a phenomenon of the sixteenth century and a product of the Italian medical schools. Most notably it seems to have been a contribution of Padua where its chief proponent was Giambattista da Monte, successively professor of the practice and of the theory of medicine from 1539 until his death in 1551. However, Andreas Vesalius informs us that he had to meet this requirement in 1537 before he could receive the degree of Doctor of Medicine so that it appears that such clinical teaching had been inaugurated earlier, possibly by Francesco Frigimeliga who had previously held the same two chairs that were later to be occupied by Da Monte. Whatever the origin, it is Da Monte who usually receives the credit for developing such clinical instruction to the point where it was remarked upon by non-Italian students as an unusual and commendable novelty of the Paduan curriculum.

It was Da Monte's method to take his students to the hospital of San Francesco, which was for long looked upon as the university's hospital, where he advised them on the procedures to be followed in the examination of patients. First the student must observe the appearance of the patient's countenance; then talk with him about his symptoms; thereafter note his pulse and observe everything necessary to gain a knowledge of the particular illness. The "everything" to be observed appears to have been pretty much in accordance with Hippocratic tradition.

After the death of Da Monte such clinical teaching continued to be carried on by his successors, and it was from Padua that Jan van Heurne of Utrecht took the idea back to the Netherlands where at Leiden clinical instruction was later to be further developed by his son Otto, still later by Sylvius de le Boë, by Boerhaave, and disseminated throughout the world.

A second result of the Greek revival of the sixteenth century was the development of medical botany. If the Greek authorities in their omniscience asserted that certain herbs or medicinal plants were required for an effective therapeutic remedy, obviously exact identification of such plants was necessary. In 1533 a chair of medical botany was established at the University of Padua, first occupied by Francesco Bonafide. However, the students still had need to make field trips in order to identify the various plants discussed, and such trips, pleasant as they might be, were time-consuming; moreover, all the medicinal plants were not to be found in a wild state in any one area. In 1542 Bonafide proposed the founding of an all-inclusive botanical garden, and with the acquiescence of the Venetian senate the garden became an actuality in 1545. At approximately the same time a second such garden was established for the medical faculty of the University of Pisa under the direction of Luca Ghini, and thereafter still further gardens were developed at Bologna, in 1568, Leiden, in 1577, Leipzig, in 1579, Paris, in 1591, Heidelberg, in 1593, and Montpellier, in 1598.

Although these botanical gardens would ultimately serve for scientific research in botany, the immediate intention was that they contain as many medicinal plants as possible in order that students and, indeed, faculty might more readily learn to recognize them. In consequence, the gardens were laid out according to the morphological characteristics of the plants and their real or assumed medicinal properties.

Unquestionably the most spectacular development in medical education during the sixteenth century concerned instruction in anatomy. This represented the first significant step towards a genuinely scientific medicine and therewith in some degree involved instruction in scientific method. It also meant that anatomy, as well as some aspects of physiology and pathology, then considered components or appendages of anatomy, advanced beyond the confines of classical Greek medicine that otherwise remained predominant.

At the opening of the century the anatomical course was still officially comprised of an annual midwinter dissection performed by a surgeon to accompany the reading of Mondino da Luzzi's fourteenth-century *Anathomia*. The purpose was illustration and verification of Galenic anatomy. After the first two decades of the sixteenth century there were some rare instances of deviation from this pattern. In Bologna, Berengario da Carpi occasionally dissected human bodies privately with his students, although he continued to conduct the official winter demonstration in the traditional manner. Although at first he related the formal demonstrations to Mondino's manual, from 1522 onwards he used his own somewhat less Galenically orientated *Isagogae breves* for that purpose, the first new dissection manual to appear in 200 years. There is also some evidence that in the third decade of the century there were a few more or less surreptitious dissections carried out in Paris, the joint effort of the more energetic and inquisitive teachers and students, and it was there, in 1536, that a second, new dissection manual was produced by Guinter of Andernach. Guinter's manual, as one might expect from the translator of a number of Galen's writings, was subservient to the current Galenic anatomy.

Despite these few exceptions, through the first four decades of the century, the course in human anatomy remained essentially limited to the formal winter dissection, often completed in three days and nights of continuous activity, the culmination of the student's study of whatever the prescribed texts, chiefly Mondino's, although sometimes in Paris the anatomical section of Guy de Chauliac's *Chirurgie*, or appropriate portions of Galen's *Use of parts*. The surgeon dissected from the most to the least rapidly putrefying parts—from abdomen to thorax, to head and limbs, while the professor, disdaining any active role in the dissection, declaimed from his *cathedra*. The students remained mere spectators, that is, those who were fortunate enough to be in a position to see what was being done. Of course, the findings were, to everyone's satisfaction, Galenic.

As with certain other innovations, the first changes in this time-honored pattern of anatomical instruction occurred at Padua where, contrary to custom, the winter dissection of 1537 was performed by the professor himself, the recently appointed Andreas Vesalius. Such personal demonstration and lecturing directly from the body were the first of the several pedagogical novelties to be introduced by Vesalius that had both appeal and instructional value for the students. To these may be added his articulation of the dissected body's skeleton as a means of revealing the relationship of the bones to one another and to other bodily structures, in contrast to the older procedure whereby the students studied the bones individually.

The final novelty of this new method was the introduction of large anatomical plates, deceptively naturalistic in appearance since in fact they revealed for the most part Galenic rather then human anatomy. At that time Vesalius was a better pedagogue than anatomist since he still gave allegiance to Galen; indeed, he had discarded Mondino's venerable *Anathomia*, actually required by the Paduan statutes, in favor of Galen's *Use of parts* and Guinter of Andernach's strongly Galenic dissection manual published in the preceding year. Despite their defects these first Vesalian illustrations were an important pedagogical contribution for the sake of their unusually large size, the extensive identifying legends keyed to details of the anatomical figures, and the presentation of anatomical terminology in Greek, Latin, Hebrew and Arabic in an effort to overcome the confusion that then existed in the meanings of such words. The *Tabulae anatomicae* were immeasurably superior to such decorative but useless anatomical illustrations as those in the *Fasciculo di medicina*, or the works of Magnus Hundt, Gregory Reisch, Lorenz Fries, or any of the so-called "fugitive sheets", and far more sophisticated then the illustrations in the *Commentaria* and *Isagogae* of Berengario da Carpi who, nevertheless, had some glimmer of the real significance of anatomical illustration.

The final development of these novelties is to be found in the *De humani corporis fabrica* of 1543, together with Vesalius' denunciation of anatomical instruction as it existed and his proposal for a new approach to anatomical study based solely upon dissection and observation of the human body. The *Fabrica* was not written as a textbook for students but rather for their teachers, and despite the outraged cries of the conservatives—that is, the former progressive Galenists—it became influential within a decade in Italy and certain other parts of Europe. It was through this work that anatomy was established as the first scientifically orientated medical discipline, although there was a considerable span of time between acceptance of the Vesalian proposals and their actual implementation in medical curricula. The medical school of Bologna was quick to follow the new proposals and, although Leonhart Fuchs, as was said, had replaced medieval medicine by that of Galen when he became dean of the medical faculty of Tübingen in 1535, he was later so persuaded by Vesalius' work as to prepare a kind of abridged and simplified edition of the *Fabrica* for the use of his own students in 1551.

By the middle of the sixteenth century anatomical demonstrations appear to have been relatively common events in Vienna, Freiburg and Heidelberg, and in France at Montpellier and Paris. How much of this activity was carried on according to Vesalian scientific principle is, however, difficult to say; at Paris, certainly, Galenic prejudice was still strong. Moreover, the Vesalian procedure whereby the professor himself performed the dissection and lectured at the cadaver remained uncommon except in Italy.

For any student interested primarily in anatomy it was essential throughout the sixteenth century that he study at Padua, if not under Vesalius, then under one of his successors such as Realdo Colombo, Gabrielle Falloppia, Fabrizi d'Acquapendente or Giulio Casserio. It was incidentally, not until 1609 that fact was translated into administrative order and the single Paduan chair of surgery and anatomy was split into two chairs, that of anatomy, now the more important, held by Fabrizi, and that of surgery, by the younger Casserio. In this matter Padua had moved more tardily than other medical faculties such as Montpellier where a chair of anatomy, although linked with botany, had been established in 1593, but certainly more rapidly than Oxford where the Tomlins Readership in anatomy was not set up until 1624.

The discipline of comparative anatomy may also be said to have had its origin at Padua where Vesalius was in the habit of dissecting various animals in conjunction with his dissection of human bodies. This practice was followed by his successors, notably by Falloppia, whose Dutch student Volcher Coiter was to carry the subject to Germany and "elevate [comparative anatomy] to the rank of an independent branch of biology", and by Fabrizi and Casserio. This last was a promoter of William Harvey's interest in the subject of comparative anatomy which played such an important role in his investigation of the movement of the blood. Harvey, it may be recalled was a student at Padua in 1600, and therefore at the very close of the period under consideration.

One of the major problems of the anatomical demonstration was that of providing a proper view for the spectators, to which may be added such details as illumination, ventilation, and sufficient space for the actual work of dissection and for whatever the number of spectators. Dissections were sometimes performed outdoors but this created the difficulty of limiting the spectators to those properly involved, and, of course, the whole demonstration was placed at the mercy of the winter elements. Indoors, too, as the Parisian faculty of medicine had discovered, the use of its own building was far from satisfactory. At least in Italy, the first solution seems to have been the use of churches, and we know, for example, that during his visit to Bologna in 1540, Vesalius presented some of his anatomical demonstrations in the church of San Francesco. It was also at this time, and in order that the students might better see certain anatomical details, that he first dissected out and revealed the structures and then arranged for the students to pass by the body in single-file in order to get a clear view. No doubt others had done or would do the same. It was, then, chiefly the problem of providing an adequate observation of the dissection that had to be solved.

The Paduan anatomist Alessandro Benedetti was the first to propose a solution in his *Anatomice* of 1502. His proposal was the erection of an anatomical theatre, presumably of wood and capable of being dismantled since the demonstrations were at that time the older sort given once a year. In essence this was a raised, well-illuminated table surrounded by tiered benches, all within a building that might serve other purposes for the remainder of the year.

There is some evidence that anatomical theatres of this nature were created in Pavia and Pisa about 1522, although no details of their construction have come down to us. The first one of which we do have any detail is that amphitheatre to be seen on the title-page of Vesalius' *Fabrica*, described in that work as a temporary wooden structure that could be built to a size that would satisfactorily accommodate 500 spectators.

In the *De dissectione* of Charles Estienne, published in 1545, there is a fairly elaborate description of a proposed anatomical amphitheatre which according to Estienne's description would have held approximately 500 spectators. There is no indication, however, that the amphitheatre constructed with royal funds in 1556 at Montpellier was based upon Estienne's plan. Guillaume Rondelet was the moving spirit behind the erection of that structure, the first of its kind in France, but no longer extant.

Curiously enough, there was no permanent anatomical theatre at Padua, the home of modern anatomy until, at the instance of Fabrizi, that theatre was erected in 1594 that still exists as the earliest surviving example of such structures. In fact, it remained in use into the second half of the nineteenth century.

It has already been mentioned that in the course of the sixteenth century, as one of the alterations that took place, anatomy usurped the position of greater importance in the medical curriculum once enjoyed by surgery. Fundamentally, however, surgical instruction did not undergo any noteworthy revision. The conflict between the physician as a man of profession and the surgeon as a craftsman has perhaps been overstated on the basis of conflicts in Paris and London. Elsewhere one finds medicine and surgery in amicable relations, and, indeed, frequent instances of physicians performing surgery. However, the whole matter is too large and too full of qualifications to be dealt with at this time. Suffice it to say that the medical schools appear normally to have provided a course in what may be called theoretical surgery for those seeking the medical degree, and sometimes, as at Paris, a second course was given in the vernacular for the budding surgeons who had not had the advantage of the university's arts course and knowledge of the all-important Latin.

The practical or clinical side of surgery was normally presented in the guild or company of surgeons. It was this kind of organization such as the United Company of Barber-Surgeons of London or the Fraternity of St. Côme in Paris that made considerable advances in the sixteenth century in the training and rigorous examination of young surgeons and generally sought to establish higher standards of practice. But, except for the official introduction of instruction in human anatomy—of which surgeons already often knew more than many of the physicians—there was merely a development of what already existed.

Finally, although ancient Greek medicine continued to dictate to sixteenth century medicine, some modifications in that control become apparent in the second half of the century. Assuming Galen's omniscience, there were, nevertheless, some matters with which it was recognized he had not chosen to deal; and with the passage of years some physicians become bold enough to suggest there were even matters of which Galen had been unaware, such as diseases and drugs of the New World and various other things that had developed since the classical era.

In the first half of the sixteenth century classical medical texts and commentaries on those texts were published, and both sorts of works had been the sources of instruction. In the second half of the century there began to appear what may be called new or modern medical treatises. Unlike the writings of Berengario da Carpi or Vesalius, these new works were not written in open defiance of Galen. Rather, they recognized a framework of Galenic medicine, but within it they elaborated beyond the established borders, reinterpreted the facts, and here and there deliberately opposed specific and hitherto accepted classical views. Such writings were the work of the *neoterici*, the "recent writers", as the conservative John Caius called them, when he declared those writings to be of no value since everything was to be found in Galen.

Nevertheless, more and more of this sort of book on what we may call internal medicine appeared, written by such men as Leonhart Fuchs and Jean Fernel. In 1573 the medical faculty of Freiburg requested permission to incorporate such newer works into its curriculum and was refused by a conservative administration. The request was an error in tactics, since elsewhere they were being quietly introduced into the curriculum of other schools thereby tempering what in some places, such as Paris, had been a fierce and almost fanatical adherence to classical medicine.

The errors of Galenic anatomy and physiology had been demonstrated, and in consequence those two disciplines were free to develop scientifically as they did in the course of the seventeenth century. Other aspects of medicine were slower to follow this course, but the new type of treatise to which I refer that was now introduced into the curriculum made the ultimate advancement of medicine inevitable. It was then in the latter sixteenth century that the medical students were exposed to the first stirrings of doubt and the first suggestions of overthrow of a complacent classical medicine, and its replacement, although in the distant future, by one more scientifically based.

BIBLIOGRAPHY

ARBER, A., *Herbals; their origin and evolution.* Cambridge, 1938.
BERTOLASO, B., Ricerche d'archivio su alcuni aspetti dell'insegnamento medico presso la Università di Padova nel cinque- e seicento. *Acta Med. Hist. Patav.,* 1959-60, **6**: 17-37.
CERVETTO, G., *Di Giambatista da Monte e della medicina italiana nel secolo XVI.* Verona, 1839.
Commentaires de la Faculté de Médecine de l'Université de Paris (1395-1516), ed. E. WICKERSHEIMER. Paris, 1915.
Commentaires de la Faculté de Médecine de l'Université de Paris (1516-1560), ed. M.-L. CONCASTY. Paris, 1964.

DULIEU, L., Guillaume Rondelet. *Clio Med.*, 1966, 1: 89-111.
DURLING, R. J., A chronological census of renaissance editions and translations of Galen. *J. Warburg Courtauld Insts.*, 1961, 24: 230-305.
HESELER, B., *Andreas Vesalius' first public anatomy at Bologna 1540. An eyewitness report,* ed. R. Eriksson. Uppsala, 1959.
KINK, R., *Geschichte der kaiserlichen Universität zu Wien.* Vol. 2, Vienna, 1854.
NAUCK, E. T., Der Ingolstädter medizinische Lehrplan aus der Mitte des 16. Jahrhunderts. *Arch. Gesch. Med.*, 1956, 40: 1-15.
O'MALLEY, C. D., *Andreas Vesalius of Brussels 1514-1564.* Univ. of California Press, Berkeley, 1964.
———, *English medical humanists. Thomas Lincre and John Caius.* Lawrence, Kansas, 1965.
PELLEGRINI, F., *La clinica medica padovana attraverso i secoli.* Verona, 1939.
PLATTER, F., *Beloved son Felix. The journal of Felix Platter a medical student in Montpellier in the sixteenth century,* trans. S. Jennett. London, 1961.
PLATTER, T., *Journal of a younger brother; the life of Thomas Platter,* trans. S. Jennett. London, 1963.
PUSCHMANN, T., *A history of medical education,* trans. E. H. Hare. London, 1891.
RADL, C., *Geschichte der Anatomie an der Universität Leipzig.* Leipzig, 1909.
RATH, G., Medical education at the south German universities in the 15th and 16th centuries. *J. Med. Educ.*, 1960, 35: 511-517.
———, *Die Entwicklung des klinischen Unterrichts.* Göttingen, 1965.
RICHTER, G., *Das anatomische Theater.* Berlin, 1936.
SHERRINGTON, C., *The endeavour of Jean Fernel.* Cambridge, 1946.
SINGER, C., AND RABIN, C., *A prelude to modern science; being a discussion of the history, sources and circumstances of the 'Tabulae Anatomicae Sex' of Vesalius.* Cambridge, 1946.
Statuta almae universitatis D. artistarum et medicorum patavini gymnasii. Venice, 1589.
STÜBLER, E., *Geschichte der medizinischen Fakultät der Universität Heidelberg 1386-1925.* Heidelberg, 1926.
———, *Leonhart Fuchs.* Munich, 1928.
SUDHOFF, K., *Die medizinische Fakultät zu Leipzig im ersten Jahrhundert der Universität.* Leipzig, 1909.
WIGHTMAN, W. P. D., Quid sit methodus? "Method" in the sixteenth century medical teaching and "discovery". *J. Hist. Med.*, 1964, 19: 360-376.

THE MODERN PERIOD

ITALIAN MEDICAL EDUCATION AFTER 1600

LUIGI BELLONI
Istituto di Storia della Medicina, Università degli Studi
Milan, Italy

The Italian Parliament is now discussing the reform of the University system. This reform will influence the teaching of medicine in the near future and therefore makes even more challenging the subject I was asked to develop, even though its importance ought properly require years of research and a much longer presentation. We must consider both the antiquity, or better, the priority of the Italian Universities and, as well, the more recent national unification which was proclaimed in 1861 and was completed, at least as far as universities are concerned, in 1866 when the Venetian region, including Padua, became a part of the Italian Kingdom. Prior to this unification the teaching of medicine assumed different characters depending on the school where it was given. This variability is to be ascribed to a difference of organization in the many small states into which Italy was divided and to the local traditions rooted in the past (Bologna, for instance, although it belonged to the Papal State from 1506, had its own character, which was different from that of Rome). To keep this subject within the bounds of a report, I intend to develop just a few points which I consider essential, and to support them with some examples. As far as these examples are concerned, I shall refer mainly to those which are better known to me and for which I can rely on a wider source of information. I refer, actually, to the teaching of medicine in the State of Milan (1), which controlled the University of Pavia from 1361. The University of Milan was founded only in 1924.

The words "after 1600" in the title seem to hint at the great scientific revolution initiated at the beginning of the seventeenth century and centered on Galileo Galilei (1564-1642), who was the champion of the experimental method and a splendid researcher himself. When and how did this methodological revolution of science succeed in imposing on the medical education which until then had followed an essentially theoretical method?

The *Memorie e documenti per la storia dell'università di Pavia* (2) gives a list of the titulars of each single lecture from the foundation of the University (1361), thus covering a period of time of more than five centuries. This list is a particularly significant document as it enables us to follow the reciprocation of the same lecturer throughout the variety of lectures given to the future *medicus physicus,* i.e., logic, astrology, mathematics and other lectures having, in our mind, a more definite medical character. We also see from this list (3) that the last *lectura Almansoris* was ascribed to Gerolamo Grazioli, who died in 1765. This gives an idea of how long the scientific revolution of the early seventeenth century had to wait before succeeding in eradicating a method of teaching which had its roots in the scholastic Middle Ages. On the other hand, we must keep in mind that, even prior to the scientific revolution, teaching methods began to approve of principles of direct observation.

Particular reference is made to: the beginning of clinical teaching with G. B. da Monte (1488/1498-1551) at Padua; to the *lectura simplicium* supported by botanical gardens; and especially to the anatomical teaching on the human body which first began with Mondino de' Liuzzi (*c.* 1275-1326) in Bologna, during the early fourteenth century, and was to be imparted from proper chairs for anatomy (or better, for surgery and anatomy) in the sixteenth century, when also the first permanent anatomical theatres were founded (G. Fabrizi d'Acquapendente, Padua 1594).

At Pavia the teaching of anatomy was illustrated in the sixteenth century by Gabriele Cuneo, who was the champion of Vesalius against the Galenist Francesco dal Pozzo (Puteus), and reached its peak in 1624 with Gaspare Aselli (1581-1625), who was the discoverer of the chyliferous vessels. Almost a century and a half of greyness followed the brief appearance of Aselli, and the same happened in the case of the other medical lectures. These lectures, except for the work of Johannes Chrisostomus Magnenus, author of *Democritus reviviscens, sive de atomis* (Pavia, 1646), had not been influenced by the innovating spirit which had been one of the aspects of Gerolamo Cardano (1501-1576).

As a general rule, medical education proceeded on the basis of traditional methods. How and when did the new experimental philosophy of Galileo Galilei start to influence medical education? In other words, when and how was medical education so thoroughly reformed as to initiate the scholar in the "new" medicine, as was clearly outlined by Marcello Malpighi (1628-1694) in his scientific testament? I refer here to the famous *Risposta* (4) published posthumously in 1697, in which Malpighi spoke in favor of the new rational medicine and defended it from the attack of Giovanni Gerolamo Sbaraglia (1641-1709) in 1689, fighting in the name of empiricism. In answer to that attack, *De recentiorum medicorum studio,* Malpighi proposed a complete system of rational medicine based on experimental philosophy, which he had so successfully cultivated, i.e. on the basis of what we would now call "basic research".

Inspired by the atomistic-mechanistic theory and provided with his microscope, Malpighi (5) structured our organism *ex novo,* detecting through microscopic anatomy, strongly based on the comparative method, the small machines executing the main functions and tried to reconstruct their operation by resort to mechanics as well as to chemistry. Humoral alterations of chemical nature played an important role in the breakdowns of these machines. Such breakdowns, even when studied only through their macroscopic equivalents, i.e., the lesions of organs detectable on the dissection table, produced alterations of the economy of the organism which developed into clinical manifestations proportional to the site and nature of the breakdowns. From this microscopic anatomy there developed in consequence the anatomical investigation of the sites and causes of the diseases, to paraphrase the title of Morgagni's work *De sedibus et causis morborum per anatomen indagatis* (1761).

The medical system thus originated was indeed open to continuous improvement, but for this very reason its was provisional and uncertain. Nevertheless it provided a method for practical medicine; Malpighi, as a practical physician, from the clinical manifestations deduced the chemical alterations of humors and the breakdowns of solids and so reached, on this basis, the therapeutic indication which he was accustomed to fulfill also with a considerable use of chemical remedies.

From a didactic point of view, we find the formulation of these instances in the opening lectures of two more-or-less direct pupils of Malpighi: *Nova institutionum medicarum idea medicum perfectissimum adumbrans*, delivered by Giovanni Battista Morgagni (1682-1771) in 1712 from the chair of theoretical medicine at the University of Padua, and *De recta medicorum studiorum ratione instituenda* delivered by Giovanni Maria Lancisi (1654-1720) in 1714 for the opening of the new medical school at the S. Spirito Hospital in Rome.

I should like to stress that Malpighi's system of rational medicine resulting from the scientific findings of the seventeenth century, was actually the same as that passed on to the following century by no less a teacher than Herman Boerhaave (1668-1738), *communis Europae praeceptor*.

It was in 1769 that the experimental mentality finally broke through in a rather dramatic way into the University of Pavia with Lazzaro Spallanzani (1729-1799), who had been appointed to the new chair of Natural History. His teaching was by no means of the notionistic type; it was, on the contrary, the best formative teaching one could think of. I wish to quote here a few lines from a report he wrote upon request of the government about his teaching method: "I most of all want to make all young people understand what I mean by *spirit of observation*. The best way to achieve this is, in my opinion, to discuss with them an object that has previously been excellently examined by a great observer such as Malpighi, Lyonet or Réaumur. In this way they easily realize that the spirit of observation is the ability to understand well an object in all its parts, to discover its different connections and to relate them to each other and to other beings, in order to attain some truth or useful consequence" (6). Spallanzani's chair—like the one of Experimental Physics to which Alessandro Volta (1745-1827) was appointed in

1778—belonged to the Faculty of Philosophy; his teaching, however, was also presented to students of medicine and surgery (7), and he therefore had a fundamental influence on the Medical School.

The appointment of Spallanzani was one of the first, perhaps the most important, fruits of the reforms being carried out at that time to develop the University of Pavia.

Already in 1730 the Senate of Milan "seemed to turn towards a revision of programs and structures of the studies, which should no longer be merely exterior and formal" and ended by taking "some partial measures" (8). In 1748, the peace of Aachen consolidated the Austrian control of Lombardy. The radical restructure of the studies began in 1765 according to an imperial decree that deprived the Senate of its authority over higher education and gave this charge to a *Deputazione* which, in 1771, was brought to the dignity of *Magistrato Generale degli Studi*. This *teaching plan* of 1771 was completed in 1773 by a *scientific plan* (9), by which the conventional lectures, i.e., theoretical medicine, practical medicine, *lectura Almansoris*, *lectura simplicium*, anatomy and surgery, were replaced by three annual propedeutic chairs and two biennial clinical chairs.

1. Anatomy and Surgical Institutions (annual course)

The subjects of this course were, in addition to the structure of the human body, "the general knowledge of the circulating, elaborated, secreted or excreted fluids"; the "changes induced by the different proportions of the fluids and solids"; "first notions of physiology" (first notions which would now be considered as pertaining to surgical pathology); and occasional information about the rare and pathological findings casually observed on dissected bodies. This was the basic element of the course: "Lectures must not be pompous, academic speeches; they should, on the contrary, be a description and a simultaneous inspection of the parts of the human body, carried out on cadavers. Students must be led to the habit of dissection, and they must practice the fine cutting or injection methods. These operations must be done by the teacher and repeated, under his control, by the most promising students, so that they will learn how to operate."

2. Surgical Operations and Obstetrics (biennial course)

This course dealt with the theoretical and systematic explanation of single operations, their practical execution on the cadaver, and finally the demonstration of these operations on the live man: "Every time an operation is carried out at the hospital, after everything has been properly set up, the students will watch its execution, so that the teacher can discuss the operation with them while explaining what is going on and the different phenomena involved in the operation." As far as the obstetrical operations were concerned, the professor had to "make use of uterine models, which he will undertake to get in the natural size, so that students become familiar with the different positions of the foetus, and thus be able to operate in the most difficult and complicated cases."

3. Medical Institutions (annual course)

During this course, taking for granted the knowledge acquired through anatomy, of "mechanical structures of the human body, it is important to acquire the knowledge of their consequences and effects, i.e., the forces and the phenomena that are the true subjects of physiology (and therefore of natural history) and of the reason of all phenomena of life." Here too, "the teacher must undertake to carry out as many experiments as possible" and "shall not be too hasty in finding a theory for everything." Another aim of the course was to "determine the signs of perfect health and the ways to preserve it. When examining a sick body the teacher will give the general description of the causes, differences and accidents of the disease, and will indicate the methods to be followed to ascertain and to distinguish them. He will finally set up the basis of the great and difficult art of prognosis while stressing its importance together with its dangers and abuses." "The major task in the education of young medical students is to endow them with the knowledge of all characteristic features of diseases and to train them to a continuous observation. It is of no less importance to free and purify the healing methods from vain hypotheses."

4. Theoretico-Practical and Clinical Medicine (biennial course)

This course followed the "Medical Institutions" and comprised "the detailed history of all diseases and their theoretical explanation." To a "theory based on facts and not on illusion of the hypotheses" and comprehensive of nosological classification, corresponded the clinical teaching in the hospital, where the teacher would see that "his students get the habit of examining the sick, recognizing the diseases by sight, distinguishing their characteristics, detecting their causes and explaining their symptoms." Clinical practice included therapeutic research: "it will be very useful indeed to treat the patients not only with known methods, but to endeavor at the same time to perfect them and to experiment with new treatments on the basis of a strict and impartial observation." Further, "it is mandatory as a rule to obtain a cadaveric section after the eventual death of these patients, and to observe carefully the alterations of all involved or suspected parts in order to find out the true causes of the diseases."

5. Chemistry, *Materia Medica* and Botany (annual course)

This course was especially intended to provide the future therapeutist with basic knowledge.

A school thus structured was actually a didactic realization of the system of medicine as proposed by Malpighi in his reply to *De recentiorum medicorum studio* (10). As mentioned before, this system was passed on to the following century by no less a teacher than Boerhaave and from Leyden it reached Pavia through Vienna, whither it had been brought by Gerard van Swieten (1700-1772), a pupil of Boerhaave.

At that time clinical teaching was conducted in the fifteenth century hospital of San Matteo, which was located close to the University. In February 1782, Simon-André Tissot (1728-1797), professor of theoretical-practical and clinical medicine for the academic biennium 1781-1783, was asked by the government to express his "ideas on the teaching of medicine and to prepare a small project of a hospital intended for that purpose." He answered by his *Essai sur le moyen de perfectionner les études de médicine,* published in Lausanne in 1785, when Tissot had already left the University of Pavia.

After a period of two years, during which the lectures were given by a temporary teacher, the clinical chair of Tissot was given to Johann-Peter Frank (1745-1821), who held it with great prestige for ten years. The author of the *System einer vollständigen medicinischen Polizey* (1779-1827) was also charged with the direction of public health of Austrian Lombardy, which he radically reformed and reorganized in 1788 on a new basis, according to the *Piano di regolamento del direttorio medico-chirurgico di Pavia* (Milano, n.d.).

On the 5 November 1785, soon after his arrival in Pavia, Frank was asked to prepare a new plan to be adopted in place of the one by Tissot which had never been realized. In 1786 he finished his *Piano degli studi per la Facoltà Medica dell'Università di Pavia* (11) and put it into practice in 1787. He later on perfected and modified his plan on the basis of the experience acquired through two successive quadrennial experiments.

In 1955, Erna Lesky (12) made a very accurate analysis of this plan and compared it to another plan made by the same Frank in 1798 for the Medical School of Vienna. It must be said that this plan is a precious document for the history of medical education and mention must be made of the great care that the author gave even to the smallest detail. If compared with the "didactic plan" of 1773, Frank's plan can be considered as a "regulation" in comparison with a "law". This does not imply that the plan did not contain any innovations, such as, for instance, the splitting of the five "big" chairs mentioned in the plan of 1773 or the addition of new disciplines such as *"Polizia Medica"*. Another remarkable achievement was the foundation, in the clinic, of a *pathological museum,* where could be collected the most meaningful postmortem findings from dissections of the bodies of those patients who had been the subjects of clinical teaching.

On the other hand, the anatomical museum was being enriched by the care of Antonio Scarpa (1752-1832), who had been called to Pavia in 1783 (13) and had opened in 1785 the new anatomical theatre: a real masterpiece of its kind, designed by Giuseppe Piermarini (1734-1808) and built by Leopold Pollak (1751-1806).

The reform of the teaching method was in fact aligned with the renewal and enlargement of the University buildings and Piermarini had been entrusted with this task. The work involved in the renewal lasted for eight years, beginning in 1771. Room was thus provided for the museum of natural history and for the laboratory of physics. The botanic garden and the chemical laboratory, previously located at the San Matteo hospital and at its pharmacy, were transferred to the former convent of Sant'Epifanio.

The period of time from the eighteenth to the nineteenth century was marked by an extreme turbulence; the Austro-Russian invasion which followed that of the French brought the closure of the University. It was only after the battle of Marengo (1800) that times were peaceful enough for the reorganization of the studies. The *Piani di studi e di disciplina per le Università Nazionali* (14), dated 31 October, 1803, assigned to the Medical School the following twelve chairs:

1. Natural History
2. Botany
3. General Chemistry
4. Pharmaceutical Chemistry
5. Human Anatomy
6. Comparative Anatomy
7. Physiology
8. Institutions of Surgery and Obstetrics
9. Pathology and Forensic Medicine
10. *Materia Medica,* Therapeutics and Public Health
11. Clinical Medicine
12. Clinical Surgery

We see from this list that the Faculty of Medicine included disciplines having a merely propedeutic character for medicine and that would now belong to the Faculty of Sciences which, on the contrary, was intended at that time as a pure "Faculty of Physics and Mathematics".

Studies for physicians and surgeons were both courses of five years. The first biennium, propedeutic, was the same for the two branches and also included humanistic disciplines:

First Year	*Second Year*
Principles of Geometry and Algebra	General Physics
Italian and Latin Eloquence	Experimental Physics
Analysis of Ideas	Botany
Greek Language and Literature	Human Anatomy
	Physiology
	General Chemistry

The following three-year course was quite different for the two branches, and the program discrepancies became even more evident with the succeeding years, as evidenced in the following listing:

Program of Studies for Physicians	Program of Studies for Surgeons
Third Year	
Materia Medica	Principles of Surgery
Principles of Surgery	Materia Medica
Pathology	Human Anatomy
Human Anatomy	Comparative Anatomy
Comparative Anatomy	Visit to the Surgical Clinic in the afternoon
Natural History	

Fourth Year

Clinical Medicine as auditors at patient's bedside	Principles of Surgery
Pathology	Clinical Surgery as auditors
Human Anatomy	Human Anatomy with surgical operations
Pharmaceutical Chemistry	Obstetrics
Visits to the Medical Clinic in the afternoon	Forensic Medicine
	Visits to the Surgical Clinic in the afternoon

Fifth Year

Clinical Medicine as practitioners	Clinical Surgery as practitioner
Human Anatomy	Human Anatomy and surgical operations as surgeons
Obstetrics	Pharmaceutical Chemistry
Forensic Medicine	Anatomical and Surgical Practice on cadavers
Visits to the Medical Clinic in the afternoon	

After the fall of Napoleon, the *Istruzioni per l'attuazione degli studj nell'I.R. Università di Pavia, pel giorno 15 Ottobre 1817, giusto le nuove Prescrizioni di S.M.I.R.A.* (15) provided twelve chairs for the Faculty of Medicine and Surgery, distributed as follows:

1. Human Anatomy, with an introduction to the study of Medicine and Surgery
2. Mineralogy and Zoology
3. Botany
4. "Sublime" Anatomy and Physiology, with Comparative Anatomy
5. General Chemistry, Animal Chemistry, and Pharmaceutical Chemistry
6. General Pathology and Semiology; Public Health and General Therapeutics; Dietetics and *Materia Medica;* Toxicology
7. Introduction to the study of Surgery and Theoretical Surgery
8. Theoretical Obstetrics and obstetric exercises for physicians, surgeons and midwives
9. Theoretical and Practical Ophthalmology
10. Special Therapy of Internal Diseases; theorico-practical teaching at patient's bedside
11. Veterinary Medicine
12. Forensic Medicine and *Polizia Medica;* principles on treatment of asphyxia and sudden threats to life

Among the innovations brought to the teaching of medicine, mention may be made of the institution of the chair of Veterinary Medicine (that led to the unification of the Veterinary School of Milan (16) with the University of Pavia, which lasted until 1858-1859) and the separation of Obstetrics and Ophthalmology as independent chairs. To these first "specializations" will be added the Dermo-syphilopathic Clinic (1860) and the Psychiatric (1864); in 1855 Morbid Anatomy had acquired an independent chair.

In 1859 Lombardy was set free from the Austrian Empire and was assigned to the Kingdom of Sardinia, which in 1861 became the Kingdom of Italy. This inherited the law on education of the 13 November 1859 (The "Casati law"). New regulations completed this law in 1862, 1866, and 1875. This last regulation was approved at the time when Ruggero Bonghi (1826-1895) was minister of public education. His book *La facoltà di medicina e il suo regolamento* (Florence, 1876) is still a precious and accurate source of historical information on the subject. In this connection it is interesting to remark the repeated references made by Bonghi to the situation of medical education in the countries of German language.

It is impossible to follow here the subsequent developments of medical education in the Italian Universities; but from a general point of view, we can notice in the successive regulations a slow but steady trend to adequate education and to the development of science and medicine. This also holds true for the "Gentile law" of 1923, which was the last important reform of public education in Italy prior to the one which is now being discussed in the Parliament.

Medical education in Bologna was associated with experimental philosophy in a different way than in Pavia; this union was the result of the Malpighian tradition, which remained alive in Bologna and led, towards the end of the eighteenth century, to the electrophysiologic discoveries of Luigi Galvani (1737-1798). Bologna was indeed witness of the following peculiar phenomenon: for the eighteenth century the experimental philosophy was kept apart from the University as an institution but was taught in a new extra-university Institution (17) by teachers who often were also lecturers at the University.

In 1689, the same year of the publication of *De recentiorum medicorum studio* by Sbaraglia (18), the archdeacon Antonio Felice Marsili (1649-1710), Chancellor Major of the University, pointed out the deplorable state in which the University had been left and suggested what was to be done in order to improve the situation. His *Memorie per riparare i pregiudizi dell'Università dello Studio di Bologna e ridurlo ad una facile e perfetta riforma* (19) did not produce any result and Marsili was removed from Bologna and appointed bishop of Perugia.

Already since the seventeenth century, the immobility of the University had found a counterbalance in the private academies, where the "neoterists" collaborated in the field of experimental philosophy. One of these academies, the Accademia degli Inquieti, was created around 1690 by Eustachio Manfredi (1647-1739). Its statutes read that "it should be considered as a fault to support those things which cannot be derived by direct observation or verified by experiments or demonstrated by positive ratiocination." The Academy developed considerably and counted among its members the most important neoterists, particularly Giovanni Battista Morgagni (1682-1771), a young man with a very mature mind who, during his stay in Bologna

(1698-1707), a few years after Malpighi's death, received his education through Malpighi's direct pupils such as Ippolito Francesco Albertini (1662-1738) and Antonio Maria Valsalva (1666-1723). It was under the presidency of Morgagni that, in 1705, the Accademia degli Inquieti—which had moved from Manfredi's home to that of Giacomo Sandri (1657-1718)— accepted the invitation of Luigi Ferdinando Marsili (1658-1730) to hold the meetings in his palace. Marsili was the brother of the aforementioned archdeacon Antonio Felice and a man of war and science who now belongs to the history of science for his many deeds and, in particular, as the initiator of oceanography. Surprising is the huge amount of natural products, scientific instruments and books that this man gathered during his life interwoven with wars and journeys. He set all this material at the disposal of the "Inquieti" and thus transformed his own palace into a workshop devoted to experimental philosophy. Is there a better model for a renewed university?

The reform of the University became a very pressing problem and it was actually planned by the same Marsili in 1709 in his *Parallelo dello stato moderno della Università di Bologna con l'altre di là de' Monti* (20). This plan was pervaded by the spirit of the man "that in the land across the mountains I heard to be called the divine Malpighi, and more than once was I asked how could it be that in Bologna there was nobody who would follow his path" (21). From Malpighi derives the definition of medicine as the "science which was born and is developing on the observation of the organic structure of man, of the nature of the fluids therein circulating, of the morbid accidents occurring in both and on the institution of the means apt to overcome the diseases caused either by a breakdown in the organic structure or by the alteration of the nature of the said fluids" (22). This detailed and modern project by Luigi Ferdinando Marsili had the negative reception that had been given twenty years before to the project of his brother.

To overcome the immobility of the authorities, however, it was necessary to find some other way out. In 1714 was inaugurated the Istituto delle Scienze, to which was annexed the Accademia degli Inquieti, thus becoming the Accademia dell'Istituto. This Institute was independent from the university and had the task of promoting and teaching the new experimental science. The Institute, to which Marsili donated all his collections and possessions, was located at the Palazzo Poggi (now seat of the University) and disposed of six scientific departments, each one of them controlled by a professor with an associate. These teachers were often lecturers at the University as well and this fact undoubtedly made easier the coordination of the two institutions which would otherwise have been very problematic, since the Institute worked on the experimental, practical and demonstrative method while the University followed the conventional method. "Teachers will have to avoid, while teaching, anything which could be like an academic lecture, but they will on the contrary have to carry out practical exercises based on observations, operations, experiments and other things of similar nature. The inobservance of this rule will be considered as a great fault" (23). At first six disciplines were taught:

1. Chemistry
2. Physics
3. Natural History
4. Military Architecture
5. Geography and Navigation
6. Astronomy

The first three disciplines were undoubtedly useful to medical students, in particular chemistry of which Marsili had complained in his *Parallelo* (24) of the absolute absence from the programs of the University. Three more specifically medical chairs—Surgical Operations, Anatomy and Obstetrics—were added under the pope Benedetto XIV (Prospero Lambertini, 1675-1758), who had actively taken care of the Institute not only during the time he was archbishop of Bologna (1731-1740), but also before, in 1726, when he succeeded in straightening up a long and dangerous quarrel. In view of its importance, it is worth recalling the source of the material used for the foundation of the three departments.

The surgical instruments had been donated to the pope by Louis XV through his first surgeon François de Lapeyronie (1678-1747), and were entrusted to the surgeon Pier Paolo Molinelli (1702-1764), so that he could demonstrate them and carry out the corresponding surgical operations on the cadavers in the hospital of Santa Maria della Vita and Santa Maria della Morte.

In 1742, the year of foundation of this chair, Benedetto XIV—who as archbishop of Bologna had had the opportunity of seeing the anatomical wax models of Ercole Lelli (1702-1766) (25)—charged Lelli with the task of preparing a collection of anatomical wax models for the anatomical department which was being set up. Lelli asked for six years time and accomplished the task in such a satisfactory manner that the pope appointed him as curator and demonstrator of the models which were set up in a room of the Institute. This magnificent collection—enriched later on by more items among which those made by Anna Morandi Manzolini (1716-1774)—belongs today to the Department for Human Anatomy of the University of Bologna (26).

The revolution in obstetrics brought about towards the middle of the eighteenth century by the diffusion of forceps, implying a profound knowledge of the normal as well as of the morbid mechanism of delivery, suggested also the execution of obstetric models. Giovanni Antonio Galli (1708-1782), who was a lecturer on surgery at the University of Bologna, had the initiative to start the manufacture, which he superintended, of "a number of models of pregnant uteri in their natural size and color, bearing foetuses in different normal and abnormal positions, so that surgeons and midwives could learn more easily how to operate on the foetus," made of different materials such as clay and wax. In 1757 Benedetto XIV bought this collection and donated it to the Institute to create an obstetrical department for the teaching of such discipline. This was the first Italian public school of obstetrics and Galli was its first titular.

After Lelli's death in 1766, the teaching of anatomy was entrusted to Luigi Galvani (1737-1798) who, in 1782, took over the chair of obstetrics inherited from Galli.

In 1803 the University moved from the ancient Archiginnasio to the premises of the Institute—where it is still located—and this move marked in a tangible way the union of the two institutions. Thanks to this union, the University was enabled to function in accordance with the *Piani di studj e di disciplina per le Università Nazionali* (27), which we have already examined in connection with the University of Pavia.

I should like to emphasize the contribution made to Italian medical education by anatomical models. Among these models stands out the "artificial anatomy" of the eye "covered only with ivory and bone" as it can still be seen at the Museum of History of Science in Florence (28). This model was the work of Giambattista Verle, who was encouraged in his activity by Niels Stensen (1638-1686) at the grand ducal court of Tuscany: "while Mr. Stensen was demonstrating to the Prince Ferdinand the anatomy of a rabbit's eye, the Prince bestowed on me the favor of his order and commanded me to make something that would closely resemble it". This story is included in the booklet *Anatomia artifiziale dell'occhio umano* (1679) by G. B. Verle who, encouraged by the success he had with his models of the eye, "was taken by the desire to imitate with an artificial anatomy not only the sense of sight, but other senses as well and all parts of the human body, using not only ivory and crystal but also wax, clay, plaster, silk, threads of wool and of linen, wool fabrics of different kinds, vellum, paper ... so to imitate them in the best possible way and with the most resembling colors. In this way the anatomists will be free from the corruption of cadavers and will be able to demonstrate in every season. I offer this souvenir so that it may be a stimulus to anyone who would like to work on Human Artificial Anatomy, when the accidents of my age will keep me from continuing my work".

We know that in 1697 Anton Maria Valsalva (1666-1723), who was carrying out the research which produced his work *De aure humana tractatus,* (Bologna 1704), asked Verle to work for him. The masterpiece of Valsalva's anatomical skill was the *Nova auris praeparatio,* i.e., the preparation of the ensemble of the whole hearing organ in a single ear, such a difficult and delicate preparation that gave birth, in the spirit of its author, to the desire of transferring this *natural* anatomy into an "*artificial* anatomy of the ear, through which it would be possible for everybody to inspect the main parts of the acoustic organ in their aspects, locations and connections." It was actually in 1697 that Valsalva instructed Verle in this matter and sent him the ossicles of the tympanum; however, even Morgagni, who was the accurate biographer of Valsalva, did not know whether this work had been completed or not, he only knew that it had reached a rather advanced stage. It would be interesting to document the connections between this work of Verle and the ivory artificial anatomy of the ear now at the Museo della Specola in Florence, labelled *Nova ostensio auris humanae* (29). It resembles, in its general features and in some smaller details, the figures of the *Nova auris praeparatio* executed by the painter Angelo Michele Cavazzoni (1672-1743) under the direction of Valsalva, who inserted them in his *De aure humana tractatus*.

The afore-mentioned "souvenir" by Verle was not fruitless, as it was not long before it was realized in Florence, thanks to another artisan who was active in that town towards the end of the seventeenth century, i.e., Giulio Gaetano Zumbo or Zummo (1656-1701) who, together with the anatomist Guillaume Desnoues, initiated the artificial anatomy with wax models (30). I have already dealt with the fruits of this cooperation and with the most important aspects of plastic anatomy in Italy in the course of the eighteenth century, an art which had its capitals in Bologna (31) (Ercole Lelli, Giovanni Manzolini and his wife Anna Morandi) and in Florence (32). Even nowadays the Museo della Specola in Florence and the Josephinum in Vienna keep a considerable number of the splendid works which Clemente Susini (1754-1814) shaped in the wax workshop of the Museum of Natural History in Florence directed by Felice Fontana (1730-1805).

NOTES

1. Luigi Belloni, La medicina a Milano fino al seicento, and La medicina a Milano dal settecento al 1915, in *Storia di Milano*. Vol. 11, Milan, 1958, pp. 595-696, Vol. 16, Milan, 1962, pp. 933-1028.
2. *Memorie e documenti per la storia dell' Università di Pavia e degli uomini più illustri che v'insegnarono*. Pt. Ia: Serie dei rettori e professori con annotazioni, Pavia 1878. This important historical source was the work of Alfonso Corradi (1833-1892), rector of the university and professor of *materia medica*, general therapeutics and experimental pharmacology. Corradi, however, is particularly known as an historian of medicine.
3. *Ibid.*, p. 144.
4. Marcello Malpighi, *Opere scelte, a cura di Luigi Belloni*. Turin, 1967, pp. 491-631. In the text of Malpighi's *Risposta* are inserted the corresponding paragraphs of Sbaraglia's *De recentiorum medicorum studio* in a contemporary Italian translation.
5. Reference is made to my introduction to the *Opere scelte* of Malpighi and to my paper, De la théorie atomistico-mécaniste à l'anatomie subtile (de Borelli à Malpighi) et de l'anatomie subtile à l'anatomie pathologique (de Malpighi à Morgagni), presented at the Symposium of the International Academy of the History of Medicine, *Theories on the structure of living matter and their influence on medicine*, Liège, 16-17 November 1967. In press.

6. This document has been reproduced in facsimile and transcribed by Pietro Vaccari, *Storia della Università di Pavia*, 2nd ed., Pavia, 1957, pp. 182-183.
7. See, for example, LUIGI BELLONI, Immatrikulations- und Textaturkunde des Tessiner Augenarztes Pietro Magistretti (1765-1837). *Gesnerus*, 1948, 5: 34-42, and pls. I-IV.
8. VACCARI, *op. cit.*, p. 147.
9. *Statuti e ordinamenti della Università di Pavia dall'anno 1361 all'anno 1859*. Pavia, 1925, pp. 243-247, 255. Frau Prof. Erna Lesky had the courtesy to let me remark the close analogy between this plan and the *Instituta Facultatis Medicae Vindobonensis curante A. Stoerck*. Vienna 1775. Cf. LESKY, *Die Wiener medizinische Schule im 19. Jahrhundert*. Graz-Cologne, 1965, pp. 16, 18.
10. Cf. notes 4 and 5.
11. *Sistema compiuto di polizia medica di G. P. Frank. Trad. dal tedesco del Dott. Gio Pozzi*, Vol. 17, Milan, 1829, pp. 159-246.
12. ERNA LESKY, Johann Peter Frank als Organisator des medizinischen Unterrichts, *Arch. Gesch. Med.*, 1955, 39: 1-29.
13. LUIGI BELLONI, "Antonio Scarpa nell'anno della chiamata a Pavia," in *Discipline e maestri dell'Ateneo Pavese*. Verona, 1961, pp. 223-233.
14. See *Statuti e ordinamenti*, quoted in note 9, pp. 277-284, 290, 303-306.
15. *Ibid.*, p. 329.
16. LUIGI BELLONI, La scuola veterinaria di Milano. Celebration speech for the 175th anniversary of the foundation of the Veterinary School, now Faculty of Veterinary Medicine, read in the Aula Magna of the Università degli Studi, 14 October 1966; in *Studium veterinarium mediolanense 1791-1966*. Milan, 1969, pp. 1-31.
17. GIUSEPPE GAETANO BOLLETTI, *Dell'origine e de' progressi dell'Istituto delle Scienze di Bologna*. Bologna, 1751. ETTORE BORTOLOTTI, La fondazione dell'Istituto e la riforma dello "Studio" di Bologna, in *R. Accademia delle Scienze dell'Istituto di Bologna. Memorie intorno a Luigi Ferdinando Marsili pubblicate nel secondo centenario della morte per cura del Comitato Marsiliano*. Bologna, 1930, pp. 383-471. PAOLO DORE, Origini e funzioni dell'Istituto e dell'Accademia delle Scienze di Bologna, *L'Archiginnasio*, Bologna, 1940, 35: 192-214.
18. See note 4.
19. BORTOLOTTI, "La Fondazione," pp. 386-403.
20. *Ibid.*, pp. 406-419.
21. *Ibid.*, p. 412.
22. *Ibid.*, p. 410: "*è una scienza che è nata e cresciuta sulle osservazioni fattesi sopra della organica struttura dell'uomo, e della natura de' fluidi, che circolano per essa, e delli accidenti morbosi, che occorrono in ambo le parti, e nell'istituzione dei medicamenti, proporzionati a quei mali che nascono dal disturbo dell'organica struttura, e dall'alterazione della natura di detti fluidi.*"
23. *Ibid.*, p. 429: "*Avranno i professori particolare avvertenza di non fare negli esercizj alcuno studio o discorso scientifico, che convenisse alla forma di una Lezione, o che si potesse chiamare una vera lezione propria delle*

cattedre del Pubblico Studio, dovendo gli esercizj versare principalmente nella pratica delle osservazioni, operazioni, esperimenti, ed altre cose di simile natura. S'imputerà a gran colpa la trascurataggine di questo articolo."

24. *Ibid.*, p. 412.
25. Luigi Belloni, Anatomia plastica: II. The Bologna wax models, *Ciba-Symposium*, 1960, 8: 84-87.
26. Where the collection is now the object of an accurate restoration thanks to the care of its present director Prof. Luigi G. Cattaneo.
27. See note 14.
28. Luigi Belloni, Stensen-Andenken in Italien: I. Stensen und die künstliche Anatomie des menschlichen Auges (Florenz 1679) von G. B. Verle; 2. Swammerdams Zeichnungen des Seidenspinners, die Malpighi im 1675 durch Vermittlung Stensens erhielt, in *The historical aspects of brain research in the 17th century*, ed. Gustav Scherz. Oxford, 1968, pp. 171-180.
29. Luigi Belloni, Suono e orecchio dal Galilei al Valsalva, *Simposi Clinici*, Milano, 1966, 3: XXXII-XLII.
30. Luigi Belloni, Anatomia plastica: I. The beginnings, *Ciba-Symposium*, 1959, 7: 229-233.
31. See note 25.
32. Luigi Belloni, Anatomia plastica: III. The wax models in Florence, *Ciba-Symposium*, 1960, 8: 129-132.

THE TEACHING OF MEDICINE IN FRANCE FROM THE BEGINNING OF THE SEVENTEENTH CENTURY

CHARLES COURY[*]
Faculté de Médecine, Université de Paris
Paris, France

The history of medical education in France from the seventeenth century onwards may be divided into two periods of almost equal length, separated by a brief but sharp division that occurred during the Revolution and gave rise to a profound tranformation in the teaching system.

The first of these two periods, closely linked to earlier times, covers the seventeenth and eighteenth centuries. In consequence, the beginning of the "classical" period in French medical education was not marked by any important change in teaching methods which differed, however, according to the different regions, but, broadly speaking, had not altered for more than four centuries.

The second or "modern" period corresponds to the nineteenth and twentieth centuries and differs markedly from the previous one. In France, perhaps more than elsewhere, it is characterized by the birth and rapid development of a uniform national method of teaching, progressively adapted to scientific procedures and to modern medicine. It is based both on university teaching and hospital training.

Medical Education in the Seventeenth and Eighteenth Centuries

During the two centuries that preceded the Revolution of 1789, medical teaching, with little change or improvement, remained under the influence of that of the Middle Ages and Renaissance, so that there is some difficulty in distinguishing a clear line of demarcation around the year 1600. The end of the religious wars, the accession of Henry IV (1589) and his promulgation of the Edict of Nantes (1598) favored limited freedom of teaching in general and of medicine in particular. Yet, although the Church became less directly absolute in the seventeenth century, the teaching of traditional doctrines predominated for at least a century and a half.

During the seventeenth and eighteenth centuries there were numerous

---

[*] The author is greatly indebted to the Wellcome Institute of the History of Medicine (London) for a preliminary translation of this text.

medical teaching bodies, but their quality, their orientation and their activity were very unequal. The official university institutions were the faculties. Considered to be the depositories and guardians of correct teaching, they jealously guarded the prerogatives they had enjoyed since the sixteenth century and their relative autonomy from ecclesiastical and civil authorities. They varied in liberality according to region, but in principle they were all hostile to innovation. They provided an almost exclusively theoretical teaching derived from tradition, and their main function, which they obstinately maintained, was that of conferring official degrees.

The national institutions outside the universities, such as the Collège Royal and the Jardin du Roi in Paris, depended directly upon the royal government; they were completely independent of the Church, and conscientious in their teaching role.

The corporate professional organizations, the guilds, communities or colleges, acquired great importance in the eighteenth century in large non-university centers such as Lyons, Tours, Clermont-Ferrand, Orléans, and even in certain "faculty" cities such as Paris, Nancy, Aix, Bordeaux, and Grenoble. They carried out research and teaching activities mainly directed towards the most neglected fields such as anatomy, surgery, physiology, epidemiology, and the growing science of chemistry. With royal support they were finally turned into academic and official societies.

Autonomous and private courses, both theoretical and practical (anatomy, surgery, obstetrics, chemistry), multiplied at the end of the eighteenth century. Organized in private premises on a very flimsy financial foundation, they were, nevertheless, successful and able to overcome the greatest deficiencies and delays of official teaching.

Hospital teaching developed at a rather late stage, chiefly during the course of the eighteenth century. It was devoted to practical teaching for young physicians but also accepted a few foreigners for study. It was especially attractive to students who were looking for professional experience that was not to be acquired in the faculties. As the old faculties became more decadent and opposed to all innovation, these other, extra-university teaching bodies became increasingly important.

University Teaching

Faculties

Under the old regime France had a total of twenty-four faculties of medicine. Four of them were dominant because of their age, activity or reputation (Montpellier, Paris, Toulouse, and Strasburg); others were also renowned for the quality of their teaching (Douai, Rheims, and Caen). On the other hand, some were progressively discredited owing to irregularities or inconsistencies of their teaching program (Orange, Avignon, Nancy, Nantes, and Angers). Finally, others which had recently disappeared, whether spontaneously or by merging with a neighboring faculty, had only a nominal existence on the eve of the Revolution. In fact, because of geographical and historical conditions, a large number of these faculties were of foreign origin and had been swallowed up in French territory only at a late stage, because of military conquest or diplomatic annexations.

The Faculty of Montpellier was constituted solely of the body of its professors, and it was under the direction and control of a Chancellor nominated by the ecclesiastical authority. The Faculty of Paris, like that of Rheims, Caen, Angers, Nantes, and a few others, was composed of the entire group of "regent" doctors whether or not they held professorial chairs. They were headed by a Dean elected in principle for two years; he was the official representative, the source of necessary power, the administrator and chief examiner. However, he did not function as a professor during the period of his authority.

An important turning point in the history of the faculties occurred at the beginning of the eighteenth century when the royal edict of Marly on the teaching of medicine was promulgated in March 1707. It attempted to standardize the statutes, organization, curricula and examinations of the different faculties. Previously each faculty had had its own statutes which, except for a few details, were very much like the statutes of other faculties in western Europe at the same period. The new regulations of 1707, despite evidence of serious intent were not, however, wholly or everywhere successful.

Uniformity in medical teaching was not achieved until a century later under the Empire (1808), when it was easier to establish national standards for newly created organizations than to impose a common framework on old ones that had arisen at different times and places and had acquired traditions and privileges.

The official foundation of the Faculty of Medicine of Paris dates from 1253; its statutes were drawn up in 1274 and remained officially effective until 1598. A parliamentary decree of 1624 recognized that it was no longer necessary to "prohibit anyone, on pain of death, from teaching or believing any maxim against old and approved authors."

During the final third of the fifteenth century the Faculty was installed in some old houses in the rue de la Bûcherie, which were altered for teaching purposes, and by the end of the sixteenth century included an assembly hall, a chapel, a library, and lodgings for the servants charged with aiding the Dean in his administrative work. With the passage of time further alterations were made, but finally, on 18 October 1775, because of the extreme age and dilapidation of the buildings, the Faculty's quarters, with the exception of the Winslow anatomical theatre, were transferred to the building of the old Law Schools in rue Saint-Jean Beauvais. There they remained until the general closure of the universities prescribed by the revolutionary law of 18 August 1792.

The Faculty was constantly troubled by the modesty of its budget, always insufficient and always overdrawn. In the middle of the seventeenth century receipts did not exceed 1621 *livres* while expenses amounted to 2363 *livres;* the situation was even worse in 1775 when receipts were only 8339 *livres* and expenses 25,213 *livres*. In fact, the budget was never balanced before the end of the nineteenth century.

The Paris Faculty's obstinately conservative, even retrograde, attitude caused, over a long period, progressive decline. Its old motto, adopted in 1640, *Urbi et orbi salus,* expressed merely a memory of the long distant past.

The Faculty of Medicine of Montpellier is the oldest in France, since its first statutes date from 1220; in fact, the city had been a focus of medical activity from 1021. Owing to the climate of southern France, the Salernitan influence, and, as well, the Spanish-Jewish-Moslem influence, the Faculty very early exploited the medical heritage of the classical world. This was expressed in the proud motto, *Olim Cous, nunc Monspelliensis Hippocrates.* The Faculty was less affected than Paris by the hold of the university and of theologians, and it always showed a spirit of intellectual independence that was favorable to medical education.

From the fourteenth century until its temporary closure, the Faculty of Medicine was housed in a modest building near the church of St. Matthew; in 1556 it acquired the first anatomical theatre in France. At the end of the sixteenth century there were six teaching chairs or "regencies", and in 1593 a botanical garden was established.

The seventeenth and eighteenth centuries represent the great period, that of Vautier, Daquin, Pecquet, Denys, Turquet de Mayerne, Barbeyrac, Bordeu, Boissier de Sauvages, Lapeyronie, Barthez, and Chaptal.

Under the old regime France had twenty-two faculties of medicine in addition to those of Paris and Montpellier. However, seven of these had disappeared before the general suppression of the universities in 1793, and all the others were closed at that date.

One of the oldest was that of Toulouse, founded in 1229, after the Albigensian religious war. Until 1520 the teaching of medicine was attached to the Faculty of Arts, and had been influenced by the doctrines of Galen; it was relatively liberal although strictly supervised by the Church which feared a survival of Catharism. Although in principle the medical course was a long and arduous one, the studies were in fact easy but did not usually lead to the doctorate.

Six faculties were founded in the fourteenth century. The Faculty of Avignon was founded by Pope Boniface VIII in 1303 and remained obedient to the pope until 1698. Its teaching was much neglected and, despite the elementary nature of the requirement, students were never numerous. It was suppressed in 1788. The Faculty of Orleans dates from 1312, but disappeared at the beginning of the seventeenth century because of lack of students. A Faculty of Medicine was established at the University of Cahors in

1469; it was much frequented in the seventeenth century because of the ease and speed with which the degree could be acquired, an abuse which led to its suppression and fusion with the faculty of Toulouse in 1751. The faculty of Grenoble, founded in 1339 by Humbert II, was united in 1565 with that of Valence which provided a more strictly orthodox teaching. The Faculty of Medicine of Orange (1365), after various closures during the sixteenth century, was reopened by William of Nassau in 1583 and reorganized in 1607. Both Protestants and Catholics took part in the teaching that led to the doctorate, but the quality was so inferior that other faculties refused to recognize this "Orange flower" degree. The university became French in 1708 and the Faculty of Medicine aligned itself with other French faculties in 1718. The Faculty of Perpignan (1379) became French in 1659, and was reorganized in 1723. Its regional recruitment, mainly Catalan, ensured that it was active up to the time of the Revolution.

Nine faculties date from the fifteenth century. That of Aix was founded in 1409 by Pope Alexander V and Louis II of Anjou, Count of Provence. It was very liberal in its outlook and soon played an important part in practical teaching. The Faculty of Dôle, founded in 1423 by Philip the Good, Duke of Burgundy, became French in 1678 and was transferred to Besançon in 1691. The Faculty of Poitiers (1431) lacked degrees, professors and students and was only a fiction by the eve of the Revolution. The Faculty of Angers was founded in 1432; its statutes date from 1484. It activities had become insignificant by the end of the eighteenth century. The Faculty of Caen was founded in 1438 by the King of England, but twelve years later it became French. Its statutes (1478) were based on those of Paris. At the end of the sixteenth century there was an influx of candidates attracted by the ease and speed with which it was possible to obtain the degree. A reform of studies in 1699 led to a relatively complete program. The Faculty of Valence dates from 1452, but students were few and the teaching insignificant. Neither new regulations in 1589 nor the royal edict of 1707 improved the situation. The practice of medicine in Brittany was reserved to graduates of Nantes (1459), although there was an exception (1653) in favor of doctors from Paris. On the whole, however, the Faculty of Nantes progressively declined during the seventeenth and eighteenth centuries. The teaching of the Faculty of Bourges (1463) was often interrupted during the wars of religion or for lack of funds. Its teaching always had limited scope, and was done in the refectory of the Jacobean convent until 1527, and then in the Hôtel-Dieu. Studies and examinations were very easy, and the Faculty of Bourges progressively declined and in fact disappeared well before the Revolution. The Faculty of Bordeaux was founded in 1480 in competition with the College of Physicians that had been founded earlier; it limited itself to the award of degrees. The edict of 1707, which ran counter to local prejudice, did little to change the situation.

Five faculties were founded in the second half of the sixteenth century.

That of Rheims (1545) inherited the medieval episcopal schools of Champagne. Its statutes date from 1680 and were similar to those of the Faculty of Paris. Its activity, judged by the number of students and the number of degrees awarded, place it among the three or four most important of the old French faculties. One of the most renowned and prosperous faculties was that of Douai, founded by Philip II of Spain in 1559; it became French in 1668 following the peace of Aix-la-Chapelle. Originally designed on the plan of Louvain, following the application of the edict of 1707, and especially after 1749, it was reorganized. It was an important center for regional recruitment because of the quality and seriousness of the studies, and it continued to give courses and diplomas up to 1799 despite the general decree of closure. The Academy of Strasburg (1566) became a university in 1621, and French in 1681. In 1602 it had some sixty students of medicine, a large number for that time. It was much frequented by foreigners and unlike the other faculties its activity progressively increased up to the Revolution. However, writing in 1736, a royal moneylender observed that there was some relaxation in teaching: "For the last seven or eight years the professors no longer have held public courses, no doubt because their private lessons are well paid and they find this method more lucrative." The faculty of Pont-à-Mousson (1598) was known for the liberality with which it awarded "extra-mural" degrees. Julien de la Mettrie (1709-1751) placed it among the "disreputable faculties where a doctor's hat is sold like a yard of cloth." After the transfer of Lorraine to France, Louis XV moved the Faculty to Nancy on 8 August 1768. The Faculty of Medicine of Besançon was founded at the end of the sixteenth century by the Emperor of Germany, but it was of no significance until after the French annexation (1678) and after it had absorbed the Faculty of Dôle (1691).

The last of the old French faculties, that of Nancy, was merely an extension of that of Pont-à-Mousson, after its transfer in 1768, and the studies were only slightly more demanding despite the favorable influence exercised by the College of Physicians of the city.

On the whole, during the eighteenth century most of the provincial faculties entered into a rapid and irreversible decline. The edict of 1767 condemned "the relaxation which has taken place in some of the faculties of medicine, either in regard to the duration and quality of the studies, or in regard to the number and nature of the examinations leading to a degree." In 1789 numerous petitions of complaint revealed the deplorable level to which the medical profession had fallen in the country as a whole. The inhabitants of Montreuil demanded "that a young man who has hardly any notions of surgery and medicine not be able, for a hundred *écus*, to obtain a licence to kill with impunity, to cripple men throughout the countryside and to take away a father, a mother from unhappy children, that he be prevented from exercising these arts without previous rigorous examinations attesting to his capacity." Numerous, similar complaints were not uncommon.

On 15 September 1793, following the principle of the suppression of all old privileged institutions that were held to be retrograde, the Convention decreed the general closure of schools and faculties.

Degrees

Under the old regime the faculties of medicine in France attempted to retain their monopoly of the teaching of medicine, an inheritance from the Middle Ages, and until the Revolution with rare exception they kept the exclusive privilege of conferrring degrees: baccalaureate, licence and doctorate. These at the same time gave the right to teach and to practice medicine.

The baccalaureate gave only a restricted right to practice and to teach under the supervision of a doctor. The bachelor could have no autonomous and independent activity, at least in principle. Teaching was limited to "lectures" given without comment.

The licence gave the right to the free exercise of medicine and teaching. In fact, however, this right was most often effective only outside the university cities, since it was not respected in those localities that had a body of doctors.

The doctorate or "mastership" conferred "the right to lecture, teach, interpret and freely exercise all medical activities, both here and in all other places [*hic et ubique terrarum*]." Nevertheless the majority of university centers refused to recognize and accept without new examinations those degrees that had been granted by other faculties, when those degrees did not clearly provide all the necessary guarantees. Thus Paris did not give the right to practice to provincial doctors, not even to those from Montpellier. However, the sovereigns and princes reserved the right to accredit "by privilege" numerous "foreign" doctors among whom were, of course, charlatans. The universality of the degree of doctors of medicine in French territory dates only from the nineteenth century.

There were, moreover, two varieties of doctorate. That called *grand ordinaire*, conferred the full right to exercise any rights acquired only in the immediate vicinity of the faculty that had awarded the degree. It implied very serious and prolonged studies, high registration fees, and difficult examinations that ended with a costly ceremony of award. Thus the number of persons receiving this degree was limited.

The doctorate called *petit ordinaire* or *per saltum* was reserved for foreign or extra-mural use. It was of value only outside the awarding faculty, abroad and in French cities without a university. It could be acquired more easily and was not regarded highly. Faculties jealous of their prestige refused to award it, notably Paris, Montpellier, Strasburg, Poitiers, and Douai. Others such as Angers, Cahors, Caen, Valence, Avignon, Pont-à-Mousson, Nancy, Rheims, and especially Orange, acquired a poor reputation because of the ridiculous ease with which they awarded successive degrees in a few weeks, even without examination.

Teaching Staff

The teaching staff included the professors, assisted by a certain number of doctors attached to the faculty, and some demonstrators. The professors were not numerous: from one to three in the less favored faculties such as Poitiers, Nantes, Orange, Besançon, Nancy, Cahors, Bourges, Valence, Avignon, Douai; seven or eight in those better endowed: Montpellier, Paris, Aix-en-Provence, Perpignan, Rheims.

The professors were generally appointed after "disputation", that is, after a competition that included two kinds of tests: prelections that were oral and lasted an hour, and the thesis from the "chair" which had to be given in public and printed. At Montpellier, after 1498, the professors were selected through competition; from the middle of the sixteenth century they were nominated for life. At Paris any doctor could be called to a chair. The election of *magis idoneos* and final selection that followed were annual events occurring on the first Saturday after All Saints' Day. Beginning in the eighteenth century, professors were selected one or two years in advance, which gave them a reasonable period in which to prepare their courses.

In the beginning professors were remunerated by their students. Later they received fixed or variable incomes that were usually not very high; in Paris at the end of the seventeenth century these were 200 *livres*, and in the middle of the eighteenth century, 600 *livres*; except for Montpellier, the income in the provinces was usually considerably less.

The number, date of foundation and designation of the professorial chairs reflect the nature, importance and development of teaching in different faculties. At the opening of the seventeenth century Paris had only two chairs: one for "natural and unnatural things", anatomy and physiology, hygiene and dietetics; the other for "things against nature", pathology and medical subjects. A chair of surgery was added in 1634, and a fourth for botany in 1646. Two further professors were appointed at the beginning of the eighteenth century, and in 1760 an independent chair of anatomy was established. After a great deal of hesitation, a course of chemistry was organized in 1770. On the eve of the Revolution the Faculty's teaching was carried on by only seven professors plus two who were entrusted with special courses.

The Faculty of Montpellier was better off. It already had four professors at the end of the fifteenth century; in 1593 a chair of anatomy and botany was added, and in 1597, a chair of surgery and pharmacy. In 1610 two further teachers were added, doctors without professorial title, and by the end of the seventeeth century there were eight chairs, two of these dedicated to practical disciplines that did not yet exist at Paris: anatomy and consultive and practical medicine.

In the other provincial faculties the number of chairs was variable. In the eighteenth century Perpignan had seven chairs; Strasburg, four; Rheims, seven; Bordeaux, four; Pont-à-Mousson, five; Nancy, four, with a fifth added for chemistry in 1776; Douai, three; Toulouse, five, and the remaining faculties more modest numbers.

There was no common rule in regard to the division of the different subjects between the members of the teaching body. In the seventeenth century the majority of faculties had a chair of botany and even a chair of anatomy, with the exception of Paris. In the eighteenth century chairs of chemistry began to be evident, while some faculties, such as that of Montpellier, had established chairs of "practical medicine", that is, clinical teaching at the bedside of the patients.

Students

Different factors determined the recruitment and mode of living of the future doctors under the old regime. The *philiatres* came for the most part from bourgeois families, professional or business families, those of magistrates, officials and physicians. Rarely did they come from noble families, and then generally from the petty nobility. Occasionally the "scholars of medicine" had very humble origins and, because of this, suffered the greatest difficulty since the studies were long and costly.

Students of medicine often had relatives who were physicians, surgeons or apothecaries, and also there were certain medical "dynasties" of influence. Such family and professional traditions were perpetuated throughout the nineteenth century and have lasted up to our day. Up to the eighteenth century the sons of physicians had numerous privileges during the course of their studies such as reduction of the length of the course, exemption from certain university taxes, favor from examiners, and the material and moral support of their families.

Some backgrounds were prohibitive. For a long time non-converted Jews could not enter the universities, bastards and sons of hangmen were also excluded. During the period of the wars of religion and even afterwards, Protestants faced serious obstacles in the provinces that were mostly Catholic.

The influx of foreign scholars, which was one of the characteristics of university life in the Middle Ages and Renaissance, began to die out in the seventeenth century. Nevertheless inside the French frontiers students passed easily from one faculty to another, carried out their studies in one city only to present themselves for examination in another, and did not hesitate to move in order to attach themselves to a local master who they had heard was a good teacher.

The life of the student could be difficult financially, but it was also turbulent and happy because of the release of youthful energy in a collective environment. The cost of being a student, especially in Paris, was always high, because of the low budget provided for the faculties. Students could register for a trimester at a cost of 6 *livres*. In the seventeenth century the cost of the examination was 48 *livres*, of which 7 *livres*, 10 *sols* were university fees; for the baccalaureate, 572 *livres* of which 36 *livres* were similar fees. These sums were increased by numerous additional expenses such as tips, banquets, presents, etc. The total reached 3500 *livres* at the beginning of the eighteenth century. In 1754 the degree of doctor of medicine cost 5614 *livres*.

Boarding-in students lodged with their teachers in chambers around the schools in special houses that more or less resembled the medieval "colleges". Because of the costliness of lighting the student's day began very early. A student's meal cost from 10 to 40 *sols* on an average, that is, from 0.5 to 2 *livres*. Complete board came to about 1200 *livres* per annum. In the provinces the costs were notably less. The few prizes and scholarships provided by rich benefactors or the public authorities were insufficient, and the more fortunate students obtained assistance by some secondary occupation or employment in the service of a rich colleague.

In order to register and follow the studies in a faculty of medicine it was necessary to satisfy a number of family and moral conditions by producing "favorable letters of reference". It was also necessary to prove that one was a Catholic by producing a certificate of baptism and by assisting regularly at the university's religious ceremonies. Certain preliminary studies were required, and many faculties demanded the master of arts degree, equivalent approximately to high-school graduation. The edict of 1707 made this last requirement obligatory in all French faculties of medicine. In consequence of such requirement the student began his medical studies towards the age of twenty.

Curricula

Over the course of time and according to the faculty, the length of studies, the teaching program and examinations could lead to three successive degrees. The duration of studies for the baccalaureate was in principle four years in Paris and three years in the majority of provincial faculties. Paris, the most demanding of all, required forty-eight months for the bachelor's degree, reduced to three years for those who held the degree of master of arts, and twenty-eight months for the sons of physicians. Studies made elsewhere were accepted for only half their actual duration.

At Montpellier, Toulouse, Cahors, Dôle, Besançon, and Poitiers, the baccalaureate could be obtained in three years. At Nantes, from 1683 this period was extended to four years but reduced to two years for masters of arts. At Caen the period was also three years, but it was doubled for students from elsewhere. Rheims insisted on fours years of study while Strasburg was satisfied with two.

Before obtaining the licence bachelors had to carry out two years of supplementary study at Paris, Angers, Nantes, Poitiers, Strasburg and Caen. Other faculties were satisfied with six months (Douai, Toulouse) or even three months (Dôle, Besançon). Finally, some like Avignon and Pont-à-Mousson often gave a licence after a few weeks or even days following the baccalaureate.

The licentiates of Paris had to carry out two years of practice before presenting themselves for the doctorate; Poitiers required three years, but in most of the faculties the degree could be obtained very soon after the licence. The total duration of studies imposed on doctors was thus eight years at Paris for most candidates, and six years for the privileged. In the provinces it did not extend beyond four to five years; at Avignon it was three years, and the time was even shorter in some faculties that tried to attract students by the brevity and ease of their curriculum.

The annual term varied from one faculty to another. In Paris the university year extended from the 13 September to the 28 June and was interrupted only for holidays, about thirty per year. The actual teaching began in the second half of November. At Douai there was daily teaching except on holy days, and at Avignon in the eighteenth century the courses began immediately after St. Luke's Day and continued until the 1 August.

Each professor normally lectured daily during the term. The lectures were in Latin, except for a course of surgery conducted in French, and lasted from an hour to an hour and a half. The use of Latin allowed the teaching to be kept on the narrow path of orthodoxy and so fend off innovating influences. In fact, the lectures were based on the classical works and were accompanied by fairly standard commentaries. These were followed by *quaestiones* and *disputationes* subject to the rules of traditional rhetoric. This sort of teaching was given *ex cathedra*, with the professor in his cap and fur-edged black robe and scarlet cloak. The bachelors took part in instruction of the younger students of the first and second years through elementary lectures that were in fact repetitions of those of the professors and took place early in the morning, at five o'clock in the summer and six o'clock in the winter. Professors and masters taught from eight to eleven in the morning and from two till four in the afternoon. The students took notes sitting side by side on wooden benches that had by now replaced the earlier bundles of straw.

The program of the four years of preparation for the baccalaureate at Paris were as follows:

1st year: *Materia medica,* pharmacy, physiology, anatomy
2nd year: Pharmacy, pathology, surgery, hygiene, obstetrics
3rd year: Physiology, *materia medica,* pathology
4th year: Physiology, surgery, pathology

Although anatomy was part of the examinations, it was the object of only summary teaching beginning with osteology and including at most two annual dissections. An important place was reserved in the curriculum for physiology and *materia medica* although these were still rudimentary.

The program was somewhat different in other faculties, notably at Montpellier where the practical orientation of studies occurred earlier than in Paris. At Dôle and at Besançon each year ended with a recapitulation examination; at Douai the students had to produce a certificate of application to the course. Elsewhere, notably at Paris, greater note was made of their presence at masses and official ceremonies than of their attendance at lectures. Surgery covered a very ill-defined area. *Quae sint chirurgicae?* was still the query of the Parisian Faculty in 1507. The surgical repertory included the works of Hippocrates, Galen, Oribasius, the sixth book of Paul Aegineta, the seventh and eighth books of Celsus, the treatise of Abulcasis, and those of Guy de Chauliac, Tagault (d. 1545), and Gourmelen (d. 1594).

The program for the licence was more developed than that of the baccalaureate; in some schools it included an obligatory period of practice. In Paris during the two years of their preparation the bachelors worked as auxiliary teachers in the morning and became students in the afternoon. From 1644 they were expected to follow regularly the free consultations which the six regent doctors offered every Saturday morning. From 1732 they were obliged to follow the practical courses of anatomy in the first year and of surgery in the second. At Caen, Nantes, and Pont-à-Mousson the candidates had to attend a practical period in hospital service or in the city as assistants to the doctors.

The doctorate program was variable. At Toulouse, up to 1707, it merged with that of the licence. The Faculty of Paris was exceptional in that it required from the licentiate a hospital period of two years; in principle this took place at the Hôtel-Dieu, at the Charité, or with doctors in the parishes.

It is very difficult to give a precise picture of the tests that the candidates had to undergo before moving on to the different degrees, since they varied a great deal from one faculty to another. In general, they were numerous, long and complicated, always touched with formalism. At Montpellier, for example, a student had to pass some fifteen tests before reaching the final doctorate. For each of the three stages it is possible to distinguish a first group of preliminary tests or "attempts" that would enable the student to go on, and a second group of "terminal" tests that led to the final nomination. This took place, as in most of the universities of Europe, at a pompous ceremony of "collation" of the degree. Interrogations as well as theses represented less an outline of the knowledge of the candidate than a greater or lesser brilliant manifestation of Latin eloquence confined within the strict limits of traditional doctrine. The reality of facts and scientific truth mattered little provided the argumentation conformed to orthodoxy and was sufficiently dogmatic and was conducted according to the rules of dialectical reasoning. Until the eve of the Revolution the faculties kept obstinately to the medieval *nego, concedo* and *distinguo* and resolutely ignored the Cartesian step of *cogito*.

The candidate for the baccalaureate had first of all to address a *supplicatio* to each of the members of the faculty, to obtain authorization to present himself for the examinations and to be *admissus ad cameram*. These oral examinations took two forms, that of general questioning, or *in communibus*, answering collectively and publicly in the presence of the whole faculty; and those of particular interrogations, that is, submitted to in private on the premises of the examiners. In Paris they lasted four days and included questions of anatomy, hygiene, pathology and botany. To these were added manual examinations of dissection and operative surgery, as well as a practical examination in botany. Between these two series of tests the candidate had to present his *probatio temporis auditionis*, in other words, the test that he had to undergo during the studies imposed by the regulations. The examination jury consisted of the Dean and four members or *tentores* of the faculty.

These interrogations were followed a few weeks later by the submission of the first thesis, called *quod libet*. In Paris this thesis, on a subject of physiology or pathology, lasted six hours, *ab aurore ad meridiem*, on a Thursday morning during the baccalaureate session held every two years. The candidate developed verbally a text of four pages in the form of five syllogistic articles with the argumentation first presented by the student's colleagues, then by a jury of nine regents; the discussion was followed by a general vote with a secret scrutiny, and a majority of two-thirds allowed him to present his final tests. At Montpellier the future bachelor also submitted a *praevium* which lasted four hours, in public, on a theme of physiology, hygiene or dietetics.

The nomination examination consisted nearly everywhere of a bachelor or "cardinal" thesis on a subject of hygiene or physiology, with or without a commentary on an aphorism of Hippocrates. This took place in public and lasted two hours at Douai, three hours at Toulouse, Dôle and Besançon, four hours at Montpellier, up to six hours at Paris and several days at Caen.

The licentiate examinations varied from one university to another. At Paris the candidate had to submit for a week to a series of detailed practical tests which were considered in a secret scrutiny. At Montpellier he had to sustain a series of four theses of four hours each. At Nantes, Angers, Strasburg, Toulouse and Poitiers these tentative tests were *quod libet* and dealt with theoretical and practical questions debated in public. Rheims required an anatomical examination, a botanical interrogation, a test of operative medicine and a thesis which was delivered from six o'clock until midday. Douai required an examination in *materia medica*.

At Paris in the eighteenth century the final admission of *licentiandus* was pronounced after the delivery of two *quod libet* theses; Montpellier also required two supplementary theses, called *points rigoureux*, which dealt with a passage from Hippocrates; at Nantes the *points rigoureux* developed a subject taken from Hippocrates or Avicenna chosen at random twenty-four hours in advance; Caen required a public thesis on a free subject, followed by a set lecture. At Poitiers, the candidate submitted to two "cardinal" tests before undergoing a *points rigoureux* on an aphorism selected at random. Toulouse, Angers, Strasburg, Rheims, Douai, and Pont-à-Mousson only required a "cardinal" dispute. Dôle, Besançon and Perpignan required a public examination.

The *grand ordinaire* doctorate also entailed two series of tests. At Paris and at Angers the examination was called *acte des Vesperies* because it took place in the afternoon; it consisted of the oral delivery of a thesis from a printed sheet, the argument of which, communicated to the candidate a few days beforehand, entailed two contradictory propositions. In other faculties the first tests lasted three days, hence their name *triduanes*; at Montpellier they consisted of six interrogations. At Strasburg the candidate underwent numerous public and private examinations of one hour each and then outlined two theoretical and practical questions each for one hour, after twenty-four hours preparation. At Douai there was a general examination of three hours on all branches of medicine. The candidate of Poitiers tackled a diurnal dispute of three hours in the morning on a theoretical question and three hours in the afternoon on a practical question. At Caen, the *inceptio* consisted of a thesis with discussion. Valence limited these initial examinations to particular tests.

The final examination consisted nearly everywhere of a thesis or a solemn discourse with great pomp. At Paris, the doctoral thesis, drawn up in Latin, was at first written, then printed from the eighteenth century after approval by the Dean; it did not exceed four folio or quarto pages; very few of these have any real scientific interest. There followed three months later the "pastillery act" in support of the thesis, during which cakes were distributed to the audience. The new doctor then received his "doctoral letters" but did not start utilizing effectively his prerogatives of *verus regens* until the end of a period of two years. In the eighteenth century in most of the faculties the final examination consisted of a public thesis on "the utility of the five parts of medicine" (Rheims), and a solemn discourse emphasizing the glory of universal medicine and the local faculty (Angers, Bourges, Caen), or on any other general theme, conventional and dithyrambic. At Valence the public thesis of the doctorate lasted three hours; it did not last less than five hours at Dôle, Besançon and Toulouse. The medical personalities and others that assisted at these long exercises in oratory well deserved the refreshments and closing repast that they received.

The doctorate examinations of *petit ordinaire*, like those of the "foreign" doctorate were much more simple. At Rheims and Caen, for example, they were reduced to the presentation of one or two theses; in other universities the degree was given after a simple examination, even without test. The final ceremony was in any case deprived of splendor and was therefore less onerous for the recipient who was to leave the city.

The conferring of each of the three degrees was marked by numerous formalities, very similar to those that took place in other universities of Europe. One should recall especially the oath of future bachelors to obey the rules and traditions of the faculty. Similarly, with the licentiates and doctors who, at least in Paris, swore to obey the faculty, to assist at the annual mass for deceased doctors and to combat the illicit exercise of medicine, meaning such exercise by doctors from the provinces as well as by true charlatans. In addition to these oaths there were religious oaths of obedience. The ceremony of conferring the baccalaureate entailed the symbolic exchange of the black student's robe for the red cloak. At Montpellier it took place with great pomp. The new bachelor received a laurel crown and the President pronounced: *Indue purpuram, conscende cathedram et gratias age quibus debes.*

At Paris the conferring of the license entailed a particularly complicated ceremonial. After having been presented by the Dean to the Chancellor-Archbishop, the candidates, dressed in red robe edged with ermine, ranged in order of merit and symbolically married the Faculty. A few days later the candidates went in procession to the archbishop to receive on their knees, and with bared head, the apostolic benediction and the prerogatives of their degree: *Licentiam legendi, interpretandi et faciendi medicinam hic et ubique terrarum.* At Montpellier the ceremony took place in the presence of the Bishop of Maguelonne and was completed by a banquet.

The "triumphal act" that marked the conferring of the degree of doctor consisted of a series of lay and religious manifestations regulated by a ceremonial full of pomp and solemnity. The essential part was the swearing of an oath followed by the donning of the doctor's cap and cloak. Accompanied by the mace-bearers, and in the presence of the regents and robed licentiates, the new doctors received the accolade of the President. The ceremony ended with an exchange of compliments, a speech of thanks, a distribution of gifts, and a banquet. This ceremony was similar in many other faculties. The insignia of the degree included a black doctor's bonnet with a crimson tassel, a gold ring, a gilded belt; the new doctor also received a copy of the works of Hippocrates.

The number of new Parisian medical students was annually about fifteen or sixteen. At the end of the eighteenth century the Faculty of Paris did not include more than eighty to 100 students of whom only about twenty were newly registered each year—a century later the total was seventy times greater. Elsewhere the number of students varied according to the reputation of the faculty. While Perpignan benefited by considerable regional recruitment and had fifty students in the eighteenth century, Valence had only two in 1791. Cahors no longer displayed that exceptional appeal of the fifteenth century when it had 100 to 150 students. The Faculty of Angers, which had annually from thirty to sixty students in the seventeenth century, had almost none at the end of the eighteenth. Avignon, where the studies were not demanding and where there were about fifteen students at the beginning of the seventeenth century, still had a dozen towards 1750, but enrollment progressively declined in the following decades. Douai maintained a vigorous program and student population, many from Flanders, but, on the other hand, at the beginning of the seventeenth century Nantes averaged only one degree every three years.

Nearly everywhere the number of annual doctoral promotions was less than that of the professors of the faculty. At Paris five to ten new doctors were nominated each year, but this recruitment was sufficient to feed the corps of Parisian regent doctors which was never very numerous (seventy-two in 1550, 113 in 1650, 163 in 1780, 144 in 1787, 139 in 1792, for a population of about 600,000). On the eve of the Revolution, however, the Faculty of Paris showed a notable reduction in activity. No doctor was nominated from 1785, and no licentiate after 13 September 1790. Among the provincial faculties, Caen received only one or two theses per year between 1660 and 1740. Although notorious for the facility with which it awarded degrees, the Faculty of Cahors did not produce many bachelors or licentiates but nominated on an average 6 doctors per year in the seventeenth century, while Rheims, which had a similar reputation nominated 13.

Teaching of Certain Subjects

It will be useful to examine briefly the teaching of some of the medical and paramedical disciplines. Although not perhaps neglected, anatomy was always badly taught in the majority of old French faculties of medicine. The rarity of theatres, the lack of cadavers, and the method of teaching help to explain this insufficiency. For a long time the teaching of anatomy had been entrusted to professors of medicine or surgery. In the seventeenth century the post of demonstrator was created in many faculties, but autonomous chairs of anatomy did not appear before the end of the century and the first half of the eighteenth, except at Aix-en-Provence which had a professor in 1462, and at Nantes where practical teaching existed from 1461. Except for Montpellier, which had a theatre in 1556, the other faculties did not have a proper place for anatomical demonstrations before the seventeenth century (Paris, 1604, Aix, c. 1670, Dôle, 1631, Avignon, 1696, Toulouse, 1785).

The first anatomical theatre in Paris was built in 1604 in the rue de la Bûcherie, a modest edifice of wood open to the elements. It is said that it was built in about a fortnight on the basis of parlementary authorization of 1526 that had not been previously implemented. The theatre was rebuilt in 1617 and opened in 1620. A later theatre, constructed in 1744 on the corner of the rue des Rats, was opened by the anatomist Jacques-Bénigne Winslow (1669-1760), for whom it was named. A circular building by the architect Barbier de Blignière, it was nine metres in diameter and could accommodate 180 spectators.

Anatomical demonstrations took place during the winter on the bodies of executed criminals distributed under the authority of the Dean of the Faculty. The number of bodies utilized did not exceed two or three each year, and the method of teaching had not improved since Vesalius's student years in Paris (1533-1536). Hence anatomy was still limited to classical lectures while an inexperienced demonstrator—often a simple barber, or sometimes a bachelor called "archdeacon of the schools"—tried mutely to illustrate the propositions of the professor on a badly prepared body. The students were forbidden to be present with sticks or swords in order to avoid bloody quarrels between physicians and surgeons. Spectators were uncomfortably packed in and often fought for the few places from which one could actually see this summary demonstration. Often the lack of cadavers led to demonstrations being made without them or, alternatively, the anatomical course was suppressed.

One of the main failures in university medical teaching up to the end of the eighteenth century was the lack of bedside teaching. In 1565, Pierre de la Ramée, lecturer at the Collège Royal, reported to Charles IX the insufficiency of practical teaching to Parisian students in comparison with that of Italy and Montpellier. The faculties had almost totally neglected this aspect of professional education; the young students at no stage received any practical experience, or at the most some faculties required of their future licentiates and doctors practical work at a hospital or with a doctor practicing in the city. The edict of 1707 generalized these measures and made them obligatory, but in practice enforcement failed because of the crowded hospitals and the reluctance of patients and of the religious staff looking after them. In any case, the hospital physicians were not well prepared for such teaching.

Nevertheless, following the foreign examples of Leyden, Edinburgh and Vienna, some praiseworthy efforts were made. In 1729 the faculty of Strasburg organized the true teaching of clinical medicine at the hospital. From the seventeenth century the students of Aix-en-Provence could follow clinical teaching at the hospital of Saint-Jacques; in the middle of the eighteenth century the students of Avignon obtained the same facilities at the hospital of Sainte-Marthe. At Angers, under the influence of Dean Pierre Hunaud II (1664-1728), from 1718 the hospital of Saint-Jean became a true annex to the Faculty. In 1760, the hospital of Saint-Eloi in Montpellier was able to accept twenty students in its clinical department. At Paris, at the end of the eighteenth century, students frequented the Hôtel-Dieu and the Charité hospital in great numbers, not so much because the university compelled them, but to follow renowned hospital teachers such as Desbois de Rochefort and Desault.

The rational organization of clinical teaching became a growing necessity. In 1778 two doctors regent of Paris, Claude-François Duchanoy (1742-1827) and Jean-Baptiste Jumelin (1745-1807) published a "Report on the utility of a clinical school in medicine", in which they suggested the nomination of two clinical professors aided by two assistants. Seven years later, Nicolas Chambon de Monteaux (1748-1828) returned to this problem in a report to the Société Royale de Médecine. The temporary closure of the faculties delayed the application of these plans which were not carried out until 1795 when they contributed in large part to the spread of French medical teaching in the first half of the nineteenth century.

Up to the beginning of the nineteenth century drugs of vegetable origin constituted the major part of the pharmacopeia. From the beginning of the sixteenth century development of relations with overseas countries led to an influx of medicinal plants which had hitherto been unknown in Europe, and these needed to be acclimatized and studied. In this there was nothing contrary to the doctrines of Hippocrates and Galen, nor to the precepts of Dioscorides or Myrepsius. Thus botanical gardens and chairs of botany arose to the benefit of teachers and students, new versions of the monastic *hortuli* of the Middle Ages. The first French botanical garden was that founded by Henry IV at Montpellier in 1598; the Faculty of Paris had its own garden in 1606, but botanical teaching given by the Faculty was rapidly eclipsed by that at the Jardin du Roi. These gardens multiplied during the eighteenth century (Avignon, 1707; Nantes, 1728; Caen, 1747; Nancy, 1751; Toulouse, 1783).

For a long time French universities neglected the teaching of the art of delivery, and male students had no training in this discipline which was reserved to midwives and to a few surgeons who, by their ability and their personal renown, could break the rules of decency.

The midwives, organized under a statute of 1560, depended upon the Fraternity of Saint-Côme. In Paris, from 1678 to the Revolution, their entry into the profession depended on successfully passing an examination before two sworn surgeons, but from 1745 they were entirely taught in a course organized for them by the Faculty. Promotions were relatively numerous, averaging about a dozen a year with a maximum of twenty-three during the academic year 1757-1758. At Strasburg theoretical and practical teaching of obstetrics developed from 1728 within the Faculty, but in the majority of other provincial schools, midwives were taught according to corporative rules, especially among the older professional bodies. Some midwives, such as Angélique Le Boursier du Coudray (1712-1789), organized a series of visits throughout the country giving public demonstrations on dummies or phantoms. In Paris during the whole of the eighteenth century the only practical center of teaching was at the Hôtel-Dieu, although some surgeons also organized private courses. The Faculty itself had only a very theoretical control over the training of midwives.

Voyages of discovery, the development of long distance navigation, and colonial expansion gave rise to certain medical problems as early as the sixteenth and especially from the eighteenth century. Nantes, which was a great seaport and point of departure to overseas regions, organized clinical teaching of colonial pathology at the Charité hospital; in 1780 it established a chair of seamen's diseases. A school of naval medicine was opened at Brest in 1783, which provided clinical teaching to young doctors who came from the interior; this specialized course lasted two to three years.

The teaching of military medicine and surgery was not within the province of the university. These subjects had had no development since the time of Ambroise Paré and remained static until the reorganization of the health service carried out by Larrey during the wars of the Empire. At the end of the eighteenth century there were only five military teaching hospitals: Lille, Metz, Strasburg (1775), Brest and Toulon (1781). In 1796 the nation had four service schools of military medicine (Paris, Lille, Metz, Toulon). The main reason for the opening of faculties of military medicine was the shortage of doctors in the armies of the Revolution.

The teaching of pharmacy remained the responsibility of the faculties of medicine until the end of the eighteenth century. The Corporation of Apothecaries of Paris acquired its statutes in 1489; these were reworded in a decree of Parlement of 3 August 1636. Future pharmacists had to follow the lectures of the Faculty at least twice a week for a year; the teaching was given by the professor of *materia medica* and by the professor of pharmacy assisted during the summer by an apothecary. Apprenticeship lasted four years and journeyman's status lasted six. The examination for master was held twice a year in the presence of at least two regent doctors and included an examination on theory, a practical test called "the herbal act", and finally the masterpiece. The candidates had to take an oath before the Faculty which ensured, among other things, the inspection of the dispensaries. In other universities the rules were very similar. At Montpellier, for example, teaching of theoretical pharmacy was ensured by the Faculty from 1532. In fact, the pharmacists taught each other, the teaching going from master to student, or by means of private courses in botany and chemistry (1650-1725).

Parisian apothecaries gradually freed themselves from the restriction of the Faculty and finally gained their autonomy by a royal declaration of 25 April 1777, which led to the official opening of a College of Pharmacy on 30 June 1780. The Schools of Pharmacy were instituted in 1803, and from 1874 some of them became mixed faculties of medicine and pharmacy. Proper faculties of pharmacy date only from 1920.

Medical Literature

In addition to oral teaching, for long the commonest form, and practical teaching that consisted of rare demonstrations and occasional hospital visits, some attention must be given to medical literature.

There were a great number of printers of medical literature in France in the seventeenth century, especially in cities such as Lyons and Paris, where they had their shops in the Latin Quarter and in the small streets neighboring the Hôtel-Dieu. In the eighteenth century a great many medical books were published: the classical works, more recent publications not without interest, as well as a great many popular books of mediocre quality by non-medical authors.

Medical publications were rarely issued for teaching purposes since students did not have the means to buy them and had to content themselves with notes taken during the course of lectures. However, the wealthier practitioners had a kind of "postgraduate" documentation that was also a means of providing scientific information which was greatly appreciated by the general public.

The medical press was not free in France. In 1536 the Faculty of Paris acquired exclusive control over all medical works, and it denounced with unfailing zeal all ideas, theories and treatments it considered to be dangerous innovations. It is sufficient to recall the violent quarrels that arose in the seventeenth century over the discovery of the circulation of the blood, the methods of bloodletting, the use of quinine and antimony, the latter formally condemned in 1615 but rehabilitated by Parlement in 1666. Those authors who wished to depart from established orthodoxy had to hide under a pseudonym or have their books published clandestinely or sent to foreign publishers.

There were not very many books available in the Faculty of Paris in the seventeenth century. These became more numerous thanks to many legacies from physicians and surgeons such as François Picoté de Belestre (d. 1733), Philippe Hacquet (1661-1737), Claude Helvetius (1685-1755) and others. The library of the Faculty, which was established in the chapel of rue de la Bûcherie was open annually from 3 March and was administered according to a regulation of 1751. In 1775 the library contained 10,000 volumes which accompanied the Faculty in its transfer to rue Saint-Jean de Beauvais, before being finally settled, in 1795, in the buildings of the old College of Surgery. The library of 15,000 volumes in 1798 was entrusted to the care of a librarian who had the rank of professor (Pierre Süe, 1739-1816). One hundred years later, it had more than 120,000 volumes which increased to 400,000 in 1930 and 800,000 in 1962, exclusive of journals. The final organization of the university libraries was established by a circular of 4 May 1878. Among the provincial libraries mention must be made of that of the Faculty of Montpellier which was opened in 1534.

National Extra-University Institutions

One of the fruits of the scientific awakening which marked the period of the Renaissance was the rapid development of anatomical studies. The seventeenth century saw the birth of modern physiology and the rise of natural sciences applied to the art of healing. From the middle of the sixteenth century it was apparent that the teaching given by the faculties of medicine, static, insufficient, and later decadent, was no longer adapted to its purpose. In consequence new organizations were founded at the instigation of royal counselors, with the object of promoting research and teaching on a national scale. These were lay and independent, compensated for the insufficiencies of the universities, and filled the existing gaps.

One of the first of these organizations was the Collège Royal founded in Paris in 1530 by Francis I under the influence of the humanist Guillaume Budé (1468-1540). The first chair of medicine was established in 1542, and at the beginning of the eighteenth century the Collège had chairs of medicine, surgery, botany and pharmacy. Their number was raised to six with the addition in 1772 and 1775 of chairs of anatomy and practical medicine.

Teaching was free and consisted of seventy to eighty lectures a year by each professor, and it is sufficient to mention the names of Magendie, Laennec, Récamier, Claude Bernard, and René Leriche to demonstrate the quality of the Collège Royal, which was later renamed the Collège de France. From the nineteenth century to the present it has never ceased to play an important part in teaching and medical research.

The Jardin du Roi, founded in 1626 at the instigation of the physician and botanist Gui de la Brosse (d. 1641), was officially organized in 1635. At first it was known as an acclimatization and botanical center, but it extended its activities to all natural and biological sciences. The first three chairs, given to royal demonstrators, were devoted to botany, chemistry, and anatomy with surgery. In 1712 an additional chair of medicine was founded that was to be held by, among others, Fourcroy and Vicq d'Azyr.

A royal declaration of 20 January 1673 guaranteed the right of the professors to carry out surgical operations and anatomical demonstrations and gave them priority for obtaining bodies of executed criminals. The teaching of botany was exemplified by Fagon, Pitton de Tournefort, and Jussieu, and it rapidly surpassed that of the Faculty begun in 1646. The chair of anatomy and surgery was particularly renowned during the time of Pierre Dionis (1650-1718), Guichard-Joseph DuVerney (1648-1730), Winslow, Ferrein, Antoine Petit, and Portal. Nicolas Lémery (1645-1715) headed one of the first courses of modern chemistry, much better than that of the Faculty.

The Jardin du Roi was renamed the National Museum of Natural History by a decree of 10 June 1793. Among its most famous directors in the nineteenth century were the chemist Eugène Chevreul (1786-1889) and the physiologist Henri Milne-Edwards (1800-1885). Human anatomy continued to be taught and studied by Flourens and Serres, but in 1855 the chairs of anatomy and natural history were merged into a single chair of anthropology, and thereafter the teaching was divorced from medicine.

Corporative Professional Organizations

Physicians had set up colleges of medicine in the numerous cities without faculties, as well as in those where the insufficiency of university teaching was most flagrant. These colleges played an important part in the development of professional activity, in the safeguarding of public health, and in the training of future practitioners. They soon became like faculties that provided good quality teaching without however being allowed to award degrees. They were more active and more liberal than the faculties, and many of them, anxious to give the art of healing its essential unity, taught on an equal basis students of medicine, surgery, and pharmacy. This system existed from 1411 in Bordeaux where the College had four elected lecturers, two physicians, a surgeon and a pharmacist. At Aix-en-Provence the union of the three disciplines was effected in 1557 under the College of Physicians of that city. Grenoble adopted a similar arrangement in 1608.

Admission to these colleges was governed by very strict rules. For example, Finot tells us that the members of the College of Nancy, founded in the first part of the eighteenth century, had to meet the following conditions: possession of the master of arts degree and a doctoral degree from an approved university; three years' practice of medicine; successful completion of a three hour examination, sustained in Latin before an assembly of the College, and an hour's explanation of an aphorism of Hippocrates; finally, payment of 300 *livres* into the funds of the College. This College provided a wide range of teaching that included anatomy, botany, and chemistry; it also concerned itself with public hygiene, gave free consultations to the poor, supervised pharmacies and the reception of masters of surgery. From 1753 it participated by association in examinations given by the neighboring Faculty of Pont-à-Mousson.

In non-university cities such as Nîmes (1397), Rouen (1538, 1651), Troyes (1539), Tours (1561), Lyons, (1577), Limoges (1646), Dijon (1654), Amiens (1656), Moulins (1657), La Rochelle (1681), Clermont-Ferrand (1681), Lille (1681), and Châlons (1685), the college represented the only medical and surgical teaching organization, although the award of degrees was necessarily made in a university city. At the end of the seventeenth century the College of Orleans displaced the old local Faculty which disappeared from lack of students, and that of Lyons, organized in the eighteenth century, quickly eclipsed the regional Faculty of Valence.

From the Middle Ages on, the surgeons of many cities in France were grouped in communities, both to affirm and to guarantee their corporate rights and to compensate for the segregation imposed on them by the faculties. In 1607 Servin could still write: "Science is not for those who have a hand; they have to allow physicians to decide." This regrettable separation between physicians and surgeons lasted almost to the Revolution.

From the middle of the thirteenth century the surgeons of Paris formed an independent Fraternity under the patronage of Saint-Côme. In 1311 they obtained a *licentia operandi* under the control of the sworn surgeons of Châtelet, and by letters patent issued at Fontainebleau in January 1544, Francis I allowed them to copy the three traditional university degrees, although this privilege was withdrawn in 1658.

Future surgeons taught one another and had to carry out from one to six years of study before attempting the tests for the "great masterpiece." This involved five examinations lasting four weeks, under the control of the Dean of the Faculty assisted by two regents, and covered osteology, anatomical dissection, the process of bloodletting and the use of medicaments. From 1616 the Fraternity had an amphitheatre, reconstructed in 1691, for its lectures and demonstrations.

Despite the stiff opposition of the Paris Faculty, in 1615 the Fraternity of Saint-Côme, with the protection of Louis XIII and the award of a shield with *fleurs de lis*, took the name of College of Surgery, reconfirmed in 1667. Conflict with the Faculty continued, but in 1699 the surgeons finally obtained the right to wear the long robe and reacquired certain lost privileges, although still required to recognize subservience to the Faculty. Gradually, however, their position improved, and by an edict of 1723 master surgeons were authorized to carry out public dissections in the absence of a physician. In 1724 Georges Mareschal (1658-1736) succeeded in gaining appointment of five royal demonstrators to the College of Surgeons which, seven years later, was reorganized as an Academy.

Anatomy and surgery were always among the less well taught disciplines in the old French faculties. The more liberal Montpellier, under Galenic influence, attempted to remedy the situation. In 1734 François Chicoyneau (1672-1752), brother-in-law of Chirac, undertook to unify teaching and to establish a common medical degree, but many other faculties, including Paris, taught these fundamental subjects less to ensure that they were known and developed than to subtract them from the rival teaching of the surgeons. At the end of a long series of stormy conflicts during the first half of the eighteenth century the monopoly of official teaching was gradually withdrawn from the faculties which retained only a nominal control, insufficient to prevent the emancipation of the surgeons.

In 1731, under the influence of François Gigot de Lapeyronie (1678-1747), the College of Surgery in Paris became an Academy, officially dedicated by Louis XV on 2 July 1748. Its statutes were rewritten in 1751 at the instigation of La Martinière, with eighty-three articles, that is, one less than the Faculty of Paris.

The Royal Academy of Surgery which was, according to the definition of Huard, "at once a corporate organization, a center of teaching and research and a consultive assembly at the service of the government," functioned until the general suppression of universities and academies in 1783. It ensured the renewal and spread of French surgery in the middle of the eighteenth century, but after having collected the best scholars of the day, it unfortunately underwent a fairly rapid decline in the last quarter of the century.

The Academy and Schools of Surgery at first occupied the old buildings of the Fraternity of Saint-Côme, reconstructed in 1711, but were removed to the site of the old College of Burgundy in the magnificent building constructed from 1769 to 1774 by the architect Jacques Gondoin (1737-1818) which now constitutes the central building of the old Faculty of Medicine.

Five chairs were founded in 1739, each with an occupant and a substitute. Professors were nominated on their qualifications for a period of three years, and this mode of recruitment ensured a quality of teaching that was very much higher than that of the Faculty. The number of chairs rose to six, shared between a professorial body of eighteen members which included the best surgeons of the day, teaching the principles of surgery, osteology and pathology, anatomy, surgical operations, *materia chirurgica*, delivery, diseases of the eye, physiology and surgical chemistry, botany and diseases of the bones.

The teaching program was spread over three years with lectures twice a week for three months. The first year was dedicated to physiology, the second to pathology and the third to treatment. As indication of the immediate success of the Academy, from the beginning the number of students was from 700 to 800.

From 1757 the teaching of anatomy took place at the school of dissection which replaced the old theatre of Saint-Côme of 1605, rebuilt in 1694 on the corner of rue de la Harpe and rue des Cordeliers; this practical school had four professors and could accomodate twenty-four students. It functioned as late as 1871.

The Academy also undertook clinical teaching partly at the free consultations of Saint-Côme which took place every Monday, and partly at the College hospital, the future clinical hospital founded in 1774, which in 1783 had thirty beds. The students were restricted to a training period of two or three years in the great hospitals such as Hôtel-Dieu, Hospice de la Vieillesse, Charité, Bicêtre, Notre Dame de la Pitié, etc. The Academy also supported one or two private courses such as those of Pelletan, Desault, Sue, Antoine Dubois and Sabatier.

From 23 April 1743 the degree of master of arts was required for the registration of all future surgeons, as it was for physicians. Officially the final degree was based not upon a thesis, an examination to which the Faculty still had exclusive rights, but upon a "public act," printed and given in Latin before eight surgeon judges in the presence of three members of the Faculty, including the Dean. This examination took place, as at the Faculty, on Thursday afternoons and lasted about five hours. If successful, the candidate became a Master Surgeon with a cap and long robe, a "free member" of the College and the Academy of Surgery. The cost of studies was much lower than for the doctorate of medicine, about 3500 *livres*. Huard has compiled a list of 238 theses submitted between 1743 and 1789, that is an annual average of five. In 1792-1793 the Parisian masters of surgery numbered 171. The receptions for masters ceased after 20 October 1789, and the College and the Academy were suppressed on 8 August 1793. Despite some attempts, the Academy was not reconstituted after the Revolution, but eight of its members became professors in the new School of Health which had twelve chairs.

Towards the end of his reign, Louis XV provided for the rest of the country the measures taken in the capital some forty years earlier. By a declaration of 12 April 1772 he officially created the Royal Schools of Surgery in many cities of the provinces.

The Royal Society of Medicine

In 1776 François de Lassone (1717-1788), physician to Louis XV, and Félix Vicq d'Azyr (1748-1794), physician to Queen Marie-Antoinette, founded the Royal Correspondence Society of Medicine, which became the Royal Society of Medicine recognized by letters patent of 1 September 1778. The aim of this organization was to establish a scientific liaison between physicians of the capital and of the provinces, and even abroad, mainly in respect to epidemiology, animal epidemics, hygiene and a knowledge of new or rare medical discoveries. It also attempted to promote research by awarding prizes and scholarships. It had no direct or immediate effect on teaching, although it produced an important idea of reform. In 1790, acting in the name of the Royal Society of Medicine, Vicq d'Azyr addressed the Constitutional Assembly on "a new plan or constitution for medicine in France." The plan envisaged the equal union of medicine and surgery, together with a reorganization of teaching by means of a fundamental change in the traditional methods so obstinately defended by the faculties.

The plan was taken up by Antoine de Fourcroy (1755-1809) and submitted by him to the Convention on 27 November 1794: teaching from books was to be replaced by a teaching system in which observation of the patient, practical anatomy and clinical nosology would become the main themes. In this respect it is not an exaggeration to say that the training of physicians in the nineteenth century was largely conceived and codified by the Royal Society of Medicine.

Private Teaching and Autonomous Courses

From the end of the seventeenth century many physicians and surgeons and others who were aware of the insufficiency of official teaching took the initiative to give separate courses. This movement, however, was less active in the provinces, even at Montpellier where the efforts of Théophile de Bordeu ran into difficulties. The faculties regarded these professors, who were completely free from their authority, as illicit competitors, all the more dangerous if their independence allowed them to pose as innovators. Lectures were announced by prospectuses distributed discreetly. More often than not the lectures were given on the teacher's premises. Indeed, this type of irregular teaching had developed from the middle of the eighteenth century, and it was successful in relation to the quality of the speakers and the program, which were better conceived and more practical than those of the faculties.

Mention may be made of the more important of those that took part in Paris. P. J. P. Chomel taught botany and pharmacology; pharmacology was also taught by Laplanche, Mitouard and Rouelle; chemistry by Nicolas

Lémery, Barbeu du Bourg, and Fourcroy; Antoine Dubois, Pierre and Jean-Joseph Sue, Petit-Radel, Goubely, Peyrilhe, Ferrand, Sabatier, Pelletan, Garangeot, Vicq d'Azyr and many other members of the Society of Medicine or of the Academy of Surgery contributed to the teaching of anatomy, physiology and surgery; Le Roy, Lebas, Lauverjat, Gaultier de Claubry taught obstetrics; and hospital doctors of the quality of Desault or Bichat enticed a large number of students from the Faculty.

According to Delaunay, some masonic lodges also organized public lectures that were well attended. Although inspired by necessity, this system had nevertheless its inconveniences. It opened the doors to questionable speakers, even empirics and charlatans who held subscription courses; this was notably true of Mesmer upon his arrival in France. Nevertheless, these free courses played a great part in the teaching of medicine and auxiliary sciences on the eve of the Revolution.

Hospital Teaching

As we have seen, the faculties of medicine almost systematically neglected the teaching resources that hospitals could offer students. The few exceptions that have been reported were rare and without great importance.

The hospitals were completely independent administratively of the faculties, but usually recruited their doctors from the teaching doctors. However, the hospitals of the great cities were important centers of professional training, reserved in fact to physicians or surgeons employed in the establishment. This meant that the hospital had its own school which functioned according to certain restrictive rules.

The Hôtel-Dieu of Paris, the oldest and for long the only great hospital of the capital, is a case in point. In the sixteenth century to have worked in this hospital was considered a notable reference, and Ambroise Paré, who had served there for three years as an assistant barber-surgeon, always spoke of this fact with pride. From the middle of the seventeenth century the administration was clearly aware of the educative role it played. It even tried to develop this within the limits of its resources and to the exclusive profit of its own surgical and medical personnel. In 1672 a regular course of surgery was organized and held twice a week for two to three months annually.

In 1726 the surgical body of the Hôtel-Dieu comprised a master surgeon, the head of the department, twelve appointed surgeons from outside, an assistant "gaining mastership", who filled the functions of the actual *chef de clinique*, twelve interns, who were given board and lodging, and seventy-four external students. From 1693 the interns were competitively selected among the most meritorious external students. Thus was outlined the basis of a hospital hierarchy that would constitute one of the most remarkable characteristics of the French medical system in the nineteenth and twentieth centuries.

The Hôtel-Dieu was also highly regarded for its school of midwifery and attracted a number of surgeons, and French and foreign obstetricians, who needed to perfect their art and were trained for a period of three months.

From 1717 the teaching of anatomy in the hospital theatre was organized on a regular basis under the leadership of Jean de Méry (1645-1723) who instituted an annual free course which continued until the Revolution.

The last years of the eighteenth century were marked by a considerable development and by a new attitude to the teaching of hospital clinical medicine. Inspired by the example of foreign countries, a few persevering and courageous pioneers anticipated official reform by some fifteen years. This radical transformation was initiated by a physician, Louis Desbois de Rochefort (1750-1796), at the Charité hospital, and a surgeon, Pierre Desault (1738-1795), at the Hôtel-Dieu. In 1780 Desbois began clinical teaching of treatment at the bedside of the patient, and may be called the true founder in France of modern hospital teaching. Attracted by this new approach, students left the Hôtel-Dieu in large numbers to go to the Charité which was less ramshackle and overcrowded, and better managed. The work of Desbois was prematurely interrupted but was followed up and amplified by his student Jean-Nicolas Corvisart (1755-1821), who was followed in turn at the Charité by the famous Laennec.

Trained at the school of the Academy of Surgery, Desault had begun his career at the Charité where he could appreciate the didactic method of his medical colleagues. He was appointed as a surgeon at the Hôtel-Dieu in 1785, and soon applied the system of Desbois to the practical teaching of surgery. His efforts were immediately crowned with success, judging by the fact that nearly 400 students or physicians went to the Hôtel-Dieu to benefit by his teaching. "Neighboring countries sent students to Paris with grants under the express condition that they followed Desault." This success led to some extent to the decline of the Academy of Surgery, but one must consider especially the personal qualities of the master and the originality of the teaching method that he employed.

A decisive impulse had started. Following the terms of the famous report addressed by Fourcroy to the Convention, 27 November 1794: "That which up to now had been lacking in schools of medicine, the practice of the art and observation at the patient's bedside, would become one of the main parts of teaching." The new methods were to be applied with the opening of the central School of Health in January 1795. Of twelve chairs, three were concerned with clinical teaching: Corvisart at the Charité and Desault at the Hôtel-Dieu were both confirmed in their functions as professor of internal and external clinical medicine which they already effectively filled. The third clinical chair, was founded at the old hospital of the College of Surgery and was given to Pelletan and then to Dubois, who had as assistants Lallement and Cabanis. Thus the history of medical teaching in France was about to enter a new era.

The Teaching of Medicine in the Nineteenth and Twentieth Centuries

In the phrase of Sabatier, the Revolution "put everything upside down, from the throne of the King of France to the humble professor's chair and the student's bench." The law of the Legislative Assembly, 18 August 1792, and the decree of the Convention, 15 September 1793, laconically put an end to the confused and moribund teaching system which had lasted for more than five hundred years: "The faculties of theology, medicine, arts and law are suppressed throughout the area of the Republic."

This radical measure, taken in the name of the ideological principles of the Revolution, has often been severely criticized. At first sight it appears to have been a measure of cultural regression, but in fact it had the happy consequence of giving France a unified system of medical education both professionally and nationally, the general principles of which are still in force. Only such a thoroughgoing eradication could have made possible the reshaping required and impatiently awaited by enlightened thinkers.

Of course, the closure of the faculties had some unwelcome immediate consequences. In the words of Bichat: "a cloud of unknown men in the theatres and schools" of which many were unwanted or charlatans, invaded the profession which was suddenly deprived of official teaching as well as all control, and practice had become free because of the degree of 2 March 1791. A message from the Directoire, dated Year VI, took a somewhat alarmed attitude: "The public is the victim of a crowd of poorly educated individuals, who on their own authority have called themselves masters of the art, distribute remedies at random and compromise the existence of many thousands of citizens. . . . A positive law should compel anyone who pretends to profess the art of healing to go through a long period of study and to be examined by a severe jury." Even more serious for a nation that was at war, there were not enough physicians for the armies.

The danger rapidly became manifest and was only dealt with at the end of fifteen months by the Convention itself. Having wiped out the past, it put into action an entirely new arrangement that took into account the judicious projects of reform set out by Vicq d'Azyr (1790) and by Fourcroy, inspired by Pinel, Cabanis, Sabatier and Guillotin (27 November 1794). As claimed by Fourcroy, one of the great reformers of the period, it was important to reestablish "in the hands of an institution worthy of the French Republic higher teaching than anything known of its kind in Europe." Eight days later, 4 December 1794, the decision was taken: three Schools of Health were to be set up in Paris, Montpellier and Strasburg, replacing the old faculties. On 20 January 1795 that of Paris, thus renovated, was reopened.

January 1795 to March 1803 marked a transitory period of profound reorganization which was hasty and incomplete. According to Fourcroy, "it was necessary to reunite and mix all children of the same family who had for too long been divided among themselves." It was also necessary to renew teaching methods and to teach future practitioners to "read little, see much and do much."

Two reforms were fundamental: (a) the unification and fusion under the same authority of the teaching of medicine and surgery. Each of these major disciplines was nevertheless distinguished by a different orientation in its tests; (b) the uniformity and the change in programs which had been brought up to date and included jointly a theoretical part taught in the schools and a practical part with laboratory exercises and obligatory periods in the hospitals.

Students no longer registered of their own free will but were recruited from all districts of the country without any conditions of entry except for their scientific aptitude and their civic spirit. From that point they became "students of the fatherland" and were to become "officers of health". They were divided among the different schools and were compelled to undergo three years of study during which they were supported by the state.

The School of Health in Paris was installed in the buildings of the old Academy of Surgery, which later became the seat of the Faculty of Medicine. The School of Montpellier had, thanks to the support of Chaptal, the old Bishop's palace of the Cathedral of Saint-Pierre which is still occupied by the Faculty.

The teaching body was constituted of professors chosen according to their reputation, each helped by an assistant; their salary was fixed at 6000 (later 10,000) *francs*, and that of the assistants at 5000 (later 6000). The School of Paris, directed by Augustin Thouret (1748-1810), had twelve chairs, only four of which were given to the regent doctors of the old Faculty: the others went to well-known professors of the former Academy of Surgery. Nine chairs of theoretical medicine were set up and three clinical hospital chairs; in addition, a professor-librarian was appointed and a professor-keeper of the collections. Montpellier was provided with eight chairs and Strasburg, six. The number of students was fixed proportionately: 300 at Paris, 150 at Montpellier, and 100 at Strasburg.

Under the Directoire, the Schools of Health were called Schools of Medicine, law of 24 October 1796, and there were also secondary schools of medicine such as that of Caen (1795).

Scholastic activity was continuous throughout the year, with the teaching program fixed by a regulation of 2 July 1796. The details may be found in Corlieu. In addition to the usual course work, the first year students, or "beginners", had a period of clinical surgery at the Hospice de l'Humanité (formerly Hôtel-Dieu) for at least four months. Second year students, or "starters", were instructed in clinical medicine at the Hospice de l'Unité (formerly the Charité) for at least four months. Third year students, or "advanced", underwent a period of clinical medicine at the Hospice de l'École (future Post-Graduate Clinic) throughout the year, with active service participation and the possibility of fulfilling paid functions equivalent to those of present day interns.

From 1797 the students were qualified by year-end examinations which differed slightly depending upon whether the candidates were going into medicine or surgery. Two years later, a terminal examination was given which did not, however, lead to a degree permitting the right of practice. All physicians who had studied during the old regime had to submit to a licencing examination.

On 7 August 1797 an École Pratique was founded, in principle with laboratories, installed at first in the old Convent of the Cordeliers. The 120 students who frequented this institution after a competition were third year students who could perfect themselves with practice in chemistry, physics, physiology, clinical medicine and pharmacy. These complementary practical studies were greatly appreciated, and candidates were numerous both from abroad as well as from France. They were qualified through a final public examination.

The major innovations of the new system were the unification of courses, the development of practical studies in the fields of fundamental and clinical sciences. The exercise of the profession itself still remained to be regulated and this was done by giving degrees following obligatory "uniform and regular" examinations. The role of reorganizer during this disturbed period was given to Fourcroy, and he drew the attention of the public authorities to the report presented on 7 Germinal, Year XI: "Those who have learned their art are confused with those who have no ideas. The life of the citizens is in the hands of greedy as well as ignorant men . . . no proof of knowledge is required." His appeal, together with that of many others, was heard, and absolute professional freedom was ended because it had become a national danger. This was recognized by the decree of 10 March 1803, by which physicians were divided into two categories: that of simple Health Officers and that of Doctors in Medicine or Surgery.

The *officiat* was instituted with the aim of acquiring quickly and at little cost a body of health auxiliaries who were to practice in the countryside. The legislators thought that this measure could be justified because "the inhabitants of the country, since they are purer [cleaner] than those of the cities, have simpler diseases which for this reason require less knowledge and less assistance." The preparation for the *officiat* required only three years of study in a school, which could be replaced by six years of practice with a doctor or five consecutive years of activity in a civil or military hospital.

The doctorate in medicine or surgery implied four years of study. The five terminal examinations took place during the last year at one of the schools; they included theoretical and practical tests of anatomy and physiology, pathology and nosology, *materia medica*, chemistry and pharmacy, examinations in hygiene and legal medicine, written and oral tests in Latin, mainly on internal and external pathology depending on the orientation of the candidate. The terminal thesis had to be printed in French or in Latin.

The legal right to the practice of medicine was strictly subordinated throughout France to the acquisition of the degree of Officer or of Doctor, and its subsequent registration with the civil and judicial administrative authorities. The distinction between the degree in medicine and in surgery lasted until 1892 when a single doctorate was instituted.

Finally, this was the period in which the typically French formula of hospital teaching with the setting up of the internship and externship was put into operation.

From then on, for more than 150 years, French medical education developed along two distinct paths: the university route, which was fundamental, official, national and open to everyone; the hospital route, which was complementary, semi-official, regional or local and reserved to an élite selected by competition. This duality was maintained until very recently when the principle of fusion and hospital-university integration was instituted (1958).

University Teaching

Faculties

The final unification of medical teaching in France under the direct authority of the state was carried out by Napoleon I at the instigation of Jean Chaptal (1766-1832). The law of 10 May 1806 set up the Imperial University of France; the decree of application that corresponded to this was dated 17 March 1808, setting up five sorts of faculties, of which that of medicine was substituted for the Schools of Paris, Montpellier and Strasburg. In other cities the Empire founded national secondary schools such as those of Clermont-Ferrand (1806), Angers (1807), Toulouse and Caen (1808). Public instruction passed from the Ministry of the Interior to the Grand Master of the University. In 1824, under the Restoration, it was placed under the charge of a special ministry which in our day has become that of National Education. The general organization of the university and its subdivision into regional circumscriptions, or academies, each directed by a rector nominated by the state, was fixed much later by the law of 10 July 1896.

After the Empire, with some variations bearing on the name and designation of the chairs, the model of the official robe (fixed in 1803), and a few details of ceremonial, all the faculties and schools of medicine obeyed the same general regulations. They provided uniform teaching which led to identical official degrees. In the pages that follow, and with such exceptions as will be noted, the Faculty of Paris will be taken as an example because it has always had the largest number of students and teachers. The regulations that apply to it, in practice, apply to all the others.

Teaching Bodies

From 1808, the official teaching of medicine in France was provided by the faculties and the Schools of Medicine, of which many, called "mixed", dealt with medicine and pharmacy but were divided into distinct sections.

The faculties provided a course of teaching that was complete, both theoretical and practical, and awarded the official degrees that were essential for the legal exercise of the profession. Each faculty was directed by a Dean nominated by the teaching body for five years—now reduced to three. Until the last third of the nineteenth century the faculties of medicine were three only: Paris, Montpellier and Strasburg.* The Faculty of Strasburg at first occupied a building near the hospital, and from 1871 developed within the site itself. This was an early anticipation of the hospital-university centers that the state has been trying to develop for some years with the application of the reform of 1958.

Five new faculties of medicine were set up at the end of the nineteenth century: Nancy (1872), Bordeaux (1874), Lille (1875),† Lyons (1877), Toulouse (1878). In 1967 a total of nineteen existed, of which ten were mixed faculties of medicine and pharmacy; to the eight preceding ones there have been added those of Aix-Marseille (1930), Clermont-Ferrand, Nantes, Rennes (1956), Grenoble, Tours 1962), Angers annexed to Nantes (1965), Rheims, Rouen (1966), Besançon (1967).‡

Most of the Schools of Medicine were founded under the Empire or under the Restoration and replaced a number of old autonomous courses that then passed under the control of the university. A number of them were at first called secondary schools: Amiens, Besançon, Clermont-Ferrand, Grenoble, Poitiers (1806); Caen, Nantes, Rennes (1808); Rouen (1814); Marseilles (1818). There were eighteen in 1820 subsidized by the municipal budget. Their field of competence was limited to the two or three first years of study and the training of "second class" health officers and pharmacists.

Others were from the beginning Imperial Schools of Medicine and Surgery. This was the case of Angers (1807) and Toulouse (1808) which were demoted under the Restoration to Preparatory Schools.

The Preparatory Schools replaced the Secondary Schools and increased in number during the century (Dijon, 1840; Clermont-Ferrand and Tours, 1841); in 1930 there were still eleven. At the end of the last century the majority were changed to National Schools with a director and about fifteen professors, equipped to provide a complete range of teaching but not allowed to proceed to the final examinations or to award doctoral degrees. In 1924 there were nine faculties in France, of which five were mixed, four National Schools, and ten Preparatory Schools. During the last few years ten schools have been turned into faculties: Amiens, Angers, Besançon, Clermont-Ferrand, Grenoble, Nantes, Rheims, Rennes, Rouen, and Tours. In 1965 two new National Schools—Brest and Nice—were founded but are not yet operative. Thus France has now twenty-five national centers for medical training, including nineteen faculties and six schools: Brest, Caen, Dijon, Limoges, Nice and Poitiers.

---

* The Faculty of Strasburg was temporarily under German control from 1870 to 1918, and 1940 to 1944.
† Lille has also a Catholic Faculty, a private and independent institution not directly controlled by the state.
‡ There are also those faculties and schools that have ceased to be French, such as Algiers (1910), and those endowed by France in countries of the Communauté, now independent: Saigon, Hanoi, Dakar, Abidjan, Tananarive, Pnom-Penh and Pondichéry.

Degrees

The separation of the medical body into medicine and surgery, established in 1803, remained in force up to the end of the nineteenth century.

As we have seen, the *officiat* constituted an inferior professional class that provided summary, rural medical services. The curriculum was modified by a decree of 22 August 1854, then again by one of 1 August 1883. Studies lasted from three to four years and the program included a clinical period and an obligatory practical studies. Teaching resembled that given to future physicians, except that theoretical teaching was limited to essential, elementary material. Practice was restricted to the region where the degree had been awarded.

This second-class degree gave rise to dissatisfaction, and the number of health officers fell from 7456 in 1847 to 4653 in 1872, while physicians remained at the level of 10,000 to 11,000. In 1881 the six faculties of that period had 756 officer of health candidates as against 3024 for the doctorate; of 632 students inscribed in all schools, 326 were to become officers of health, and 306 doctors. Less than a quarter of all the students in medicine in France were prepared to be content with a minor position in the medical hierarchy, and the rank of officer of health was finally suppressed by a law of 30 November 1892.

The doctorate gave the right to practice without restrictions throughout French territory. Towards the end of their studies candidates decided if they wished to take the degree of medicine or that of surgery or even both. However, a law of 30 November 1892 suppressed the degree of doctor of surgery. From then onwards, only one degree existed, that of doctor of medicine, required of every physician and surgeon, and obtained on average at the age of twenty-eight years. Apart from special cases of temporary replacements given, under supervision, to hospital interns and students that had finished their period of study, the practice of medicine, with all the rights and duties that it entails, is rigorously subordinated to the possession of a state degree duly registered with the administrative authorities and the Medical Council. Any infraction of this regulation constitutes the crime of illegal practice. Under certain conditions, foreign students may obtain the doctoral degree of the university which does not, however, confer on them the right to practice in France.

Teaching Staff

At the beginning of the nineteenth century the teaching body of the university was constituted exclusively of professors. Later it was augmented by the *agrégés* and a number of categories of auxiliaries, *chefs de clinique*, *chefs de travaux*, prosectors, faculty aides, etc.

The professors constituted the permanent titular staff of the faculties and schools, their numbers varying considerably from one faculty to another but always increasing. Thus in Paris there were twelve in 1795, twenty-three in 1823, twenty-nine in 1875, thirty-four in 1894; the number reached 152 in 1967, of which forty professors had a personal title. Montpellier, which had six professors at the time of the Revolution, now has fifty, of whom five have the personal title.

The recruitment and control of professors was initially declared by the decrees of 7 March 1795, which were for the most part confirmed by the decree of 17 March 1808. Under the decree of 31 July 1810 professors were nominated by competition; examinations, taken before a jury of seven, and later eight members of the faculty, included written compositions and lectures presented after a preparatory period of twenty-four hours. Under the Restoration, from 17 February 1815 to 5 October 1830, the competition was abolished and nominations were made by the faculty. The competitive system was thereafter reestablished with tests including a list of publications and titles, a printed dissertation, a written answer to a question posed by the faculty, and two teaching lectures of one hour each. The competition was finally suppressed by a decree of 9 March 1852, and from then on professors of medicine were nominated by the Chief of State on the recommendation of the Ministry. These recommendations were drawn from a double list prepared by the Council of the faculty on the one hand, and on the other by the Higher Council of Public Instruction, later known as the Consultative Committee of Universities. This system is equivalent to a double election of a new professor by his senior peers, both at the local level and at the national level; the choice is made almost exclusively from *agrégés* or former *agrégés*, at least in the clinical disciplines. Candidates need only present an outline of their qualifications and publications. However, the government reserves the right to decide in case of disagreement between the two lists; it also intervenes in the choice of the holder of a newly created chair. Professors may be designated by chair or by personal title reflecting their work or renown.

The French system has certain details that need to be noted. A professor may apply to the Ministry to be shifted from one chair to another, following the agreement of the interested consultative bodies; such changes, officially permitted since 1796, have been very numerous and have been the object of severe criticism, notably on the part of Bouillaud in 1830 and by many others later. On the other hand, alterations or changes from one faculty to another are exceptional.

From 1920, the honorary title of "professor without chair" could be awarded to certain members of the staff which the faculty wished to recognize as possessing special merit. The temporary title of "associate professor" may also be awarded.

The *agrégés* and lecturers may replace the professors in case of absence of the latter; they participate in the teaching under the direction of the professors and also participate in examinations. The post of *agrégé* was established by a royal ordinance of 2 February 1823, and the first competition for thirty-six places in Paris occurred on 20 November 1823. At present nomination is made by the Minister after competition for the hospital and university places that are vacant or have been newly created. The number of *agrégés* and lecturers in practice varied according to faculties or schools. In Paris it was twenty-four in 1823, twenty-eight in 1875, thirty-five in 1894. It reached 389 in 1967, a number three times greater than that of the professors. At Lyons, it has increased from thirty-nine for twenty-five professors in 1884 to sixty for fifty-three professors at the present day. At Montpellier there are only forty-three for fifty professors.

Over the last century and a half a very important role has been played by auxiliary personnel. The *chefs de clinique,* nominated every four years, must help the professors of surgery, obstetrics and clinical medicine in their hospital teaching. The first regulation governing them dates from 6 May 1813. A regulation of 2 February 1823 fixed their number for Paris at four. Thereafter their numbers were much increased and it is now 481. In the beginning recruitment was by simple nomination from former hospital interns. From 23 August 1862 this post was gained by competition; from 14 August 1863 it was opened, at least in principle, to all doctors of medicine under the age of thirty-four. In the new hospital-university hierarchy the functions of *chef de clinique* are linked with those of hospital assistant.

The *chefs de travaux* must assist the professors of basic sciences and, under their supervision, direct the practical work carried out in the faculty. The *chefs de travaux* for anatomy have existed since 1880. At present nearly all the chairs of basic sciences have one or more such aides and the number reaches 149 in the Faculty of Paris. They are helped by numerous assistants and monitors (656 in Paris in 1967).

The practical teaching of anatomy is carried out by a body of prosectors, governed by a regulation of 1795. Their strength, which was initially six in Paris, was reduced to four in 1801, after which time they were given eight anatomy aides. The prosectors were nominated by competition for a maximum of six years and could not occupy any other public salaried position. From 1823 to 1839 the number of prosectors, as that of the aides, was reduced to three, and the duration of their functions was limited to three years. In 1879 the number of prosectors was increased to eight, and an engagement of four years; the number of aides reached twenty-four, and an engagement of three years. The number of prosectors in Paris is now ten; that of anatomy aides is thirty-five.

Chairs

The total body of teaching personnel, professors and *agrégés*, has considerably increased in the past eighty years. In most of the provincial faculties it almost doubled between 1884 and 1966; thus it has gone from sixty-four to 124 at Lyons, from fifty to 106 at Bordeaux, from forty-five to eighty-three at Lille, from forty-three to ninety-three at Montpellier, and from forty-one to seventy-nine at Nancy. This increase has been even more marked in Paris where the staff, which was 120 in 1884, is now 539 (1967). The number of chairs has progressively increased, in step with the progress of medicine and as new subjects or disciplines have taken their places in the curriculum.

Initially the Faculty of Paris had only twelve chairs, created in 1795. Under the Restoration it went through a serious political crisis which led to individual disciplinary dismissals and an authoritarian and conservative reorganization on the part of the state. However, by a royal ordinance of 2 February 1823 the number of chairs increased to twenty-three, of which fifteen were for theoretical medicine; chairs of clinical medicine increased from three to eight. The chair of the history of medicine, created under the Revolution, was suppressed, and only reestablished in 1870.

Three supplementary chairs were organized between 1829 and 1835. A fourth chair of surgical clinical medicine (1829), a chair of pathology and general therapeutics (1831) and a chair of pathological anatomy (1835).

Between 1850 and 1870 the Faculty obtained three new professorships: histology (1852), comparative medicine and experimental pathology (1869), history of medicine and surgery (1870). The last named chair, which brought the total number to twenty-nine, was the result of a generous legacy from a non-medical personality, Salmon de Champotran.

From 1872 to 1892 a number of new chairs of clinical medicine was instituted in different specialities, most of them going to the most eminent members of clinical medicine.

At the end of the nineteenth century the total number of professorial chairs in Paris was thirty-five. It has since tripled because of the increase in the number of students and the importance of certain new disciplines. In 1967 the Faculty of Paris had 110 organic teaching chairs, of which twenty-three are for basic sciences, thirteen for medical clinical sciences, six for surgical clinical sciences, thirty-nine for hospital clinical medicine and twenty-nine for hospital clinical surgery. In addition, there are forty professors with personal titles, of whom twelve are in the basic disciplines and twenty-eight are in the clinical, medical and surgical disciplines.

The faculties in the provinces, even the least important, have a dozen chairs and the more active have fifty or more (forty-five at Montpellier, fifty at Marseilles, fifty-three at Lyons and Lille, fifty-six at Bordeaux). The majority of other faculties have twenty-five to thirty-five chairs.

Students

The Revolution and the first Empire notably enlarged the recruitment of students of medicine, less for idealistic principles than because of the needs of the civil population and of the armed forces.

The law of 22nd Ventôse, Year X, sought to make medicine accessible to everyone. Courses were free, and the degree no longer entailed the heavy expenses of the old regime. During the nineteenth century the cost of studies, in the strict sense of the word, registration fees, examination fees, etc., was markedly reduced. Although they had exceeded 6000 *livres* before the Revolution, they were only 1000 *francs* at the beginning of the nineteenth century, 1260 *francs* at the middle of the century, and 1360 *francs* towards the end. At present they are 290 new *francs*. Thus the state assumes much of the cost of medical education, and numerous university scholarships have been established for the more meritorious and necessitous students.

Because of their duration, medical studies were still very onerous throughout the nineteenth century, and the centralization of teaching centers in a few large cities was expensive for students who had to live away from their homes. In addition to the often mentioned cases of Bichat, Larrey, Laennec and Claude Bernard, there are many other instances of future doctors who had to carry on their studies under severe financial handicaps. According to a remark of Lasègue in the final quarter of the last century: "As in nearly all parts of the world, our medical students belong to the second rank of the bourgeoisie. I have collected the professions of their parents; they are small proprietors, employed in modest jobs, shopkeepers, pharmacists, or, in a minority of cases, physicians. The father has often wished his son to succeed him, but his business is a burdensome one, very time-consuming and not well remunerated."

The number of students has increased markedly. At the School of Health in Paris it reached 896 in 1797, increasing to 1390 in 1801; between 1825 and 1830 the number of doctoral candidates inscribed in the Faculty was between 1800 and 2000; it reached 3855 in 1877 and 5144 in 1894—of which number more than 1000 were foreigners, and 195, women. In 1964 the Faculty of Paris had 7690 students with 4050 inscribed in the preliminary year; in 1966 the number in student courses reached 9481, and 4903 were taking the preparatory year. 5300 were registered in specialized courses.

The attendance at provincial faculties has always been less notable. In 1881, for example, Lyons had 165 medical students, Bordeaux, 155, Montpellier, 154, Nancy, eighty-three, and Lille, fifty-four, while Paris attracted 2413.

The number of doctoral degrees awarded has increased in proportion: 300 per year at Paris under the Restoration, and 1055 in 1959; for the whole of France it was 1128 in 1900, and 2123 in 1959.

Registration requirements have often varied since the Revolution. Although since the beginning of the eighteenth century the master of arts degree has always been required, no prior university qualification was imposed until 1808. After that all medical students had to have the arts degree of the secondary school. In addition, from 1823—except between 1831-1836—the science baccalaureate was also required. This latter requirement was suppressed in 1852 but reestablished in 1858.

In 1893 the scientific training of medical students was reinforced by the further requirement of a year's attendance at the faculty of sciences leading to a certificate in physics, chemistry and natural sciences (P.C.N.). In 1935 this requirement was modified to a certificate in physics, chemistry and biology (P.C.B.). In 1960 the preparatory year was temporarily fused with the first year of medicine, which led to a reduction in the length of study but an increased burden for the beginning medical student, and the preparatory year was reestablished in 1963 with a Preparatory Certificate for Medical Studies (C.P.E.M.).

Curricula

The medical curriculum has been frequently modified over the last 160 years, with the most notable changes taking place in 1878, 1893, 1911, 1934, 1958, and 1966. It is not easy to summarize the different regulations that have succeeded one another and have been set aside. The details may be found in the thesis of our pupil, Madame Picard. We shall limit ourselves here to recalling the tenor and the orientation of the most striking reforms.

The first half of the nineteenth century was characterized by the clear preeminence accorded to clinical teaching which has for long constituted one of the main titles to fame of French medicine. For a hundred years the reforming tendencies in this field have oscillated between two opposite poles: to develop above all the aptitudes and the quality of the practitioner beginning with a clinical teaching considered to be among the best in Europe, or to follow the requirements of modern medicine and utilize intensively the possibilities offered by the development of the auxiliary sciences, which have become "basic", essential to work and research in a laboratory. The conflict which has periodically opposed these two extreme tendencies and the compromises which have been adopted to maintain the unity of the teaching system, have not existed without provoking a certain amount of excitement. There has often occurred progressive burdening of courses, which has resulted through the rapid accumulation of new information and the increase of specialities.

Blindly confident of the value of the anatomical and clinical method, many French medical educators have not recognized at first the place due to medical biology which was born towards the middle of the nineteenth century. In the introduction to his "Clinical Lectures at the Hôtel-Dieu" (1857), Armand Trousseau (1801-1867) was the eloquent interpreter of the traditional concept which had been turned into a piece of crude mysticism: "The small amount of time that you dedicate to medicine makes it very difficult for you to study auxiliary sciences. You must have sufficient notion of chemistry and physics to be able to understand the application of these sciences to medicine. But I should profoundly deplore the time that you might lose in order to acquire a more extended knowledge of chemistry . . . So, gentlemen, let us have a little less science, and a little more art." Ackerknecht believes that this static tendency marked the beginning of an *impasse* in which French medicine found itself after 1850. From that time foreign students who had hitherto been attracted by the Parisian school turned to the German universities that were providing a much more developed physical, chemical and microbiological training.

Drawn in the wake of Claude Bernard and Pasteur, other French teachers tried to revise the educational system and "put it on a more forward-looking basis". They insisted that more room must be found for biological and laboratory techniques. They succeeded in obtaining the teaching of various obligatory practical courses (1878) and numerous new courses. A further decisive step was taken (1893) by the institution of a preparatory scientific year. In our day divergence of opinion may be seen in the timetables that give time to clinical sciences and other subjects, in the recent reform of medical studies. The formula which consists of proposing to students two distinct orientations, one towards practical medicine and the other towards laboratory and research medicine, was not put into effect until recently for fear of compromising the principle of unity of the doctorate of medicine. The creation of a cycle of studies and of research in human biology (Decree of 12 August 1966) will perhaps bring about a satisfactory solution to this difficult problem on which to some extent the future of French medicine depends.

The duration of studies leading to the doctorate of medicine has been on the increase. Up to the end of the nineteenth century the total period of study was only four years. In 1911 it was increased by one year, and since 1934, by two, the sixth year being exclusively in a hospital. On the other hand, since 1893 medical studies have been compulsorily preceded by a preparatory year.

The courses have naturally been subject to numerous modifications, both in the nature of the subjects and the time devoted to them. They were in fact not precisely fixed until 1893, and thereafter altered in 1911 and 1927.

In 1934 teaching was somewhat arbitrarily subdivided into three parts: theory, practice and clinical medicine. The first two years were given to the basic sciences, and the hospital stages were carried out in medical and surgical departments. The courses of internal and external pathology were split between the third and fourth years. The third year also included theoretical and practical teaching of pathological anatomy, bacteriology, parasitology, obstetrics and operative medicine. The fourth year included medical and surgical anatomy and operative medicine.

Practical work was carried out in medicine, surgery, obstetrics and certain compulsory specialities such as neuropsychiatry and pediatrics, or optional ones like dermatology and syphilology, ophthalmology, ear, nose and throat, or contagious diseases. The theoretical and practical course of the fifth year included hygiene, forensic medicine and deontology, treatment and hydrology, and pharmacology. During the sixth year students were required to carry out compulsory periods of duty in medicine, surgery, obstetrics and certain specialized services or laboratory work.

By 1960 there was an evident overloading of the curriculum: 920 hours of work in the first year, 1090 hours in the second, 1520 in the third, 1760 in the fourth, and 1640 in the fifth. The sixth year was reserved for compulsory clinical or laboratory sessions.

Reorganization has led to the following less exacting program (1966-1967): First year: anatomy (180 hours), histology and embryology (80 hours), physiology (100 hours), metabolic biochemistry (80 hours), biophysics (70 hours), semeiology (in the morning), a total of 510 hours, excluding hospital sessions.

Second year: anatomy (100 hours), histology and embryology (40 hours), physiology (100 hours), chemistry in relation to physiology and semeiology (35 hours), biophysics (50 hours), bacteriology, virology and immunology (90 hours), medical and surgical semeiology (hospital in the morning), a total of 415 hours, excluding hospital sessions.

Third year: pathological anatomy (60 hours), general pathology (35 hours), pathological physiology and experimental medicine (35 hours), medical pathology, pathological and physical chemistry applied to pathology (120 hours), surgical pathology (60 hours), hematology and parasitology (60 hours), radiology (30 hours), hospital sessions of medicine and surgery (in the morning), a total of 400 hours, excluding hospital sessions.

Fourth year: pharmacology (100 hours), pathological anatomy (60 hours), medical pathology, physical and chemical pathology (100 hours), surgical pathology (60 hours), obstetrical pathology (60 hours), radiology (20 hours), hospital sessions of medicine and surgery (in the morning), a total of 400 hours excluding hospital sessions.

Fifth year: therapeutics and hydrology (130 hours), forensic medicine and deontology (50 hours) preventive medicine and hygiene (60 hours), occupational medicine and rehabilitation (60 hours), applied genetics (15 hours), applied psychology (25 hours), hospital sessions (in the morning), a total of 340 hours, excluding hospital sessions.

Sixth year: hospital sessions as an intern.

Examinations also underwent numerous modifications bearing on the nature, rules, chronology and respective importance of the different examinations. In 1798 students had to undergo three examinations at the end of the fourth year, the last of them a printed thesis on a subject chosen by the candidate. From 1829 there were five examinations begun in the third year. They included a thesis and a dissertation on an aphorism of Hippocrates— the last suppressed in 1831. The use of Latin in interrogation, discussion and argumentation was abolished in 1862. In 1849 it was decided that five examinations should take place at the end of the first year, but from 1878 they were spread throughout the period of study. In 1893 the five examinations were extended to include practical work and hospital sessions, and with the completion of studies there were three examinations of recapitulation called "clinical" examinations (medicine, surgery and obstetrics), and a printed thesis on a freely chosen subject.

At the present time examinations take place at the end of each year of study; they include written, oral and practical tests dealing with the different subjects taught, with a ratio varying from 0.5 to 2. Written tests are all handled anonymously. No student may register at the beginning of the year if he has not successfully passed the practical and theoretical tests that closed the previous year. At the end of the hospital period of the sixth year students submit to three "clinical" and recapitulative examinations in which they must examine a patient under the supervision of one of three members of a jury. The inaugural thesis is submitted in public before four professors, and at Paris, and in many other French faculties, it is followed by the swearing of the Hippocratic Oath.

Numerous modifications in the qualifications of students, courses and tests have been particularly frequent during the last twenty years, and they have not failed to create a certain amount of confusion in the minds of the administrators, and certainly in those of professors and students. Such changes have not yet led to the results that were expected. The quality of teaching has much improved in the sense that teaching technique has been modernized and there is a stricter focus on the subject of teaching. But the level of knowledge acquired by students has not, unhappily, risen in proportion, no doubt because of the growing requirements of the course of basic sciences which exercise a detrimental effect on clinical training.

Practical Teaching of Clinical and Basic Sciences

Clinical teaching came to official notice at the opening of the Schools of Health in 1795. This was initially limited to the creation of a chair of internal clinical medicine at the Hôtel-Dieu, a chair of postgraduate clinical medicine at the Hospice de l'École, and the post of assistant professor in clinical surgery at the Charité. Students had to attend three sessions of four months each. The number of clinical chairs reached eight in 1823, with the foundation of three new chairs of clinical medicine—at the Hôtel-Dieu, Charité, and Salpêtrière—and a chair of obstetrics at the Maternity Hospital. The Pitié Hospital became the seat of a fourth chair of surgical clinical medicine in 1830. The number of chairs has since increased considerably.

Each clinical chair within the hospital buildings had an operating theatre for teaching purposes. From 1813 professors were assisted by *chefs de clinique,* but it was not enough for the professor to have a place to teach and assistants; it was necessary to have the clinical sessions compulsory and subject to certain rules.

By a decree of 30 October 1841, beginning in the third year, every student was obliged to attend "hospital clinic for at least a year, either as an extern or as a medical student". A decree of 8 June 1862 increased the period to two years and extended the requirement as well to fourth year students, who could not take the examinations of the Faculty without a certificate attesting faithful attendance. Later, participation was required from the first year.

In most of the clinical departments the method of teaching has remained almost the same from the beginning. At the beginning of the morning the *chefs de clinique* teach the rudiments of semeiology and pathology; they then lead the students to the ward and teach them to ask questions and to examine the patients and, in particular cases, they carry out demonstrations of clinical application and, if necessary, an autopsy with commentary. Once or twice a week the professor gives a lecture in the theatre on a subject drawn from one of the cases observed.

In the first half of the nineteenth century the number of students was strictly limited to fifty for each department. In fact, reputable clinics soon had an excessive number of students who gathered round the beds of the patients to detriment of the latter and that of the quality of teaching. The recent reform envisages that the number of students attached to each clinical head will not exceed twenty-five.

Practical work in the basic sciences was made compulsory in the first four years by a decree of 20 June 1878, completed by a ministerial circular of 20 December 1879. Each series of practical sessions ended with an obligatory examination that led, as did the theoretical tests, to admission or failure at the end of the year. Lectures and practical demonstrations also took place, under the control of the Faculty, in the hospital laboratories attached to the clinical chairs, as well as in other laboratories, institutes and research centers.

Teaching of Certain Special Subjects

The rapid and considerable progress achieved in medicine over 150 years has led to the subdivision of the profession into numerous specialties. The official teaching of specialities was begun late, by a ministerial circular of 14 August 1862. At the request of Dean Rayer (1793-1867) supplementary courses were organized and the first specialized clinical chairs were founded. In fact, these administrative dispositions only recognized what was happening; they generalized the individual initiative already taken for some time by professors and by department heads in the functions of many specialized hospitals such as the Necker Hospital for urology, the St. Louis Hospital for venereal and cutaneous diseases, Bicêtre, Salpêtrière and Sainte-Anne for nervous and mental diseases.

In the first years of the twentieth century this so-called second cycle teaching took a new turn with the founding in many faculties of a certain number of certificates or specialty diplomas, such as that of colonial medicine at Marseilles (1902), forensic medicine and psychiatry at Paris (1903), hygiene at Toulouse and Lyons (1905) and Lille (1907). In 1906 there was instituted a certificate of Higher Medical Studies dealing almost exclusively with basic sciences, which to some extent anticipated the preparatory certificate for the diploma and doctorate in human biology (3rd cycle) that was set up recently under the decree of 12 August 1966, and which will be required of future teachers and research workers.

At the same time the number of specialized subjects for teaching has been considerably increased to the benefit of the most varied disciplines. The term "diploma" has been replaced by that of "certificate of special studies" (C.E.S.) to avoid any ambiguity and to safeguard the fundamental principle of the unity of the degree of doctor of medicine. Preparatory teaching for these different national certificates is generally spread over three years leading to a final theoretical and practical examination.

These certificates, the organization of which tends to multiply in the different faculties, give their holders the qualification of "specialists", and the importance is reflected in the increasing number of students regularly registered: 5300 at the Faculty of Paris in 1967.

The Faculty now controls twelve "institutes" which participate in teaching and favor research. Other official organizations play an active part in the training of specialists and research workers under the control of or with the support of the National Center for Scientific Research (C.N.R.S.), the National Institute of Health and Medical Research (I.N.S.E.R.M.). Numerous physicians have perfected their theoretical and practical training in the Collège de France or other organizations such as the Institut Pasteur, Radium Institute (Pierre-Curie Foundation), Cancer Institute (Gustave Roussy Foundation), etc.

The training of military physicians first took place at Strasburg, then from 1883, at Bordeaux and Nancy. Later, the Military Health Service School was established at Lyons; it was reorganized in 1919. Students were chosen competitively and then followed five years of general teaching at the Faculty of Medicine before completing their education at a military academy or in one of three centers designated for this purpose in Paris, Bordeaux and Marseilles.

Under the Revolution midwives had to follow two courses of one year and carry out a practical stage of nine months. The law of 10 March 1803 instituted an obligatory stage of at least one year at the delivery clinic of the Maternity Hospital. The law of November 1892 and the decree of 25 July 1893 required two years of study, the first taking place in a preparatory school designated for this purpose, and the second in a faculty of medicine which awarded the required diploma for exercising the profession. The duration of studies is now three years.

Hospital Teaching

Hospital teaching which is independent of that of the faculties has for long been the second pillar of medical instruction in France. Although it has progressed on a parallel route to that of university teaching, it has been in constant close relations, and the reform of 1958 tends progressively to abolish the distinction between the two separate currents and the two hierarchies.

Hospital teaching, which is a feature of French tradition, has very old origins. It also responded to the necessity that became urgent in the seventeenth and eighteenth centuries to give future physicians the practical training that was lacking in the faculties. Physicians and surgeons carrying out functions as department heads in the hospitals were the teachers and their students were the aides and assistants attached to those establishments, it being understood that the older students, while continuing to learn, would contribute to the training of the younger ones.

This kind of activity was increasing in hospitals from the end of the old regime. During the Revolution, the Empire, and the Restoration, those physicians who were not heads of one of the rare official clinical departments, used their service and talent, on their own initiative, for teaching; despite the protestations of the faculty, jealous of its prerogatives, they were supported by the General Council of Hospitals. In 1817 Joseph Récamier (1774-1852) and his coworkers organized "clinical courses", anticipating by six years the foundation of a chair of clinical medicine at the Hôtel-Dieu. At the instigation of the universities the Ministry of the Interior raised objections, and a medical and surgical commission, formed to deal with this, replied in the name of Marie-Antoine Petit (1762-1840): "that the right to teach belongs to all physicians without distinction, and one can have no possible doubt about the right of hospital physicians to give public clinical instruction to their students; it is this practical instruction in science that has contributed to their progress and their students benefit by special means of instruction that neither theoretical lessons, nor ordinary clinical teaching in the faculty can give them."

Napoleon established the *externat* and the *internat* of hospitals and this justified and gave an impulse to hospital teaching. Under the direction of the *médecins des hôpitaux,* that is, those with the highest degrees and the most envied in the hospital hierarchy, hospital teaching has not ceased to develop in our day at the side of the teaching given from the clinical chairs. It has attracted considerable French and foreign interest and has always had a rare quality, in particular in the field of specialities. Conflicts with the faculty have progressively diminished and finally disappeared; a high proportion of the hospital staff also belongs to the staff of the faculty, and the latter without distinction sends their students into the hospital departments attached to the university as well as into other hospital departments.

An effort to coordinate individual hospital initiative was attempted in 1907 with the setting up of the Association of Medical Hospital Teachers at Paris. This finally led to the foundation in 1952 of the College of Hospital Physicians in Paris, which tries to organize a more homogeneous course for the teaching of specialities and postgraduate teaching, a distinct need at present.

The Externat and the Internat

One of the outstanding features of the organization of French hospitals is the externs and interns, recruited by competition, who are at the same time students and collaborators with the heads of departments, the oldest among them also acting as auxiliary instructors.

In the seventeenth and eighteenth centuries some hospitals already employed internal assistants or boy-surgeons, who were boarded in the hospital, and external apprentices. They were mentioned in 1613 at the Hôtel-Dieu of Paris, where their functions were regulated by a decree of 14 June 1655, completed by that of 14 May 1749. These students questioned and supervised patients, carried out necessary care prescribed by the chief surgeon, bandaged, participated in bloodletting and other matters. In the beginning, interns were chosen from externs as the result of examination before six physicians and a surgeon of the hospital. This plan of recruitment first appeared in the "new constitutional plan for medicine in France", set out in 1788 by the Royal Society of Medicine. In the revolutionary period, the School of Health envisaged, in article 13, that some students proposed by professors and designated by a commission should be taken by the hospitals to fulfill the functions that would later be carried out by externs: to keep note of visits, supervise administration of medicaments and their effects, and distribute meals; to keep records of diseases, draw up observations, collect weather observations daily, assist the professors in postmortem examinations and carry out any dissection or research entrusted to them.

The body of externs and interns as it exists today was established under the Consulate at the instigation of Chaptal, then Minister of the Interior, a result of the "General Regulation for the Health Service of the Hospitals of Paris," 23 February 1802. The major significance of this plan was the division of students into externs and interns, meaning two periods of service, the second not being reached until the first had been completed. It was decided that nominations would be made by competition through examinations, and similar regulations were later adopted by most of the provincial hospital administrations.

The first intern competition, for twenty-four places, took place on 13 September 1802, and the first extern competition the following year. The regulations and details of these two competitions have often varied since their origin, and they were notably modified in 1904, 1921, 1927, and 1933; however, the principle of a single annual competition has been maintained for all places declared vacant in the hospitals of the same city, or even the same region; and, in addition to the system of anonymous written examination, for the internship, these have always been preceded or followed by an oral test. From 1964 externs have no longer been recruited by an open competition, but as a result of the records of university examinations at the end of the first two years of study at the Faculty; selection is therefore made according to knowledge of basic sciences as well as of medical and surgical semiology. This break with tradition, although favored by some, has been strongly criticized by clinicians.

The number of places put to competition has progressively increased because of the opening of hospitals and new services. At Paris it was thirty-eight in 1853, seventy in 1902, ninety in 1939, and reached 257 in 1966. Competition has always been keen; the proportion of interns nominated in relation to registered candidates is about one in eight. At the present time (1967) there are 2904 externs and 761 interns for nearly 10,000 students.

The system of the *internat* has been the object of both high praise and sharp criticism. It has been objected that it obliges candidates to undergo a long and sterile bookish preparation, opening the way to favoritism of judges and "patrons" and creates a kind of closed and proud professional aristocracy. It is nonetheless true that the list of annual promotions during the nineteenth and twentieth centuries contains nearly all the greatest names in French medicine. As Sir William Osler wrote, from the point of view of a foreigner:

The Interne is a special French product, unlike anything else in the medical world. He is still a student, yet he has all the responsibility of a practitioner and he is house surgeon, house physician, clinical assistant, laboratory assistant, special research student rolled into one. He lives in the hospital for four years, a sufficient length of time to give him an exceptionally good education and a large experience. He comes into delightful relations with his chiefs, he lives in charming comradeship with his fellows, and if there is anything in him he finishes his term with an admirable bit of original work which appears as his thesis for the M.D. . . . A first-class Interne is about the best hospital product with which I am acquainted, and it is no wonder that as a body the "Internat" is looked on as the special glory of French medicine.

MEDICAL SOCIETIES, PUBLICATIONS, LIBRARIES AND MUSEUMS

Medical societies arose from the splitting up of the old Academy of Surgery and the Royal Society of Medicine and rapidly multiplied from the end of the eighteenth century. Here mention will be made of the oldest and most prominent. Two years after the foundation of the Society of Medicine of Paris, Bichat founded the Emulation Medical Society (1798) which attracted the most active young physicians in Paris. This was followed by the Society of Medical Instruction (1801), the Galvanic Society (1802), the Anatomical Society, founded by Dupuytren in 1803, closed in 1807, and reconstituted by Cruveilhier in 1820. The Society of Practical Medicine (1805-1808) was renamed (1893) the Society of Medicine and Practical Surgery.

The Society of the School of Medicine (1800), which brought together the teachers of the Faculty, opened the way for the Royal Academy—later the National Academy—of Medicine, founded in 1820 by Antoine Portal, and now an advisory organization at the disposal of the public authorities. It has no direct effect on teaching, although it has an official influence on the education of the public and the orientation of medicine in France.

The middle of the last century saw the birth of a series of learned societies that have since then occupied a considerable place in French medicine: the Society of Surgery (1843), an anticipation of the present Academy of Surgery; the Society of Biology (1848), the Medical Society of the Hospitals of Paris (1849), the societies of Medico-Psychology (1855), Anthropology (1859), Legal Medicine (1860), Therapeutics and Pharmacodynamics (1866), etc. For the last 100 years a large number of national and local societies have been formed, both in Paris and in the provinces, which dedicate their activity to different specialities. They exercise directly or indirectly a great influence on the training of young physicians, notably through bulletins, annual proceedings, notes and reviews which they publish. The French Society of the History of Medicine was founded in 1902.

Because of its nature and inevitably fragmentary character, oral teaching is not entirely suited to a complete medical training, and publications, notably periodicals, are playing an increasingly important part. On the eve of the Revolution and in the first years of the nineteenth century, medical periodicals were not very numerous in France. One of the oldest was the *Recueil Périodique d'Observations* (1754), to become the *Journal Général de Médecine, Chirurgie et Pharmacie;* it twice ceased to appear (1794-1801, 1817-1822). The renowned *Journal de Chirurgie,* founded by Desault in 1791, lasted only a short time; and a similar fate overtook a number of medical journals during the last century. The *Gazette de Santé* founded in 1773, on the other hand, had a long existence; in 1830, under the editorship of Jules Guérin, it became the *Gazette Médicale.* About 1850 there were already in France forty-nine medical periodicals published in Paris and nine in other cities in the provinces. Nevertheless written teaching could not be truly effective until it had been given to the largest number of readers, the task of the libraries.

Each French faculty possesses its own library, open to faculty, students and research workers. The most important is without doubt that of the Faculty of Paris, the origins of which have already been mentioned, and about 1900 it already had 120,000 books and periodicals. It is administratively attached to the Central Library of the University and directed by a Chief Curator. It has today about a million publications including 15,000 French and foreign journals. Mention should also be made of the Central Medical Library of the Paris Hospitals, initially founded for interns and now attached to the College of Physicians. There is also the C.N.R.S. library, and among others those of the Institute Pasteur, the Public Assistance Documentation Center, and the Military Hospital in Val-de-Grâce. The Academy of

Medicine possess a private library of about 200,000 volumes and receives regularly more than 3500 journals.

Finally, mention should also be made of the medical museums, generally specialized but unhappily not much frequented, such as the Dupuytren, Orfila, History of Medicine, Legal Medicine at the Faculty of Paris, Hôpital Saint-Louis, Assistance Publique, Val-de-Grâce, Institut Pasteur, and Palais de la Découverte at Paris, the museums of the Faculty and of the Civil Hospital at Lyons, and many others.

Summary

During the two centuries that preceded the Revolution medical education in France still belonged to the past. Teaching was almost exclusively theoretical, rendered static by a common university tradition, except for a few local variations stemming from the individuality of some faculties. Practical training was slight and elementary.

The modern period began in 1795. The long hibernation that had prevailed since the Middle Ages gave place to a sudden and radical change. The national reorganization of teaching and its unification opened the way to a kind of *duumvirate* between renovated theoretical teaching by the faculties and practical teaching with a double and conflicting orientation. Clinical training, a particularly French development, consisted of early hospital periods and a hierarchical selection by means of competition. The practical introduction of the basic sciences was made by means of demonstrations and obligatory exercises organized in the faculty laboratories, with a more or less open conflict between the holders of a belief in theoretical teaching and the supporters of a practical education. This resulted especially from a confrontation between the partisans of a mainly clinical training for future practitioners, that is, the majority of physicians, and the promoters of biological education as a means to engage in laboratory research and scientific progress. In order to avoid a divorce between these two tendencies, attempts at compromise have been sought none of which has been entirely satisfactory.

The contemporary period is based on a profound change, the principles of which were laid down about ten years ago. The new system is an attempt to develop two teaching methods which are not as yet very widespread in France: that of supervised teaching, more interesting and more effective than collective lectures given in theatres; and that of integrated teaching, which aims to suppress all separation between the different disciplines. Both these methods aim to contribute jointly to the envisaged teaching objective. At the same time, the teaching of biological or clinical specialities and postgraduate training play an increasing part. These new measures are linked with the administrative fusion of the university and hospital sectors and form a single system that tends to abolish the traditional difference in teaching, healing and research. This is a recent reform with many facets with very wide prospects and it is as yet impossible to appreciate or foresee the results.

Many French apologists, critics and reformers have made varied and divergent comments on the development of medical education during the last four centuries. Nevertheless, it is interesting to recall some of the views of foreign observers, who, more objective in their attitude, temper the individual and national attitudes.

During his practical training of fifteen months, carried out at Montpellier at the end of the seventeenth century, John Locke, the English physician and philosopher, had every opportunity to observe in his Journal of 1678—in terms to which Molière would not have objected—the peculiar manner of making a doctor of medicine.

A century and a half later other English-speaking visitors sounded a rather different note. The Revolution had destroyed in order to renew, and the Empire had set up the structure of a new national university. This was the peak period of French medical teaching, owing especially to the quality of the teachers and the orientation of the anatomical and clinical teaching, and exercised a powerful attraction on foreign observers. In 1843, the American Ferdinand Campbell Stewart (1815-1899), although putting the young physicians of his country above those of France, considered Paris to be the best teaching center in the world.

In his fundamental work on the history of medical teaching, Theodor Puschmann, a professor at Vienna, summed up the situation in 1890:

The medical schools of France are richly supplied with teachers, and the government spares no expense in providing them. . . . The medical schools of Paris and Lyons, of which I can speak from personal experience, are arranged in a most exemplary manner. . . . The abundant supply of material for instruction, the strict manner in which the attendance and work of the students are looked after, the close connection between theory and practice, the application of anatomical facts to practical medical science and especially to surgery, lead to remarkable results. . . . The system of medical education in France possesses along with many advantages . . . also certain lamentable defects.

Among these defects Puschmann stressed the late beginning of practical teaching and the undesirable aspects of the system for recruiting teachers, especially professors, which rested on selection by competition and was too subordinate to considerations of local opportunity to allow individual aptitudes to be displayed and developed on a national level.

The great medical educator Sir William Osler, giving his impressions of Paris in 1908-1909, had a great deal of sympathy for the French teaching system, although he also had some criticisms:

As elsewhere the Paris medical students are in three groups—good, indifferent and bad. A casual visitor to the laboratory and the hospitals gets only a general impression, and that given to me was of a very industrious hard-working set of men. From the start the student knows that success depends on his brains, or on a facility to use them in a certain way. . . . He is early made to realize that every single step in his career until he reaches an "agrégé" professorship depends on how he conducts himself at the "concours." This must have a very steadying effect on a young fellow. . . . One advantage the French medical student has over all others. . . . in Paris, the hospital is his home. . . . The hospital is everything; the medical school is well, quite a secondary consideration.

Osler shared the activities of a Parisian medical student and gave an excellent picture of hospital teaching in Paris at the beginning of this century; he emphasized its quality but he criticized its precociousness as well as the practical arrangements that led to an excessive influx of spectators and the tendency on the part of teachers to develop in an exaggerated manner their ability as orators. He has given us a penetrating study of the specifically French institutions of the *internat* and *externat,* which he particularly praised. He concluded his account with a tribute to France, eloquently concise, addressed to teachers and students: "After a stay of three and a half months, I am leaving Paris with many regrets. I am sorry not to be a member of the Faculty of Medicine." One may hope that an equally illustrious contemporary medical personality will give an equally flattering opinion in the years to come.

BIBLIOGRAPHY

ACKERNECHT, E. H., *Medicine at the Paris hospital 1794-1848.* Baltimore, 1967.

AMZALAC, J. C., *Réflexions sur l'enseignement de la médecine en France des origines à la revolution.* Paris, 1967. Thesis.

ASTRUC, J., *Mémoires pour servir à l'histoire de la Faculté de Médecine de Montpellier.* Paris, 1767.

BARIÉTY, M., and COURY, C., La Faculdad de Medicina de Paris desde sus orígines hasta nuestros dias. *Arch. Fac. Med. Madrid.* 1965, 8: 87-88.

——— *Histoirie de la médecine.* Paris, 1963.

BARON, H.-TH., *Questionum medicarum series chronologica.* Paris, 1730.

BINET, L., and VALLERY-RADOT, P., *La Faculté de Paris—cinq siècles d'art et d'histoire,* Paris, 1952.

BOUCHET, A., Parallèle entre l'Hôtel-Dieu de Paris et celui de Lyon vers 1750. *Cah. Lyonn. Hist. Méd.,* 1966, 10: 3-28.

BOYER DE CHOISY, M.DE, *Les étudiants en médecine de Paris au XVIe siècle.* Versailles, 1905.

BULLOUGH, V., *The development of medicine as a profession.* Basel, 1966.

CAUBET, C., *La Faculté de Médecine de Toulouse.* n.p., 1887.

CHOMEL, J.-B., *Essai historique sur la médecine en France.* Paris, 1762.

Commentaires de la Faculté de Médecine de l'Université de Paris (1396-1792). 25 vols. in fol. manuscript. Bibliothèque de la Faculté de Médecine, Paris.

CORLIEU, A., *L'ancienne Faculté de Médecine de Paris.* Paris, 1877.

——— *L'enseignement au Collège de Chirurgie depuis son origine jusqu'à la révolution.* Paris, 1890.

——— *Centenaire de la Faculté de Médecine de Paris (1794-1894).* Paris, 1896.

COURY, C., L'Ecole chirurgicale de l'Hôtel-Dieu de Paris au XXe siècle. *Episteme.* 1967, 1: 153-164.

——— L'Ecole médicale de l'Hôtel-Dieu de Paris au XIXe siècle. *Clio Med.*, 1967, **2**: 307-326.

——— Sir William Osler and French medicine. *Med. Hist.*, 1967, **11**: 111-127.

——— *L'Hôtel-Dieu de Paris.—Treize siècles de soins, d'enseignement et de recherche.* Paris, 1969.

COURY, C., and PECKER, A., Aperçus sur l'enseignement de l'obstétrique en France au XVIIIe siècle; Angélique de Coudray à Grenoble, in *XXe Congr. int. Hist. méd.*, Berlin, 1968.

DECHAMBRE, A., Écoles de médecine, in *Dictionnaire Encyclopédique des Sciences Médicales.* 1st ser., Vol. 32, Paris, 1885, pp. 323-400.

DELAUNAY, P., *La vie médicale aux XVIe XVIIe et XVIIIe siècles.* Paris, 1935.

DELMAS, P., Les étapes de l'enseignement clinique à Montpellier. *Chanteclair*, 1929, no. 261 (November).

DEWHURST, K., John Locke à Montpellier; notes médicales tirées de son journal (1676-1678). *Monspel. Hippocrates*, 1967, **10**: 9-20.

DUBARRY, J., and BEAUSOLEIL, C., La cérémonie du doctorat en médecine avant Molière. *Presse Méd.*, 1957, **65**: 57-58.

FABRE, R., and DILLEMANN, G., *Histoire de la pharmacie.* Paris, 1963.

FAUVELLE, R., *Les étudiants en médecine de Paris sous le Grand Roi.* Paris, 1899.

FINOT, A., *Les facultés de médecine de Provence avant la révolution.* Paris, 1958.

——— *Notes sur l'histoire de la clinique médicale et de son ensignement (d'Imhotep à Trousseau).* Paris, 1958.

FOURCROY, A. F., *Rapport et projet de décret sur l'établissement d'une école centrale de santé à Paris, fait au cours des comités de salut publique et d'instruction publique. 7 primaire an III.* Archives Nationales, A.D. VIII, 30.

FOURNIER, M., *Les statuts et privilèges des universités françaises depuis leur fondation jusqu'en 1789.* Paris, 1890-1894. 4 vols.

FRANKLIN, A., *Recherches sur la Bibliothèque de la Faculté de Médecine de Paris.* Paris, 1864.

GENTY, M., Un médecin de la Charité avant la revolution: Desbois de Rochefort (1750-1786). *Progrès Méd.*, 1935, **12**: 54-55. Supplement.

HAHN, A., and DUMAITRE, P. A propos de l'histoire de la Faculté de Médecine de Paris; les "Commentaires" (1395-1792). *Ann. Univ. Paris*, 1954, no. 4, 498-507.

——— *Histoire de la médecine et du livre médical.* Paris, 1962.

——— Les manuscrits à la bibliothèque de la Faculté de Médecine de Paris. *Sem. Hôp. Paris*, 1954, **30**: no. 79.

HAZON, J. A., *Notice des hommes les plus célèbres à la Faculté de Médecine en l'Université de Paris depuis 1110 jusqu'en 1750.* Paris, 1778.

HUARD, P., *L'Académie Royale de Chirurgie. Conférence donnée au Palais de la Découverte le 5 Novembre 1966.* Paris, 1967.

——— La presse médicale française dans la première moitié du XIXe siècle. *Concours Méd.*, 1967, **89**: 8228-8238.

JONES, H. W., The Faculty of Medicine at Paris. *Ann. Med. Hist.*, 1959, **1** (ser. 3): 1-29.

LACASSAGNE, J., *Histoire de l'internat des hôpitaux de Lyon (1520-1900)*. Lyons, 1900.

LASÈGUE, C., *Études médicales*. Vol. 1, Paris, 1884.

LEFRANC, A., *Histoire du Collège de France*. Paris, 1893.

LÉVY-VALENSI, J., *La médecine et les médecins français au XVIIe siècle*. Paris, 1935.

MILLEPIERRES, F., *La vie quotidienne des médecins au temps de Molière*.

MONTANIER, H., *Les facultés de médecine de Provence avant la révolution*. Saint-Germain, 1868.

MONTARIOL, *Sur l'étude de la médecine depuis le XIVe siècle jusqu'à la révolution*. Toulouse, 1912. Thesis.

N., and N., *L'enseignement et la diffusion des sciences en France au XVIIIe siècle*. Paris, 1964.

OSLER, W., Impressions of Paris; teachers and students. *J. Am. Med. Ass.*, 1909, **52**: 771-774.

PÉRY, L., *Histoire de la Faculté de Médecine de Bordeaux et de l'enseignement médical dans cette ville* (1441-1788). Paris, 1888.

PICARD, G., *La règlementation des études médicales en France. Son évolution de la révolution à nos jours*. Paris, 1967. Thesis.

PIZON, P., Un étudiant en médecine du premier empire. *Presse Méd.*, 1954, **62**: 1259-1261.

PUSCHMANN, T., *Geschichte des medicinischer Unterrichts von den ältesten Zeiten bis zur Gegenwart*. Leipzig, 1889.

RAYNAUD, M. A., *Les médecins au temps de Molière*. Paris, 1862.

ROLAND, R., *Les médecins et la loi du 19 Ventôse, an XI*. Paris, 1883.

ROULE, L., *Les médecins du Jardin du Roi aux XVIIe et XVIIIe siècles*. Paris, 1942. Thesis.

SABATIER, J., *Recherches historiques sur la Faculté de Médecine de Paris*, 1837.

SHRYOCK, R. H., *Histoire de la médecine moderne*. Paris, 1956.

TURCHINI, J., *La Faculté de Médecine de Montpellier. Aperçu historique*. Montpellier, 1963.

VALLERY-RADOT, P., *La Faculté de Médecine de Paris. Ses origines. Ses richesses*. Paris, 1945.

——— Les formalités de la thèse de maîtrise en chirurgie au 17e siècle. *Sem. Hôp. Paris*, 1946, **22**: 25-26.

——— L'internat des hôpitaux de Paris; ses origines; son prestige. *Presse Méd.*, 1952, **60**: 1069-1070.

——— La thèse de doctorat en chirurgie au 19e siècle. *Presse Méd.*, 1942, **50**: 441.

VARNIER, M., and STEINHEIL, G., *Documents pour l'histoire de l'Université de Paris. Commentaires de la Faculté de Médecine de Paris. (1777-1778)*. Paris, 1903.

VAUTIER, R., La vie des étudiants en médecine à travers les âges. *Presse Méd.*, 1954, **62**: 431-432.

VICQ D'AZYR, F., *Nouveau plan de constitution pour la médecine en France*. Paris, 1790.

PRINCIPLES AND PROBLEMS OF MEDICAL UNDERGRADUATE EDUCATION IN GERMANY DURING THE NINETEENTH AND EARLY TWENTIETH CENTURIES

HANS H. SIMMER
UCLA School of Medicine
Los Angeles, California

It is often implied that the rise of medicine in Germany during the second half of the nineteenth century and its fall in the current century not only reflected medical science but the teaching of medicine as well. During the latter part of the last century and the first decade of ours Germany reputedly offered the best teaching of medicine (1, 2, 3). A thorough evaluation of this phenomenon has not yet been attempted, in particular not for undergraduate medical education.

In 1876 Theodor Billroth (1829-1894) published his classic work *Über das Lehren und Lernen der medizinischen Wissenschaften an dem Universitäten der deutschen Nation* (4), more a stock-taking than a history. The *Geschichte des medicinischen Unterrichts von den ältesten Zeiten bis zur Gegenwart* (5, 6) by Theodor Puschmann (1844-1899) followed in 1889, but as far as Germany of the nineteenth century is concerned, it was more a history of medicine than of medical teaching. Both authors lived and worked in the midst of the new development. The only extensive assessment of medical teaching in Germany in the early twentieth century is found in the publications of Abraham Flexner (1866-1959) (7, 8). There are histories of medical faculties, like that of Eberhard Stübler (1891-1960) for Heidelberg, with some discussion of medical teaching (9). There are the meticulous collections of facts and statistics about students, assistants, *Privatdozenten* and curricula in Freiburg i.Br. by Ernst Theodor Nauck (1896-) (10, 11, 12). In addition there is a recent abortive attempt to describe in a thesis the history of medical teaching and the examining system in Germany from 1240 to the present (13). On the other hand, we find numerous pamphlets, articles and printed lectures about the history and problems of medical teaching (14-30), some of which are of recent origin (24-30). Quite surprisingly, there is no definite work on the modern history of medical teaching in Germany.

Before such a work could be attempted, material would have to be provided first for each specialty as well as for each university in a fashion similar to that exemplified in Nauck's monographs. This material is not now available, thus my task has to be a different one. Aware of the dangers of writing on a subject without such a foundation I shall, nevertheless, try to delineate some principles and problems of the German system of medical education during the last century. The origin of such principles and problems in earlier centuries cannot be dealt with in this presentation.

My discussion will be restricted mainly to the undergraduate clinical education. Far from being complete, it will concentrate on four subjects:

1. The Germany secondary schools as preparatory schools for the study of medicine.
2. The principles of solitude and freedom in learning medicine.
3. The principle of freedom of teaching medicine.
4. The principle of the unity of research and teaching for the professor and that of research and learning for the student.

The Secondary School as Preparatory School for the Study of Medicine

A specific kind of secondary school education was to become an essential condition for the quality of doctors in Germany in the nineteenth century: the *Gymnasium* (31, 32). Wilhelm von Humboldt (1767-1835) was the founder of its neohumanistic form.

After elementary schooling a pupil would enter the *Gymnasium* at the age of nine or ten. In a formal education he would learn how to learn (32). Emphasis was laid upon the mastering of Latin and Greek which in von Humboldt's view included psychology and philosophy of language in addition to grammar. Languages as "the cast of the world" were to educate the whole human being. After nine years the student could take a final examination, introduced in Prussia in 1788, but enforced first in the early nineteenth century. If he passed he received a certificate of maturity called the *Abitur*. That was all he needed to be admitted to the university. In Prussia, from 1827, this certificate was an absolute requirement for the study of medicine (33).

The value of Latin for medical students was soon to become a major subject of discussion. Since during the first half of the nineteenth century Latin was still the language used by some professors in medical faculties, it was, of course, necessary for the future student of medicine to understand it. However, fewer and fewer students were able to understand it sufficiently, let alone speak it fluently. By the middle of the nineteenth century, Latin was replaced by German in teaching, in written theses and in disputations, but at different times in different medical faculties and even departments. As a last step in this development the aspirant for *Habilitation* was permitted to dispute in German, in Freiburg i.Br., for example, in 1845 (34) and in Heidelberg, in 1855 (35). A conversation in pseudo-Latin or just Latin phrases continued to be used by some clinicians in the presence of patients, who were thus prevented from gaining information about their diseases. As recently as 1921 students could be asked "to use the *lingua latina,* while they hardly commanded their mother-language" (36).

Latin had been given up in medicine officially long before that, not only because of the growing inability to speak it but also for other reasons. For one thing, Latin had often been misused. As Friedrich Hufeland (1774-1839) pointed out, though scholarship could only be based on knowledge of older and newer languages, in spite of that knowledge one could be an insipid theoretician and an incompetent practitioner (37). Otto von Bollinger (1843-1909) described it later in this way: the clinicians made a show of philosophical self-conceit, and the classical Latin they spoke was only a disguise for the vacuity of their clinical lectures (38). This complaint surprises. Wasn't one advantage of Latin considered to be a training in systematic thinking and mental discipline? Quite obviously such exercise might aim at the form and less at the content of expression. Indeed, the teaching of old languages often deteriorated into monotonous logical gymnastics.

If Latin was not officially used any longer in medical teaching, should then the premedical education in classical languages remain a requirement? Of course, as long as the *Gymnasium* existed as the only kind of secondary school, the student had no other choice, but this sitiatuion was about to change. Since the *Gymnasium* did not allow sufficient training in mathematics and natural sciences, since it mainly trained the memory and to a lesser degree the senses, new schools developed (39). From the older *Realschulen,* schools to educate merchants and craftsmen, a *Realschule I. Ordnung* was developed by 1859, later called *Realgymnasium.* Latin was taught, but instead of Greek more mathematics and natural sciences were substituted. From 1870 on graduates of these schools were admitted to Prussian universities, but not yet to medical faculties. The rapid progression of technics and industrial arts necessitated even more concessions in 1882: *Oberrealschulen* were founded which emphasized natural sciences and modern languages, giving up classical languages completely (40).

Shortly after the founding of the *Realgymnasium* and even more so after the establishment of the *Oberrealschulen* the question arose which of these school systems was the more suitable for a student of medicine. It has been one of the most disputed questions in medical education in Germany ever since.

In 1869 the Prussian government consulted the medical faculties on this subject; four out of nine voted in favor of admitting students leaving the *Realgymnasium* to medical study, not enough to bring about a change (41).

Seven years later, Billroth emphasized the intellectual and disciplinary value of the study of Latin and Greek as leading to a degree of elasticity of thought attained by no other method, let alone the pedagogic importance of ideal and beautiful concepts of the antique (42). The medical student particularly needed Latin and Greek, otherwise he would have been in a chronic state of mild despair not knowing the spelling of every tenth or twentieth word in a medical work and, of course, not understanding the meaning of the terms derived from Latin and Greek. As a rule, the Germans used the etymological spelling rather than the phonetic. When German began to be used for lectures, no effort was made to invent a German scientific nomenclature, according to Billroth, a thoroughly German mistake (43). With the advance of science, many new terms had to be coined, in particular for pathology; the influence of Rudolf Virchow (1821-1902) caused a preference for Greek words provided with Latin endings. Billroth (44) admitted that this *mélange* of Greek and Latin was an evil, but was inevitable and had to be accepted. By the way, in 1876, he did not find that mathematics and the basic concepts of logic stood in any especially close relation to the study of medicine and to medical practice. In his opinion, imagination was more important than logic. Be it noted that Wilhelm von Humboldt had just recommended mathematics to train the mind and to make understanding, knowledge and mental life attractive not "by external circumstances, but by their inner precision, harmony and beauty" (45).

In 1876, Ernst von Leyden (1832-1910), though considering the classical education to be the best, urged that the graduates of the *Realgymnasium* be given a chance (46); but a strong request to maintain Latin and Greek came soon thereafter from the practicing physicians. In 1879, 163 medical societies of Germany were requested to pronounce opinion upon the matter of *Gymnasium* versus *Realgymnasium* (47). The majority voted against the *Realgymnasium* as a preparatory school for future physicians though many of them recommended a reform of the classical *Gymnasium*. Apparently, there was fear that the new secondary schools would be inferior to the *Gymnasium* still required for theology, law and philosophy. There was a desire to preserve the social position of the medical profession, and it was generally believed that the doctor must possess that measure of general knowledge which satisfies the highest demands in the country he lives in. The classical languages were considered to be an essential part of that knowledge (48).

Among those in favor of the dual secondary school education enjoying equal rights—not identical, but equivalent—was also Theodor Puschmann. In his "History of Medical Education" he devoted about one third of his final considerations to this subject (49). Unlike Billroth in 1876, Puschmann was an eager advocate of the *Realgymnasium*. How could it be justified, he argued, that a stumbling block be thrown in the path of a student, requiring him to study Latin and Greek, for which he was not gifted, when he showed a remarkable aptitude for natural sciences and promised to make an excellent doctor? And so the discussions went on and on.

In 1890 it was reported that 411 German professors had stated that the preparation of the *Gymnasium* was insufficient to be the basis for the study of medicine and natural sciences (50). On one occasion or another almost every outstanding professor voiced his opinion. Recently Liselotte Buchheim (1914-1966) gave examples of different opinions of professors at Leipzig (51). Among other famous discussants of this subject was Rudolf Virchow. In his *Rektoratsrede,* 1892, he required two things from his first year students: delight in study and ability to do independent work (52). He no longer considered the *Gymnasium* with its discipline in grammar to be an instrument for developing such attitudes and ability. but whether or not mathematics and natural sciences would compensate for deficiencies in the classical education, he left to the future (53).

Less than ten years later, in 1900, Kaiser Wilhelm II (1859-1941), himself bitterly recalling the insufficiency of his classic education (54), ordered the equivalency of all three secondary schools, the *Gymnasium,* the *Realgymnasium* and the *Oberrealschule.* Medical associations were still opposed (55) but by 1901 could not prevent the admittance of graduates of all three schools to medical faculties. Due to their influence, however, students of *Oberrealschulen* had first to pass courses in Latin.

The argument that Latin helps students to understand medical terminology has continued to be used in recent times (56), but Greek would also have to be required were one to aim at any understanding of medical terms at all. Considering the limited time students of *Oberrealschulen* do spend with Latin, it has further been questioned whether it can provide them with a training of the intellect and a superior cultural background for which Greek would, in fact, be preferable.

When Abraham Flexner visited Germany in 1910, he was somewhat baffled by the many discussions still going on (57). He made a remarkable proposal (58). "Something would be proved for the humanistic cause, if it could be established that students of mediocre capacity from the classical schools do better in medicine than students of mediocre capacity from the *Realschulen*. . . . While extraordinary ability attains its objects, mediocrity can noticeably be helped by propitious and hindered by unpropitious surroundings. . . " But on this point statistics were not available and still are not.

The discussions going on in these more recent times did not greatly affect the course of events. According to Nauck's statistics for the medical faculty in Freiburg i.Br., more and more students from all three secondary schools were admitted (59). While in 1900 seven graduates of the *Realgymnasium* were admitted in addition to 120 from the *Gymnasium,* later, when students of *Oberrealschulen* were also admitted, the percentage of *Gymnasium* graduates dropped further. In 1934, among the first year medical students, sixty-two came from the *Gymnasium,* fifty-eight from the *Realgymnasium* and thirty-three from the *Oberrealschule.*

Nevertheless, the German secondary schools usually permitted only students of higher quality to pass and thus to receive the certificate of maturity without which they could not enroll in the study of medicine and without which they did not have to pass an additional university entrance or admission examination. During the nineteenth century the content of secondary school education began to change. The shift from the classical languages to natural sciences reflected the "transition from the philosophical to the scientific age" (60). A thorough and long school education remained, however, a prerequisite for students of medicine. What Wilhelm von Humboldt had stressed as early as 1809 also remained basically valid: "Medicine is not only a technical discipline. . . . but a rational science which can only be studied in connection with historical, mathematical and philosophical sciences which are the propedeutics of all rational education (*Bildung*). With rare exceptions, medicine as a rational science prospers only in minds which have gone through long and laborious exercises in schools and universities" (61).

THE PRINCIPLES OF SOLITUDE AND FREEDOM IN
LEARNING OF MEDICINE

The transition from the secondary school to a medical faculty of the German university was at the same time easy and not at all easy. It was easy for the student because he was used to hard brainwork. He was, however, not prepared to learn on his own, without daily assignments; and this made the transition difficult for him. Wilhelm von Humboldt had requested the *Gymnasium* to prepare the student in such a way that "he can be left physically, ethically, and intellectually to freedom and independent work and so that he, free of pressure, does not turn to idleness or a practical life, but has the intense longing for sciences" (62). It remains to be elucidated why the *Gymnasium* often failed to achieve this aim. Complaints about the students misuse of his freedom are threaded throughout the modern history of medical education in Germany. Towards the end of the nineteenth century Virchow expressed it this way. "The temptation to enjoy academic liberty first of all by neglect of study is doubtless great for a young student. Stepping out from under the rigid discipline of the *Gymnasium* into the golden sunlight of academic liberty, he feels sensuous delight in stretching his limbs and acting in an irresponsible, improvident way" (63).

Why was the student given so much freedom? The answer lay in the distinctiveness of the German University after 1810. In that year the University of Berlin was founded, its beginning being marked by an idealistic neohumanism. Friedrich von Schiller (1759-1805), Friedrich Wilhelm Josef Schelling (1775-1854), Johann Gottlieb Fichte (1767-1814), Friedrich Schleiermacher (1768-1834) and especially Wilhelm von Humboldt contributed to the new ideal (64, 65, 66, 67, 68). For only sixteen months in 1809 and 1810 von Humboldt was head of the Prussian school system, as director of the section for culture and education in the Department of the Interior. It was his greatest merit that during this short period he succeeded in obtaining the state's sanction of a new type of university (32). Among others, Eduard Spranger (1882-1963) (65) and Helmut Schelsky (1912-) (67) have discussed the historical background of von Humboldt's reform. We cannot go into the long history of leading ideas such as academic freedom nor can we discuss the events of the actual founding of the University of Berlin and its medical faculty in particular, but we would like to restate von Humboldt's idea of the University. It is best expressed in his memorandum on the inner and external organization of higher scientific institutions, which though left unfinished, is a hallmark in the history of modern education (69). We referred to this *exposé* twice above in the context of secondary schools, and will return to it later on.

In this chapter it suffices to quote von Humboldt's leading thought: "Everything depends on upholding the principle of looking upon science as something not yet quite discovered and never completely solved, and unceasingly to seek it as such." (70) Such science should remain the major point of view of scientific institutions. "Since these institutions answer their purpose only if they, as much as possible, face the pure idea of science, *solitude* and *freedom* are the prevailing principles within them." (71) *Einsamkeit und Freiheit*, solitude and freedom, were to become, as Schelsky expressed it (72) the *"soziale Leitbild"*, the leading social idea of the new University of Berlin. They were also the principles on which German universities arose in the nineteenth century.

Medical education never quite fitted into this concept, but it benefited greatly from it, though not as much and as evidently as medical science. Von Humboldt's influence on medical education was twofold: a direct one through his two memoranda of 1809 (73, 74) and an indirect one through the partial, though often unconscious, acceptance of his ideas and principles by medical educators later on when his memoranda had been long forgotten. His influence upon medical teaching has received little, if any attention in the past (75), and certainly deserves more consideration and elucidation than can be given in this discussion.

Von Humboldt's memoranda of 1809 consisted mainly of detailed proposals on how to organize a section for medicine in the State Department of the Interior. As he expressed in another context (76), the State government, as a necessary evil, has to provide means and forms to allow science to prosper, but it should not forget that science would advance much better without any interference. Max Lenz (1850-1932) has pointed out the analogy between this concept and the theory of Martin Luther (1483-1546) regarding the relation between state and church (77).

In any case, in his medical memoranda, von Humboldt attempted just that: to propose a state medical administration which would guarantee the best possible flowering of the medical sciences. Although dwelling mainly on technical details, he also presented basic concepts, influenced by Johann Christian Reil (1759-1813), as described by Lenz (78). As to medical education, von Humboldt's basic concept was the following: there should be only one form of education, the study of medicine in a medical faculty of a university under the administrative control of the state. The study should be followed by a postgraduate practical training in a large non-university hospital. Such an internship, though in a modified way, first became obligatory for all German graduates in 1901. The second proposal gained recognition in a shorter time, but nevertheless more than forty years passed before it was put into practice: after 1825 the universities remained the only schools for the study of medicine. Thus medical teaching became "especially theoretical, a university education oriented towards an all encompassing science, and isolated and withdrawn from the common ties of science" (79); and thus medical students were exposed to solitude and freedom.

Schelsky has recently pointed out that the concept of solitude did not have a long life (80). In von Humboldt's view the continuous search for truth isolated the individual. "It is only in the university that the individual can discover through and within himself, the insight into pure science. For this self-act liberty and helpful solitude in the truest sense are necessary." (81) Science, of course, can only prosper if scientists discuss matters and stimulate one another (82). A delicate balance, indeed, had to be found: the right proportion of solitude and intellectual intercourse. This applied to teacher and student as well.

Nevertheless the freedom to choose solitude and, at other times, fellowship was characteristic for students and professors during most of the nineteenth century. It was also characteristic for medical faculties, but it appears that the principle of solitude disappeared earliest from the medical faculties. Medical professors found less and less solitude while the ever increasing demands on their time were changing them into managers of large enterprises called university hospitals and institutes. This development, by the way, had not been foreseen by von Humboldt and was certainly not favored by him. Without giving his reasons, he categorically declared that the university ought always to prefer a small hospital (61).

The student of medicine fitted even less into von Humboldt's ideal. How could he follow this advice: "Attendance at lectures is an incidental matter: it is essential that, for a number of years, one . . . live for oneself and for science?" (83) The student of medicine had to gain a certain amount of knowledge in order to be able to take care of patients. During the semesters he had not much time to indulge in science, he was hardly able to digest what he heard or saw.

It was not exceptional for a student to attend six to eight lectures per day. He could attend a lecture several times, not being bound to an obligatory curriculum; but he also had to attend more and more lectures and courses. In Freiburg i.Br., for example, during the summer semester of 1808, the medical faculty offered sixteen lectures and six courses; by 1910 the numbers had increased to seventy-three and thirty-eight respectively (84). Though lectures and courses differed in length and their demands upon the student, and although not all offerings were obligatory, the student's time was increasingly taken up by teaching sessions. If solitude was sought and utilized, it was mainly for grinding out what was thought to be necessary for the examinations. Admittedly the student had considerable time for himself during the vacations which lasted about five months each year, but were at different times and in different degrees occupied partly by military service or partly by clinical clerkships. Nevertheless, some students used this time as von Humboldt wished it to be used. However, many students had neither the talent nor the inclination, even though they had the freedom.

The principle of freedom of learning has been a continuously recurring topic of endless discussions (85). Should there be freedom at all for medical students? If so, shouldn't it be a limited freedom? And if so, in what way and to what degree should freedom be limited? Von Humboldt, in his reemphasizing the principle of freedom and giving it new content, did not express himself on the problems that the principle of freedom of learning had to encounter in the medical faculty. He had been opposed to technical schools for surgeons as they existed up to the middle of the nineteenth century (86). Such schools should only train those who were not gifted for study and would later be medical auxiliaries. Von Humboldt denied the army the right to train special military doctors. Such training was going on in the Pepinière, the *Militärärztliche Bildungsanstalten* in Berlin which one might call an army medical college. He proposed that the better students of the Pepinière obtain all their medical training in the medical faculty, while the others be trained only as medical corpsmen. Spranger has reported on the controversy between von Humboldt and Johann Goercke (1750-1822), the director of the Pepinière, to illustrate von Humboldt's contempt for *Fachschulen,* technical schools (87). The rigid military drill in the Pepinière, of course, did not agree at all with von Humboldt's ideas. Prussian discipline applied to studies and forced upon the students by their superiors was diametrically opposed to the discipline coming from within the academically oriented student.

But could one expect the future physician to have this inner determination? Apparently, one expected too much of him. Again and again, professors complained about the students wasting valuable time or not using their time optimally. An early complaint is to be found in a letter to Johann Friedrich Meckel (1781-1833) (88). In 1827 Meckel sharply criticized the students for attending the clinical lectures two to four semesters earlier than they were supposed to do, thus being insufficiently prepared. In 1840 the medical faculty of Berlin complained that the students were against theoretical lectures and preferred to watch operations and to participate in clinical lectures (89). Meckel demanded that students and also teachers should be ordered to comply strictly with a logical curriculum. Nineteen years later, Joseph Hermann Schmidt (1804-1852) called for the same referring to the requirements of the law faculty, among other reasons (90). Puschmann also advocated rigorous measures and hardly concealed his approval of a school-like system with frequent discussions and examinations to rectify misconceptions and clarify things not understood, a system, as he stated, preserved in military medical schools (91). Others like Theodor Ludwig Wilhelm Bischoff (1807-1882) (92) and Billroth (93) were opposed to any obligatory curriculum. Karl Freiherr Stein zum Altenstein (1770-1840), since 1817 Prussian's "secretary for religious, educational and medical affairs", had in 1826 ordered a quadrennium and suggested a curriculum. Schmidt thought this to be in error. One should order what the student has to learn and recommend a quadrennium (94). Not only should lectures and courses be obligatory but also the sequence in which they were to be taken had to be ordered. Recommended curricula, however, remained the characteristic feature of German medical faculties later on, while the time of required study changed to five years in 1901 and, in 1927 to five and a half years or eleven semesters.

Many curricula did exist during the nineteenth century. They differed from university to university even within one state, as documented by Stübler (95) for Heidelberg and Nauck (96) for Freiburg i.Br. Not even within a medical faculty was full agreement always reached among professors as one can read in Bischoff's pamphlet of 1848 (97). Basically, all curricula agreed in as far as they proposed basic sciences first, followed by pathology and pharmacology, and finally clinical specialties. In a way, the students were ordered what to learn since they had to pass examinations: the *tentamen philosophicum* in basic sciences, in Prussia changed in 1861 to *tentamen physicum*, and the faculty and state examinations at the end of the study. But during the years prior to these examinations the students always had considerable liberty in spite of the demands of Schmidt and others to curtail freedom. Wrote Schmidt about his own proposals: "Should the goddess of [academic] freedom not consent to that, she will have to be ignored. This is discourteous, it is true, but most practical. *Mulier taceat in republica.*" (98).

In an attempt to redefine academic freedom, Schmidt declared science to be truly free if pursued for itself, for mankind, and only thirdly for breadwinning purposes (99). How little was changed during the following decades is evident from Virchow's cutting remark in 1892, "Academic liberty is not the liberty to do nothing, nor liberty to pursue pleasure, but liberty to study" (100).

Under the auspices of the goddess of academic freedom the student continued to take many liberties, one of which was to change medical schools several times during the course of his study. There had been restrictions against migration prior to 1871; thereafter, the student could attend a medical faculty of his choice and often switched from one university to another. Further, the minimum time of study was often extended for reasons ranging from the distraction of pleasures to that of research. For the top student the system provided an excellent opportunity to seek the teachers of his choice and to broaden his horizon.

The student had the further freedom of attending a lecture several times if he thought it benefitted him, or, after having received certification of his first attendance in a special booklet, he could skip all the other lectures, until the last one which at various times also had to be certified. By 1901 the misuse of academic freedom was such (101) that a new regulation of examination was ordered.

In addition to signing up for certain compulsory lectures the student thereafter, at the time of examination, had to present certificates which attested regular participation in other compulsory courses and lectures. Such a certificate was called a *Schein,* a word which in German is also used for pretense. As it turned out, it was not possible to prevent students from pretending that they had attended lectures and courses regularly and also successfully, as was required from 1924. Even this system was misused at times (102), since it was not always rigidly applied by professors who assumed the students to be mature and honest and often considered themselves above the unpleasant task of policing them. Of course, such control became increasingly difficult with the tremendous increase in students. Between 1830 and 1890 the number of medical students at twenty-one German universities had tripled from about 3000 to more than 9000 (103). After a transitory drop the number rose again in the twentieth century. At the universities in the area of the Federal German Republic the total number of medical students increased from about 7000 in 1910 to more than 32,000 in 1963 (104). Without an appropriate increase in professors the student became anonymous. It was not a rare event that he spoke with the chairman of a clinical department for the first time during the final examination.

At the turn of the century, as von Strümpell expressed it clearly in 1901, the problem of freedom of learning had boiled down to its excellence for top students and its inadequacy for and misuse by the masses of average students, let alone the below average students (105). Essentially, nothing changed during the following decades.

In 1965, Thauer summarized the development and present situation by stating: "We still hate to abolish it [the privilege of freedom of learning] because for our very best students the atmosphere of complete freedom seems to be an indispensable condition for an optimal development of their mental abilities.... With the growing masses of students and our failing system of selection, however, the old and unique privilege unfortunately has proved to be insufficient.... Influenced by Humboldt's ideas in this field, we failed to recognize this development in its early stage, and we are guilty of not having adapted our system in time to the changing attitude and abilities of our average student" (106).

The Principle of Freedom of Teaching

Academic freedom consisted not only of liberty of learning, but also of *Lehrfreiheit,* freedom of teaching. Both were old principles, and both were given new content in as far as they were newly oriented to "living science" by von Humboldt (107). As freedom of learning had its counterpart in obligatory curricula, freedom of teaching had its antipole in ordered teaching norms.

It is very remarkable that the dual control of teaching exerted by the state government from outside the universities early in the nineteenth century, was almost entirely delegated to the medical faculties themselves by the end of that century. The two controlling mechanisms were state board examinations and the vocation of professors as chairmen.

Various boards had been established, such as *Medicinal-Kollegien* or *Sanitäts-Kommissionen* to examine students before licensing them to practice medicine. Some professors had become lost in theoretical systems, their teaching was not oriented towards the needs of medical practice. The state, however, as Schmidt expressed it, had to be granted the right to distinguish between liberty of *Lehre* and that of *Leere*, both being pronounced equally, but one meaning teaching, the other vacuity (108). Since, moreover, great laxness prevailed in granting medical degrees, and protecting physicians from "foreign" competition, examing boards were set up usually consisting of medical officials and distinguished practicing physicians. Prussia had a state board of examiners from 1725. Such a system was necessary, and yet it created new problems. How many physicians could be found outside the medical faculties who were not only competent practitioners, but also men characterized by "love of science and pure devotion to the advancement of the commonwealth" as von Humboldt had requested in 1809? (109). He gave this board of examiners the name *wissenschaftliche Deputation*, scientific deputation, not specifying whether its members were professors or not. If we interpret the term "scientific" correctly, he did not exclude professors. Indeed, professors were admitted to examining boards and again in 1837 confirmed in this role by von Altenstein (110). He requested, however, that they not examine on those subjects they had treated in their lectures (111). With the development of scientific medicine in the nineteenth century the exclusion of professors or the restrictions imposed upon them became untenable, if not ridiculous, as described by Schmidt (112) and Karl Otto Weber (1827-1867) (113).

The necessary change came in 1846 in Prussia when mainly professors were appointed to the *Prüfungskomission*. The state of Baden first acknowledged clinical professors after 1871 when this state became part of the German empire (114). Professors and *Privatdozenten* were to become a permanent state board of examiners in the twentieth century, non-university physicians being excluded entirely.

The history of the university and state examinations from 1800 is too intricate and involved to be considered here. What matters in this discussion is the fact that the states eventually gave up the prerogative of checking on the quality of graduating medical students. This was not basically changed when in 1901 the government introduced a practical year, a kind of internship in a university hospital, in another qualified hospital or in the office of a qualified physician. It could hardly be considered to be a control of university teaching since after this year no further examination was required to obtain a license.

One might speculate why the government finally left the state examinations to medical faculties. Practical problems such as finding enough and suitable examiners outside the university might have been one factor. More important perhaps was the arrival of German scientists at the forefront of the world's medicine during the nineteenth century. The success of modern medicine could not be questioned by administrators. World authorities, and there were many in almost every medical school of that time, could not be denied the authority to examine in their respective fields.

This was granted even though the vocation of a chairman, the other governmental control over medical education, had lost much of its effectiveness. Election of a new professor by the members of the faculty without any governmental involvement had often led to nepotism and continuing mediocracy. Von Humboldt simply questioned the objectivity of the professors. Though admitting that, of course, "a reasonable and fair curatorium" will allow the faculties to have some influence, he demanded that the state have the exclusive right to appoint professors (115). It was of utmost importance to choose the right men and thus provide "wealth, strength and variety of intellectual vigour" (116). In what way and to what degree the faculties should influence the call to a chair, how then the government would go about finding the best available candidate, to these eminently important questions von Humboldt gave no answers.

Although early in the century the government often appointed professors against the wishes of the medical faculties, it became customary later in the century to appoint the candidate who had been nominated *primo loco* by the medical faculty (117). In Prussia, for example, as early as 1834, the medical faculty was permitted to nominate three candidates for a vacant professorship. It appears that in 1861 in Greifswald this permission by the state was first expressed as a right and obligation of the medical faculty (118). Once a professor had been appointed as a chairman he had the liberty to do or to stimulate research in whatever area he preferred, and he had the liberty to teach his specialty in whatever form and content he determined to be suitable.

How then could a system work that was based on liberty of teaching and lack of suitable controls from without? One answer probably lies in the highly competitive system of the academic career in Germany. For many years a future chairman was assistant, *Privatdozent* and professor. During those years, while aiming at a chair the candidate had to excel in research and teaching, though the latter was usually not valued equally with research achievements. And there was more at stake after he became chairman: his career usually began at a small university with a small number of students and thus a relatively small income and a relatively slight reputation. This was considered to be a stepping stone. Rarely did a professor stay at the university of his first chair. So, for many years the professor was under constant demand from within his profession. One might say that he had the freedom to be excellent. Indeed, he tried to excel in his field of science. Many professors had been trained in physiology, biochemistry or other basic sciences, but for teaching no special training was available, and no guidance was offered as to the purpose of teaching.

The freedom of teaching allowed the professor to aim at whatever he thought to be worthwhile. Thus, the ways to teach and the aims of teaching were manifold. And yet, general characteristics can be recognized.

During the last century the states expected the students to be ready to practice medicine at the end of their university study. The state examination could be taken immediately after completion of the university studies; if passed, the student was able to practice medicine. During the latter part of the century it became apparent that the university education was no longer sufficient, though many attempts were made by professors to help the student to gain practical experience. Work in the *Poliklinik*, the outpatient department, was required although its value was disputed for many years (119); in several hospitals students were accepted for work on the wards under close supervision, but not many places were available. As a rule only the best students were accepted; and the professors regretted that they were not able to offer this education to all (120). Here, too, the best student was well taken care of.

For the majority of students another solution had to be found. A practical year, a kind of internship, as von Humboldt had proposed, was more and more considered and disputed. It already existed in 1839 in Württemberg, but not in Baden and Prussia. When in 1882 new regulations were formulated for all states of the German Empire, a practical year was omitted. And yet, there was growing unrest about this. Practicing physicians wrote on the subject, all complaining, though not all as acidly and sarcastically as Mandel in 1892 (121). "To memorize and recite in the examination", he wrote, "is transitory like wind . . . one learns something only by practicing it until one masters it" (122). A practical year was finally introduced in 1901, extended by an additional three months of supervised work in a rural area in 1939 and, finally, in 1953 extended to two years.

With this development it was gradually admitted that the university education prepared the graduate less and less to practice medicine. What, then, was the aim of university study? Was it the education of future scientists in medicine with purposeful neglect of practical training? This was the tenor of many compliants (123). In part, such complaints originated in a misunderstanding of scientific medicine or in a failure to grasp and apply the principles of new medicine.

Under the influence of the medical sciences teaching changed from transmission and apprenticeship to the application of the so-called scientific method. Education in natural sciences, although existing earlier, became absolutely indispensable when these sciences were systematically applied to medicine in the nineteenth century.

It was not only that natural sciences provided the student with specific tools for diagnosis and treatment. More than that, as Phoebus wrote in 1849 (124), "the practical work in natural sciences, particularly if it consists of independent observations, sharpens the senses and the mind of the medical student, gives him practice not only in formal, but also in materially correct judgment and conclusions, and finally develops his ability to observe critically."

The new attitude in clinical medicine was then an unprejudiced and critical scrutiny applied deliberately, consciously, and as consistently as mental powers can make it. The student was exposed to practical problems, but the emphasis was on theoretical education. He was taught the fundamental problems of medicine and he was intellectually trained to be mentally prepared for the ever increasing knowledge in medicine. The ideal developed not to set up training camps where the student mainly learned the techniques and the tricks of the medical profession, but to expose him to the scientific method by allowing him to think first with the professor and, hopefully, later independently. The aim was no encyclopedic knowledge of facts and tools, but the ability to cope with ever increasing problems, and the ability to learn new details whenever necessary. This approach required close contact between teacher and student and between student and patient, but the freedom of teaching was such that the professor was even permitted to give up the close contact with his students. Good contact existed earlier in the nineteenth century, but became less and less towards its end especially in the larger universities. The transition is best illustrated by the bedside teaching of Johann Lucas Schoenlein (1793-1864), and the auditorium teaching still going on now. Bedside teaching has a long history (125). Schoenlein was its last great representative in Berlin. There from 1840 he gave his clinical lectures on the ward, examining the patient and then sitting in a comfortable armchair beside the patient's bed (126). He truly had a *Lehrstuhl*, a teaching-chair. He stressed the method of critical observation. He made the patients available to the student in such a way that everyone could gain personal experience. The *Klinik*, as it was called, was not a lecture with demonstrations, but theoretical and practical guidance of the individual student. The student saw the patient daily and followed the course of disease. In case of death, Schoenlein meticulously used the postmortem findings to judge retrospectively whether and where diagnosis, therapy and prognosis had been right or wrong (127, 128).

This was time-consuming and laborious, but rewarding. Schoenlein, who was also personal physician to the King and counselor for medical affairs in the ministry of education, found less and less time for teaching in his later years. Other professors had other demands on their time. Furthermore, the number of students increased considerably. In Berlin, between 1819 and 1909 the student body increased from 397 to 1568, and the teaching staff from twenty-eight to 202; the number of chairmen (*Ordinarien*), however, grew only from ten to eighteen (129). Another form of teaching had to be chosen for the chairman if he insisted on teaching all students. The form of the grand lecture (*Hauptvorlesung*), a class-room lecture, was introduced.

Schoenlein's successor in Berlin, Friedrich Theodor Frerichs (1819-1885), in 1859 built a lecture hall for more than 200 students (130) and taught in a way much different from that of Schoenlein. "The patient in his bed was brought into the lecture room and was now the strict object of the lecture, only from time to time somewhat unfeelingly looked at, during which time Frerichs clearly discussed all pertinent points. When Frerichs was in the right mood to examine thoroughly, his discussions were instructive and precise. Very often, however, he didn't go into details, made a snap diagnosis and a routine prescription" (131). Others let the students participate, as *Praktikanten*. Flexner has given a description of it in 1912: "In the arena of the amphitheatre two *Praktikanten* appear, after a dozen names have been called. The professor sets them to examining the patient. They are completely nonplussed. He directs their attention to a certain spot. 'Can't you feel a cyst?' They are not quite sure. With some hesitation one of them ventures a timid 'Yes.' It is not quite convincing. 'Really?' the professor inquires. There can apparently be no doubt it is there. 'Yes' they both reply with emphasis and that is all. By the time the second patient is brought in, the professor has wholly forgotten about his student assistants. He examines, describes, expounds. A third case is brought in—typhoid, it proves. The professor does everything. Twice only are the *Praktikanten* addressed . . . thence forward they retreat ever further into the background, differing from the students in the benches only in being a little closer—and much more uncomfortable" (132).

Of course, among the professors, there were magnificent lecturers, like Friedrich von Müller (133). Two points were made in favor of the grand lecture (134): A master mind at work was exhibited daily to two hundred students or more who would not be influenced by the great man otherwise, even if they were on the wards of his 200-to-400 bed hospital.

It was further pointed out that the lecture constituted an object lesson in scientific method, but it was also stated that one could not follow such lectures with much profit unless a considerable personal experience had preceded and that one could not learn to master medicine by merely looking or listening. It had to be learned, if at all, by doing (135). According to Flexner, the technique of the grand lecture erred in substituting at the start a related or described or exhibited experience for a personal one (136). And, furthermore, since the student heard a succession of such discourses daily, his mind was becoming overwhelmed with facts, theories and ideas delivered to him far more rapidly than his mental processes operated: they simply stopped.

Flexner's observations could still be confirmed in recent years; his criticism is still valid. It appears that with the growing disproportion between professors (*Ordinarien*) and students the grand lecture became a necessity. It appears further that such lectures were in line with the university concept of training students mainly mentally, and less practically. It will, however, need further study to elucidate how and why the grand lecture developed and why it was continued at a time when its disadvantages were clearly recognized. Whatever the reasons are, the principle of freedom of teaching made such development possible.

The Unity of Research and Teaching for the Professor and That of Research and Learning for the Student

In his memorandum on the organization of scientific institutions, von Humboldt refused to accept the concept according to which the university's task was to teach science, while research was the domain of academies (137). "Science can not be really presented as science without its independent comprehension each time, and it would be inconceivable that in so doing one would not even often, make discoveries. University teaching, furthermore, is not such a troublesome affair that one should consider it as an interruption of the leisure to study, but rather as a contrivance of science" (138). In essence this meant the unity of research and teaching. For this reason, von Humboldt vehemently fought against medical schools separated from universities as planned around 1800. We owe it mainly to him, at least for Berlin, that medical education remained within the university and thus gained from and took part in the scientific revolution of the nineteenth century.

Already in the universities of Halle and Göttingen, which were founded in 1694 and 1737 respectively in the age of enlightenment, when philosophy replaced theology, teachers were expected to be researchers (139). But as in Göttingen, research and teaching in medical specialties were not too closely related (140). Furthermore, during the same century many academies of sciences were founded as research institutions outside the universities. Consequently, there was a demand for closure of the existing universities and the founding of special technical schools (141). In 1810, through Wilhelm von Humboldt, the unity of research and teaching was reinstituted in the University of Berlin and became exemplary for other German universities.

A new concept of research developed, involving clinical as well as laboratory work. There was at times a misconception, that only laboratory methods are scientific while clinical study at the bedside is empirical. However, as Abraham Flexner pointed out, "science is indifferent as to where or how observations are made: It is only concerned with the vigor, precision and consistency of observation" (142). There were only differences in the tools to be used. Another attitude characterized the new scientific physician. As Phoebus had expressed it in 1849: "Even the best physician cannot always arrive at a diagnosis, but there is a tremendous difference between the doctor who in objectively obscure cases is aware to what degree and why diagnostic procedures are insufficient, and that doctor to whom also objectively clear cases remain subjectively obscure." (143). "The sense of limitation with the cool, persistent effort to pierce the barrier at its weakest spot" (144) was now the attitude of the clinical and non-clinical scientist as well.

This approach can be found earlier, no question, but it was to become the consistent and most typical approach in the nineteenth century. Though Romantic medicine led astray temporarily, by the middle of the century most survivors of the predominantly deductive method were replaced by exponents of the predominantly inductive, the so-called scientific method. This development did force differentiation since one man could master only a small limited field. Specialties developed which only later became independent departments: physiology, physiological chemistry, pediatrics and dermatology are examples (145). While lectures and courses in these specialties were later to become compulsory, in Prussia in 1861, formal logic and philosophical psychology were finally dropped out of the curriculum. The *tentamen philosophicum* was changed to the *tentamen physicum*.

It was of utmost importance for the unity of research and teaching that laboratories for the preclinical sciences and hospitals for the clinical sciences were within the university. Even more, laboratories were set up in the hospital. As early as 1802, Reil demanded that each hospital have a chemist and an anatomist (pathologist) as associates, but it took long until laboratories were established for them in hospitals (146). In Berlin, Frerichs started with a one room laboratory (147). Von Ziemssen's hospital laboratories, established in 1884 in Munich, were to become the model for the future (148).

Laboratory and clinical chairs were of the same status. Basic sciences influenced clinical sciences and *vice versa*. Prolonged training and some more or less independent research in physiology, physiological chemistry or pathology became the normal basis of a clinical career. Johannes Müller (1801-1858), Karl Ludwig (1816-1895), Rudolf Virchow, Rudolph Heidenhain (1834-1897), Felix Hoppe-Seyler (1825-1895) and others were the teachers to mention only a few. Ludwig Traube (1818-1876), Hugo von Ziemssen (1829-1902), Adolf von Strümpell (1853-1925) and many others were their students.

This contiguity of the sciences and the clinic in the modernized university, brought about modern medicine. Highest demands were put on the clinical teacher. Could he really be an active researcher in all the fields of internal medicine of which he was the chairman? Could he also be a good teacher of those branches he had never worked in as a researcher? As long as he did active research in one field he probably would be able to grasp facts and problems of a closely related field and deal with patients of that field scientifically, but the development of new specialties was inevitable. How long in his academic life could a professor himself conduct research? When did administrative duties and clinical burdens of a private practice reach such a degree that he could not attend the laboratory any longer? Research had to be delegated to younger coworkers sooner or later, thus disrupting the personal unification of research and teaching for the professor. What effect had this on his teaching?

These questions will suffice to indicate that the history of the concept of the unity of research and teaching for clinical professors deserves special treatment not yet attempted. In any event, since about 1860 the German medical student was consistently exposed to the "scientific method" of medicine. As pointed out, he could not acquire this method by merely listening to the professor. His practical education on the other hand was far from ideal.

Did then the medical faculty provide the scientific method, though less related to a patient, but within the framework of biomedical research? The professor, at least in his younger years, was teacher and researcher. With his help, the student would go into the depth of a problem and learn how to approach it. Besides obtaining an answer to a specific question he would learn the complexity of medical problems and he would also learn general methodology, but supposedly influencing him as a physician.

The arrangement by which the student could learn through research was the requirement of a doctoral thesis without which a German student could not receive the title *Doktor*.

Little attention has been paid to the basic principle of the doctoral thesis: the unity of research and learning. In the concept of von Humboldt (149) the student was supposed to learn by active participation in the scientific process of gaining new knowledge. Both professor and student were aiming at cognition, the former to advance science, the latter to learn. "Both live for science" (150). This ideal was never fully realized (151), least of all in the medical faculty.

Here, the customs of the past were a heavy burden. Lenz has described the controversy of Reil, Hufeland and Karl Asmund Rudolphi (1771-1833) over thesis and promotion—a controversy which threatened to disrupt the new medical faculty in Berlin directly after its founding (152). The questions raised not only reflected the past, but also indicated problems which continued to exist throughout the nineteenth century. Should the promotion replace the state examination? Who should write the thesis, the professor or the student? Should the content of the thesis be clinical or experimental? Should the thesis be disputed *coram publico*? Should one insist on a high quality of the thesis? A temporary solution was found for Berlin (153), the details of which are less important than the question which led to it.

As to the regulations for promotion great differences existed in different states early in the century. In the state of Baden, for example, not many students graduated as doctors, since they could practice without the degree if they passed only the state board examinations. And many students were not able or willing to spend the money required for the doctorate. In Prussia, on the other hand, until 1869, promotion by dissertation was a condition for the state license. Though students did not need the MD degree to practice after 1869, almost all underwent the additional effort and expense. Billroth, who received his degree in 1825 in Berlin, pointed out that the title raised the social position of a physician (154).

Society, indeed, exerted pressure, even more so since the unauthorized use of the title was punishable. The title determined the rank in society and also was a kind of advertisement. The medical faculties tried to see to it that the student not only succumbed to the demands of society, but also achieved a scientific piece of work, which found its expression in different grades like *summa cum laude* and *magna cum laude*, a difference, by the way, which usually was not known to the patient and also had little effect on the doctor's career.

While Billroth was clearly in favor of the doctoral thesis, Puschmann in 1889, also in Vienna, was decidedly against it (155). Puschmann had completed his thesis in 1869 in Marburg. According to him the doctorate was nothing more than an ancient and superfluous custom since it neither conferred the right to practice nor was it a condition upon which admission to the State examination depended. He proposed to grant the title to everyone who had passed the State examination, as had been done in Vienna since 1848. Or, if the title should really distinguish physicians, the theses should be of higher quality.

Even more of a surprise than Puschmann's negative attitude was Abraham Flexner's evaluation of the German doctorate in 1912. In his extremely thorough assessment of the German medical faculties he devotes a little less than half of a page and two short footnotes to this subject (156).

Flexner considered the MD examination "little more than a costly formality". Usually compiled according to formula, the dissertation might acquaint the student with the literature and might even be the beginning of an academic career. Flexner also briefly mentioned the procedure of that time: the thesis accepted, the candidate appeared before a committee of the faculty, usually the dean and two other professors, "by whom he was colloquially interrogated for fifteen minutes each". Later this was different in different universities. In Tübingen after World War II, for example, from all the original requirements only the acceptance of a written thesis remained and, of course, the expense, about which Flexner stated in conclusion of the subject, that it is "perhaps the most formidable aspect of the ordeal" (156).

In general, it appears that when the non-professional state board examiners disappeared during the second half of the last century the university professors had ample time to test the students in the state examinations.

Thus, an extensive general faculty examination for the doctorate became unnecessary, and was finally, as in Tübingen, given up completely.

The history of the medical doctoral theses in modern Germany has still to be written. It will be necessary to evaluate how often the ideal of the unity of research and learning was achieved, and why it apparently failed in so many instances. Lack of proper guidance and facilities were factors. Supervision shifted from professor to *Privatdozent* and, unofficially, even to assistants not yet distinguished as researchers. Nevertheless, some students certainly did gain more than a title for society and more than Flexner saw. Under good guidance and with appropriate facilities they did research historical subjects; they provided valuable clinical statistics; and they performed excellent experimental work. By doing so, they not only learned the scientific method, but also, though rarely, contributed to the progress of the medical sciences. In the field of endocrinology, the case of Paul Langerhans (157) is often cited. Of course, such discoveries as his were rare, but more important, students were specifically influenced by participating in scientific work. Here again, the system favored the top students who more easily received good guidance and benefitted more from their work on the thesis.

Concluding Remarks

Many aspects of undergraduate medical education have not been discussed such as more recent attempts to limit the number of admissions. Neither the role of textbooks nor the use of technical aids in lectures and courses nor the problems caused by the development of new specialties (158) have been described. Neither the influence of the *Privatdozent* (159) in medical education has been sufficiently dealt with nor the development of the preclinical education in medical faculties. A discussion of general principles has prevailed. More questions have been raised than answers could be given: it might be called preliminary thoughts to a history of modern medical education in Germany. Nevertheless, it has become evident that the strength of the German system was also its weakness. Guided by the principle of freedom of teaching and of the unity of research and teaching the professor taught students whose ideal was freedom of learning and, supposedly, the unity of research and learning. Such principles had existed earlier; through the efforts of Wilhelm von Humboldt, they were to become the leading ideals of the German university of the nineteenth century, reaching into our time; admittedly, they were basically theoretical in character.

Ralf Dahrendorf recently stressed the theoretical character of Wilhelm von Humboldt's university: "The tradition of the University of Berlin in the history of the German Universities is theoretical in at least three regards: it emphasizes research and not teaching in its didactic problems, it stresses science and not its application to practical questions of life or professions, for which these sciences qualify, and, even within scientific research, it accentuates the reflective philosophical element over the experimental" (160).

Be it put into proper perspective for medicine: the emphasis on research led to modern scientific medicine which put education of medical students on a much sounder basis than had ever existed before. Secondly, the stress on medical and premedical science and not on its application led to numerous discoveries which brought about modern medicine as it is now applied to sick patients. Medical teaching, at least since the later nineteenth century, did not dare any more to neglect new knowledge as applied to diagnostics and therapy. Finally, as far as the character of medical research is concerned, the philosophical element remained of utmost importance if integrated into experimental research (161, 162). For the student, participation in induction and deduction was as important as conducting experiments and examinations.

The real problems appear to have been of another nature. First, the medical professor was not able any longer to represent continuously the concept of the unity of research and teaching. This concept, sooner or later in different specialties and sooner or later in the life of a professor, stiffened into a *Leerformel*, a formula of vacuity, as Schelsky has expressed it (163). Under the pretext of freedom of teaching, overburdened by mainly administrative duties and in a "self-delusion of the abstract authorities" (164) the professor did not devote himself sufficiently and efficiently to undergraduate teaching. The principle of freedom of teaching then was not properly restricted in its realization for undergraduates. On the other hand, the same principle proved to be eminently effective for the best students and particularly for postgraduate training.

Another development additionally made the philosophical and also the experimental element in education more or less ineffective: the loss of personal contact between professor and student. Good contact usually existed between the few highly gifted students and their teachers, especially in the smaller universities, but with the ever increasing number of students, the average and below average students no longer received proper guidance and training. Thus, it is not surprising that William Gilman Thompson (1856-1927), returning from Germany to the United States, wrote in 1906: "I do not see that there is anything in the scheme of German undergraduate instruction that we need adopt for the betterment of our own" (165). It should be stated, however, that Thompson's verdict was mainly directed against clinical undergraduate education. The preclinical training, as advanced in German universities in the nineteenth century was to become examplary for the world. This subject will deserve special future treatment. It also should be added to Thompson's statement that it was written at a time when German medical research was leading in the world. The ones who gained greatly from the advances in medical sciences were not so much the undergraduates, with the exception of top students, but the post-graduate students, be they clinically or research oriented specialists. They were the ones who carried the fame of Germany abroad (1, 2, 3).

While all reforms in undergraduate education in the past were half-hearted and inconsequential, drastic changes have been introduced recently by the founding of a new biomedical university in Ulm (166, 167). The future will show whether this endeavor marks the beginning of a new era of medical education in Germany.

NOTES

1. WARTMAN, W. B., *Medical teaching in western civilization.* Chicago, 1961.
2. BOMMER, THOMAS NEVILLE, *American doctors and German universities. A chapter in international intellectual relations 1870-1914.* Lincoln, Nebraska, 1963.
3. WELCH, WILLIAM H., Introduction, in *The medical sciences in the German universities; a study in the history of civilization.* Translated from German of Theodor Billroth. New York, 1924.
4. BILLROTH, T., *Über das Lehren und Lernen der medicinischen Wissenschaften an den Universitäten der deutschen Nation nebst allgemeinen Bermerkungen über Universitäten. Eine kulturhistorische Studie.* Vienna, 1876.
5. PUSCHMANN, THEODOR, *Geschichte des medicinischen Unterrichtes von den ältesten Zeiten bis zur Gegenwart.* Leipzig, 1889.
6. ACKERKNECHT, ERWIN H., Introduction to THEODOR PUSCHMANN, *A history of medical education,* trans. E. H. HARE. New York, 1966.
7. FLEXNER, ABRAHAM, *Medical education in Europe.* Boston, 1912.
8. ───── *Medical education, a comparative study.* New York, 1925.
9. STÜBLER, EBERHARD, *Geschichte der medizinischen Fakultät der Universität Heidelberg 1386-1925.* Heidelberg, 1926.
10. NAUCK, E. TH., *Zur Geschichte des medizinischen Lehrplans und Unterrichtes der Universität Freiburg i. Br.* Freiburg i. Br., 1952.
11. ───── *Studenten und Assistenten der Freiburger medizinischen Fakultät. Ein geschichtlicher Rückblick.* Freiburg i. B., 1955.
12. ───── *Die Privatdozenten der Universität Freiburg i. Br. 1818-1955.* Freiburg i. Br., 1956.
13. BOHNER, H., *Die Geschichte des medizinischen Ausbildungs-und Prüfungswesens in Deutschland von 1240 n. Chr. bis heute.* Köln, 1962. Inaugural-Dissertation.
14. SCHMIDT, JOSEPH HERMANN, *Die Reform der Medicinal-Verfassung Preussens.* Berlin, 1846.
15. BISCHOFF, THEOD. LUDW. WILH., *Beleuchtung der Bemerkungen eines Grossh. Hess. Arztes Dr. * über die neue Grossherzogl. Hess. Prüfungsordnung für Mediziner.* Giessen, 1848.
16. LEYDEN, E., *Ueber die Entwicklung des medizinischen Studiums. Rede gehalten zur Feier des Stiftungstages der militärärztlichen Bildungsanstalten.* Berlin, 1878.
17. ZIEMSSEN VON, Ueber die Aufgabe des klinischen Unterrichts und der klinischen Institute; *Dt. Arch klin. Med.,* 1878, 23: 1-30.

18. Gusserow, Zur Geschichte und Methode des klinischen Unterrichtes. Berlin, 1879.
19. Recklinghausen, Friedr. von, Die historische Entwicklung des medicinischen Unterrichts, seine Vorbedingungen und seine Aufgaben. Strassburg, 1883.
20. Ebstein, Wilhelm, Über die Entwicklung des klinischen Unterrichts an der Göttinger Hochschule und über die heutigen Aufgaben der medizinischen Klinik. Klin. Jb., 1889, 1: 67-109.
21. Virchow, Rudolf, Lernen und Forschen. Berlin 1892; translated, "Study and research" in Rep. Smithson. Instn., 1892, pp. 653-665.
22. Orth, Johannes, Medizinischer Unterricht und ärztliche Praxis. Wiesbaden, 1898.
23. Strümpell, Adolf von, Über den medizinisch-klinischen Unterricht. Erfahrungen und Vorschläge; Erlangen and Leipzig, 1901.
24. Heischkel, E., Die Entwicklung des medizinischen Unterrichtes, Medsche Welt, 1939, 13: 1238-1241; 1267-1269.
25. Buchheim, Liselotte, Mitglieder der Leipziger Medizinischen Fakultät über medizinisch-pädagogische Fragen des ausgehenden 19. Jahrhunderts. Wiss. Z. Karl-Marx Univ. Lpz., 1955-56, 5: 39-42.
26. Ackerknecht, E. H., Typen der medizinischen Ausbildung im 19. Jahrhundert. Schweiz. Med. Wschr., 1957, 87: 1361-1366.
27. Rath, Gernot, Advances in medical education in Germany, J. Med. Educ., 1961, 36: 1092-1101.
28. ——— Die Entwicklung des klinischen Unterrichts. Göttinger Universitäts-Reden No. 47. Göttingen, 1965.
29. Hamperl, H., Das Medizinstudium heute, Medsche. Welt, 1963, 17: 982-996.
30. Thauer, Rudolf, German medical education today and tomorrow, J. Med. Educ., 1965, 40: 343-350.
31. Paulson, Friedrich, Geschichte des gelehrten Unterrichts. 3. ed., vol. 2, Berlin, Leipzig, 1921.
32. Spranger, Eduard, Wilhelm von Humboldt und die Reform des Bildungswesens. 3 ed., Tübingen, 1965; see also op. cit., Nos. 45, 81.
33. Seligman, Herbert, Ueber die Erziehung und Vorbildung der Dozenten der Berliner Medizinischen Fakultät in der ersten Hälfte des 19. Jahrhunderts. Berlin, 1936. Inaugural-Dissertation.
34. Nauck, E. Th., op. cit., No. 12, pp. 28, 130 ff.
35. Stübler, E., op. cit., No. 9, p. 289.
36. Hueck, Walter, Wie man anno 1921 in München Medizin studierte. Medsche Klin., 1964, 59: 2035-2038.
37. Hufeland, Friedrich, Ueber das Verhältniss der theoretischen zu der praktischen Bildung des Arztes. C. W. Hufeland's J. prakt. Heilk., 1839, pp. 4 ff.
38. Von Bollinger, O., Wandlungen der Medizin und des Ärztestandes in den letzten 50 Jahren. Munich, 1909.
39. Paulson, Friedrich, op. cit., No. 32, pp. 544-576.
40. Ibid. p. 581.

41. PUSCHMANN, T., *op. cit.*, No. 5, p. 497.
42. BILLROTH, TH., *op. cit.*, No. 3, pp. 139-147.
43. *Ibid.*, p. 141.
44. *Ibid.* p. 142.
45. HUMBOLDT, WILHELM VON, "Über die innere und äussere Organisation der höheren wissenschaftlichen Anstalten in Berlin, 1810?" In: *Wilhelm von Humboldt's gesammelte Schriften.* ed. B. GEBHARDT. Vol. 10, pt. 2, p. 256. Berlin, 1903.
46. LEYDEN, E., *op. cit.*, No. 16, p. 32.
47. PUSCHMANN, T., *op. cit.*, No. 5, p. 497.
48. ―― *op. cit.*, No. 5, p. 497; and LEXIS, W., *Die Universitäten im Deutschen Reich.* Berlin, 1904, p. 128.
49. PUSCHMANN, T., *op. cit.*, No. 5, pp. 495-501.
50. NAGEL, ALBRECHT, *Die Vorbildung zum Medizinischen Studium und die Frage der Schulreform.* Tübingen, 1890, p. 5.
51. BUCHHEIM, LISELOTTE, *op. cit.*, No. 25, p. 40.
52. VIRCHOW, RUDOLPH, *op. cit.*, No. 21, pp. 11 ff.
53. *Ibid.* p. 19.
54. FLEXNER, ABRAHAM, *op. cit.*, No. 7, p. 33.
55. PAULSEN, FRIEDRICH, *op. cit.*, No. 29, p. 747.
56. RATH, GERNOT, *op. cit.*, No. 27, p. 1093.
57. Eigener Bericht des ärztlichen Vereins München: Welche Mittelschulvorbildung ist für das Studium der Medizin wünschenswert? *Münch. med. Wschr.*, 1910, **57**: 1045-1046.
58. FLEXNER, ABRAHAM, *op. cit.*, No. 7, p. 38.
59. NAUCK, E. TH., *op. cit.*, No. 11, p. 59.
60. VIRCHOW, RUDOLPH, The founding of the Berlin university and the transition from the philosophic to the scientific age. (Rektoratsrede 1893). *Rep. Smithson Instn.*, 1896, pp. 651-695.
61. HUMBOLDT, WILHELM VON, Denkschrift über die Organisation des Medizinalwesens, 25. Juli 1809. In: *Wilhelm von Humboldt's gesammelte Schriften*, ed. BRUNO GEBHARDT. Vol. 10, Pt. 2, p. 127, Berlin, 1903; see also p. 126.
62. ―― *op. cit.*, No. 45, pp. 255-256.
63. VIRCHOW, RUDOLF, *op. cit.*, No. 21, p. 9.
64. SPRANGER, E., *Wilhelm von Humboldt und die Humanitätsidee.* Berlin, 1909.
65. ―― *op. cit.*, No. 32.
66. ―― *Gedenkrede zur 150-Jahrfeier der Gründung der Friedrich-Wilhelms-Universität in Berlin.* Tübingen, 1960.
67. SCHELSKY, HELMUT, *Einsamkeit und Freiheit, Idee und Gestalt der deutschen Universität und ihrer Reformen.* Reinbeck bei Hamburg, 1963.
68. DAHRENDORF, RALF, Traditionen der deutschen Universität. In: *Hochschulführer*, ed. by P. KIPPHOFF, T. VON RANDOW, and D. E. ZIMMER. Hamburg, 1965, pp. 11-23.
69. HUMBOLDT, WILHELM VON, *op. cit.*, No. 45, pp. 250-260.
70. *Ibid.* p. 253; similarly expressed on p. 251.

71. *Ibid.* p. 251.
72. SCHELSKY, H., *op. cit.*, No. 67, p. 68.
73. HUMBOLDT, WILHELM VON, *op. cit.*, No. 61, pp. 123-128.
74. ——— Plan zur Organisirung der Medicinal-Section im Minsterio des Innern, in *op. cit.*, No. 61, pp. 128-138.
75. Neither BILLROTH (*op. cit.*, No. 4) nor PUSHMANN (*op. cit.*, No. 5) mentioned Wilhelm von Humboldt. FLEXNER (*op. cit.*, No. 7, pp. 7, 8) briefly mentioned his "brilliant achievement", the founding of the University of Berlin, stressing its importance for German universities and also for medicine without, however, delineating the latter influence.
76. HUMBOLDT, WILHELM VON, *op. cit.*, No. 45, p. 252.
77. LENZ, MAX, *Geschichte der Königlichen Friedrich-Wilhelms-Universität zu Berlin.* Vol. 1, p. 187, Halle, 1910.
78. *Ibid.* pp. 199 ff.
79. HUMBOLDT, WILHELM VON, *op. cit.*, No. 74, p. 132.
80. SCHELSKY, H., *op. cit.*, No. 57, pp. 118 ff.
81. HUMBOLDT, WILHELM VON, "Der königsberger und der litauische Schulplan". In: *Wilhelm von Humboldts gesammelte Schriften,* Vol. 13, Berlin, 1920, p. 279.
82. ——— *op. cit.*, No. 45, p. 251.
83. ——— *op. cit.*, No. 81, pp. 279-280.
84. NAUCK, E. TH., *op. cit.*, No. 10, p. 59.
85. e.g., *op. cit.*, Nos. 14, 15, 16, 19, 21, 29, 30.
86. An example is the "Medizinisch-chirurgische Lehranstalt" in Münster. See: FRATZ, PAUL, Das Matrikelbuch der Münsterschen Medizinisch-Chirurgischen Lehranstalt. *Arch. Gesch. Med.* 1942, **35**: 68-97.
87. SPRANGER, EDUARD, *op. cit.*, No. 65, pp. 142-143.
88. J. F. Meckel to Dr. Schulze on 15 May 1827; letter quoted by SCHMIDT, J. H. *op. cit.*, No. 14, pp. 41-43.
89. HEISCHKEL, E., *op. cit.*, No. 24, p. 1269.
90. SCHMIDT, J. H., *op. cit.*, No. 14, p. 47.
91. PUSCHMANN, T., *op. cit.*, No. 6, pp. 627-629.
92. BISCHOFF, T. L. W., *op. cit.*, No. 15, p. 10.
93. BILLROTH, T., *op. cit.*, No. 4, p. 134.
94. SCHMIDT, J. H., *op. cit.*, No. 14, p. 38.
95. STÜBLER, E., *op. cit.*, No. 9, pp. 192-196.
96. NAUCK, E. TH., *op. cit.*, No. 10, pp. 53-55.
97. BISCHOFF, T. L. W., *op. cit.*, No. 15.
98. SCHMIDT, J. H., *op. cit.*, No. 14, p. 58.
99. *Ibid.*
100. VIRCHOW, RUDOLF, *op. cit.*, No. 21, p. 8.
101. STRÜMPELL, A. VON, *op. cit.*, No. 23, pp. 3 ff.
102. HAMPERL, H., *op. cit.*, No. 9, p. 982.
103. EULENBERG, FRANZ, *Die Frequenz der deutschen Universitäten von ihrer Gründung bis zur Gegenwart.* Leipzig, 1904, p. 255.
104. THAUER, R., *op. cit.*, No. 30, p. 344.
105. STRÜMPELL, A. VON, *op. cit.*, No. 23, p. 4.

106. THAUER, R., *op. cit.*, No. 30, pp. 347-348.
107. SCHELSKY, H., *op. cit.*, No. 67.
108. SCHMIDT, J. H., *op. cit.*, No. 14, p. 33.
109. HUMBOLDT, WILHELM VON, *op. cit.*, No. 74, p. 131.
110. BILLROTH, T., *op. cit.*, No. 4, pp. 159 ff.
111. *Ibid.*
112. SCHMIDT, J. H., *op. cit.*, No. 14, pp. 94 ff.
113. WEBER, KARL OTTO, Memorandum, quoted by STÜBLER, E., *op. cit.*, No. 9, pp. 282-287.
114. *Ibid.* p. 287.
115. HUMBOLDT, WILHELM VON, *op. cit.*, No. 45, p. 259.
116. *Ibid.* p. 254.
117. BILLROTH, T., *op. cit.*, No. 4, pp. 280-307.
118. *Ibid.* p. 289.
119. See, for example, for 1888: HENOCH, "Ueber den poliklinischen Unterricht" *Dt. med. Wschr.*, 1888, **14**: 53; VON DUSCH, T., "Ueber den poliklinischen Unterricht" *Dt. med. Wschr.*, 1888, **14**: 69; VON JURGENSEN, T.: "Ueber den Unterricht in der Poliklinik," *Dt. med. Wschr.*, 1888, **14**: 75.
120. V. ZIEMSSEN, H., "Ueber den klinischen Unterricht in Deutschland" *Dt. Arch. klin. Med.*, 1874, **13**: 1-20.
121. MANDEL-FORBACH, *Die Ausbildung des Arztes*. Hamburg, 1892.
122. *Ibid.* pp. 3, 9.
123. See for example HELLERMANN, *Ueber die heutige Ausbildung der Mediziner und deren Wirkung auf die ärztlichen Verhältnisse*. Leipzig, 1890; GUTTSTADT, A., *Über die praktische Ausbildung der Ärzte in den Kliniken*. Berlin, 1892; ISRAEL, O., "Zur praktischen Ausbildung der Aerzte." *Berl. klin. Wschr.*, 1892, **29**: 1291; also ORTH, I., *op. cit.*, No. 22.
124. PHOEBUS, PHILIPP, *Ueber die Naturwissenschaften als Gegenstand des Studiums, des Unterrichts und der Prüfung angehender Ärzte*. Nordhausen, 1849, pp. 2 ff.
125. LESKY, ERNA, The development of bedside teaching at the Vienna Medical School. pp. 217-234 in this volume.
126. ARTELT, WALTER, Die Berliner Medizinische Fakultät. *Ciba Ztschr.* 1956, **7**:25-87.
127. VIRCHOW, R., *Gedächtnisrede auf Julius Schoenlein*. Berlin, 1865.
128. ACKERNECHT, ERWIN, H., Johann Lucas Schoenlein 1793-1864, *J. Hist. Med.*, 1964, **19**: 131-138.
129. LENZ, MAX, *Geschichte der königlichen Friedrich-Wilhelms-Universität zu Berlin*. Vol. 3, Halle a.d.S., 1910, pp. 490, 492, 493 and 498: also see FERBER, C. VON, *Die Entwicklung des Lehrkörpers der deutschen Universitäten und Hochschulen 1864-1954*. Göttingen, 1956.
130. *Ibid.* p. 106.
131. KISCH, E. H. quoted by ARTELT, W., *op. cit.*, No. 78, p. 2595.
132. FLEXNER, A., *op. cit.*, No. 3, pp. 175 ff.
133. ——— *An Autobiography*. New York, 1960, p. 102.
134. ——— *op. cit.*, No. 7, p. 171.
135. *Ibid.* pp. 171 ff.

136. *Ibid.* p. 172.
137. HUMBOLDT, WILHELM VON, *op. cit.*, No. 45, pp. 256 ff.
138. *Ibid.* p. 257.
139. ACKERKNECHT, E. H., *op. cit.*, No. 26, p. 1364.
140. EBSTEIN, W., *op. cit.*, No. 20, pp. 69 ff.
141. SCHELSKY, H., *op. cit.*, No. 67, pp. 37 ff.
142. FLEXNER, A., *op. cit.*, No. 7, p. 6.
143. PHOEBUS, P., *op. cit.*, No. 124, pp. 4 ff.
144. FLEXNER, A., *op. cit.*, No. 7, p. 7.
145. EULNER, HANS-HEINZ, *Die Entwicklung der Medizinischen Specialfächer an den Universitäten des deutschen Sprachgebietes.* Frankfurt a. M., 1963. Med. Habilitations-Shrift.
146. LENZ, M., *op. cit.*, No. 77, p. 49.
147. ARTELT, W., *op. cit.*, No. 78, p. 2595.
148. ZIEMSSEN, VON, *op. cit.*, No. 17.
149. HUMBOLDT, W. v., *op. cit.*, No. 45, pp. 251-252.
150. *Ibid.* p. 252.
151. DAHRENDORF, RALF, *op. cit.*, No. 68, p. 16.
152. LENZ, M., *op. cit.*, pp. 373-379.
153. *Ibid.* pp. 379-381.
154. BILLROTH, T., *op. cit.*, No. 3, pp. 232-233.
155. PUSCHMANN, T., *op. cit.*, No. 5, p. 481.
156. FLEXNER, A., *op. cit.*, No. 7, p. 258.
157. LANGERHANS, PAUL, *Beiträge zur mikroskopischen Anatomie der Bauchspeideldrüse.* Berlin, 1869. Inaugural-Dissertation.
158. EULNER, HANS-HEINZ, *op. cit.*, No. 145.
159. BUSCH, ALEXANDER, *Die Geschichte der Privatdozenten. Eine soziologische Studie zur grossbetrieblichen Entwicklung der deutschen Universitäten.* Stuttgart, 1959.
160. DAHRENDORF, RALF, *op. cit.*, No. 68, p. 17.
161. HARTMAN, MAX, *Die philosophischen Grundlagen der Naturwissenschaften. Erkenntnistheorie und Methodologie.* 2 ed. Stuttgart, 1959.
162. VAN'T HOFF, J. H., *Imagination in science. Translation and introduction by G. F. Springer.* Berlin, Heidelberg, New York 1967.
163. SCHELSKY, HELMUT, Wer herrscht in der Universität? *Christ und Welt,* 1968, **21** (No. 23): 17.
164. *Ibid.* The German term is "Vertrautheitsselbsttäuschung der abstrakten Autoritäten."
165. THOMPSON, WILLIAM GILMAN, *Medical education in Germany, Austria, Bohemia and France.* Unpublished manuscript in the New York Academy of Medicine Library, 1906, p. 11.
166. *Bericht des Gründungsausschusses über eine Medizinisch-Naturwissenschaftliche Hochschule in Ulm.* Ulm, 1965.
167. ROBERG, NORMAN B. Medical reform in West Germany. *J. Am. Med. Assoc.,* 1967, **200**: 603-608.

MEDICAL EDUCATION IN THE NETHERLANDS 1575-1750

G. A. LINDEBOOM
Free University
Amsterdam, The Netherlands

Considered in historical perspective the beginning of medical education at university level in the Netherlands coincided with the birth of a new state, the Republic of the Seven United Provinces, and it was marked by the foundation of the University of Leyden.

Before that date there were in the Netherlands only a few doctors of medicine. Most of these men had not taken their degrees in Louvain (1), which was at the time the only university in the Low Countries, but in France or, preferably, in Italy. The care of the sick was at a low level and was mostly provided by surgeons and unqualified medical practitioners.

Yet, in the first half of the sixteenth century there was already some medical teaching in Holland, but only for surgeons and midwives.

In this period of municipal particularism this teaching was provided by the guilds in the major towns. As early as 1555—twelve years after the publication of the *Fabrica* of Vesalius—at the request of the Amsterdam Surgeons' Guild, King Philip II of Spain, Count of Holland, granted it a charter. By that charter the Guild obtained permission to dissect annually the body of an executed criminal for the purpose of instruction of surgeons and barbers, so that (as it runs in the text of the charter) "they would not cut veins instead of nerves, or nerves instead of veins, and would not work as the blind work in wood, as Galen says." Soon afterwards the post of *praelector anatomiae* was created in Amsterdam, and in the two succeeding centuries a series of outstanding physicians maintained this teaching at a high level. The praelectors, with the governing board of the Guild, had themselves painted by the best artists of their times. The most famous example of such a practice is of course Rembrandt's portrait of Doctor Nicolaas Tulp in *The Anatomy Lesson*.

However, this was the teaching of one medical subject, and was not medical education in its broad sense. Strangely enough, medical education was a side-effect of the war with Spain. The rebellion against the Spanish regime began in 1568, and led to a war which was to last eighty years. In the beginning the Spaniards occupied the greater part of Holland, and in 1573 they began a siege of the city of Leyden which lasted nearly a year. Thousands of the inhabitants died from starvation. In the end the Dutch invoked the assistance of the water, which was Holland's enemy as well as its friend. The dykes were cut, but only after many weeks of anxious expectation did the wind turn and the rising water dissipate the besiegers on 3 October 1574.

William the Silent intended the University of Leyden to be not only a reward to the city for its courageous defense, but also a diplomatic move in the struggle with Spain. In his letter of 28 December 1574 to the States of Holland and Zeeland, in which he advised the foundation of a university, he explained that it ought to be a "strong blockhouse" of resistance to the enemy and should also further serve the unity of the nation. In the light of the approaching negotiations with Spain, it was necessary that the university should be founded as soon as possible, and indeed, six weeks later, on 8 February 1575 it was opened.

Three years later (1578) the University of Leyden, unlike many other universities, was, with the exception of the Faculty of Divinity, opened to students of all religious faiths—a resolution which greatly favored the influx of foreign students.

Though, in the following decades, four other provinces also founded universities, namely Franeker (1585), Groningen (1614), Utrecht (1636), and Harderwijk (1648), Leyden, the oldest, remained the best and the most renowned—not least for the medical education it provided.

From 1581 the University was housed in the church of the former convent of the White Nuns. It was administered by a Board consisting of three Governors, called the Curators, who were appointed by the States for life, and of the four Burgomasters of the city who were changed every year.

From the outset the Curators took a broad view and tried by every means to attract famous scientists; they sent a warship to fetch the famous man-of-letters Joseph Scaliger from France, in order to protect their future professor from the dangers of a journey through troubled countries.

For the Medical Faculty it soon appeared that special arrangements were necessary for the teaching of the basic sciences, botany and anatomy.

At that time botany was a purely medical discipline; it fostered the knowledge of medical drugs and, in fact, it was taught only in the Faculty of Medicine. Though the famous botanist Rembert Dodonaeus (Dodoens, 1517-1585) was for some years professor of medicine at Leyden, he never had a good opportunity for his teaching of botany because there was no garden at his disposal. The University Botanic Garden was founded in 1587 (2), and was thus one of the oldest in Europe; its supervision was entrusted to Gerardt de Bondt (1537-1599), the first professor of medicine (3).

Neither were there, in the beginning, facilities for the teaching of anatomy, a still more important basic science (4). De Bondt lectured on that subject also, but his lectures were purely theoretical. However, in 1589 young Pieter Pauw (1564-1617), a skilled anatomist, was appointed extraordinary professor of anatomy. In the same year Pauw performed a dissection; the account of his expenses, amounting to more than eleven Dutch guilders, dated 29 December 1589, is still preserved in the Archives (5). In 1597 the anatomical theatre was built (6).

There was every reason for building a theatre for the number of medical students was steadily increasing. The Curators stated with some pride in 1589 that the number was already greater than it had ever been in the University of Louvain, even in its most flourishing period (7).

At Leyden many of the medical students came from abroad, especially from England. Indeed, the very first medical student at Leyden was an Englishman named Jacobus James, who matriculated on 21 September 1578 and took his degree in 1581. In doing so he became the second graduate in the Medical Faculty.

From 1580, the year in which the first medical graduation took place (8), up to 1625 one hundred and thirty-seven students graduated, that is, about three a year (9).

It appears that the medical course at Leyden could be completed in about two years. A proposal of a professor of divinity, made as early as 1575, that no one should be permitted to graduate as a doctor of medicine until he had been apprenticed to some very learned physician, was not accepted (10).

According to the original regulations, which remained operative during the two centuries that followed (11), the student was required to have held two public disputations *exercitii causa* before he was allowed to pass the final examination.

Such public disputations, which served to give the students practice in speaking and in repeating those things that they had been taught, were held monthly. Though they occurred regularly in the Faculty of Philosophy during Boerhaave's time, there is not sufficient evidence that this custom was then followed strictly in the Medical Faculty (12).

Before a student was allowed to graduate he had to pass a preliminary examination privately (13). Then, on a following day, he had to explain in the afternoon an aphorism of Hippocrates that had been communicated to him at 8 o'clock that morning.

Finally, the candidate had to explain, against the objections of the professors, the case-history of a patient and to suggest the detailed treatment. This disputation lasted an hour (14).

After that the candidate was entitled to have his thesis printed and to request from the Rector a date for his graduation. At the graduation ceremony the candidate had to defend his dissertation and a few additional short theses. The disputation lasted an hour, the end of which was marked by the beadle's exclamation *hora*. Then the graduand received his doctor's diploma and took an oath (15).

Except for the teaching of botany and anatomy, all medical teaching was originally theoretical. At Leyden lectures were held only during four days of the week, Wednesday and Saturday being free (16). The programme was the same for every day. The first syllabus—which was in manuscript—dates from October 1587 (17).

We learn from it that at 8 o'clock in the morning Jan van Heurne (1543-1601), who was appointed in 1581, lectured on *praxis medicinae*, using a textbook composed by himself. At 2 o'clock in the afternoon Gerardt de Bondt spoke on the *Prognostics* of Hippocrates and the *Physiology* of Fernel alternately.

On 1 March 1592 the first printed syllabus appeared for the summer months (18). Now Van Heurne lectured on the *Diet in acute diseases* of Hippocrates and De Bondt on the third book of Paul of Aegina. Since all this teaching was theoretical, the professors may have felt that something was lacking. The first Leyden professors of medicine (19), Gerardt de Bondt and Jan van Heurne, were both graduates of the University of Padua, which was administered by the Venetian Republic. The latter, Jan van Heurne, is sometimes erroneously credited with the first bedside teaching in Leyden (20).

Now it appears that in Padua the first bedside instruction to classes of students was given by Giovanni Baptista da Monte (Montanus) (1498-1552), who was appointed professor in 1543, the year in which Vesalius left Padua. De Bondt and Van Heurne must have studied under Degli Oddi and Bottoni, the successors of Montanus. They would not have forgotten the bedside teaching they had enjoyed in Italy.

It is a fact that, at the meeting of the Curators held on 4 December 1591, there was a proposal from the Medical Faculty that the professors should instruct the students in the hospitals (21). It is not clear whether this plan was put forward by De Bondt or by Van Heurne, or by both (22). No names are mentioned in the notes, but, in any case, at this time an attempt was made to introduce bedside teaching into the medical curriculum—a novelty north of the Alps. However, the Curators resolved to postpone further consideration of the proposal till their next meeting, but as the later minutes are silent on this point, it is probable that the plan was never realized.

However, forty-five years later, the son of Jan van Heurne, Otto van Heurne (1577-1652) who in 1601 succeeded his father as professor of medicine, took a similar initiative, but with more success.

It must be conceded that on this occasion the stimulus came from outside. In 1635 the activities of the Medical Faculty of Leyden were at a low level; a serious epidemic of plague had driven the professors and the students out of the city and all lectures had been suspended. Meanwhile, in all secrecy, the foundation of a university in Utrecht, at a distance of less than thirty miles, had been planned and the fact was made known in 1636. This news came as a real shock to Leyden University, and especially to the Medical Faculty—the more so as the newly-appointed professor of medicine in Utrecht, Willem van der Straeten (1593-1681), announced in his inaugural oration on 17 March 1636 his intention to teach the students also in the

Nosocomium in which the care of the patients had been entrusted to him by the states (23).

These facts were the inducement for Otto van Heurne to write without delay an address to the Curators requesting facilities for clinical instruction. It was discussed by the Curators at their meeting on 13 May 1636 (24). Van Heurne pointed to the possibility that the foundation of the University of Utrecht might lead to a falling-off in the number of students at Leyden, and furthermore that the provision of clinical instruction would mean that the students would no longer have to travel to Italy or France to visit patients in the hospitals with the doctors. Therefore, he offered himself to provide, together with the two city physicians, clinical teaching in the public hospitals. Incidentally he made a discreet allusion to an increase in his own salary.

The wise Curators did not entertain the proposal immediately, but sought the advice of the Medical Faculty. The Faculty unanimously wrote a favorable recommendation on 29 September (25), and so the Curators, at their meeting of 4 November 1636, resolved to institute clinical instruction, the "collegium medico-practicum," also called practical exercises (26). This instruction was to be given twice a week, on Wednesdays and Saturdays, by two professors in turn, each during three months.

The patients in the hospital were treated by the city doctors, and it was resolved that one of the city physicians, as well as the surgeon, had to be present when the professor presented his bedside teaching. The town physician had to fetch the professor from his home so that they made the journey to the hospital on foot together (27). Soon it was necessary to decide that the surgeon, unless he was ill, be not allowed to send his apprentice in his place.

The first professors to be charged with this teaching were Otto van Heurne and Ewald Screvelius. Probably from the very beginning the bedside instruction was given in the Caecilia Hospital, an old lunatic asylum. At a later date two wards, each of six beds, one for men and one for women, were reserved in this hospital and were adapted for the purpose of erection of some kind of gallery behind which the students stood (28).

The clinical lectures started in the following year. At his demonstrations Otto van Heurne began enthusiastically to interrogate the students about signs and symptoms, diagnosis and prognosis, but soon he had to give up this practical method as it displeased most of the students (29).

It seems that in the beginning all did not go well, and that some criticism of the behavior of the professors during their hospital lectures reached the ears of the Curators. In any case, on 3 June 1637 the Curators received the surgeon at their meeting and interrogated him on this point (30). It was decided to hear also professor Screvelius, who explained his method of clinical instruction and requested that he might go on in the same manner. There may have remained some doubt in the minds of the Curators, for they recommended him to follow the method which was the best and the most profitable to the students (31).

If anyone were to think that the story of the clinical lectures at Leyden is a story of unremitting success, he would be badly mistaken. In several periods clinical instruction was neglected by the professors charged with it, and again and again we see the Curators taking action in order to restore it. Of course, there were some professors who performed their task faithfully, the most brilliant among them, without doubt, was Franciscus de le Boë Sylvius (1616-1672) (32). Born in Hanau, he was a descendant of a noble French Huguenot family. After his graduation at Bâle in 1637 he gave at Leyden for some time private lectures and demonstrations on the newly discovered circulation of the blood, but, seeing no possibility of an academic career, he went to Amsterdam, where he soon built up a large practice.

However, in 1658 Sylvius was called to Leyden as professor of medicine, where he occupied the chair for fourteen years till his death. He was not only an experienced anatomist and an outstanding clinician, but also a gifted teacher and an eloquent orator. Soon his fame spread and attracted many foreign students from Great Britain as well as from Northern and Central Europe.

With great enthusiasm he devoted himself also to clinical teaching. He even asked permission from the Curators to give clinical lectures every day. He himself made the postmortem examination of patients who had died in the hospital, but the audience was so numerous that he asked and obtained permission for two additional galleries to be constructed in the rather small room situated next to the entrance to the hospital, which was reserved for this purpose. Theoretically, Sylvius was a convinced iatrochemist, and, in fact, he was the most famous exponent of the Iatrochemical School. He believed in treatment based on his theoretical system, and he prescribed so many expensive drugs not included in the official list of medicaments to be administered that the regents of the hospital complained to the Curators of the University (1656). However, the Curators shielded their prominent professor and asked the regents not to discourage Sylvius in his activities in the hospital: they agreed to pay an additional sum of 120 Dutch guilders yearly to cover the extra cost of his drugs.

So Sylvius went his way; he reintroduced the practice of interrogating the students at the bedside, and he is said to have done this in a true Socratic manner, leading the student to diagnosis and prognosis step by step.

During the last years (1670-1672) of his life Sylvius gave the clinical instruction jointly with his former pupil, now his colleague, Lucas Schacht (1634-1689). Schacht acquitted himself of this duty in a praiseworthy manner. However, after his death in 1689, a period of decline and great difficulties for the faculty set in. At that time the man who was to bring new fame to the University and who was to establish the medical curriculum on modern lines, was still a modest student of philosophy and divinity. That man was Boerhaave.

In fact, Herman Boerhaave (1668-1738), the son of a minister, born in the parsonage of the small village of Voorhout near Leyden, descendant of a South-Flemish Huguenot family, was originally destined to follow the wishes of his father, who had intended that his gifted son should become a clergyman. However, Boerhaave conceived his studies broadly. A graduate in philosophy in 1689, on the advice of a fatherly friend, the secretary to the Curators, Jan van den Berg, he applied himself to the study of medicine in order to become later a preacher and at the same time a physician, as was not very unusual at that time, especially among the Mennonites. Well-trained in independent studies, he attended no medical lectures, but only the public dissections. After two and half years at Leyden he graduated in 1693 at Harderwijk, where the cost of graduating was lower. An incident in a canal boat, as a result of which he was suspected of Spinozism, made him give up a career in the church and he settled as a physician at Leyden. After eight years, in 1701, the Curators, who had not succeeded in finding a worthy successor to Drélincourt, appointed Boerhaave as a "lector"—a reader—for three years.

This appointment, and that of Albinus the Elder in the following year, brought about a turn in the low tide of the Faculty's existence, and students entered in increasing numbers. The teaching post of "lector" was next in importance to that of professor, and a "lector" delivered a public address. In his address Boerhaave recommended the study of Hippocrates. His duty was to teach the Institutes of Medicine, an introduction into medicine on the basis of anatomy, physiology and pathology.

From the first Boerhaave showed himself a born teacher. In the years following his appointment foreign students asked him to deliver lectures on anatomy and chemistry, subjects which were both neglected at that time. Bidloo, the anatomist, was often absent in pursuance of his duties as physician to the king-stadtholder William III, and Le Mort, the chemist, had a grievance against the University and often did not give his lectures. Boerhaave, well versed in these fields and convinced of their importance as basic sciences to medicine, gladly accepted. He gave up lecturing on anatomy when Bidloo returned to his academic duties after the death of William III (1702). However, he continued private lectures on chemistry until the death of Le Mort in 1718, after which he was appointed as his successor.

In 1703 the University of Groningen invited Boerhaave to accept a post as professor. The Leyden Curators, afraid to lose their gifted "lector", but unable to appoint him as professor, since the statutory complement of professors of the Medical Faculty (five) was complete, promised him the first vacancy in the Faculty. Probably in order to seal their promise and to show their intention to the world, Boerhaave was allowed to deliver an academic oration. It was at the end of this formal address that the young "lector" unfolded the scheme of the medical curriculum as it is still carried on.

Boerhaave, now nearly thirty-five years old, spoke on the use of mechanistic reasoning in medicine. In the controversy between the Iatrochemical and the Iatrophysical Schools he was definitely on the side of the latter. Physics at that time was a more developed and mature science than chemistry, which was still mainly practiced in apothecaries' shops. Boerhaave argued that "of two physicians, both equally practised in their profession, the one more fitted to advance his science is that one who, more than the other, is experienced in the principles of mechanics." And then he described the ideal physician and the steps of an ideal education for a physician.

The first stage was entirely dedicated to mathematics and natural science. Boerhaave saw the young student "absorbed in the contemplation of geometric figures, weights and velocity, the construction of machines and their effects on bodies around them." He studies "the various characteristics of fluids: their viscosity, elasticity, volatility, weight and adhesion; as described in the discipline of Hydrostatics." He makes experiments in Hydraulics, Mechanics and Chemistry.

"A second scene unfolds him to us, in which he is already within the sacred halls where medicine itself is practised." "There we see him silently direct his gaze, sharpened by the light of geometric studies, on opened corpses and the bodies of animals opened during life." "With the help of discoveries made by the great efforts of others, he forms a clear conception of the structure of the human body."

"Next, he sets himself to studying the vital fluids, which, within the body as well as extracorporally, he submits to a careful examination, using all the means made available to him by anatomy, chemistry, and hydrostatics, and with the aid of the microscope." In this way "he can make a complete survey of all the phenomena which the body in normal condition displays."

"With the aid of these data . . . he becomes able, slowly but surely, to disclose truths which, although embodied in those data, are not perceptible to mere senses, but are only revealed by the processes of logical deduction."

In this manner he traces the immediate causes of every effect, and as a result of all his investigations "he forms a clear and complete conception of the nature of these causes."

After having faithfully completed these studies, we see our physician "striving after the ultimate goal." "Now he advances to the Holy of Holies: to the inner chamber of the Aesculapian temple." From the Greek medical writers "he garners the precious facts with which their works abound."

He performs post-mortem examinations, produces diseases in animals artificially: "gathers, from personal experience, a variety of data concerning the effects of diseases and medicines; at other times he supplements the knowledge thus obtained, by consulting the best writers in the field; finally he arranges all these data in a carefully considered pattern, which compares with what Theory has taught him, so that ultimately he acquires a good understanding of the course of various disesases and their treatment."

In this way Boerhaave concluded his portrait of the "perfect medical man", and he believed that he had demonstrated adequately that this level could not possibly be attained without the study of mechanics, to which he, the iatrophysicist, attached undue importance.

We discern clearly the stages of the modern medical curriculum: the propedeutics of mathematics and natural science, the study of normal anatomy and physiology, and finally the study of pathology and therapy.

Undoubtedly this scheme was new. It was of paramount importance that Boerhaave demanded a firm basis of natural science as the first stage in the study of medicine. He saw clearly that only in this way could knowledge of the normal and pathological processes in the human body be extended, and he declared "that even the slightest progress made is a decisive step forward, and a firm basis for further advances."

However, when this scheme is carefully considered it is seen to be an image of Boerhaave's own medical studies. Nowhere is there any mention of a teacher; this ideal physician finds his own way just as Boerhaave did himself!

Moreover, though there is mention of personal experience of diseases and medicine, the patients and the bedside training and teaching are not mentioned *expressis verbis*. This is the more striking as one of Boerhaave's greatest merits was that he gave its proper place to clinical instruction, in which he excelled above all his contemporaries.

But Boerhaave, as far as we know, did not himself receive any clinical instruction, since it was given very poorly, if at all, when he was a medical student. When he delivered this oration he was still developing mentally and he probably had not yet obtained a clear conception of its irreplaceable value.

As for myself, I believe it to be true that, if Boerhaave had drawn up the medical curriculum twenty years later, he would not have omitted to stress the importance of clinical instruction. It was only after he had been a professor for some years that he had the right idea.

In January 1709 the severe cold of the winter was fatal, not only to many plants, but also to the professor of botany, Petrus Hotton (1648-1709). So, six years after this oration, the Curators had the opportunity to redeem their promise, and they did so without delay. Boerhaave was appointed professor of botany and medicine. By that time he was already making an international name for himself not least by his two textbooks, the *Institutiones medicinae* and the *Aphorismi de cognoscendis et curandis morbis*.

The appointment forced him to find his way unaided in botany, a science that he had neglected for some sixteen years. Five years later we see him as Rector Magnificus of the University, as the Vice-Chancellor is called in Holland. In this year the Curators, after the death of Bidloo, entrusted the clinical teaching to Boerhaave, and appointed him also professor of the *collegium medicopracticum*. It is highly probably that this occurred at his own request. His biographer Matthieu Maty (1718-1776) relates that Boerhaave obtained the reopening of a hospital which had long been closed to the students (33). The truth seems to lie in the fact that in previous years bedside teaching had again been neglected by the professors. After the death of Bidloo in the spring of 1713 only Frederik Dekkers (1644-1720) remained to give bedside teaching, but apparently the old man had not carried it out, for, at their meeting on 8 August 1714, the Curators stated with regret that after Bidloo's death the *exercitia,* that is, clinical instruction, had remained completely stagnant (34).

It is to the lasting merit of Boerhaave that he revived bedside teaching and brought it to an unprecedented level of excellence. Till his last illness, for nearly a quarter of a century, he taught clinical medicine in the Caecilia Hospital, and it is a testimony to his diligence than on 16 April 1738, during such a lecture, an attack of dyspnoea forced him to interrupt it and thus to end his long term of bedside instruction.

From 1719 Boerhaave was assisted faithfully by Herman Oosterdijk Schacht, a stepson of Lucas Schacht (35).

A historian (36) gives a vivid impression of such a clinical lecture. The audience stood behind a balustrade in front of a slightly elevated gallery. The professor made his ward round, but lectured especially on one patient; in the following lectures he discussed the change in signs or symptoms and the course of the disease of that patient. The students wrote down rapidly the words of the professor, and Gerard van Swieten using his own shorthand did this for some twenty years. After the lecture Boerhaave, followed by some pupils, used to go to the room of the master and of the matron of the hospital to give prescriptions for drugs and his instructions relating to the diet. When a patient had died, he often performed the post-mortem examination himself. How skilled he was in this way may be gathered from his excellent description of his autopsy on the Count of Wassenaer, in which Boerhaave, for the first time, observed the presence of a spontaneous rupture of the oesophagus (37).

We may wonder how a clinical teacher, who could not of course at that time employ percussion, or auscultation, or laboratory tests, could talk for an hour about one and the same patient. Yet, anyone who reads the students' notes of Boerhaave's lecture on the *cachecticus* soon learns how he taught his pupils to observe exactly the patient and his symptoms. He was a master at describing the external appearance of the patient, his skin, his breathing, his pulse, and the sounds that he made on breathing. And further, he was by no means a therapeutic nihilist, for on the contrary he had belief in his own prescriptions. With great care he ordered the diet, physical methods such as massage and foot-baths, and a choice of drugs, especially medical herbs which he himself also took in his illnesses.

I do not believe that Boerhaave's influence on the medical curriculum is to be derived from or can be explained by, his description of the ideal physician, as given in his oration of 1703. It was the man behind the description who influenced his pupils to follow the brilliant example he himself gave (38).

Indeed, for many years Boerhaave himself taught enthusiastically at the different levels of medical education that he had designed. For some ten years, from 1718 until 1729, when the state of his health caused him to resign from his professorships of botany and chemistry, he filled three chairs simultaneously. He lectured every day from three to four hours, on botany, chemistry, the institutes of medicine, and on *praxis medicinae,* and he also gave clinical instruction. In those years Boerhaave was a Faculty in himself. It is curious that, though he was a convinced iatrophysicist and opposed the Iatrochemical School, he applied himself to chemistry day and night, and in his course of chemical demonstrations he was the first to introduce biochemistry in the education of the students.

In this way he attained to a balanced scheme of medical education even more completely than he had envisaged and described it long before. By his living example and never failing industry he maintained his teaching at a high level, and he showed to the world what medical education could and ought to be.

His extraordinary gift for teaching and his eloquence made an unforgettable impression upon his numerous pupils, who came from all parts of Europe.

Burton (39), in his biography written five years after Boerhaave's death and some twenty years after his own period at Leyden, repeatedly speaks of "our Professor". Haller, who came to Leyden in 1725, confessed that he was unbelievably delighted when he first heard Boerhaave lecturing (40); and even when Haller had long been a professor himself, he still used to refer to Boerhaave as his "Praeceptor".

Haller once called Boerhaave *communis Europae Praeceptor*—the common teacher of the whole of Europe—and this certainly was not saying too much. It is also true that Boerhaave himself never intentionally sought to be the preceptor of Europe, for he never left his own country, and his farthest journey was to Harderwijk. But his textbooks went to Europe; translated or in the original Latin, they were surreptitiously reprinted in several countries such as England, France, Germany and Italy. And further —Europe came to him. His audience was, to a great extent, international.

Till the end of his academic activities he always had about a hundred students at his lectures, more than half of whom were foreigners. During the years of his teaching, 1701 to 1738, more than two thousand medical students enrolled at Leyden (41). Among them there were gifted young men who came especially to Leyden in order to hear Boerhaave and many of them later became leading figures in their own countries (42).

It is the tragedy of medical education in the Netherlands in the eighteenth century that after Boerhaave's death clinical teaching soon collapsed. Oosterdijk Schact carried it on for a time until his death in 1744; thereafter no bedside demonstrations were given, although they continued to be announced in the syllabus, but with the significant restriction *quousque licet*. Neither Van Royen (1704-1779) nor Gaubius (1705-1780) was in the least interested in clinical instruction. Johann Georg Zimmerman (1728-1795) found neither clinical professors nor patients in the Caecilia Hospital in 1753.

This period of decline lasted more than forty years. The 'cri de coeur' of the historian (43) who described the glorious days of the flourishing period of bedside teaching ended in a public appeal to the Curators of the University. In fact, it was the Curators who, on 1 June 1786, resolved that clinical teaching should be reintroduced. In the following year the old Caecilia Hospital was restored for the nursing of twenty patients, and before the end of the century a small university hospital was fitted up in an old patrician house near St. Pieter's Church.

However, in the first decades of the nineteenth century the spirit of medicine, in Holland as in other European countries, was inclined to speculation and was divorced from observation and experiment. Neither the microscope nor the clinical thermometer was used. In fact, this period was to last until, under the influence of renewed interest in the natural sciences, medical education also experienced a new birth and clinical instruction found again the place in medical education to which it had been first raised by Boerhaave (44).

NOTES

1. J. E. KROON, *Bijdragen tot de Geschiedenis van het Geneeskundig Onderwijs aan de Leidsche Universiteit 1575-1625.* Leyden, 1911, p. 23.
2. P. C. MOLHUYSEN. *Bronnen tot de Geschiedenis der Leidsche Universiteit.* Vol. 1, The Hague, 1913, pp. 51, 140*. The Garden was, and still is, situated on a plot of ground behind the University Building.
3. In 1592 Pieter Pauw was appointed ordinary professor of medicine and botany, and was charged with the supervision of the Garden. However, soon afterwards, in 1593, the Curators succeeded in attracting the famous Carolus Clusius (1526-1609) to Leyden. Though he was a cripple, because of a dislocation or a fracture of the femur, he did much for the enrichment of the Garden. He was assisted by Dirck Outgerszoon Cluyt, who was a pharmacist, first at Delft and later at Leyden, and had a great knowledge of medical herbs.
4. On 24 August 1587 the Vice-Chancellor drew the attention of the Curators to this fact.
5. MOLHUYSEN, *op. cit.*, p. 166*.
6. The anatomical theatre was erected in the old Faliede Bagijnen-Church, where the University had been housed from 1577 until 1581. After 1591 it served as a library and for the teaching of anatomy.
7. This was said in the resolution appointing Peter Pauw, 8 February 1589. KROON, *op. cit.*, Bijlage VI, p. 131. In 1589 the number of medical students at Leyden could not have exceeded twenty. In 1586 eight students matriculated, in 1587 not a single one, in 1588 six and in 1589 five students.

8. The first medical graduate was Joachimus Simonis from Friesland; he took his degree on 12 February 1580.
9. In 1589 the Curators decided that no one should receive the doctor's degree who had not reached the age of 28 years. Later this age was lowered to 24 years. It is possible that this regulation explains the fact that several students of Leyden went to France (for example to Rheims) to take their degree.
10. This proposal was continued in a comprehensive programme of the studies in the various Faculties, which was drawn up by Guilielmus Feugueraeus, professor of Theology, MOLHUYSEN, I, *op. cit.*, pp. 39*-42*. In it the medical curriculum is summarized as follows: "*Restat fidelissimus ille naturae administer medicus qui paulo diversa a reliquis duobus ratione a nobis informatur instituiturque: auditioni quidem plurimum temporis tribuit toto studii medici curriculo; sed primis eius annis declamationi et disputationi aliquantulum, corporum vitalium, vegetabilium et metallicorum inspectioni, dissectioni, dissolutioni ac transmutationi multum, postremis vero tanti exercitii annis, tribus prioribus sudoribus, quartum longe uberioris fructus adiungit, nempe peritissimi simul et doctissimi medici individuam acoluthiam, quo duce Hippocrates, Galenos sua cuique morbo remedia adhibentes videt, admiratur, imitatur: ac denique tum demum doctoris insigni titulo ornatur, cum non aegrotantium periculo, sed magistri censorio labore dignum se in conservanda aut recuperanda valetudine naturae ministrum probet et patefaciat.*"
11. These regulations were drawn up by the Senate on 20 July 1576. The medical examinations were described as follows: Molhuysen I, p. 48*:
    "*Forma Examinis, qua Doctor Medicinae titulo donandus creabitur et in collegium Doctorum adsciscetur.*

    Qui in nostra academia instituti laureae testimonium petunt, specimen suae eruditionis edent, binis publicis disputationibus ante promotionem habitis. Quibus absolutis, tandem Hippocratis Aphorismus privatim interpretandus proponetur ut intelligatur quam habeant in universalibus methodum. Post vel ex Hippocrate vel ex Galeno alicuius morbi curatio explicanda praescribetur, ut intelligatur quem usum in singularibus habeant, praemissa oratione rei institutae accommodata. Post interpretationem nonnulla ab examinatoribus obiicientur ad quae illis respondendum erit. Si satisfacient eligant thesim publice defendendum. Qui docti accedunt, se primum probent Aphorismi interpretatione, morbique particularis explicatione, eo quem paulo ante diximus modo, deinde publica disputatione delecti thematis."
12. It is certain that in the first half of the seventeenth century disputations were also held in the Medical Faculty. On 4 February 1640 the English student ROGER DRAKE defended *Theses de circulatione naturali seu cordis et sanguinis motu circulari pro Cl. Harveio*, with Professor Walaeus in the chair, and in Utrecht, in June 1640 there was also a disputation on the circulation of the blood by Johannes Haymann, with Professor Regius (de Roy) in the chair. The disputations or theses were printed as broadsheets.

13. This examination occurred secretly after Professor Pieter Pauw had made objection to the promotion of Laurens Brant publicly on 6 July 1592. The professors were obliged by oath to keep secret an unfavorable result of this first examination. The student was interrogated on the *Institutiones Medicinae* and further on the whole of medicine, especially therapeutics (*methodus medendi*), "Aengaende de medicynen in institutionibus medicinae, voorts Summarie per totam Medicinam ende bysonder in Methodo medendi" KROON, *op cit.*, p. 116, MOLHUYSEN, *op. cit.*, I, p. 215*.
14. In his diary Haller deals with the graduation of his countryman Waldkirch on 19 July 1725, but he gives a rather different account of the two final examinations. It would seem that Haller's account is incorrect. See G. A. LINDEBOOM. *Haller in Holland.* Delft, 1958, pp. 49-50.
15. Originally every newly created doctor received an authoritative book (the Bible for graduands in divinity, the Corpus Iuris for graduands in law, a book of Aristotle for graduands in philosophy, and a book of Hippocrates for graduands in medicine), a golden ring and a silk hat (*holosericus pileus*). See MOLHUYSEN I, p. 48*. How long this custom was carried on is not known.
16. Wednesday was reserved for public disputations and for graduations.
17. MOLHUYSEN, *op. cit.*, I, p. 157*.
18. *Ibid.*, I, p. 191*.
19. Pieter van Foreest (1522-1597) walked during the opening ceremony in the cortège of professors, but, after pronouncing a public address, he returned to his practice in Delft within a few days. Quite probably no more had been expected of him.
20. E. GURLT, *Geschichte der Chirurgie und ihre Ausübung*, III, p. 287. Perhaps Gurlt borrowed this information from G. D. J. SCHOTEL, *De academie te Leiden.* Haarlem, 1875, who, on p. 145, apparently confused Jan van Heurne with his son Otto (see later); cf. KROON, *op. cit.* p. 44.
21. KROON, *op. cit.*, p. 45.
22. The young Pieter Pauw, being only an extraordinary professor of anatomy, may be excluded as the originator of the plan.
23. Willem van der Straeten (Stratenus) had been physician to the municipal hospital at Utrecht since 1621, and he lectured on anatomy to the surgeons. That he did indeed give bedside teaching appears from ALBERT KYPER, *Medicinam rite discendi et exercendi methodus.* Leyden, 1643, p. 215. Kyper states that Van der Straeten "... *diligentem curam in Nosocomiis aegris suis praestat, atque diaetae accuratissimam rationem observat, verum etiam ubi aegrotum astante Studiosorum corona examinaverit, post ex historia istius aegri solvit quaestionem, quis sit morbus, quae ejus causae, ut & symptomatum, quae prognosis, quae eis debeatur curatio, atque historiam ita tractatam publica disputatione ventilari permittit.*" So Van der Straeten allowed the students to use the case-history of a patient whom he had demonstrated for a public disputation—certainly an excellent method of providing clinical instruction.
24. MOLHUYSEN, *op. cit.*, II, p. 312*.

25. *Ibid.*, II, p. 314*.
26. *Ibid.*, II, p. 209.
27. *Ibid.*, II, p. 260. From this resolution of the Curators of 13 August 1641 it appears that, at that time, the clinical lectures were held at 9 and 3 o'clock. It is not clear whether on Wednesday and Saturday two lectures were held, or that for example, the lecture on Wednesday was held at 9 o'clock and on Saturday at 3 o'clock.
28. *Ibid.*, IV, p. 16*. The resolution of the Curators dates from 18 February 1686, and was taken on the advice of Lucas Schacht.
29. Kyper, *op. cit.*, p. 256. "*celeberrimus Dn Ottho Heurnius, Med. Practicae, Anatom. ac Chirurg. Professor, ac Praeses Collegii Practici publici, initio cum Nosocomio praeficeretur, ut discentium profectibus melius consuleret, ipsis interrogandos aegros exhibuit, atque ordine ipsorum sententiam de morbo, causis ejus & Symptomatum, prognosi, ac curatione quaesivit, suamque sententiam ultimo loco exposuit, sed quoniam iste mos plerisque non placebat, prudenter eo abstinuit, atque suam tantum sententiam de morbo cum curatione nunc exponit: tamen certus sum, eundem Clarissimum virum adhuc ejus animi esse, ut si Studiosorum vota priorem methodum rursus expetierint, promptissime ipsis gratificaturus sit.*"
30. MOLHUYSEN, *op. cit.*, II, p. 215.
31. *Ibid.*
32. See E. D. BAUMANN. *François de le Boë Sylvius.* Leyden, 1949.
33. M. MATY. *Essai sur le charactère du grand médecin ou éloge critique de Mr. Herman Boerhaave.* Cologne, 1747, p. 131 "*Il obtient que l'on rouvrit un hôpital de malades, qui avoit longtems été fermé aux Etudians.*" Maty studied at Leyden from 1732 until 1740; he took, as Boerhaave had done, degrees in philosophy and medicine.
34. G. C. B. SURINGAR. De Leidsche Geneeskundige Faculteit in het begin der achttiende eeuw. *Ned. Tijdschr. Geneesk.* 1886 (2), p. 1-39, note 20.
35. Herman Oosterdijk Schacht (1672-1744) lost his father (named Oosterdijk) at the age of three months; his mother, Anna van Pool, married Lucas Schacht as her second husband in 1675. The name Schacht was added to Herman Oosterdijk's original name.
36. DANIEL VAN ALPHEN, in FRANS VAN MIERIS, *Beschryving der stad Leyden.* III, Leyden, 1784. Aanhangzel p. 47*.
37. H. BOERHAAVE. *Atrocis, nec descripti morbi, historia.* Leyden, 1724, Photostatic reprint. Introduction by G. A. Lindeboom and French translation. Nieuwkoop, 1964. (Dutch Classics on History of Science, Vol. IX).
38. See also for Boerhaave's influence I. SNAPPER, *Meditations on medicine and medical education.* New York, 1956, pp. 50-85.
39. WILLIAM BURTON matriculated at Leyden on 6 September 1724, when aged 21 years. He published anonymously. *An account of the life and works of Herman Boerhaave*, London, 1743; he died at Yarmouth in 1756.
40. "*. . . incredibili voluptate memini me perfusum esse, quando primum veriorem medicinam amoenissima eloquentia ornatum proponentem*

audivi", HALLER, *Bibliotheca medicopractica*, IV, p. 142.
41. J. E. KROON, Boerhaave as professor-promotor, *Janus*, 1918, 23: 291. According to my count, in total 2358 medical students enrolled at Leyden from 1701 to 1738 (from 1701 to 1708: 401).
42. Of the five Americans who studied at Leyden in Boerhaave's time, only two practiced medicine in America, but without developing much influence. Samuel Nicholson remained in Europe. Isaac Dubois died after a few years, and William Bull gave up medicine. Lewis Johnston, who matriculated at Leyden in 1729, took his degree at Rheims (1732) and practiced later in Savannah, Georgia; he became a member of the King's Council. John van Bueren, who must have studied at Leyden in the early 1730's practiced some 30 years in New York; in 1736 he became physician to the Infirmary of the Public Work House. See H. E. SIGERIST, Boerhaave's influence upon American medicine, *Memorialia Herman Boerhaave*, Haarlem, 1939, p. 40.
43. DANIEL VAN ALPHEN, *op. cit.*
44. G. A. LINDEBOOM, *Herman Boerhaave. The man and his work.* London, 1968.

THE DEVELOPMENT OF BEDSIDE TEACHING AT THE VIENNA MEDICAL SCHOOL FROM SCHOLASTIC TIMES TO SPECIAL CLINICS

ERNA LESKY
Universität Wien
Vienna, Austria

There are two reasons for my choosing bedside teaching as the topic of my paper at this symposium on the History of Medical Education. First, the perfection of this very important form of medical instruction is one of the culmination points in the rich history of the Vienna School of Medicine. We may justly say that the type of clinical instruction developed by that School during the second half of the eighteenth century became the general pattern for European clinics. As it is indeed of more than merely regional interest, the emergence of this widely imitated model may claim the attention of an international forum. All the more so as from the last decades of the nineteenth century onwards thousands of American physicians have profited from the final outcome of this development, the formation of specialists, which has thus had stimulating effects in the United States. Secondly, it seems to me that up to now little has been known of the local effort of the Vienna Medical School to establish bedside teaching even before the time of van Swieten (1700-1772). It is of considerable interest and offers new insights into the structure of late scholastic teaching.

Turning now to those earlier stages of practical teaching in Vienna, we must, however, take account of the fact that the development was neither steadily nor continuously aimed towards a clearly conceived goal. Surely enough, external circumstances produced the obstacles. War, plague, and religious struggles more than once endangered the very existence of the university after its establishment in 1365 (1). Instead of a uniform development, there were three phases of practical instruction to be distinguished, each reflecting the impulses active in the European history of science.

The first of these preliminary phases coincided with the late Middle Ages, when the universities were still dominated by scholasticism and instruction at the medical schools was directed by an essentially uniform canon derived from classical and Moslem tradition (2). Usually our handbooks of medical history distinguish this time of authoritative, dialectical teaching centered on the textbook as "book-medicine" from the later periods of "empirical medicine" based upon observation and experience.

However, the statutes of the medical schools of Montpellier (3) and Paris (4) of 1240 and 1325 respectively reveal a factor difficult to integrate into this picture of medieval university medicine. E. Wickersheimer (5) has already pointed out that even then candidates for the licenciate or doctorate were required to prove that they had undergone practical medical apprenticeship for a prescribed period of time. E. Th. Nauck (6) found similar articles in the statutes of most German universities of the fourteenth, fifteenth, and sixteenth centuries. Wickersheimer (7) concludes that since they are of constant presence and general validity: "There can be no doubt that the Middle Ages realized the importance of practical teaching for the training of physicians." Nauck arrives at a similar conclusion. Both scholars demand a revision of the opinions current since the well-known books of Th. Billroth and Th. Puschmann (8) in which, in their extreme formulation, it is maintained that "practical medical instruction was done independently of the universities."

The historian will ask if these articles in medieval university statutes remained a theoretical demand or whether they reflected actual conditions. Source material for investigation is scarce, yet it is available. In 1909 Wickersheimer (9) chanced to obtain the account of a one-time Parisian student in Wolfenbüttel, from which he was able to prove that around 1400 the Paris masters took their students to the bedside.

What about Vienna? Also in Vienna we find the following passage (10) on practical instruction already in the first statutes of the Medical School of 1389—obviously an imitation of the Paris model: "Item, we decree that everybody who wants to attain the licence or doctorate of medicine has to visit the sick for the practice of medicine during at least one year, and this in the company of a doctor of the same school." Against other statements to the contrary that question the connection between the university and practical teaching, this Viennese formula clearly states: Bedside teaching is an essential element of the university instruction. A student has to devote to it a certain stage of his studies, and it is to be offered not by some practicing physician but by a teacher of the medical school itself.

As to the question whether or not practical courses actually took place, there is that dramatic incident of 1455 preserved in the Acts of the Medical School (11). Already at that time, the students were apparently displeased with compulsory lectures on the *Articella* or the *Canon* of Avicenna and the commentaries on it. They would rather practice at the bedsides, *visitare practicam*, as we read again and again in the Acts. On 30 April 1455, the faculty decreed (12) by a majority vote "that no student should attend practical classes before having finished with all the lectures prescribed for the baccalaureate." At a later meeting they defended their resolution professing (13) that they feared lest students, by neglecting classroom lectures, should after graduation "practice like quacks, to the distress of the school, without due system and in ignorance of the canon of dogmata." The students were on the brink of revolt against the faculty, holding the unanimous opinion that their written privileges must be respected, and they even appealed to the university authorities. In the end, the faculty and the students agreed to a compromise.

The Viennese demand of 1455 stipulating *visitare practicam* had its exact parallel in the often discussed incident at the University of Padua (14). After the death of Giovanni Battista da Monte (1498-1551), at Padua, too, the students did not want to miss bedside teaching and fought valiantly for its reestablishment. This struggle, however, did not take place at a conservative, medieval university, but a full century later (1578) at the very school in Italy which was mothering the new impulses of renaissance medicine. The singular significance of the Viennese incident thus lies in the fact that it bears witness to impulses inherent in scholasticism itself which—generally speaking—tended towards observational medicine.

The Vienna material contains hints which allow clearer definition of these impulses. Doctor Martin Stainpeis (1450/60-1527)(15), whose *Liber de modo studendi seu legendi in medicina* (16) we wish to discuss, actually belonged to the next generation. Around the turn of the century, when New Learning entered the medical school in the persons of the physicians, Bartholomaeus Steber, Vadian, and Cuspinian were teaching at the Vienna School (17). However, in his book, written in the second decade of the new century (18), Stainpeis leaves no doubt, when he argues with the modernists (19), that he himself was still rooted in the old teaching method as practiced in Vienna around the middle of the fifteenth century. In this connection he speaks (20) about *practicam visitare* during the fourth year of the curriculum as a matter of course. That this does not mean a revolution or innovation in medical education in the sense that it should be founded on observation or investigation, and that Stainpeis was indeed in full accord with the scholastic tradition, is evident from two facts: (*a*) Stainpeis's authorities were not only Galen, who was held in high esteem by the philologically minded physicians of the Renaissance, but also Rhazes, Mesue and Avicenna, the main figures of scholastic medicine. (*b*) Stainpeis's statement that the student "could see with his own eyes what he has read in the textbooks" proves that he still considered the book as the core and in the first place, while bedside teaching existed only to illustrate what the student had read. With this interpretation of scholastic bedside teaching, a phenomenon which has so far not been placed within its historical context, we are in agreement with H. Grundmann (21), who says of dissection as it had been practiced since the beginning of the fourteenth century: ". . . dissection, however, did not serve to investigate unknown phenomena but only to find confirmation and to demonstrate what had been learned and gleaned from books."

The second phase in Viennese bedside teaching was completely overshadowed by the new Paduan clinical concept (22). Giovanni Battista da Monte developed this new method of looking for the *particulare*, the single phenomenon in the individual case history, which he demonstrated at the bedsides in the Hospital of San Francesco before students from all parts of Europe. For a short space of time it seemed as if Vienna instead of Leyden was destined to be the immediate heir and administrator of the Paduan tradition of bedside teaching. The Vienna of 1550 offered ideal conditions: Ferdinand I was a sovereign sympathetic to reform. There was the reform-minded pupil of da Monte, Franz Emerich (1496-1560) (23) and the Civic Hospital (24), where in 1554 another pupil of da Monte, Martin Stopius (25), had become chief physician. In the same year of 1554 the *Renovatio nova* of Ferdinand I was issued (26), which supplied the university with a new set of guiding rules and was in force until the days of Maria Theresa. It also contained new regulations for clinical instruction, and as L. Senfelder (27) has pointed out they clearly bear the stamp of Emerich's personality, who then held the chair of practical medicine. We read in these regulations (28) that according to the intention of the legislator "it is the task of the professors of the School of Medicine to have the students participate not only in their visits to the hospital, but also to private patients—as far as the convenience of the latter allows for this. The purpose is to put the students in a position to combine theory and practice so that they might thereby become experienced physicians."

The great aim of uniting medical theory and practice had been envisaged and decreed by law. Ferdinand I, however, failed to supply the means necessary for realizing his reform. At first, Franz Emerich, the direct pupil of da Monte, was so sure of his mission that he tried to reach his goal even despite inadequate means. We know from one of Emerich's contemporaries, the Reichshofrat Georg Eder (29), that Emerich did take his students to bedsides. Eder credits Emerich with having for the first time introduced practical instruction in Vienna. We should forgive this error of the lawyer ignorant of earlier efforts in the field of medical instruction. On the basis of the previous remarks historians will correct once and forever this longstanding (30) error.

In 1560 Emerich died. He did not live to witness the decline of his school, which in succeeding times degenerated into a mere preparatory school for Padua (31), while Leyden took over the mission which Emerich had envisaged for Vienna, namely to promote bedside teaching. And it was from Leyden that Gerard van Swieten 200 years later came to establish Vienna as the center of clinical education. But for all the glory that is duly attached to his name and work it should not pass unnoted that thirty years before his arrival the faculty had made another effort to introduce modern methods of teaching in all fields of medicine.

The initiative for the reform program of 1718 (32)—as we may call it—

was assumed by the government, which had become aware of the development of new teaching methods of demonstration at the thriving universities of Germany and above all in the Netherlands in the course of the seventeenth century. The same government that for more than one and a half centuries had done nothing for the faculty and its professors now reproachfully demanded information as to why bedside teaching in Vienna was behind the times. The neglected faculty retaliated with urgent and vociferous demands: Supply us with an anatomical theatre, botanical gardens, chemical laboratories. Hire experienced physicians and assistants for the richly endowed hospitals and, above all, pay adequate salaries to the professors and do not believe that they can subsist on starvation salaries of 110 guldens per annum (33), while the rich funds of the university are amassed at the imperial treasury without serving their proper purpose. If you answer to our demand you will have public bedside teaching with the *status morbi* first explained to the students in a lecture, the appropriate measures deliberated, and finally the patient publicly operated upon by physicians and surgeons in the presence of all concerned. Nor was pathological anatomy neglected in this widely inclusive reform project, when the faculty suggested that "the dead body should be cut open" to demonstrate to the students "what might have been the internal trouble".

In 1843 Anton von Rosas, the chronicler of the Vienna Medical School, deplored the fact that not a single proposal of 1718 was realized. We may, however, assume that when the Hospital of the Holy Trinity was founded in 1741 (34), which 35 years later under Maximilian Stoll became the famous center of clinical education, the proposals of 1718 of the faculty were remembered. The new hospital was not only supplied with qualified chief physicians and assistants, but there were as well *famuli*. These were medical students who practiced under the special supervision of the chief physician for internal diseases, who was to give them "adequate instruction, i.e., to spare no effort for their better instruction, and to offer all the necessary explanations of the important cases for the treatment of patients as well as autopsies or dissections of criminals." Besides, the students themselves had to examine patients, to draw up case histories and to take turns in nursing those who were put under their charge, day and night. It is self-evident that this well-organized system of *famuli* at the Trinity Hospital, which was then the best-equipped hospital in Vienna, marked a definite advancement over the casual visiting of patients practiced before. It was soon to show its beneficial results. Leopold Auenbrugger, who invented the technique of percussion, had been *famulus* at Trinity Hospital in 1746 (35). His was the most important medical achievement accomplished in Vienna during the eighteenth century.

The instruction of *famuli* at Trinity Hospital, it is true, was modern as well as efficient, but it did not as yet represent the type of academic clinical studies developed later. The contact with the Faculty and with the proper science was missing, and there was no professor with special appointment to instruct the students in the practice of medicine. Peter Quarin, at that time professor of practical medicine, the father of the first director of the General Hospital, seems to have been aware of this lack. As a young man, in 1718, he had been among the faculty members who had signed the great reform project. He did the best possible with a government that had not yet realized the necessity of a university hospital. As he was at the same time physician at the Hospitallers he took his students there and managed to give them practical lessons as well as he could (36).

Such was the state of affairs when Gerard van Swieten (1700-1772) arrived from Leyden in 1745. We thought it essential to dwell upon these details as they imply the necessary revision of the conventional picture of medical history (37). Van Swieten's achievement will hereby not lose in greatness, but it will be evident that to the Vienna physicians he did not at all appear as the *deux ex machina* at the scene of bedside teaching. The first decree issued according to his directions in 1747 (38) contained nothing new. It laid down that "the professors should bring their students to the hospitals and almshouses for a period of two years, where they should observe the cases, and by examining two or three of them at a time and by prescribing medicine should instruct the students on diseases and their cures." All this did not progress much beyond the practice at Trinity Hospital and at the Hospitallers and what had already been suggested by the passage on bedside teaching contained in the *Renovatio nova* of 1554. Even in the great reform plan for the School of Medicine of 17 January 1749 (39) van Swieten still kept to this cursory form of bedside teaching. When he defined the new aims as "offering evidence in the practice of medicine for the truth of what the students had learnt theoretically", it seems as if the scholastic Stainpeis were speaking rather than van Swieten, the enlightened pupil of Boerhaave, who in other parts of his program declared that he intended to make all branches of contemporary medical science, i.e., anatomy, physiology, botany, chemistry, and *materia medica*, serve practical medicine as the culmination of medical art. Indeed, during the succeeding years van Swieten was, by energetic efforts, to catch up with the times.

In 1749 van Swieten even contemplated that he himself might conduct practical classes in the above sense (40). It was only in 1753 that he seems to have realized that this was out of the question because of his other duties as physician-in-ordinary to the imperial family, president of the Medical Faculty, director of medical and surgical studies, and, furthermore, entrusted with all matters concerning public health (41).

Now the time had come for carrying out what Ferdinand I and his physician Franz Emerich had planned, but not realized, in 1554: the union of theory and practice in medical instruction.

In 1754 Maria Theresa and van Swieten in a unique way put forward this union, in the *exercitatio clinica viva*. When they founded a lying-in clinic at the Civic Hospital and called Anton de Haen (1704-1776) to Vienna, a man passionately devoted to the cause of medicine, they set an impressive example of how much a country could do for medical education through the authority of absolute monarchy combined with the insight of a great physician with organizing talent. With de Haen's clinic, Vienna became the center of clinical instruction for Europe of the eighteenth century. From numerous contemporary descriptions (42) we are familiar with the clinic's equipment and the manner in which de Haen conducted his practical classes from 8 to 10 in the morning. His institution is usually regarded as an exact copy of the Leyden model, and emphasis is usually laid upon the fact that there were twelve beds, six for male and six for female patients, at which de Haen practiced the subtle examination of patients just as Boerhaave had done with his students. At times historians have overlooked the fact that in some respects the new clinic went beyond Leyden and was the soil in which seeds for further development were planted. This must be attributed to two measures of van Swieten, both of them originating from the century's concern with demonstration and the direct appeal to the senses inherent in contemporary observational and experimental medicine. The first of van Swieten's measures aimed at confronting the students with a variety of pathological phenomena chosen from the greatest possible number of cases. There were only twelve beds at de Haen's clinic, it is true, but their function may best be compared to that of exchangeable frames in a picture gallery. De Haen had been authorized by van Swieten to select from all Viennese hospitals the cases which supplied the most typical examples of diseases and were thus best suited for the instruction of the students (43). In addition, there were plenty of cases to be observed daily in the dispensary of the Civic Hospital, which served the poor. With the second measure, namely the regulation that all Viennese hospitals should supply cadavers for postmortem examinations (44), van Swieten in 1749 prepared the way for the later outstanding development of pathological anatomy. We need only open de Haen's *Ratio medendi* (45), that famous handbook of clinical case studies, to realize that in Vienna pathological anatomy had become a well integrated part of clinical instruction.

The 1754 innovations at the Civic Hospital have still another significance besides the fact that bedside teaching was made an academic subject and that pathological anatomy (=*Anatomia practica*) was added to the curriculum. It meant at the same time the full emancipation of internal medicine as a special field in its own right. When Boerhaave, the teacher of van Swieten and de Haen, started his academic career in 1701, he taught as professor of theoretical medicine. Later on, for years he taught botany and chemistry as well as practical medicine, and only twice a week devoted himself to clinical instruction. De Haen, on the other hand, during the whole time of his activities at Vienna until 1776, had no other function but that of a clinical internist, i.e., professor of practical medicine. He lived as well as taught at the clinic, and instead of teaching there twice a week he did so twice a day having classes in the morning as well as in the afternoon. In his inaugural lecture (46) he declared before the students: "I give myself up to you, and all my being will be devoted and sacrificed to you", a promise which was to inspire his work day after day for 22 years.

What was the personality of this unprecedented clinical teacher at Vienna, de Haen, who rightly deserves this distinction? The fifteen volumes of his *Ratio medendi* supply the answer. Seeker of the truth, a man who passionately strove to settle the problems disturbing his mind, a teacher who in front of his students attempted to distill the generally valid from a variety of cases and, at the same time, to detect the specific characteristics of the individual case, the final aim being the appropriate diagnosis and the effective therapy by using the thermometer, the scalpel and even animal experiments. Given his inborn restlessness, he himself confessed that "in all matters he was asystematic" (47); it was only natural that results assured one day were questioned the next, and who dares blame him for this. Willful, high-spirited, contradictory and erratic, de Haen still succeeded in making his bedside teaching an *exercitatio clinica viva* in the best sense of the word. But because of his asystematic nature the students found it very difficult to follow his extravagant gushes of inspiration, couched, furthermore, in antithetical language. Thus we may understand today the moving apology of his *Ratio medendi continuata* (48) as a sort of testament written four years before his death. He complains about not being understood and in spite of all his efforts being rewarded only with the blame of self-contradiction.

De Haen died in 1776. At the Vienna clinic he was succeeded by his pupil Maximilian Stoll (1742-1787) (49), who had an entirely different character. By the gentleness of his nature, his humanitarian tolerance, and, above all, the clarity of his teaching, Stoll inspired the enthusiasm of hundreds of pupils from all parts of Europe—and also an American from Philadelphia with the name of Franz Joseph Kaufmann (50). He taught at Trinity Hospital, which he turned into the model clinic of the late eighteenth century (51) as regards didactic matters and administration. In 1776, the practical clinic had been moved from the Civic Hospital to Trinity Hospital, which offered better-lighted and larger rooms. Besides, there were already visible the traces of the enlightened spirit characteristic of the rule of Joseph II, who aimed at strict regulation and rationalization in all fields. I should like to give some examples: the technique of examining patients became more refined (52), and instruction centered around six focal points, i.e., anamnesis, diagnosis, prognosis, therapy, and *decursus morbi*, which might be followed by postmortem examination; compared with the practices of de Haen, the progress lay in the stipulation that students must write down their observations in the form of case histories (53). The second significant feature to be noted is the fact that in 1774, for the first time, it was decreed by special regulations that medical studies should be focussed on bedside teaching

(54), and bedside teaching was to become an integral part of the syllabus.

In 1785 a new syllabus decreed by emperor Joseph II himself was issued (55), revolutionizing medical studies by abolishing the doctoral disputations and dissertations and instead introducing bedside examinations and the recording of case histories. The new curriculum was rooted deeply in the utilitarianism of Joseph's time, favoring knowledge that could be put to immediate practical use. For the scientist Stoll, however, the dominating role the case history had come to play in the training of physicians implied something else. He saw new possibilities for medical research in the collection of many descriptions of one and the same disease, and by comparing them he hoped to extract generally valid rules for medical practice (56). When he defined his programme in the words *a particularibus ad universalia* (57) Stoll brought to perfection the method which da Monte had long before envisaged in his *Methodus resolutiva*, namely using inductive methods at the bedside. At the same time, Stoll's words like pointed epigrams implied the complete contrast to the postulations of scholastic medicine, which took the students to the bedside only to offer them the illustrations and confirmation of what had been taught by the theory of *universalia*.

Having surveyed the methodical development of bedside teaching at Vienna, starting with the initial impulses emanating from da Monte and concluding with Stoll's final achievement, it is time to survey in brief the extensive effects. It was a significant feature of the Austrian administration that as a matter of course it permitted the countries under its rule to profit from the cultural progress that resulted from the successful establishment of model institutions in the metropolis. Thus bedside teaching modelled on that of the Vienna clinic was introduced at Pavia in Lombardy in 1770, at Freiburg im Breisgau in 1780, at Prague in 1784 and at the Hungarian university of Pest in 1784 (58). We should also keep in mind in this connection that Corvisart in Paris introduced percussion, as it had been invented by Auenbrugger, the most outstanding achievement of the Vienna school of bedside teaching. Moreover, we must not forget that when in 1790 Fourcroy (59) suggested that clinical instruction should be instituted at Parisian hospitals, he confessed to having been impressed by the model of Vienna, and especially that of Pavia, which we know was patterned after Vienna.

In 1785, at Stoll's suggestion, Joseph II sent Johann Peter Frank (1745-1821) to Pavia to organize bedside teaching (60). At Pavia, Frank further developed the method of Stoll, and in this form it was to return to the metropolis from the Lombard *filia Vindobonensis*, when in 1795, Frank took up Stoll's work at the General Hospital of Vienna. Again Vienna was fortunate enough to gain a clinical teacher of the highest distinction as to pedagogic qualifications and as an organizer, and, in addition, he was the third of this sort within half a century. Ten years earlier the clinic had moved to the General Hospital, founded in 1784 (61). As is well-known, the removal from Trinity Hospital to the new building on Alser Strasse marked a step backwards. Stoll had no longer the energy to make a stand against red tape and an overpowering hospital director, Joseph Quarin.

After Quarin had abdicated, Frank very soon made up for this retrogression. He enlarged the clinic so that it could bed twenty-four patients, and in his dual capacity of professor of practical medicine and director of the General Hospital he was able to exploit for clinical instruction the entire capacity of this 2000-bed city of the sick, the center of clinical instruction. These facts are known well enough from Frank's autobiography (62) and equally that he gained Rudolph Aloys Vetter (1763-1806) as a pathological prosector for his clinic. This marked a decisive step beyond Stoll, who, as de Haen before him, had been his own prosector. Now the emancipation of pathological anatomy within the framework of internal clinical studies was possible due to the demand for a more differentiated correlation of clinical and anatomical findings.

Another instance of specialization within clinical instruction is of equal importance. The syllabuses of 1774 and 1785 contained as a mere suggestion that advanced students as well as young doctors should by themselves treat patients under the supervision of their professor. Frank in his project for the reform of medical studies in Vienna, dating from 31 October 1798 (63), and covering 515 folio pages, made it compulsory and accordingly proposed a five years' study instead of four years, as fixed in 1785. Thus the two last years at school were to be reserved exclusively for clinical instruction, students attending bedside instruction as mere observers at first, then themselves treating the sick under the direction of the professor (64). Frank's proposals remained a draft only. It was not until nineteen years later, in 1817, that he published in the sixth volume of his "System of a Complete Medical Policy" a modified and enlarged version of the principles contained in his project of 1798 (65). In the meantime, in 1804, his principles of clinical education had been introduced into the universities of the Monarchy. What irony that it had been the most fervent opponent of Frank, Joseph Andreas Stifft (1760-1836), the powerful physician-in-ordinary to Francis II, who had realized the essential proposals of Frank's reform project in the very year of 1804, in which he had succeeded in exiling Frank his rival from Vienna (66).

With this reception by Stifft into the medical syllabus of 1804, internal medicine reached a stage in its development of bedside teaching, which represents in a certain respect a culmination point. The regulations of 1810 and 1833 contained modifications and expansions for clinical instruction in surgery and obstetrics, but as to internal medicine, no changes were to be effected until a final phase, which presently will concern us (67).

In dealing with this last phase of Viennese medical history, during the second half of the nineteenth century, we shall concentrate on two major issues. The first concerns the problems in connection with rapidly increasing numbers of students, and the second, the training of specialists. Neither of these problems to be discussed has so far been solved satisfactorily, and they represent today the most controversial points of modern medical education.

In 1798 Frank by reason of his 200 students, had demanded a second clinic for internal diseases at Vienna, and, as a second measure, had introduced classroom lectures. Under the title of "Special Pathology and Therapy of Internal Disease" it survived up to the days of Oppolzer, i.e., 1871, offering not much more than supplements to practical instruction which, as it had been under de Haen, Stoll and Frank, remained centered on the ward and the bedside (68). I quote from a pupil of Oppolzer to give some impression of the teaching method that for more than a century established the fame of the Vienna clinical school: "Oppolzer had all his classes in the ward at the bedside. The students crowded around him hardly giving him enough elbow-room" (69). The rising number of students since the 60's ruled out this private and immediate form of instruction. The syllabus of 1872 consequently introduced the classroom lecture as the center of clinical education (70). As a meagre substitute for the former morning and afternoon instruction at the bedside the following practice was introduced: students were questioned about the cases presented to them. Wagner von Jauregg (71) informs us how little intensive and effective this practice turned out to be: "Heinrich von Bamberger (72) used to present one or two cases in every lecture, asking for volunteers to examine the patient. After having been told the case history the student was supposed to formulate the diagnosis. As most medical students were afraid of giving themselves away or revealing their lack of knowledge it was always the same few volunteers, and as an advanced student I myself belonged to this group."

There is no need for translation into modern terms. The reader will be quite aware of the fact that we are in the midst of the harassing problems of our own days. When in Europe criticism of the great lecture is constantly growing and the demand raised for the substitution of teaching in small groups at the bedside—which has already been realized in the United States —this practice implies nothing else but a return to the sound principles of de Haen's *exercitatio clinica viva,* the development of which has already been outlined in this lecture.

We mentioned the description given by Wagner von Jauregg—and there is a similar one by Billroth—of compulsory classroom lectures in the last three decades of the nineteenth century (73). We should not forget, however, that at the same time there were fields in which the traditional bedside teaching of Vienna reached its climax and exerted its most far-reaching influence. We are thinking of the special fields of ophthalmology, otology, rhino-laryngology, dermatology, and urology. It was the time when every year the professors of the Vienna special clinics together with their assistants held about 275 courses of five to six weeks' duration, the time when Vienna was the Mecca for thousands of young American doctors. It would be necessary to devote an entire lecture to the impact of the Viennese specialist training upon American medicine. Here I must confine myself to a few hints, but I may safely do so as there is the excellent book of Thomas Neville Bonner on "American doctors and German universities" (74) published five years ago, which contains a detailed discussion of this development. All I want to do now is to take from this book one instance that should suffice as an illustration of the wide scope of Viennese specialist training. Because of the increasing number of American medical men that flocked to Vienna, it became necessary in 1904 to found an organization that would take care of them. This was the American Medical Association of Vienna, which still exists. Its records lists the names of 21,000 American physicians (75). In view of these figures we ask ourselves how it was possible and what there was about Vienna which, as if magnetically, attracted thousands of American physicians. An answer may be offered from a statement by the otologist Adam Politzer. When he wrote down the following sentences he was fully aware that with his teaching method he stood within the great tradition of clinical instruction from de Haen to Oppolzer: "This is teaching directly based upon the observation of the sick and free from subsidiary theoretical considerations. By it the symptoms of diseases are indelibly impressed upon the memory of the students, which is absolutely necessary as a factual basis for their later practice. This method should, however, not be regarded as a mere unsystematic description of case histories since it means the critical analysis of symptoms and processes on a strictly scientific basis challenging students to think independently and, if conditions are favorable, even to do research of their own" (76).

It was this method of scientific bedside teaching that made Vienna the "first European place for medical instruction" (Gower) and it also explains what Bonner says, namely that "of all the medical schools of Europe, none had a more widespread and enduring attraction for Americans than Vienna" (77).

BIBLIOGRAPHY

ASCHBACH, JOSEPH, *Geschichte der Wiener Universität*, Vol. III: *Die Wiener Universität und ihre Gelehrten 1520 bis 1565*. Vienna, 1888.

BILLROTH, THEODOR., *Über das Lehren und Lernen der medizinischen Wissenschaften an den Universitäten der deutschen Nation*. Vienna, 1876.

BONNER, THOMAS NEVILLE, *American doctors and German universities. A chapter in international intellectual relations, 1870-1914*. Lincoln, Nebraska, 1963.

FLEXNER, ABRAHAM, *Die Ausbildung des Mediziners. Eine vergleichende Untersuchung, ins Deutsche übertr. von Walther Fischer*. Berlin, 1927.

FRANK, JOHANN PETER, *System einer vollständigen medicinischen Polizey*. Vol. VI, Vienna, 1817. *Supplementbände zur med. Polizei, herausgeg. von G. Voigt*. Tübingen, 1812-1827. 3 vols.

HECKER, J. F. C., *Geschichte der neuren Heilkunde*. Berlin, 1839.

KINK, RUDOLF, *Geschichte der kaiserlichen Universität zu Wien.* Vienna, 1854, 2 vols.

LAIN ENTRALGO, PEDRO, *La historia clínica. Historia y teoria del relato patográfico.* Madrid, 1950.

LERMOYEZ, MARCEL, *Rhinologie, otologie, laryngologie. Enseignement et pratique de la Faculté de Médecine de Vienne.* Paris, 1894.

LESKY, ERNA, Johann Peter Frank als Organisator des medizinischen Unterrichts, *Arch. Gesch. Med.*, 1955, **39**: 1-29.

――― Die Wiener medizinische Schule im 19. Jahrhundert, in *Studien zur Geschichte der Universität Wien*, Vol. VI, Graz-Cologne, 1965.

――― Das Wiener allgemeine Krankenhaus; seine Gründung und Wirkung auf deutsche Spitäler. *Clio Med.*, 1967, **2**: 23-37.

LÖBEL, GUSTAV, Geschichtliche Notizen über das medizinische Clinicum der Wiener Universität. *Wien. med. Wchschr.*, 1871, **21**: 27-29, 31, 32, 36, 38, 39.

NEUBURGER, MAX, *Das alte medizinische Wien in zeitgenössischen Schilderungen.* Vienna, 1921. (Cited as Neuburger I.)

――― *Die Wiener medizinische Schule im Vormärz.* Vienna, 1921.

――― *British medicine and the Vienna school; contacts and parallels.* London, 1943.

PETERSEN, JULIUS, *Hauptmomente in der älteren Geschichte der medicinischen Klinik.* Copenhagen, 1890.

PUSCHMANN, THEODOR., *Geschichte des medizinischen Unterrichtes von den ältesten Zeiten bis zur Gegenwart.* Leipzig, 1889.

ROSAS, ANTON VON, *Kurzgefasste Geschichte der Wiener Hochschule im Allgemeinen und der medicinischen Facultät derselben insbesondere.* Vienna, 1843-1847. 3 pts.

SCHRAUF, KARL, *Acta facultatis medicae universitatis vindobonensis*, Vols. II-III, Vienna, 1899-1904.

SENFELDER, LEOPOLD, *Öffentliche Gesundheitspflege und Heilkunde*, Vols. I-II, Vienna, 1904-1916.

TEMKIN, OWSEI, Studien zum "Sinn"-Begriff in der Medizin. *Kyklos*, 1929, **2**: 21-105.

WARTMAN, WILLIAM B., *Medical teaching in western civilization.* Chicago, 1961.

WICKERSHEIMER, ERNEST, *La médecine et les médecins en France à l'époque de la renaissance.* Paris, 1905.

WIGHTMAN, W. P. D., *Science and the renaissance. An introduction to the study of the emergence of the sciences in the sixteenth century*, Vol. I. Edinburgh, 1962.

NOTES

1. R. KINK, *op. cit.*, Vol. I, pp. 253 ff., 340.
2. HERBERT GRUNDMANN, Naturwissenschaft und Medizin in mittelalterlichen Schulen und Universitäten, *Abh. Ber. dt. Mus.*, 1960, **28** (pt. 2): 24 ff. Also HEINRICH SCHIPPERGES, Medizinischer Unterricht im Mittelalter, *Dt. med. Wschr.*, 1960, **85**: 856-861; J. ASCHBACH, *op. cit.*, I, pp. 319 ff.; E. WICKERSHEIMER, *op. cit.*, pp. 40 f.

3. MARCEL FOURNIER, *Les statuts et privilèges des universités françaises.* Vol. II, Paris, 1891, p. 7.
4. E. WICKERSHEIMER, *Commentaires de la Faculté de Médecine de l'Université de Paris (1365-1516),* Paris, 1915, p. XXXIV: ". . . *quod nullus deinceps ad magisterium valeret promoveri nisi per duas estates practicaverit extra Parisius, vel continaverit per duos annos practicam Parisius in comitatu alterius magistri."*
5. ———— La clinique de l'Hôpital de Strasbourg au XVIIIe siècle. *Arch. Int. Hist. Sci.,* 1963, **16:** 3 f.
6. ERNST THEODOR NAUCK, Zur Geschichte des medizinischen Lehrplans und Unterrichts der Universität Freiburg i. Br. *Beitr. freiburg. Wiss.-u. UnivGesch.,* 1952, **2:** 16, n. 28; Cf. also GERNOT RATH, Die Entwicklung des klinischen Unterrichts, *Göttingen Univ. Reden,* 1965, **47:** 7 f.
7. E. WICKERSHEIMER, *supra,* n. 5, p. 3; also WICKERSHEIMER, *op. cit.,* 1905, p. 42.
8. TH. PUSCHMANN, *op. cit.,* p. 213; also NAUCK, *supra* n. 6, p. 17; W. B. WARTMAN, *op. cit.,* p. 61, even in our time clings to Billroth's error about practical teaching at the university: "that there was as yet no practical instruction in anatomy and at the bedside."
9. E. WICKERSHEIMER, Les secrets et les conseils de maître Guillaume Boucher et de ses confrères; contribution à l'histoire de la médecine à Paris vers 1400. *Bull. Soc. Fr. Hist. Méd.,* 1909, **8:** 199-305.
10. R. KINK, *op. cit.,* Vol. II, p. 162.
11. K. SCHRAUF, *op. cit.,* Vol. II, pp. 79 ff.
12. *Ibid.:* "*quod nullus scolaris, antequam audivit lectiones ad gradum baccalareatus requisitas, visitet practicam."*
13. *Ibid.,* p. 81: ". . . *quod scolares visitando cum doctoribus practicas negligerent lectiones, ad quas obligantur, et neglig[erent] studia in sciencia medicine et post, cum ad gradum promoventur, in scandalum facultatis ut emperici sine debito ordine et sine doctrina canonum in practica procedunt."*
14. ARTURO CASTIGLIONI, Una pagina di storia dell'insegnamento clinico (da Padova a Leida), *Bijdr. Gesch. Geneesk.,* 1938, **18:** 246-258; partly in English translation in W. B. WARTMAN, *op. cit.,* pp. 54 f. See also PETERSEN, *op. cit.,* pp. 41 f.; L. PREMUDA, Die Natio Germanica an der Universität Padua, *Arch. Gesch. Med.,* 1963, **47:** 97-105.
15. K. SCHRAUF, *op. cit.,* Vol. III, pp. IX ff.; R. KINK, *op. cit.,* Vol. I, p. 222; A. v. ROSAS, *op. cit.,* Vol. I, pp. 149-164.
16. This book is referred to in the literature as a very rare one. Its contents are given in great detail in the Berlin medical thesis of KARL-FRIEDRICH MERKER, *Ueber das Studieren in der Medizin nach dem Werke eines Wiener Hochschullehrers zu Beginn des 16. Jahrhunderts,* Berlin, 1930.
17. Cf. J. ASCHBACH, *op. cit.,* I, pp. 1, 89; K. SCHRAUF, *op. cit.,* Vol. III, Introd.; K. GROSSMANN, Die Frühzeit des Humanismus in Wien bis zu Celtis Berufung, *Jb. Landesk. Niederöst.,* 1929, **22** (new ser.): 306.
18. The postscript of the book is dated 21 February 1517. Bound with it is an *Antidotale praeservacionis* (*Finis, anno* 1520) so that for this reason

PUSCHMANN, *op. cit.*, p. 202, gives the year of publication as 1520.
19. STAINPEIS, *op. cit.*, fol. 136v: "*Neque hoc dixi, nisi ut me ipsum consolarer; sunt enim inimici mei obloquentes mihi et detrahentes, sunt alii boni faventes, sunt tertii dicentes: Modus et processus* [i.e., the method of his book] *nihil valet, cum tota intentio sit scriptis vacare doctorum (ut moris est universitatis Viennensis), sed talis superfluus et inutilis est, tum quia textui insistere satis esse videtur, tum etiam quia solius temporis conssumptioneum inutilem affert.*" Cf. K. SCHRAUF, *op. cit.*, Vol. III, p. XIII; O. TEMKIN, *op. cit.*, p. 79.
20. STAINPEIS, *op. cit.*, fol. 17v: "*Practicam visitare. Anno quarto incipiat, ita ut ea quae legerit, ad experientiam etiam oculis videat.*"
21. H. GRUNDMANN, *supra*, n. 2, p. 32. Cf. also the differentiation between scholastic medicine and renaissance medicine which is along the same line, W. P. D. WIGHTMANN, *op. cit.*, p. 223.
22. On the new methodological thinking cf. O. TEMKIN, *op. cit.*, p. 78. Also see V. L. BULLOUGH, *The development of medicine as a profession*. Basel-New York, 1966, pp. 75 f. The *methodus resolutiva et divisiva* is especially treated by GIAMBATTISTA DA MONTE, *Opuscula varia*, ed. HIERONYMUS DONZELLINUS, Basel, 1558, Vol. I, p. 19.
23. See J. ASCHBACH, *op. cit.*, III, pp. 183 f.; A. v. ROSAS, *op. cit.*, Vol. II, p. 70; K. SCHRAUF, *op. cit.*, Vol. IV, pp. XXVII f.; R. KINK, *op. cit.*, Vol. 1, p. 273, n. 327; L. SENFELDER, Franz Emerich, 1496-1560; ein Reformator des medizinischen Unterrichtes in Wien, *Kultur*, 1907, 8: 61-74.
24. L. NOVAK, *Das Bürgerspital*. Vienna, 1820; A. HAUSER, *Das Bürgerspital in Wort und Bild*, Vienna, 1888; K. WEISS, *Geschichte der öffentlichen Anstalten, Fonde und Stiftungen für die Armenversorgung in Wien*, Vienna, 1867, p. 83 ff.
25. L. SENFELDER, *op. cit.*, p. 46; J. ASCHBACH, *op. cit.*, III, p. 275.
26. R. KINK, *op. cit.*, Vol. I, p. 257 ff.
27. L. SENFELDER, *supra*, n. 23, p. 69.
28. R. KINK, *op. cit.*, Vol. II, pp. 379 f.: "*. . . Id quod maxime Medicae Facultatis Professoribus iniunctum esse Volumus, ut nedum in hospitalium, sed et aliorum aegrotorum visitationibus Studiosos eatenus admittant, quatenus hoc ipsum absque infirmorum gravamine fieri potest, quo facilius Theoricae practicam iungere adeoque periti Medici fieri queant.*"
29. G. EDER, *Catalogus Rectorum ad annum* 1538, p. 74, cited by J. ASCHBACH, *op. cit.*, III, p. 185: "*Celeberrimus hic medicus experientia summus, eruditione in sua professione nulli secundus, omnes medicinae partes in hac schola per annos XXV continuo maxima cum laude docuit. . . . Primus veram hujus artis methodum ex Galeno in hanc scholam introduxit; primus optimos quosque authores ipsi Galeno pro luce accomodare coepit; primus auditores ad aegrotos in praxi secum circumduxit.*"
30. This is also true for my information given by E. WICKERSHEIMER, *supra*, n. 5, p. 258, only on account of the literature.
31. K. SCHRAUF, *op. cit.*, Vol. IV, p. XXVII.

32. The original text is printed in A. v. Rosas, *op. cit.*, Vol. II, pp. 232-235.
33. For the sake of comparison it may be noted that Anton de Haen—invited from the Netherlands to Vienna by van Swieten—received a salary of 5000 fl. per annum. The salary of the other professors was limited by van Swieten to 2000 fl. Cf. R. Kink, Vol. I, p. 453, n. 585.
34. Cf. "Die Nachricht von dem Krankenspitale zur allerheiligsten Dreifaltigkeit," printed in A. v. Rosas, Vol. III, pp. 22-64; also J. Knolz, *Darstellung der Heil- und Humanitätsanstalten*, Vienna, 1840, pp. 219-230; K. Weiss, *supra* n. 23, pp. 158 ff.
35. L. Auenbrugger, *Experimentum nascens de remedio specifico sub signo specifico in mania virorum.* Vienna, 1776, p. 19.
36. Valentin ab Hildenbrand, *Ratio medendi in Schola Practica Vindobonensi*, Pt. I. Vienna, 1809, p. 1.
37. From J. F. C. Hecker, *op. cit.*, p. 366, via Th. Billroth, *op. cit.*, pp. 32 ff., to the much used book of J. Petersen, *op. cit.*, pp. 125 f., this description has remained authoritative.
38. G. Löbel, *op. cit.*, p. 283.
39. R. Kink, *op. cit.*, Vol. I, pt. 2, p. 257: "*Pour perfectionner après les estudiants et mesme les jeunes medecines, rien de plus propre que de leur montrer dans un hospital deux ou trois malades à la fois, pas plus, et leur prouver par l'exercise de la medecine la vérité de ce qui on leur aura appris.*"
40. *Ibid.*: "*D'abord que j'auray fini mon ouvrage* [reference to *Commentaria in Hermanni Boerhaavii aphorismos de cognoscendis et curandis morbis*, of which in 1749 only two of the five volumes had appeared], *je veux bien encore me charger de ce travail, surtout pendant les mois d'hyver, que suis en ville.*" During the summer van Swieten was with the court at Schönbrunn Palace.
41. On 27 April 1753 van Swieten advised the Empress to call de Haen from The Hague, cf. H. Wyklicky, Zur Kenntnis des Wiener Klinikers Anton de Haen, *Wien. med. Wschr.*, 1958, **108**: 596-598; J. Boersma, *Antonius de Haen, 1704-1776, Leben en Werk*, Amsterdam [1963], p. 37.
42. M. Neuburger I, p. 5; G. Löbel, *op. cit.*, pp. 662, 687; J. Boersma, *supra*, n. 41, p. 24.
43. J. P. Frank, *op. cit.*, Vol. VI, pt. 2, p. 273, talks in this sense about hospitals as "storehouses of clinical schools without which one cannot do."
44. A. v. Rosas, *op. cit.*, Vol. II, p. 285.
45. Anton de Haen, *Ratio medendi in nosocomio practico, quod in gratiam et emolumentum medicinae studiosorum condidit Maria Theresia.* Vienna, 1758-1773. 15 vols.
46. ——— *Praelectiones in Hermanni Boerhaavii Institutiones pathologicas*, ed. F. de Wasserberg. Cologne, 1784, Vol. I, "Exordium"; translated into German by G. Löbel, *op. cit.* pp. 688 ff.
47. After de Haen had argued against the hypothetical and systematic medicine of the late Baroque period, he presented himself as an observer of nature in the following words cited from J. Petersen, *op. cit.*,

p. 149: "*Interea dum id ab illorum ingenuitate expectamus, in id ego sedulus annitar, ut ad lectores aegrorum, una cum circumstantium Discipulorum Corona, simplicem Naturam contemplando magis intelligam magisque, in omnibus plane Asystematicus.*"

48. Vol. I, Vienna, 1772, "praefatio," translated into Dutch by J. BOERSMA, supra, n. 41, pp. 28 f.
49. J. F. C. HECKER, op. cit., pp. 500 ff.; G. LÖBEL, op. cit., pp. 711 f., 735 f., 762 f.; V. FOSSEL, *Studien zur Geschichte der Medizin*, Stuttgart, 1909, pp. 153-191; P. LAIN ENTRALGO, op. cit., pp. 258 ff.; E. LESKY, op. cit., 1967, pp. 24 f.
50. Franz Joseph Kaufmann from Philadelphia was even Stoll's assistant in 1776-1778, cf. G. LÖBEL, op. cit., p. 857.
51. JOHN HOWARD, *Nachrichten von den vorzüglichsten Krankenhäusern und Pesthäusern in Europa*, trans. annot. by Christian Friedrich Ludwig, Leipzig, 1791, p. 479. FRIEDRICH NICOLAI, *Beschreibung einer Reise durch Deutschland und die Schweiz im Jahre 1781*, Vol. III, Berlin-Stettin, 1784, Beylage VI, 4, pp. 69-78.
52. M. STOLL, *Orationes academicae*, ed. J. EYEREL. Vienna, 1815, p. 58; *De methodo examinandi aegros*, in *Ratio medendi*, ed. J. EYEREL, Vol. VI, Vienna, 1790, pp. 269 ff. According to P. LAIN ENTRALGO, op. cit., p. 263, subjective and objective were more distinctly differentiated by Stoll, although the anamnesis does not purely refer to questioning the sick. Cf. O. TEMKIN, op. cit., p. 55.
53. Cf. the reproach referring thereto of the Anonymus of 1774 against de Haen in M. NEUBURGER I, p. 5, with Stoll's strict demand, *Orationes academicae*, cited in n. 52 supra, pp. 16, 59, and the already exactly defined six rules for writing down anamneses by his pupil V. v. HILDENBRAND, *Initia institutionum clinicarum seu prolegomena in praxin clinicam*, Vienna, 1807, pp. 178 ff.
54. *Instituta facultatis medicae vindobonensis*, ed. ANTONIUS STÖRCK, Vienna, 1775; cf. to this A. v. ROSAS, op. cit., Vol. III, pp. 204 ff., 223 ff.
55. J. P. FERRO, *Einrichtung der medizinischen Fakultät zu Wien*. Vienna, 1785.
56. M. STOLL, *Ratio medendi*, ed. Eyerel, Vol. VI, Vienna, 1790, pp. 291 f.
57. ——— *Orationes academicae*, cited supra n. 52, p. 38.
58. J. F. C. HECKER, op. cit., p. 369.
59. A.-F. FOURCROY, *Nouveau plan de constitution pour la médecine en France presenté à l'Assemblée Nationale par la Société Royale de Médecine*. Paris, 1790, p. 158 f. Even before the revolution, in 1784, WÜRTZ, *Mémoires sur l'établissement des écoles de médecine pratique à former dans les principaux hôpitaux civils de la France à l'instar de Vienne*, Paris, 1784, had vigorously pointed to the example of the Viennese clinical school.
60. J. P. FRANK, op. cit., Supplement, Vol. 1, pp. 162 f. Cf. E. LESKY, op. cit., 1955, pp. 1 ff.
61. E. LESKY, op. cit., 1967, pp. 23 ff.
62. *Biographie des D. Johann Peter Frank von ihm selbst geschrieben*, Vienna, 1802, pp. 47 ff.; M. NEUBURGER I, pp. 119-125; E. LESKY, op. cit., 1955, p. 8.

63. Preserved in the Haus- Hof- und Staatsarchiv, Studien-Revisions-Hof-Commission. Kart. No. 3864 (1801). *Ibid.*, fol. 340.
64. E. LESKY, *op. cit.*, 1955, p. 14. In the article of E. HEISCHKEL, Die Entwicklung des medizinischen Unterrichtes, *Medsche Welt*, Berlin, 1939, **13**: nos. 35-36, one may read how the same principles of teaching affected the surgical-clinical teaching of Berlin University founded in 1810.
65. For the internal clinic cf. especially J. P. FRANK, *op. cit.*, Vol. VI, pt. 2, pp. 258 ff.
66. E. LESKY, *op. cit.*, 1955, p. 14; 1965, p. 37.
67. *Ibid.*, p. 14.
68. *Wien. med. Wschr.*, 1871, **21**: 380.
69. TH. BILLROTH, *Aphorismen zum "Lehren und Lernen der medizinischen Wissenschaften"*, Vienna, 1886, pp. 24 ff., and the diagram on p. 69. In the winter term of 1866 there were 1105 medical students at Vienna University, in 1886 there were 2673. Cf. E. LESKY, Probleme medizinische Unterrichts in der Zeit Billroths, *Wien. klin. Wschr.*, 1963, **75**: 529-532. *Ibid.*, 1965, pp. 293-306.
70. As to the same development at the German universities, cf. E. HEISCHKEL, cited *supra* n. 64, p. 17 of the offprint.
71. J. WAGNER-JAUREGG, *Lebenserinnerungen*, ed. and completed by L. Schönbauer and M. Jantsch. Vienna, 1950, p. 11. A similar opinion on this way of clinical "practitioning" is found in A. FLEXNER, *op. cit.*, pp. 219 f.
72. Heinrich von Bamberger (1822-1888), the successor of Oppolzer and teacher of Wagner-Jauregg in internal medicine was head of the II. Medizinische Klinik that had been founded in 1849—51 years after Frank had suggested it.
73. TH. BILLROTH, cited *supra* n. 69, p. 23.
74. Lincoln, Nebraska, 1963.
75. *Ibid.*, p. 79.
76. A. POLITZER, Die Aufgabe des otiatrischen Unterrichtes, *Wien. med. Wschr.*, 1899, **49**: 377.
77. BONNER, *op. cit.*, p. 69.

MEDICAL EDUCATION IN ENGLAND SINCE 1600*

F. N. L. POYNTER
Wellcome Institute of the History of Medicine
London, England

There has been a good deal of discussion in Britain lately about the so-called "brain drain", through which many hundreds of young British doctors are escaping each year to find more satisfying and better paid employment in the United States. The fact that they are so easily recruited for posts abroad argues that the present system of British medical education is a good one which succeeds in producing competent doctors.

Some British medical teachers might take the view that it is the material rather than the system which is good and that with a better planned and integrated system of medical education these young doctors would be even better. Britain is at present going through one of its periods of doubt and self-criticism. One of the good things which can be said of our democracy is that whatever criticism may be levelled at us from abroad there are always many at home who are ready to go farther. Medical education, from the sixteenth century onwards, has been almost constantly a source of such criticism and dissatisfaction. Over a century ago, when a great deal of time and energy was spent on planning and petitioning for medical reform, there was much talk of a "single portal of entry" by State examination. Today there are still twenty-four licensing authorities, ranging from universities to an ancient guild of the city of London. The medical profession in England, as in most countries, is intensely conservative, with a great regard for its vested interests and its institutions. It has never been clear that criticism made of some sections of it was altogether disinterested or whether it was the attacks of those outside trying to get in, or really, as is often claimed, in the best interests of patients.

Medical education reflects the organization of the profession and its institutions and just as vestigial features are very prominent in the profession in England, so they may be clearly seen in the system of education. This has never been designed and planned as a whole for its purpose. Indeed, no-

---
* N. B. Since the completion of this paper, the Report of the Royal Commission on Medical Education (Chairman, Lord Todd) has been published and has aroused a great deal of discussion by its radical proposals, especially those suggesting the amalgamation of several of the London medical schools.

body is agreed on the purpose, that is, what type of doctor the system is intended to produce. To take an industrial analogy, it is rather as if a great variety of machine tools were assembled from a number of different car factories and linked together in the belief that the ultimate product would be a motor car of some kind, although nobody was at all sure what it would look like or how it would perform. The product does indeed work and does indeed pass the different kinds of inspectors, each of whom is supplied with a different blueprint for his tests.

It is easy to trace three particular lines of continuity in the profession in England from medieval times to the present day. Originally the physician, the surgeon, and the apothecary, the last of the three has now become the general practitioner. Overlaying the three and often interweaving with them are now many other patterns, representing technical and academic specialists, scientific research workers and administrators. Although the consultant physician may become a professor, a research worker or an administrator, there is little chance today of such mobility for a general practitioner, although it is worth remembering that Harvey, Jenner, Hunter, and James Mackenzie were all in general practice. For a hundred years following the Medical Act of 1858 the system of medical education in England was in fact, as Charles Newman has pointed out, designed to produce the "omni-competent general practitioner" (1). Now that the general practitioner represents only a minority of the profession, this pattern is beginning to break down in order to allow early streaming into a great range of specialized courses. Just as in the past however, it is not by curricula alone that we must judge the practical efficacy of medical education, but by how the curriculum is interpreted and what may be done to supplement it in the medical schools.

Certain historical factors have contributed to give medical education in England a specific character which has proved very resistant to change. For many centuries the only universities in England were those at Oxford and Cambridge, where Medicine was not highly regarded or vigorously pursued as an academic training. Both entered the seventeenth century with Regius professors of Medicine and Linacre lecturers; both had statutory rights to claim the bodies of executed criminals for anatomical study; both required the study of some medicine and some science in their Arts degree, followed by the textual study of books by Hippocrates, Galen, and Avicenna for the specifically medical degrees of Bachelor and Doctor of Medicine respectively. Neither had hospitals in which clinical instruction could be given and no formal arrangements had been made allowing students to obtain clinical experience with local practitioners. At Oxford all students entering upon study for medical degrees were required to hold the Bachelor's degree in Arts, so that the complete course, from entry to the university to the award of the doctorate, lasted thirteen years. At Cambridge the requirement of an Arts degree had recently been rescinded, but this encouraged more foreign students to apply for incorporation as Cambridge M.D.'s rather than any marked improvement in medical teaching at Cambridge. However, it was possible for Bachelors of Arts intending to take up medicine to apply for a licence to practise from the university. While getting clinical experience as independent practitioners in the town, they could also study for and obtain their medical degrees. Some successful practitioners (of whom John Symcotts (2) is an example) did not bother to obtain these academic qualifications until very much later when it was to provide the crown rather than the foundation of their careers.

It was only doctors who qualified at Oxford or Cambridge who were eligible for election to the College of Physicians in London (3). This was a disciplinary rather than an educational body and it prosecuted vigorously unlicensed practitioners, who may well have held qualifications of equal value, in order to protect its monopoly of the London practice. As the supreme medical body it was authorised to supervise the activities of the apothecaries and to challenge surgeons who in their practice went beyond their strict limits of treating "outward diseases". In the England of that time—and for long after—social rank and order was everything. The physicians were gentlemen who took a pride in their learning and their university education. The surgeons were not trained at a university. Their formal education, usually obtained in a grammar school, ended at fourteen when they were apprenticed for seven years to a master surgeon, at the end of which term they were admitted to the freedom of their company and could begin to practice on their own account. Apothecaries were also apprenticed and when qualified acted as the intermediaries between the physician and the patient. It was his close contact with patients and his familiarity with the physician's usual prescriptions which became the basis for his own independent general practice, a role which became more and more important throughout the seventeenth century and was eventually given legal sanction in 1704 (4).

Surgeons and apothecaries may also have been granted the Bishop's licence to practice, a qualification dating from the first Medical Act of Henry VIII in 1511-1512, when the Church was made responsible for seeing that only respectable and trustworthy individuals were allowed to treat sick patients. Candidates for this licence were examined by qualified physicians and surgeons in a brief interview. Many holders of this licence were simple and honest empirics who were well known in their neighborhood for their efficacious treatment. Some were women.

The predominant pattern then was one of practical training and experience rather than a university education. The obvious place where both could be combined was the capital city, London, where the royal hospitals, the medieval foundations of St. Bartholomew's, St. Thomas's and Bethlem could have provided clinical training. Unfortunately, London had no university, although Camden and others put forward plans for such an institution right at the beginning of the seventeenth century. The so-called Inns-of-

Court in London were in all but name the only Law University in the country and it would have seemed logical to follow the precedent of Bologna and to add a Medical Faculty to it. The fact that the plans for a university of London were frustrated at that time has left its mark on English medical education up to the present day. When the theory taught in the universities was divorced from practice in the hospitals would-be practitioners knew which was more useful and in the eighteenth and nineteenth centuries were to turn to the hospitals in great numbers.

Meantime, Oxford enjoyed a few decades of greater promise for the future. For the first time, in 1611, the university got a Regius professor of Medicine who really wished to teach medicine. Thomas Clayton was only thirty-six when he took the chair and before his death in 1647 he infused new life into the Oxford medical school. A physic garden was established in 1621 and the new Tomlins lecture in anatomy in 1624. Bartholin's *Anatomy* was printed in Oxford to serve as a textbook and the lecturer was given the right to claim the bodies of executed criminals. Clayton encouraged medical students to become apprenticed to local practitioners and even urged them to carry out experiments. Dr. Robb-Smith (5), the most recent historian of the school, suggests that this temporary revival provided the right setting for the brilliant work that was done in Oxford in the middle decades of the century, when Petty, Wren, Boyle, Lower and Willis carried out work of lasting value and blazed the trail for the foundation of the Royal Society. It was in London, however, and not in Oxford, that this celebrated scientific society made its home, and there followed in Oxford a period of torpor that was only exceeded by the apathy in the Cambridge medical faculty. Many of the Cambridge M.D.'s continued to be incorporated degrees based on study and qualification at a foreign university, usually Leyden.

In the early eighteenth century there was a new trend when many Cambridge students attended St. Thomas's Hospital in London to study under the celebrated Richard Mead. This link between Cambridge and St. Thomas's has continued to the present day. New chairs of anatomy and botany were established at Cambridge but made little impression. Cambridge's long record of success in physiology however has been traced to this same period when Vigani's chemistry laboratory became in effect the first physiological laboratory in England. There Stephen Hales, a Fellow of Corpus Christi College, applied the scientific principles he had learned under Isaac Newton to the problems of physiology and so was able to make the first estimation of the pressure of the blood in the arteries. William Heberden was also a Cambridge teacher for a time, but his outstanding qualities as a medical teacher were not appreciated nor given any scope so that he moved to London. While at Cambridge he wrote the essay on the study of medicine which is regarded as a classic in the history of medical education. A more promising development in both universities was the foundation of voluntary hospitals which were available for use as clinical schools. Boerhaave's methods at

Leyden and their triumphant vindication at Edinburgh had even stirred the complacency of these ancient universities. Addenbrooke's Hospital was built at Cambridge in 1766 and the Radcliffe Infirmary at Oxford in 1770. In order to encourage the use of the Radcliffe as a medical school the university went so far as to alter its statutes, reducing from three years to one the interval between the B.A. and the Bachelor of Medicine degree, with a further three years instead of six for the M.D. In fact this had little effect, for simultaneously the majority of medical students were enrolled at Edinburgh or as private pupils of the leading physicians and surgeons in the London hospitals.

These hospitals trebled in number in the eighteenth century with the foundations of the Westminster, St. George's, Guy's, the London and the Middlesex Hospitals. Mead was already teaching Cambridge students at St. Thomas's when William Cheselden (6), the leading surgeon of his time, began regular courses of instruction in anatomy and physiology. When the continuance of these courses was threatened by the Barber-Surgeons Company, who claimed that dissection could be performed only in its own hall, Cheselden began his campaign for the separation of the Barbers and Surgeons which in 1745 led to the separation of the two by Parliament and the building of a Surgeons' Hall in which teaching could be given. Cheselden's lectures were in fact the forerunner of a long series of distinguished private medical schools, the best known of which was the Windmill Street School (7) founded by William Hunter and where instruction was given by his celebrated brother John Hunter, William Hewson, and William C. Cruikshank.

These schools were in essence an extension of the old apprenticeship system. The voluntary hospitals, founded mostly by laymen, paid no salaries to their medical or surgical staff. Yet these honorary appointments were eagerly sought after, for they brought not only patients but also pupils. These latter were either apprenticed for seven years, with a handsome fee on the signing of the indentures, or enrolled as hospital pupils paying annual fees. The lay governors of the hospitals at first objected that the presence of many pupils interfered with the treatment of patients, but soon came to realise that they acted as unpaid auxiliaries and, in their senior years, as additional medical staff, as well as providing successors to the honoraries. This last fact provided an additional inducement to parents to send their sons to the leading hospital consultants, for if they did well they might inherit not only the hospital appointment but the consultant's fashionable and lucrative practice. This is one of the facts of life in medical education which everybody knows about but never mentions. As such it has persisted to the present time.

Realising that ward rounds or attendance at surgical operations had to be supplemented by formal lectures, the consultant staff of a hospital agreed together to pool their intellectual and financial resources and provide regular courses, either in the hospital itself, if the governors provided a room, or in neighbouring premises. It was in this way that the London Hospital Medical College (8) came into existence in 1786, a school which has been described as the first "medical school of university type" in London, for it provided a comprehensive education in the whole field of medicine and surgery and the ancillary sciences as they were then recognised. On completion of the course, students were provided with certificates of attendance and/or proficiency and could then present themselves for an M.D. at Cambridge or some other university. The Scottish university of St. Andrews was much favoured for this purpose because the M.D. required no residence or even attendance for an examination, but merely letters testimonial from two practitioners of standing certifying the candidate's knowledge and abilities. It was in fact the ease with which medical qualifications and licences to practice were obtained that led at this time to a campaign for medical reform. Young doctors who had worked hard at their studies and whose parents had laid out a good deal of money for their education felt it was unfair that a charlatan with a good bedside manner could compete successfully with them for patients and suffer no penalty. From the middle of the eighteenth century it had also been possible for anybody who had served three years with the armed forces in a medical capacity to set up in practice on returning to civilian life without examination or certificates. The reformers wanted the licence to practice linked with regular training and tests of proficiency and their efforts were rewarded in 1815 by an Act of Parliament—the Apothecaries Act—which gave the London Society of Apothecaries the right to examine and licence general practitioners throughout the country.

This Act was to prove a turning point in medical education in England, for the Apothecaries took their new responsibility very seriously and gave long consideration to the new type of training required of candidates at its examinations. This included not only apprenticeship to a recognised general practitioner for five years, but also attendance at lectures and a period of hospital practice. The Society was also one of the first bodies to institute a written examination, as opposed to the oral question and answer and disputation which had persisted in the universities since the Middle Ages. The existing private medical schools were greatly encouraged by these requirements and new schools were established both in London and the provinces. The *Lancet* was started with the avowed object of reporting the lectures given in the medical schools and teaching hospitals—a procedure which aroused the indignation of the teachers but pleased many students who did not wish to pay lecture fees, or to attend the lectures themselves if they were already paid for by their parents. In the year 1800 the Surgeons' Company had become the Royal College of Surgeons and its membership (M.R.C.S.), obtained by examination in surgery, but not midwifery, was the passport to general surgical practice, so that many general practitioners, especially those intending to practise as surgeon-apothecaries in the provinces, sought the double qualifications of L.S.A. (Licentiate of the Society of Apothecaries) and M.R.C.S. Since anatomy was the basis of surgical teaching there was great demand for opportunities of dissection. As these were still very restricted in England, despite the activities of the so-called "resurrectionists", a great number of medical students, after the end of the wars in 1815, flocked to Paris for instruction in anatomy. This migration of students to another country, as well as the crimes of Burke and Hare, were among the arguments used in petitioning for new regulations for the provision of bodies for dissection. In 1832 the Anatomy Act made it possible for the bodies of any unclaimed paupers who died in hospital to be used for this purpose. Teachers in English medical schools may have lost some income through the transfer of students to Paris, but the practice of scientific medicine gained, for the type of clinical instruction, with careful case histories and record of physical signs, then developed in France was something of a revelation to English students. And not only did they bring back new ideas; some brought back with them Laennec's new stethoscope.

The success of the new system of training and qualification introduced by the Apothecaries Act of 1815 was never in doubt. It was already acknowledged in the report of the Select Committee (of the House of Commons) on Medical Education in 1834 (9). For the greater part of the nineteenth century the vast majority of family doctors in England qualified in this way. By comparison the contribution of the universities was negligible and even that of Edinburgh was dwarfed.

Many candidates for the new examinations came from the provinces and it was a result of their treatment by the Royal College of Surgeons that the parliamentary Select Committee on Medical Education was appointed in 1834. The provincial cities had begun to establish their own medical schools, the first being opened in Manchester in 1824, when the famous John Dalton agreed to give the lectures on chemistry. Others soon followed at Birmingham, Sheffield, Leeds, Newcastle, Bristol, Liverpool and Exeter. Their courses included anatomy and physiology, chemistry, botany, *materia medica* and therapeutics, the theory and practice of medicine, medical jurisprudence, the principles and practice of surgery, midwifery, and the diseases of women and children. Fees ranged from £30 to £50 for the whole course. The instructors were, as they were in London too, busy practitioners who had to give their lectures in the early morning or in the evening. Although they were inspected and approved by the Apothecaries, these schools were for many years denied recognition by the College of Surgeons, who insisted that all candidates for their membership should study anatomy and surgery in the London schools. It was the outcry consequent upon this which led to the appointment of the Committee. Evidence given before it revealed—what everybody already knew—that the examiners of the College of Surgeons were also the teachers in the London schools who feared a drop in their income from fees if the provincial students stayed at home and studied locally. Their opposition was eventually overcome and from 1839 study in provincial schools was recognized by the College.

In 1856 the Newcastle school became the Medical Faculty of the University of Durham, which awarded its first M.B. and M.D. degrees in 1858. Throughout the second half of the nineteenth century the drive for improved education, especially in scientific subjects, led to the foundation of Colleges of Sciences and Arts in the provincial cities and in 1873, despite some opposition, based on a belief that medical students were immoral and that contact with them would corrupt the Arts students, the medical schools were amalgamated with these new colleges. These have now developed into the well known universities of Manchester, Liverpool, Birmingham, and so on.

The two other medical schools outside London—and I speak of those of Oxford and Cambridge, which do not like to be called "provincial"—also went through considerable changes in the nineteenth century. In the first half of the century Cambridge really began to pay some attention to Medicine. The Downing bequest in 1800 led to the foundation of another college where medical studies could be pursued and to a new chair of medicine. The school was fortunate in the appointment of Dr. John Haviland as professor of anatomy. He has been called the real creator of the modern medical school at Cambridge. When he was made Regius professor of medicine in 1817 he gave regular weekly lectures and twelve years later attendance at lectures was made compulsory. At the same time the examination for the M.B. included for the first time written papers in English in place of the brief oral examination in Latin which had formerly been considered adequate. The Downing professor was also required to give weekly lectures throughout the year and in 1842 there was introduced for the first time as part of the examination for M.B. the clinical examination of a hospital patient, an innovation which was readily taken up by other schools and is now a traditional part of all examinations for medical degrees in England. By the time Haviland died in 1851 the school was already well on the way to the high position it attained under the leadership of George Paget, George Humphry and Michael Foster.

At Oxford the school was less fortunate and in mid-century the Royal Commission on the University of Oxford reported that "Oxford has ceased altogether to be a school of medicine. Those few persons who take medical degrees there, do so with a view to the social considerations which these degrees give or the preferments in the University for which they are needed but study their profession elsewhere." However, Henry Acland had already begun his long association with the school which was to lead to important developments. It was his idea that the B.M. should be granted for study in the preliminary medical sciences at Oxford; that students should then carry out their clinical studies elsewhere (usually St. Bartholomew's Hospital) and return to take the D.M. to set the seal on their complete medical education. Although this suited the peculiar conditions at Oxford it was unfortunately adopted by the General Medical Council for the whole kingdom and so brought about the divorce between theory and practice which has lasted to the present day.

Meantime, the number of students attending the teachers in the London Hospitals was increasing rapidly. The private medical schools which had grown out of ward rounds and surgical attendance by apprentices in the hospitals were by mid-century dying out as more and more of the well-known hospitals took over the schools as an official part of their organization. In 1827 a notable development in the history of medical education in England was the foundation of the University of London. At that time it was still impossible to enter the universities of Oxford or Cambridge without subscribing to the articles of the official State Church, the Church of England. With the spread of the noncomformist sects, Quakers, Baptists, Methodists, and so on, had come the establishment of the so-called Dissenting Academies, some of them universities in all but name, such as the celebrated Academy at Warrington, and some with distinguished men among their teachers. Inspired by the principles of Jeremy Bentham, a group of men led by Lord Brougham decided that the time was ripe for the foundation of an officially recognized university where nonconformists could send their sons for their higher education. It was to be relatively inexpensive and it was to concentrate on science and medicine, two fields in which nonconformists had achieved distinction. In order to provide its medical students with clinical instruction the North London Dispensary which became within a year or two the North London Hospital and later University College Hospital, was opened in the immediate vicinity of Gower Street. A year after the institution of the university another rival college was founded by the Church party to hold the balance against what they called "the godless institution in Gower Street." This was King's College, and as a result the university was forced to change its name to University College and both colleges became part of the University of London, which was for long simply an examining and not a teaching body (10). Advantage was taken of this special feature to allow students of the several provincial colleges founded later in the century to take the University of London degrees and this continued well into the present century when the so-called university colleges became universities granting their own degrees. This led to a concentration of examining bodies in London, for here too were the royal colleges of physicians and surgeons and the Society of Apothecaries. It was no wonder then that London became the most important centre of medical education in England. Here too was located the General Council of Medical Education and Registration of the United Kingdom, established by the Medical Act of 1858, a body which later changed its name to the General Medical Council and is commonly referred to as the G.M.C.

The Medical Act embodies the laws by which the medical profession in

Britain is still controlled. The G.M.C. is a branch of the Privy Council and therefore an organ of government. It has nothing to do with the British Medical Association which had to agitate for nearly thirty years before it succeeded in obtaining a representative on the G.M.C. The Council represents a typical British compromise, for it brought together the ancient corporations and royal colleges with representatives of the universities and of government. The objectives which were entrusted to it were the establishment of as uniform a system of medical education and examination as could be achieved without abolishing the existing system and beginning again, and also with the registration of all those individuals who succeeded in satisfying the examiners of one or other of these bodies so that they could be enrolled as "legally registered general practitioners." The liberty of the sick patient to seek the help of whom he pleased was safeguarded by allowing freedom to anyone to practise medicine, but only registered practitioners could sign a death certificate or hold State office (11).

For over a century the G.M.C. has exerted a steadily increasing influence on medical education in England. For the greater part of that time it had legal authority to inspect only the various types of examinations by which candidates entered the profession, but in recent years it has also inspected the schools themselves. Its weakness has always been a direct consequence of its composite nature, for the bodies represented on it were originally competitors in a very competitive world, each resentful of any attempt to curtail its individual privileges. The official policy which emerged from their deliberations therefore tended to be neutral in character and likely to offend those who were most active and most impatient in the cause of reform. It remained a fact that for many decades after the foundation of the Council the majority of students presenting themselves for examination and registration had not enjoyed a university education so that the curriculum was necessarily of a very practical character. In 1867, the subjects to be studied were: general anatomy, physiology, chemistry, *materia medica*, practical pharmacy, medicine, surgery, midwifery, and forensic medicine. What this meant in practice can best be judged by glancing at some of the examination questions of the time, or at one of the many little "crammers' books" which were so much favored by medical students. A little later the Council announced that "some limit must be assigned to the amount of knowledge which can be fitly exacted." In 1885 it became clear that teachers themselves were to be left to assign these limits and to choose which of the growing number of subjects should be included. A policy of this kind places a great responsibility upon the teachers who collectively form a vested interest to preserve the *status quo* and make it very difficult for any new scientific subject to find adequate time in the syllabus. Nevertheless, the G.M.C.'s powers to control the examination did offer an opportunity of influencing the actual teaching. In 1885 chemistry and physics were placed at the head of the pre-clinical subjects, and hygiene and mental diseases

Figure 1. First page of *al-Fakhir*, on diseases, their causes and treatment, by Abu Bakr Muh. al-Razi (Rhazes, 865-925). Courtesy of the Ahmadiyah Library, Aleppo.

Figure 2. Latin translation of Ibn Sīnā's *al-Qānūn fī al-Ṭibb* (*Prima quarti canonis Avicene principis in explanatione*, 1498). The commentary surrounding the centrally placed Latin translation indicates the great attention given to the *Canon* in the Latin West during the 15th and 16th centuries. Courtesy of the University of California Medical Center, San Francisco, California.

Figure 3. Chapter 31, Section 2, of the surgery of Abu al-Qasim al-Zahrawi (Abulcasis, d. *c.* 1013), shows forceps used as stump extractor and elevator to help in extraction of teeth. Photograph by courtesy of Kh. Bakshsh O. P. Library of Patna, Bihar, India.

Figure 4. Mosaic design of the "Tree of Life" on a marble wall of the Ẓāhirīyah's domed structure. Muslims believed in the healing powers of medicinal plants. Courtesy of the Ẓāhirīyah Library.

Figure 5. Title-page of Vesalius' *Fabrica* (1543), displaying the new method of anatomical instruction and the first representation of an anatomical theatre.

Figure 6. Bedside Teaching in Scholastic Times. From *Hortus Sanitatis*, Mainz, 1485.

Figure 7. The university quarter. At right, the old university; at left, the Dominican church. Engraving by Jacob Hufnagel. Vienna, 1609.

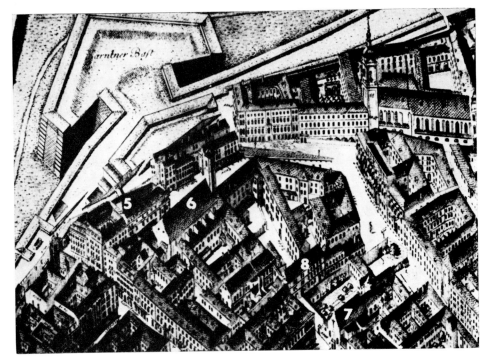

Figure 8. The Bürgerspital. Engraving from A. Huber, *Vogelschau der Stadt Wien,* 1769-1774.

Figure 9. Engraved portrait of Maximilian Stoll (1742-1787) by I. E. Mansfeld. n.p., n.d.

A DESCRIPTION OF THE Western Islands OF SCOTLAND.

CONTAINING

A Full Account of their Situation, Extent, Soils, Product, Harbours, Bays, Tides, Anchoring-Places, and Fisheries.

The Antient and Modern Government, Religion and Customs of the Inhabitants; particularly of their Druids, Heathen Temples, Monasteries, Churches, Chappels, Antiquities, Monuments, Forts, Caves, and other Curiosities of Art and Nature: Of their Admirable and Expeditious Way of Curing most Diseases by Simples of their own Product.

A Particular Account of the *Second Sight*, or Faculty of foreseeing things to come, by way of Vision, so common among them.

A Brief Hint of Methods to improve Trade in that Country, both by Sea and Land.

With a New MAP of the Whole, describing the Harbours, Anchoring-Places, and dangerous Rocks, for the benefit of Sailors.

To which is added, A Brief Description of the Isles of *Orkney* and *Schetland*.

By *M. MARTIN*, Gent.

The SECOND EDITION, very much Corrected.

LONDON,

Printed for A. BELL at the Cross-Keys and Bible in *Cornhill*; T. VARNAM and J. OSBORN in *Lombard-street*; W. TAYLOR at the Ship, and J. BAKER and T. WARNER at the Black Boy in *Paternoster-Row*. M.DCC.XVI.

It was this book that first induced Johnson to take his celebrated Tour.

Figure 10. Title-page of the second edition of Martin Martin's *A description of the western islands of Scotland, &c.*

Figure 11. Print by John Kay, depicting the hostility of the Seven Professors of King's College, Aberdeen, to a suggestion (1786) that King's College and Marischal College be united. Dr. Skene Ogilvy, dubbed by Kay as "the beauty of holiness" (he was a singularly ugly man) is in the pulpit lecturing his colleagues on the duty of returning good for evil. Dr. William Chalmers, Professor of Medicine, is shown as a skeleton-like figure bearing a scythe, and, according to Kay's notation, saying "Degrees Male and Female in Medicine and Midwifery sold here for ready money."

Figure 12. Olof Rudbeck *the Elder*. Painting by M. Mijtens the Elder.

Figure 13. The Anatomical Theatre in Uppsala. Drawing by Olof Rudbeck.

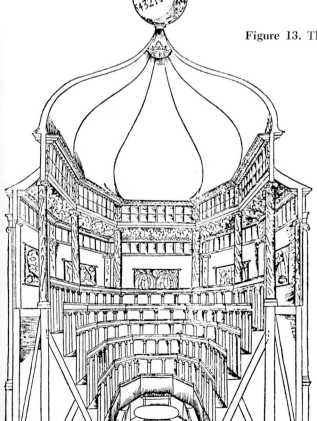

Figure 14. Olof af Acrel. Pastel by Gustaf Lundberg, 1776.

Figure 15. Gripenhielm House, later The Serafimer Hospital.

Figure 16. The Karolinska Medico-Surgical Institute, 1815.

Figure 17. Domus Anatomica and Theatrum Anatomicum, Copenhagen.

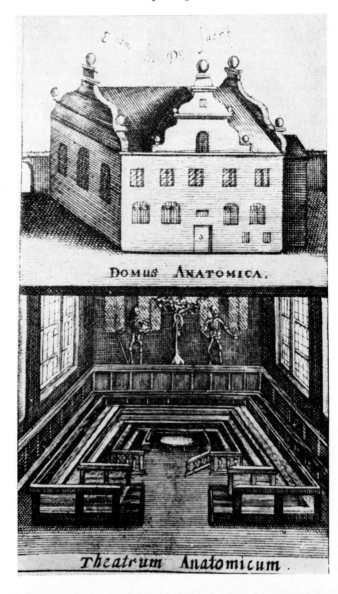

Figure 18. Israel Hwasser.

Figure 19. Michael Skjelderup. Painting by Johan Gørbitz, 1846.

Figure 20. Steatite seal from Mohanjo-Daro showing human figure in Yogic posture, wearing trident-like head dress. Surrounded by various animals, the figure bears resemblance to Shiva Pashupati, "The lord of animals," who in the *Purānās* is considered as the first propounder of the *Āyur Veda*. Copyright: Archaeological Survey of India.

Figure 21. Brahmā, one of the Trinity, considered as the author of the *Āyur Veda*. In left upper hand he holds a *kalasha* (vase) containing *amrita* (nectar of life), which continued as a symbol of the physician, divine or mortal, through the ages. From Aiholi (7th century A.D.), now in Prince of Wales Museum, Bombay. Copyright: Archaeological Survey of India.

Figure 22. Kāshīnareśōh Divodāsa (Dhanwantari) teaching Āyurvedic science to Sushruta and other trainees. Dhanwantari holds the symbolic *amrita kalasha* in his left hand. Before him lie a manuscript and a few bottles in a basin. From a 19th century Āyurvedic text. Courtesy of Dr. D. V. Subba Reddy.

Figure 23. The *marmās* (vital points) on the surface of the upper and lower extremities as described by Sushruta. These have no relation to the Chinese points for acupuncture, but are the points which the physician is warned to approach with caution since important neurovascular structures and joint cavities lie in close proximity to them. From *Sushruta Samhitā*, trans. by Kaviraj Kunja Lal Bhishagratna, Vol. 2 (1963). Courtesy of the Chowkhamba Sanskrit Series Office, Varanasi.

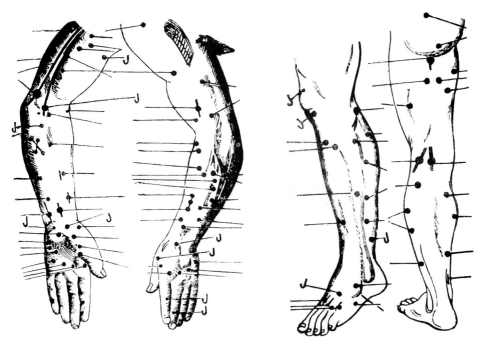

Figure 24. Painting from the island of Bali, depicting the *Pancha mahābhūta* (five elements) which, according to Āyurveda, form the human body. In the center is *tejas* (fire), below is *prithvī* (earth), at left, *ap* (water), above, *ākāsha* (ether), and at right, *vāyu* (air). Courtesy of Dr. Lokesh Chanra, Director, International Academy of Indian Culture, New Delhi.

Figure 25. Terracotta female figure depicting laparotomy, from ruins of Kaushāmbī (2nd century B.C.). Copyright: Allahabad Museum.

Figure 26. Part of facade of Royal Hospital, Goa, showing the Portuguese Royal Arms and the inscription "Hospital Real". Courtesy of Mr. V. T. Gune, Director of Historical Archives, Panjim, Goa.

AN ATLAS

OF

ANATOMICAL PLATES

OF THE

HUMAN BODY,

ACCOMPANIED WITH DESCRIPTIONS IN HINDUSTANI.

BY FRED. J. MOUAT, M. D.

FELLOW OF THE ROYAL COLLEGE OF SURGEONS OF ENGLAND—ASST. SURGEON, BENGAL ARMY—MEMBER OF, AND SECT. TO, THE COUNCIL OF EDUCATION OF BENGAL.—PROFESSOR OF MATERIA MEDICA AND MEDICAL JURISPRUDENCE IN THE BENGAL MEDICAL COLLEGE, ETC. ETC. ETC.

ASSISTED BY

MOONSHI NUSSEERUDIN AHMUD,

LATE OF THE CALCUTTA MADRUSSA.

THE DRAWINGS ON STONE BY C. GRANT, ESQ.

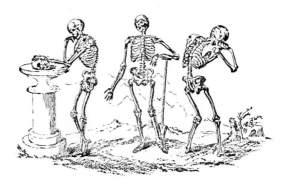

CALCUTTA:

Bishop's College Press.

M.DCCC.XLVI.

Figure 27. Title-page of the first Atlas of anatomy in English to be published in India. Courtesy of Dr. Herman Henkel, Director, John Crerar Library, Chicago.

Figure 28. The Lady Hardinge Medical College, New Delhi, established in 1916 for the training of women physicians. Courtesy of Dr. M. Chaudhuri, Principal, Lady Hardinge Medical College and Hospital.

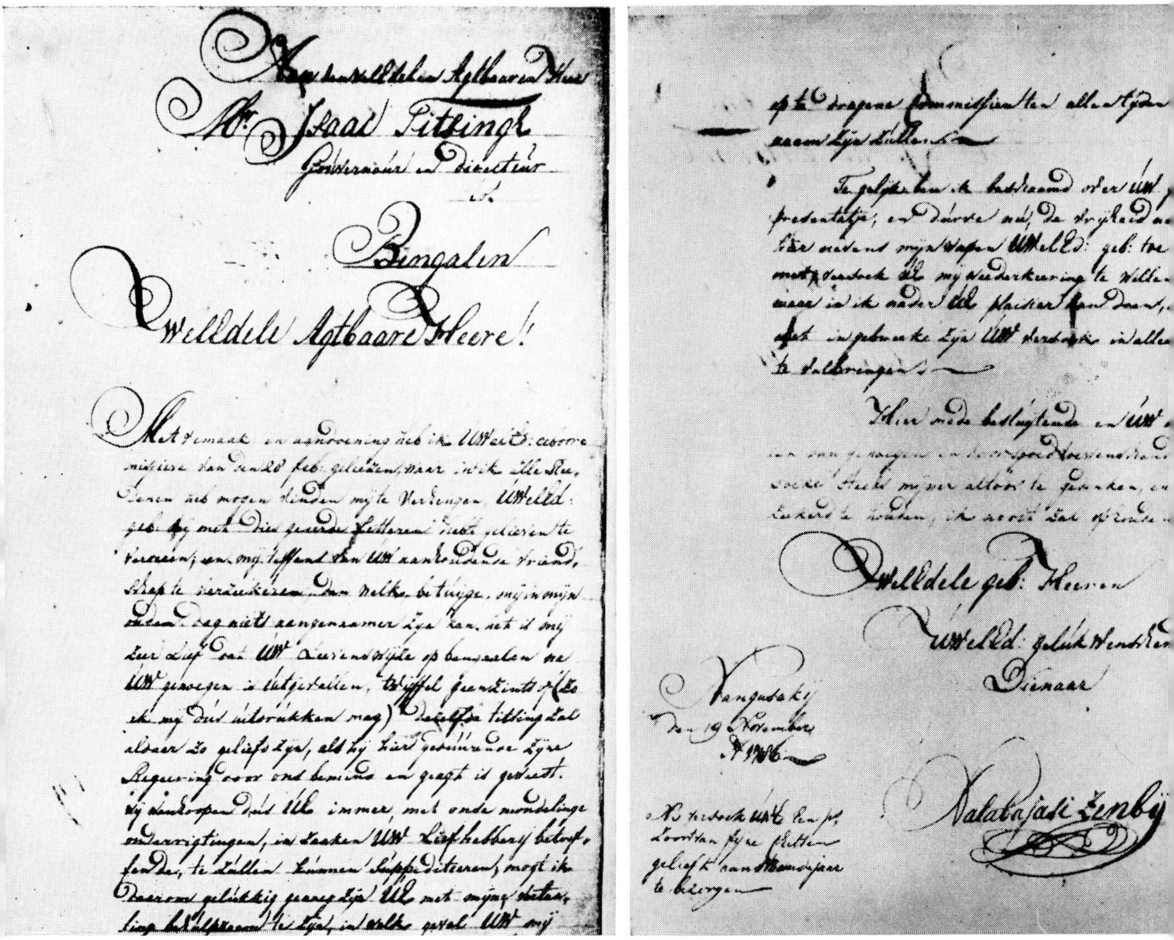

Figure 29-31. Letters in Dutch from Japanese physicians to Isaac Titsingh in 1786. From Kyoto National University, Library of Faculty of Medicine.

Figure 32. Specimen page from Rangaku Kaitei, published in 1788 by Ōtsuki Gentaku.

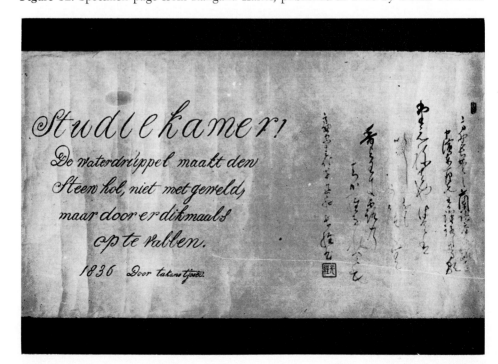

Figure 33. Pencil sketch of Philipp Franz Balthasar von Siebold by a Japanese student.

Figure 34. Siebold at Edo in 1826. Sketch by his student Iwasaki Kanyen.

Figure 35. The teaching hospital of Pompe van Meerdervoort at Nagasaki with the Japanese flag on one wing and the flag of the Netherlands on the other.

ESTATVTOS
HECHOS POR LA MVY
INSIGNE VNIVERSIDAD
DE SALAMANCA.

Año.

M. D. LXI.

EN SALAMANCA.
en casa de Iuan Maria de Terranoua.
M. D. LXI.

Figure 36. Title-page of the Statutes of the University of Salamanca, 1561.

CONSTITVTIO
NES TAM COMMODAE
APTÆQVE, QVAM SANCTÆ
Almæ Salmanticensis Academiæ toto terrarum orbe flo-
rentissimæ.

SALMANTICÆ
Excudebat Ioannes Maria à Terranoua.
M. D. LXII.

Figure 37. Title-page of the Constitutions of the University of Salamanca, 1562.

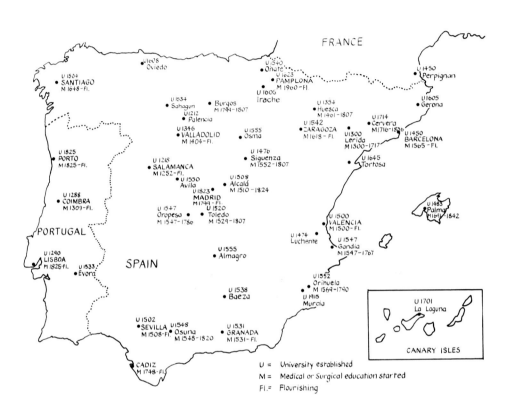

Figure 38. Universities and medical schools of Spain and Portugal, 1250-1950.

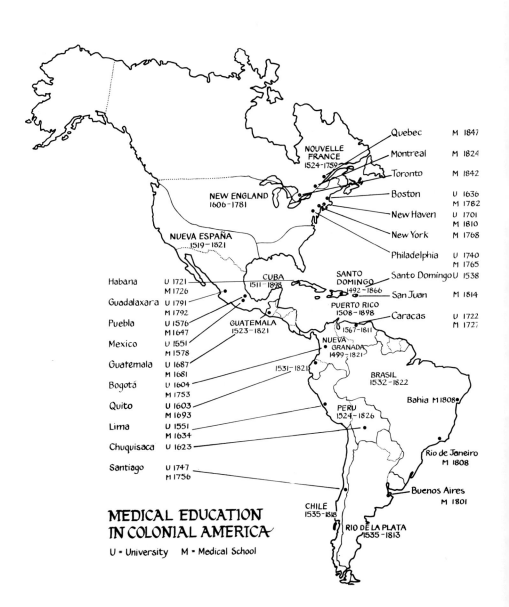

Figure 39. Medical education in colonial America.

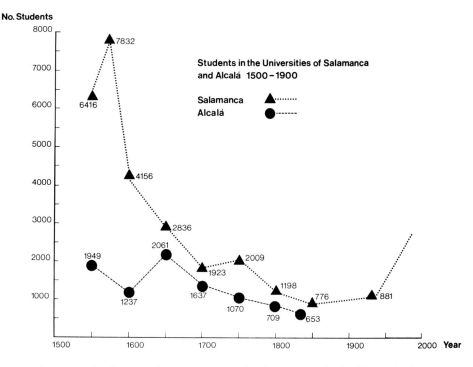

Figure 40. Students in the Universities of Salamanca and Alcalá, 1500-1900.

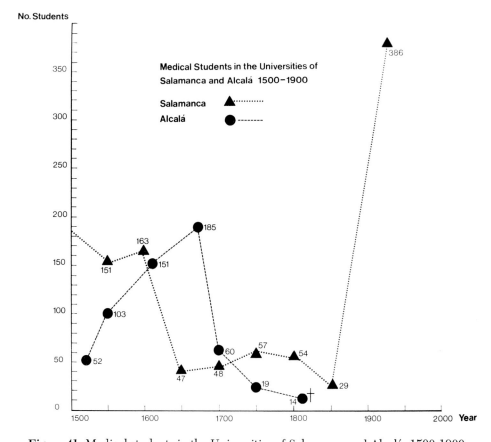

Figure 41. Medical students in the Universities of Salamanca and Alcalá, 1500-1900.

were added. After a further revision in 1890 the length of the course was extended from four to five years (12). In 1922 drastic revision attempted to pull together the dozen or so apparently unrelated subjects into a coherent whole with a pervading purpose of scientific preventive medicine. Halfway between these two efforts to provide a more reasonable curriculum came the Haldane Commission of 1909 whose final report (13), published in 1912-1913, revealed the great influence of the Flexner Report (14) in its findings and recommendations. When Abraham Flexner turned his attention from American to European medical schools in 1912 he found what he considered the ideal model in Germany rather than in Britain. The highly organized scientific departments of the German universities made a powerful contrast to the casual and amateurish medical schools he inspected in England. Here, if the pre-clinical subjects were taught at all, they were entrusted to the most junior of the clinical staff. More often they were taught elsewhere by teachers who were training candidates for disciplines other than medicine, so that medical students often found it difficult to keep their goal in sight. When they transferred to medical school for clinical training they were often urged by the consultants who gave them part-time teaching to put behind them all the theory they had learned in their preclinical science classes. There was not only a split in their education but almost a dichotomy. The universities had little or no influence over clinical teaching, so that while Flexner admired the bedside teaching in English hospitals, he strongly advocated the institution of the German method, and Haldane followed this recommendation closely in his own report.

An Act passed through Parliament in 1900 empowering the University of London to incorporate other institutions, and all the medical schools attached to the teaching hospitals agreed to go into the university. Unfortunately, this had little immediate practical effect, so that when Flexner saw them in 1912 there was still little apparent connection with the university. The schools still depended on student fees and on payment by the hospital for services rendered. The hospital staff continued to teach without pay and lecturers in non-clinical subjects were appointed and paid by the school. Even the students were not required to register with the university and less than a third of them took the degrees of Bachelor of Medicine and Surgery of the University of London. Many were still from Oxford and Cambridge and returned there for their examinations. Others took the conjoint examination of the Royal Colleges of Physicians and Surgeons (L.R.C.P., M.R.C.S.), while a few still qualified with the Licence of the Society of Apothecaries (L.S.A.).

Three new and important factors now acted on the schools to bring about a change: (*i*) the advance of the basic sciences and the growth of specialization with the consequent need for scientific equipment and laboratories; (*ii*) the outbreak of the first world war; and (*iii*) economic stringency. The Haldane Report's recommendations were designed to help the university to break the "closed shop" of the hospital schools by establishing in every hospital "professorial units" in medicine and surgery. The report quoted Osler to the effect that "a professor of medicine requires the organization of a hospital unit, if he is to carry out his three-fold duty of curing the sick, studying the problems of disease, and not only training his students in the technique of their art, but giving them university instruction in the science of their profession." This department should be controlled entirely by the university, or with the cooperation of the hospital.

Nothing was done about the Commission's recommendations during the first war. In the economic troubles which followed that war the financial difficulties in which the universities found themselves led to the establishment of the governmental University Grants Committee to advise the Treasury on the distribution of funds to the universities and university colleges. The teaching hospitals, which were of course the old voluntary or charity hospitals, also had more and more difficulty in attracting contributions while their costs soared. It became clear that the medical schools associated with them would have to depend more and more on the universities for financial aid. Once it came to have a strong financial stake in the schools the university was able—gradually—to insist on the institution of the professorial units envisaged by the Report. How the term "gradually" is understood in Britain may be gathered from the fact that it took another world war and a nationalized Health Service to bring about any substantial speeding up of this historical process.

One notable advance between the wars was in the provision of facilities for postgraduate education. In 1921 the Athlone Committee urged the establishment within the University of London of an Institute of State Medicine where special training could be given in such subjects as public health, industrial medicine and forensic medicine, as well as medical ethics and economics. As Britain was at that time still the heart of a great colonial empire the science and practice of medicine in the tropics was also important and in 1929 the London School of Hygiene and Tropical Medicine was opened, its first Director being Sir Andrew Balfour who had for nearly twenty years directed the research in tropical medicine sponsored by Henry Wellcome. This same Athlone Committee also urged the establishment of a school where advanced medical education could be given to already qualified practitioners. Although efforts were made to locate such a school in one of the well known teaching hospitals none could be found that was willing to give it a home. It opened in 1931 as the British (now the Royal) Postgraduate Medical School (15) in a suburb of London where a former workhouse infirmary, the Hammersmith Hospital became its home. The Greater London Council supports the hospital, the university maintains the school, and here was rapidly built up under a succession of first class men the most advanced and successful medical school in the British Commonwealth. As an example of what can be achieved when the restrictions imposed by tradition and vested interests are loosened this institution with its brilliant record has been of tremendous value in radically changing the attitude of British doctors to medical education and the way it should be organized.

In 1944 another government-appointed committee, the Goodenough Committee (16), made its recommendations for the medical schools and the place of teaching hospitals in the pattern of nationalised medicine that was then taking shape. Consequently, when the National Health Service was launched in 1948 the teaching hospitals were left outside the regional hospital network, but were placed under the control of independent boards of governors responsible directly to the Minister of Health who provides the funds for running the hospital. Representatives of the medical staff, the schools and the university sit on this board. The schools themselves are controlled by another independent board of governors responsible to the university which provides the funds for running the school, and representatives of the hospital sit on this board. The staff is paid by the Ministry for treating patients and by the university for teaching students. Outwardly, this dualism may be seen as the culmination of the symbiosis of consultants and hospital pupils which goes back to the early eighteenth century, but on closer examination we see that here, at last, is broken the financial link between clinical teachers and students which had been for so long an invisible barrier between clinical teaching and the university system.

Other developments in the last twenty years, mostly as a result of the Goodenough Report, include the establishment of professorial units in pathology, obstetrics and gynaecology as well as medicine and surgery, and a postponement of registration as a "qualified practitioner" until the student who has passed all his examinations has served a period as "house physician" or "house surgeon", appointments corresponding to the American "interneships". The value of the Postgraduate Medical School has been so universally recognized that it has now become the basis of a British Postgraduate Medical Federation embracing a score or so of Postgraduate Institutes associated with the hospitals where the medical specialties are practised (17).

I have said nothing so far of the medical education of women. Their struggle to enter the profession has been told in the admirable biographies of the pioneer medical women, especially in the biography of Elizabeth Garret Anderson by Jo Manton (18). Forced to go abroad to take the M.D., which she did most successfully at Paris, she finally qualified for the British Medical Register by becoming a licentiate of the Society of Apothecaries in London. From her efforts arose the London School of Medicine for Women located at the Royal Free Hospital. Among the recommendations of the Goodenough Committee was one that treasury funds for the medical schools should be made conditional on their admitting a certain percentage of women students. It is interesting to note that women often carry off the prizes in the schools which were formerly strictly guarded male preserves. So great have been the social changes in Britain since the second world war that most medical students are now supported at least in part by State grants. The preclinical subjects are now taught in the ordinary secondary schools and pupils intending to take up a medical career have to make up their minds very early in school life, at the age of thirteen or fourteen, in order to pass the necessary subjects before they leave school, for without them they will not be admitted to medical school. The abandonment of the liberal arts at such an early age is a source of anxiety to all educationists. Senior consultants, who once studied for an arts degree at Oxford or Cambridge before even deciding to study medicine—and they are now a rapidly diminishing number—wonder how long a learned profession which has been graced by a long succession of medical scholars and men of wide culture can resist the growing opinion that it is one of a number of specialties covered by the comprehensive term "technology". The present system reinforces this trend. Those who have designed the system feel confident that the final product will work, but few would like to hazard a guess as to the kind of person the English doctor of the next century will be.

NOTES

1. NEWMAN, CHARLES, *The evolution of medical education in the nineteenth century.* London, 1957.
2. POYNTER, F. N. L., and BISHOP, W. J., *A seventeenth century doctor and his patients: John Symcotts, 1592-1662.* Streatley, Beds. 1951. (The introduction contains an account of the education of a physician in the seventeenth century.)
3. CLARK, Sir GEORGE, *A history of the Royal College of Physicians of London,* Vol. I. London, 1964.
4. WALL, C., and CAMERON, H. C., *A history of the Worshipful Society of Apothecaries of London,* ed. E. A. UNDERWOOD, Vol. 1. London, 1963.
5. ROBB-SMITH, A. H. T., Medical education at Oxford and Cambridge prior to 1850, in *The Evolution of Medical Education in Britain,* ed. F. N. L. POYNTER. London, Pitman Medical Publishing Co. Ltd., 1966, pp. 19-52.
6. COPE, Sir ZACHARY, *William Cheselden.* Edinburgh & London, 1953.
7. ―― The Great Windmill Street School, in his *Famous General Practitioners.* London, 1961.
8. CLARK-KENNEDY, A. E., *The London: a study in the voluntary hospital system.* London, 1962-3. 2 vols.
9. Great Britain, Parliament, House of Commons, Select Committee on Medical Education, *Report . . . with minutes of evidence, and appendix,* 4 pts. London, 1834.
10. BELLOT, H. H., *University College, London 1826-1926.* London, 1929.
11. See the chapter on Education and the General Medical Council and the references given there in *The Evolution of Medical Education in Britain,* ed. F. N. L. POYNTER. London, 1966.
12. NEWMAN, Sir GEORGE, *Some notes on medical education in England,* London, 1918; *Recent advances in medical education in England,* London, 1923.

13. Great Britain, Royal Commission on University Education in London, (Chairman: Lord Haldane) *Final Report*. 3 pts. London, 1913.
14. FLEXNER, A., *Medical education in Europe*, New York, 1912. Carnegie Foundation, Bulletin No. 6.
15. NEWMAN, C., A brief history of the Postgraduate Medical School, *Med. Hist.*, 1966, **10**: 285-288.
16. Great Britain, Ministry of Health ("Goodenough Committee") *Report of the Interdepartmental Committee on Medical Schools*. London, 1944.
17. FRASER, Sir FRANCIS, *The British Postgraduate Medical Federation; the first fifteen years*. London, 1967.
18. MANTON, Jo, *Elizabeth Garrett Anderson*. London, 1965.

SCOTTISH MEDICAL EDUCATION

L. R. C. AGNEW
UCLA School of Medicine
Los Angeles, California

There are two main types of medical education—formal and informal—and this paper is mainly concerned with formal medical education as conducted by the Scottish universities and by certain other institutions. Now, although formal Scottish medical education is our main topic, we should at least recognize the existence of informal Scottish medical education—indeed, there were even families of healers in the West of Scotland who were active long before any of the Scottish medical schools made any significant contributions to medical education. The best known of these families were the M'Beaths (later Macbeth, and, in Latin, Betonus or Beaton) and several of their manuscripts, written in Gaelic, still survive. One such manuscript, in the National Library of Scotland, is that of John Beaton (15th century) in which there are mentions, *inter alios*, of Hippocrates, Dioscorides, Galen, Rhazes and Avicenna; and this suggests that some, at least, of the hereditary Highland healers had access to classical medical manuscripts.

The simplest system of informal medical education, of course, is folk medicine itself, with a predominantly oral transmission of information from generation to generation—and Scotland has a rich heritage of curious medical practices. The literature in this field is extensive; and while it is perhaps invidious to select a single volume, Martin Martin's *A Description of the Western Islands of Scotland, &c.* (London, 1703) is worth mentioning as a useful source for details of some of the more curious beliefs and practices. Parenthetically, it is interesting to note that it was the second edition ("very much corrected") of this work (London, 1716) which aroused Dr. Johnson's curiosity in the Western Islands of Scotland, and thus indirectly brought about his celebrated trip to Scotland in 1773 accompanied by the engaging, albeit sycophantic, James Boswell. Martin Martin was struck by the innocence of the islanders, and observed that "they are to this day happily ignorant of many Vices that are practised in the Learned and Polite World: I could mention several, for which they have not as yet got a Name, or so much as a Notion of them." And we cannot leave Martin without reference to what can only be dubbed "hammer therapy":

There is a Smith in the Parish of Kilmartin, who is reckon'd a Doctor for curing Faintness of the Spirits. This he performs in the following manner: The Patient being laid on the Anvil with his Face uppermost, the Smith takes a big Hammer in both his hands, and making his Face all Grimace, he approaches his Patient; and then drawing his Hammer from the Ground, as if he design'd to hit him with his full Strength on the Forehead, he ends in a Feint, else he would be sure to cure the Patient of all Diseases: but the Smith being accustom'd to the Performance, has a Dexterity of managing his Hammer with Discretion; tho at the same time he must do it so as to strike Terror in the Patient: and this they say has always the design'd Effect.

Before dealing with Scottish medical education in the post-1600 period, two important sixteenth century developments in the field of formal medical education are worth noting. In 1505 the barbers and the surgeons of Edinburgh petitioned the Town Council to be enrolled among the Incorporated Crafts of the Burgh and to be granted a "Seal of Cause" or Charter of Privileges. The petition was successful, and the Seal of Cause was granted on 1 July 1505—and, as Creswell has observed, "to this union the Royal College of Surgeons of Edinburgh traces its foundation as a corporate body" (1). Provision was made in the petition for appropriate training of the surgeons and their apprentices—for example, the apprentices should be able to read and write; also, an executed criminal's body should be made available once a year for anatomical study and instruction (". . . and that we may have anis in the yeir ane condampnit man efter he be deid to make anatomea of quhairthrow we may haif experience"). There was a curious clause in this seal of petition which read ". . . that na persoun man nor woman within this burgh mak nor sell ony aquavitae within the samyn. Except the saidis maisteris brether and friemen of the saidis craftis." Alas, the present Fellows of the Royal College of Surgeons of Edinburgh no longer enjoy this most glorious and lucrative of privileges.

Towards the close of the sixteenth century another important development occurred in early Scottish medical education—specifically the formation of the Faculty of Physicians and Surgeons of Glasgow by Dr. Peter Lowe (1550-1612), one of Scotland's greatest surgeons and medical educators (2). Dr.— or Maister, as he usually signed his name—Peter Lowe had spent most of his professional life as a surgeon in France, and much of it on the battlefield, and returned to England around 1596 and to Glasgow in 1598. By November of 1599 he had obtained a Charter from James VI which established the Faculty of Physicians and Surgeons of Glasgow. Lowe, incidentally, had published in 1597 a surgical text written in English, and in the second edition of this work (1612) he mentions the provisions of the Royal Charter he obtained for the Faculty of Physicians and Surgeons of Glasgow. The Faculty, later the Royal Faculty and nowadays known as the Royal College of Physicians and Surgeons of Glasgow, was empowered to control both surgical and general medical practice in the Glasgow area by conducting examinations and by issuing appropriate licences. This they continued to do until the Medical Act of 1858, although their virtual monopoly in medical education was latterly challenged by the University of Glasgow. "It is a pity, however," bemoaned one writer as late as 1827, "that so much hostility and jealousy should exist between the University and [the] Faculty of Physicians and Surgeons, who, instead of prosecuting each other's candidates, and calling in question their power to practise within a certain district of country, ought to unite in every liberal object which could enhance the respectability of their common profession" (3).

Indeed it was a pity, this bickering between two great institutions, for certainly the Faculty of Physicians and Surgeons had done a fine job throughout its long and proud history—not least in uniting the physicians with the surgeons, a truly rare accomplishment. Perhaps a true *rapprochement* can never occur between surgeons and physicians—for reasons which are known to all who have practised medicine—but the Glasgow Faculty had from its start put these gentlemen in close professional association with one another, and this undoubtedly has led to a greater tolerance and understanding than is usually found elsewhere.

Although we have not yet discussed medical education at Glasgow University, it is appropriate to note here that this University in 1817 became the first British university to introduce a separate degree in surgery—specifically the C. M. This was an attempt, of course, to widen the professional horizon of the Glasgow University medical student who wished to practise surgery within the area patrolled, so to speak, by the Faculty of Physicians and Surgeons. But it was more than that—it was a tacit recognition of the fact that the Glasgow University M.D. in surgery left much to be desired. Indeed, the M.D. degree itself was a less rigorous offering in 1817, so it is not surprising that "for the first ten years after its institution the degree of C.M. was in greater demand than the degree of M.D., 202 obtaining the former and 146 the latter" (4).

The C.M. degree originally took two years, but in 1826 the curriculum was extended to three years, and in 1839 to four. By 1839 two years of hospital attendance were required, half the time being devoted to surgery and half to medicine; and courses—each lasting six months—were required in anatomy, anatomical dissections, surgery, chemistry, physiology, practice of medicine, midwifery, *materia medica* and pharmacy; a three months' course was required in forensic medicine and, finally, the candidates "were to be examined in all the subjects of the curriculum, to submit a thesis in English on a surgical subject, and to give evidence of a competent knowledge of Latin" (5).

The Faculty of Physicians and Surgeons were hardly enthusiastic about this new degree in surgery and fought tooth and nail in the law courts against any implied or actual trespassing on their medical property by those failing to hold a Faculty diploma. Fortunately the Medical Act of 1858 brought such niggling to an end, other British universities introduced degrees in surgery, and latterly joint degrees in medicine and surgery (e.g., M.B., Ch.B.), were generally offered, and the C.M. today is a higher surgical degree in its own right.

The Scottish Universities

St. Andrews

The four original Scottish universities—all of which still exist—were St. Andrews, founded in 1411, Glasgow (1451), Aberdeen (1494) and Edinburgh (1583). Although St. Andrews is the oldest of the Scottish universities, its influence on medical education was minimal until the end of the nineteenth century For centuries medicine was regarded as a sort of liberal arts subject for the well-rounded student rather than as a formal academic offering in the modern sense. After the Reformation, John Knox in *The Book of Discipline* (1560) suggested that St. Andrews concentrate on teaching medicine, while Glasgow handle arts, and Aberdeen law and divinity, but this was not done. There was a Chair of Medicine in 1772 but no medical school until 1897 when St. Andrews amalgamated with Dundee. But the absence of a true medical school until 1897 did not prevent St. Andrews from offering the degree of M.D. *in absentia* or *honoris causa*, and there was much criticism of this practice although among the M.D.'s of St. Andrews were such distinguished men as Edward Jenner and Jean Paul Marat.

Aberdeen

Aberdeen recognized medicine right from the start by appointing professors of medicine, or "mediciners" as they were more commonly known. One early mediciner was Gilbert Skeyne (*c*. 1522-1599) who had the distinction of writing Scotland's first printed medical book (*Ane Breve Descriptioun of the Pest*. Edinburgh, 1568). As at St. Andrews, medicine was but part of the offerings in the field of liberal arts, and for practical medicine students had to serve an apprenticeship with a practitioner. Although Aberdeen University today is a single institution, before 1860 there were two separate, indeed rival, colleges—King's College (founded in 1494) and Marischal College (founded in 1593). King's was the first to award the M.D. degree, but not until the year 1630; and in 1637 Mr. William Broad of Berwick was to become the writer of Scotland's first M.D. thesis. King's College awarded its first M.B. degree in 1685 and Marischal College awarded its first M.D. in 1713. Often these early Aberdeen medical degrees were given, again like St. Andrews, *in absentia* or *honoris causa*. There was much rivalry between King's College and Marischal College, and even as late as 1850 King's College was still protesting that Marischal College had no right to grant medical degrees; but by 1860 peace was established by the formation of a Joint University, and with medicine being the responsibility of Marischal College alone. There had been much dissatisfaction with the old bipartite state of medical affairs—for example, in 1789 the students of Marischal College had improved medical teaching by getting extra-mural classes started by the simple device of founding the Medical (Medico-Chirurgical) Society—and after the union of the Colleges the prestige of Aberdeen's medical degrees was enhanced.

Glasgow

Although Glasgow University was founded in 1451 its encouragement of medicine was sporadic and unsystematic until the eighteenth century, although, to be sure, the first Professor of Medicine (Robert Mayne) was appointed in 1637. William Cullen (1712-1790), however, was the real founder, in 1744, of the Glasgow University School of Medicine. Cullen had been in general practice in Hamilton with William Hunter (1718-1783); and when Hunter, like so many talented young Scots both then and now, found the high road to London irresistible, Cullen turned his attention to medical education, and specifically to Glasgow University. He lectured at first on medicine, and latterly chemistry and *materia medica*, and by 1751 was appointed Professor of Medicine. On his appointment to Edinburgh University in 1755 he was succeeded by his distinguished pupil, the great chemist Joseph Black (1728-1799).

In addition to men of the caliber of Cullen and Black, a great medical school needs a great teaching hospital; and this was forthcoming in 1794 with the opening of the Royal Infirmary (6). Additional clinical appointments were made, either by the University or on an extramural basis; many of the extramural teachers were affiliated with Anderson's University (founded in 1796) and, later on, Anderson's College Medical School (founded in 1887 but no longer in existence), and with the Portland Street School (*c.* 1827-1844). There was much rivalry—indeed at times frank hostility—between the University medical faculty and the extramural teachers; part of this rivalry undoubtedly had its roots in the fact that many of the extramural teachers were more stimulating than their University counterparts, who tended to be conservative and status-minded (7). One disenchanted medical student has left us the following frank description of Robert Freer, who was Professor of Medicine at Glasgow University from 1796 to 1815:

At all times, and on every occasion, the Dr. seemed the same dull heavy headed man, and during the several years we attended him, both during his medical lectures in College and while he was one of the attending physicians in the infirmary, we never upon any one occasion saw the expression of his face alter. It was the same every day dull heavy countenance without the slightest variation, as if a coloured marble bust had been set upon a pair of living shoulders . . . [He] never was, we are sorry to say, much esteemed as a popular lecturer, his dogmas and doctrines being deemed stale and uninteresting, and might have passed current about a century ago, but certainly his lectures were by no means adapted to fascinate the attention of the student of 1813, who expected he would have attended more to the changes and improvements progressively taking place in medicine at the present day, but the old man's mind was obstinately running back upon the vista of years gone by.

The medical student who wrote this (8) was also critical of the teaching of anatomy during the period (1812-1815) of his attendance at Glasgow University:

At the present time the Anatomical Schools of Scotland are in a most deplorable state from the want of subjects, it now being quite impossible to procure more than two or three bodies in the course of the year by exhumation. In Glasgow, bodies are so scarce that they are salted in summer and hung up to dry like Yarmouth herrings (9).

. . . This dissecting room attendance was in sooth a most heartless and discouraging one to us, for we too soon found that anyone who wished here to benefit and improve himself was so beset by a crowd of youngsters bent upon idleness and mischief that he soon was forced to retire and court quietness elsewhere. This place that ought to have been of the utmost importance to the student in a practical point of view was neither more nor less than a den of tumult and uproar, of wrangling and brawling and every species of schoolboy mischief. . . . The Dissector, Mr. Jardine, might have managed matters otherwise had he attended in the least degree to the well-being of his pupils, but he either had or pretended to have too much business on hand elsewhere, and left them to do the best they could without his aid after having swindled them out of their money.

More than a dozen years later there were still students' complaints about the teaching of anatomy; one such complaint was directed against the poor use made for teaching purposes of William Hunter's fine museum:

I cannot, at the present time, however, allow myself to dismiss the affairs of this College, without expressing my regret that the doors of the Hunterian museum should be so completely and obstinately closed against the entrance of the students. It is peculiarly rich in morbid specimens of the human body, and was given by its noble founder, the late Dr. William Hunter, chiefly as an important and valuable adjunct to the medical school. Yet the many valuable lessons, which a careful study of its contents could not fail to produce, are unfortunately lost to the students, from the illiberality of its management. We are only permitted for *an hour in the year* (ex gratia) to perambulate its apartments, but not to handle or too minutely inspect any of the morbid preparations (10).

Until the nineteenth century the regulations governing the award of medical degrees at Glasgow University were, as Coutts elegantly observes, "regulated partly by use and wont and partly by express enactment". But on 30 March 1802, the Senate decreed that henceforth a curriculum of three years would be needed, and that candidates would be required to attend courses in anatomy, surgery, chemistry, pharmacy, theory and practice of physic, *materia medica*, and botany; three examinations (in Latin) would also be required in these subjects. "If the examiners reported favourably, the next step was to propound to the candidate an Aphorism of Hippocrates and a case of disease, on each of which he had to compose a Latin commentary to be read before the Senate on an appointed day. On that day he was also liable to be further questioned before the Senate on any branch of medical science. If he passed this ordeal successfully, he might graduate there and then, unless he had published a thesis. Candidates who published theses were required to defend them in the *Comitia*, which included the Senate, the students, and other members of the University. It was laid down that no thesis should be received if it contained illiberal reflections on the writings or lectures of any medical teachers or writers" (11).

Edinburgh

Although the University was established in 1583 its medical school did not get under way until 1726. There was a medical chair established in 1685, however, but this was brought about by the influence of the Royal College of Physicians of Edinburgh. The history of this important body need not detain us (12)—suffice it to say that it was founded in 1681 by Dr. (later Sir) Robert Sibbald (1641-1722). Sibbald had studied for a time at Leiden (although he obtained his M.D. at Angers), and both he and Archibald Pitcairne (1652-1713) were pioneer professors of medicine in the so-called Town's College; Pitcairne was later (1692) to be appointed for a short time to Leiden's chair of medicine (13). But despite these promising local developments in medical education, the real power, albeit indirectly—a medical *eminence grise*—was Leiden's renowned Hermann Boerhaave (1668-1738). This great clinical teacher attracted young men from all over Europe, and among these were five who were destined to form the original faculty of the Edinburgh School of Medicine. The first of the five was Alexander Monro (1697-1767), whose father—a surgeon in the army of William of Orange—was so impressed with the Leiden medical school that he sent his son to study there; and when the young man returned home to Edinburgh he was appointed, in 1720, to the chair of anatomy—the new medical school's first professorial appointment. Later on, in 1726, four additional professorial appointments were made; this quartet, like Alexander Monro, were all Leiden-trained Edinburgh men—John Rutherford, Alexander St. Clair, John Innes and Andrew Plummer. These appointments (together with the earlier one of Monro)—of Rutherford and St. Clair to chairs of the theory and practice of physic, and Innes and Plummer to chairs of medicine and chemistry—gave the Edinburgh School of Medicine (which opened in 1726) a fine start; and within a few years—specifically after the opening of the Royal Infirmary of Edinburgh in 1738 (14)—the Edinburgh School of Medicine became the most famous in Europe, a happy state of affairs which lasted until the rise of the great French clinical school in Paris in the early years of the nineteenth century.

During the second part of the eighteenth century—Edinburgh's greatest years—among the many students from abroad were several from America who were destined to make distinguished contributions to medical education on their return. The names of William Shippen (1736-1808) and John Morgan (1735-1789) come immediately to mind—the cofounders (15) in 1765 of America's first medical school (the University of Pennsylvania School of Medicine, Philadelphia); Shippen obtained the Edinburgh M.D. in 1761 and Morgan in 1763. And there was the great Benjamin Rush (1745-1813), who obtained the Edinburgh M.D. in 1769 and who returned to America to join the original medical faculty of the University of Pennsylvania School of Medicine, to be signer of his country's Declaration of Independence, and later on (1812) to write America's first text on mental diseases. Of a slightly later vintage was Philip Syng Physick (1768-1837)—the so-called Father of American Surgery—who obtained the Edinburgh M.D. in 1792, and who became in 1805 the first Professor of Surgery in the University of Pennsylvania School of Medicine. Still later, Edinburgh's influence was to be felt in Canada, for in 1823 four young Edinburgh-trained physicians (William Caldwell, Andrew F. Holmes, William Robertson, and John Stephenson) opened the Montreal Medical Institution; this was Canada's first medical school, and in 1829 it became the Medical Faculty of McGill University.

Table 1 shows the number of M.D. degrees granted by the University of Edinburgh from 1726 until 1840. One cannot, of course—except perhaps nowadays—expect the number of graduates to reflect the actual number of students attending a medical school. In fact, only a minority of medical students attending the Edinburgh School of Medicine in its early years obtained the M.D. degree. This is well seen from Table 2, which shows the actual number of medical students at Edinburgh during the decade 1790-1800; during this period only 229 M.D. degrees were granted. Table 2 also shows the preponderance of medical students over those of other faculties.

Two of the most important questions in medical education are: how may a school achieve greatness, and how may that greatness be maintained? The factors responsible for Edinburgh's greatness during the eighteenth century were undoubtedly her stimulating teachers, imbued directly or indirectly

TABLE 1
NUMBER OF M.D. DEGREES GRANTED BY THE UNIVERSITY OF EDINBURGH, 1726–1840*

Years	Degrees	Years	Degrees	Years	Degrees
1726–1738	21	1771–1780	213	1811–1820	831
1740–1750	53	1781–1790	280	1821–1830	923
1751–1760	105	1791–1800	197	1831–1840	1124
1761–1770	103	1801–1810	493		

* Based on figures obtained from *Nomina Eorum qui Gradum Medicinae Doctoris in Academia Jacobi Sexti Scotorum Regis quae Edinburgi est, Adepti Sunt.* Edinburgh, 1846.

TABLE 2
Comparative Numbers of Medical (Physic) Students at Edinburgh, 1790–1800*

Years	Literature and Philosophy	Divinity	Law	Physic	Total
1790–1791	426	128	129	510	1193
1791–1792	473	131	129	547	1280
1792–1793	453	133	142	581	1309
1793–1794	464	137	124	527	1252
1794–1795	470	146	154	525	1295
1795–1796	427	140	143	508	1218
1796–1797	496	130	156	577	1359
1797–1798	471	125	154	591	1341
1798–1799	461	125	124	592	1302
1799–1800	446	127	97	634	1304
TOTALS	4587	1322	1352	5592	12,853

* From data provided by A. Duncan, Sr., and A. Duncan, Jr., in the *Annals of Medicine for the Year 1799*, Edinburgh, 1800, Vol. IV, p. 537.

with the Boerhaave tradition—and also the inspired use of the clinical facilities provided by the Royal Infirmary. Edinburgh's decline in fame during the early years of the nineteenth century is less easy to pinpoint, but among the precipitating factors were such things as inferior academic appointments, nepotism (16), faculty squabbles (17), and the competition provided by extramural schools and teachers; and there were important outside factors such as the Apothecaries Act of 1815 which meant that Scottish medical degrees were not recognized in England until the Medical Act of 1858 abolished regional licenses.

The so-called brain drain currently arouses a great deal of concern in Britain, but is no new phenomenon regarding Scottish medical graduates—Scotland's medical brains have been drained off for more than two centuries. Scotland indeed has been—and still is—one of the greatest suppliers of trained medical men in the world. For example, in 1750 Scottish university medical graduates comprised forty-one per cent of all university medical graduates practising in London, and sixty-one per cent of those in the Provinces; and by 1850—despite the Apothecaries Act—the figures had risen to sixty-five per cent and seventy-nine per cent respectively (18). As regards actual numbers of medical graduates in Britain, Table 3 shows the dramatic rise in the number of Scottish medical graduates from 1601 to 1850 (19).

In the eighteenth century—and indeed well into the nineteenth century—Scottish medical education was far superior to its English counterpart; but, as Newman has observed, "It is extraordinary not only that it was not more imitated, but also that the infiltration of Scottish teachers into England, which is no new thing, did not lead to a much wider adoption of Scottish

TABLE 3
Rise in Number of British and Scottish Medical Graduates, 1601–1850

Year	Oxford and Cambridge	Continental Universities (not incorporated at Oxford and Cambridge)	Scottish	Total
1601–1650	599	36	None	635
1651–1700	933	197	38	1168
1701–1750	617	385	406	1408
1751–1800	246	194	2594	3034
1801–1850	273	29	7989	8291

methods. The English system must have been extremely tough to have resisted it, but resist it it did, until in the present century many features more Scottish than English have been adopted" (20).

Much could be written concerning Scottish postgraduate medical education; suffice it to say that the four Scottish medical schools have fine programmes (Edinburgh and Glasgow being particularly active) and attract students from all over the world.

For a country of such small size Scotland's contributions to the history of medical education have been little short of prodigious; and with each of her original medical schools flourishing vigorously at the present time Scotland can confidently be expected to contribute as significantly in the future as she has done so many times in the past.

NOTES

1. C. H. CRESWELL, *The Royal College of Surgeons of Edinburgh.* Edinburgh, 1926, p. 4.
2. For full details of this important institution, ALEXANDER DUNCAN's *Memorials of the Faculty of Physicians and Surgeons of Glasgow, 1599-1850*, Glasgow, 1896, should be consulted.
3. *Lancet*, 1827, 12: 344.
4. J. COUTTS, *A history of the University of Glasgow.* Glasgow, 1909, p. 546.
5. *Ibid.*, p. 545.
6. For the early history of this institution see M. S. BUCHANAN's *History of the Glasgow Royal Infirmary.* Glasgow, 1832.
7. There is need for a comprehensive survey of the contributions of Scotland's extramural schools and teachers to Scottish medical education. Part of this need has been met recently by the publication of DOUGLAS GUTHRIE's short volume on *Extramural medical education in Edinburgh*, Edinburgh, 1965.
8. THOMAS LYLE (1792-1859), surgeon and songwriter (see *Dictionary of National Biography*), wrote these passages; I have transcribed his *University reminiscences* from his unpublished manuscript which I hope to publish *in extenso* in due course.
9. Until the passage of Warburton's Anatomy Act (1832)—which was un-

doubtedly hastened by the lamentable exploits of Burke and Hare in Edinburgh in 1828—the obtaining of enough subjects for dissection remained a difficult problem for British medical schools. The specific complaints of leading teachers of anatomy before the passage of the Anatomy Act are convincingly presented in the *Report from the Select Committee on Anatomy* (House of Commons, 22 July 1828).

10. *Lancet*, 1827, **12**: 343.
11. J. COUTTS, *op.cit.*, p. 541.
12. For full details see R. P. RITCHIE's *The early days of the Royall Colledge of Phisitians, Edinburgh,* Edinburgh, 1899.
13. This interesting appointment has been the subject of a recent article by G. A. LINDEBOOM, *Ned. Tijdschr. Geneesk.*, 1966, **110**: 20-26.
14. For the history of this institution see A. L. TURNER's *Story of a great hospital; the Royal Infirmary of Edinburgh 1729-1929,* Edinburgh, 1937.
15. Strictly speaking, Morgan was the promoter of the University of Pennsylvania School of Medicine; however, since Shippen had previously started his own medical school in Philadelphia but had abandoned the project in favor of Morgan's more ambitious scheme, I think the description of these men as co-founders of the University of Pennsylvania School of Medicine is reasonable.
16. Alexander Monro *tertius* (1773-1859) is an oft-quoted example. Monro *tertius'* father and grandfather had brought fame to Edinburgh in general and specifically to the teaching of anatomy, but Monro *tertius* was less gifted; Charles Darwin, who was a medical student at Edinburgh during the years 1825-1827, observed in his *Autobiography* that Monro *tertius* "made his lectures on human anatomy as dull, as he was himself, and the subject disgusted me."
17. The academic vendetta between James Gregory (1753-1821) and John Bell (1763-1820) is well known and resulted in several bitter publications; less well known, perhaps, is Gregory's physical attack on James Hamilton who held the Chair of Midwifery. Gregory beat Hamilton vigorously with his cane and was fined £100 for the offence—nothing daunted, he was heard to observe that he would pay double if he could do it again.
18. Figures derived from Table VI of A. H. T. ROBB-SMITH's chapter on Medical Education at Oxford and Cambridge Prior to 1850, in *The evolution of medical education in Britain* ed. F. N. L. POYNTER. London, 1966.
19. Figures derived from ROBB-SMITH, *op.cit.*, Table V.
20. C. NEWMAN, *The evolution of medical education in the nineteenth century.* London, 1957, pp. 12-13.

MEDICAL EDUCATION IN SCANDINAVIA SINCE 1600

WOLFRAM KOCK
Medicinhistoriska Museet
Stockholm, Sweden

In describing the history of medical education in Scandinavia it is natural for a Swede to begin with, and to concentrate upon, that of his own country. This is the more natural as the premier university in Sweden (Uppsala) is the oldest university in Scandinavia (1477) (1), and developments in the northern countries proceeded on very similar lines.

Sweden

CONDITIONS IN SWEDEN UP TO THE MIDDLE OF THE EIGHTEENTH CENTURY

It was not until relatively late that medical education became available in Sweden. Until the seventeenth century medical care was dominated by the poorly trained barber-surgeons (23) who in Stockholm had been united in a guild since 1496. Even the king, Gustav Vasa, was unable on his deathbed (1560) to have treatment from a physician, despite the fact that physicians had been periodically summoned to the court from abroad.

Scientific medical training is said to have been given for a few years in the 1580's at the *collegium* established by the culture-loving King Johan III on the island of Gråmunkeholmen off Stockholm. However, it was of little practical importance, and the first professorship of medicine in Sweden did not come into existence until 1595 at the University of Uppsala. However, the first holder of this chair, Johannes Chesnecopherus (1581-1635), did not enter upon his professorial duties until 1613. He had been mainly trained abroad, where he spent four years of medical study at the University of Marburg and took his doctorate at Basel with a thesis entitled *De hydrope* (45).

Even before Chesnecopherus and two colleagues formed a faculty of medicine at Uppsala in the 1620's, temporary medical training had come into existence at the university, where Johannes Johannis Rudbeck *the elder* (1581-1646), professor successively of mathematics, Hebrew and theology, lectured on wild herbs and their uses. His brother and fellow-professor Peter Johannis Rudbeck *the elder* (1578-1629), professor of oratory, "explained diseases, their causes and occasions" (20).

Chesnecopherus's colleagues were Johannes Franck (Franckenius, 1590-

1661), who was appointed in 1624 to lecture on physics, anatomy and botany, and Johannes Raicus (c. 1580-1632), who became a professor in 1627. When three years later Raicus moved to Dorpat (Tartu) in Estonia, he was succeeded in 1636 by Petrus Kirstenius (1577-1640), a philologist and medical man, who had attended Gustav Adolf II as physician-in-ordinary (33).

Under the university's constitutions of 1625 and 1626, one professor of medicine lectured on theoretical and practical medicine and therapy, and the other, on medicine, botany and anatomy, and also performed an annual anatomical dissection. Both professors also lectured on Hippocrates and Galen.

As in other civilized countries, so in Sweden, even during the first half of the seventeenth century, homage was paid to scholastic medicine based upon the works of Hippocrates, Galen and the Moslem writers, especially Avicenna. However, Swedish students were more attracted by studies in theology, languages, mathematics and natural science than by medical studies which held only uncertain prospects for the future. At times medical studies at Uppsala were pursued by only one or two students.

For a short time Stockholm seems to have offered serious competition to the Faculty of Medicine of Uppsala. The learned Grégoire François du Rietz (1607-1682), summoned to Sweden in 1642 to be physician-in-ordinary to Queen Christina, gave medical lectures in Stockholm for several years in the 1640's. This competition led the faculty at Uppsala to find a pretext for forbidding Du Rietz's teaching, and in the rigidly Lutheran Sweden of that time this was easily accomplished by calling attention to the fact that Du Rietz was a Roman Catholic. Nevertheless, he was of great importance as the founder in 1663 of the Collegium Medicorum, later the Collegium Medicum, a forerunner of the present National Board of Health (10, 34).

The regulations for medical education issued in the University of Uppsala's new constitutions (1655) were in force down to 1852. These regulations required that one professor be concerned with the means of safeguarding and restoring health: therapy, dietetics, surgery, and pharmacy; and that the other lecture on pathology, anatomy, botany and chemistry. The latter professor was also required to perform an anatomical dissection at least once every other year. As formerly, both professors lectured on Hippocrates and Galen. Those who wished to study medicine were first required to obtain a broad humanistic education and were compelled to sit for a theological examination in the faculty of theology and philosophical examination in the faculty of arts and sciences (1).

Under the direction of Olof Rudbeck *the elder* (1630-1702) and Petrus Hoffvenius (1630-1682) (11), medicine in Sweden underwent a period of expansion. But still no Swedish medical man had had the doctorate conferred on him in his own country, and Swedish medical education usually concluded with travel abroad for the purpose of further study.

After Rudbeck had completed his training, including a journey to Leiden for study under Van Horne, he returned to Uppsala and worked there as professor of medicine from 1660 onwards. He literally created a memorial for himself in the magnificent anatomical theatre which he built from his own drawings and partly with his own hands in the Gustavianum at Uppsala. In this theatre four public anatomical demonstrations took place during Rudbeck's time. Later such demonstrations also took place in other towns in which medical training was given, such as Stockholm, Lund, Dorpat (Tartu), Åbo (Turku), and possibly Greifswald, whose university was under the jurisdiction of Sweden during the period 1637-1815 (20).

The already-mentioned Collegium Medicum, founded by Du Rietz and three other doctors practicing in Stockholm, was intended to be a counterpart to the medical societies that existed in many places abroad. The founders seem to have thought of forming a faculty of medicine in Stockholm, together with a hospital. However, Olof Rudbeck considered that this proposal "would be the greatest disaster for the Faculty of the University". The Stockholm faculty did not come into existence, probably largely through Rudbeck's influence with Magnus Gabriel de la Gardie, the Lord High Chancellor and Regent. But on 16 May 1663 the Collegium was granted the right of inspection, and in certain cases the right of examination, of physicians and surgeons. These privileges were confirmed in 1680 and further extended by a Royal letter in 1684. In the Medical Ordinances issued on 30 October 1688, but not printed until 1699, it was declared that prospective district medical officers and municipal medical officers in the provinces must, like the Stockholm physicians, undergo the *examen publicum practicum* before the Collegium. The surgeons, on pain of fine, were forbidden to treat internal diseases in Stockholm or other towns that had physicians and must, if possible, have a physician in attendance at all serious surgical operations—which, however, the physician could not perform in place of the surgeon.

However, by a Royal decision on 26 April 1737 the Collegium was deprived of the right to examine and licence physicians with Swedish doctoral degrees. From this time an increased number of Swedish medical men began to take the doctorate, not only as formerly in Holland and Germany, but also at Swedish universities. With about a million inhabitants in Sweden proper and 2.5 millions in the whole kingdom during the period 1668-1710, Sweden had no less than five universities: Uppsala, Dorpat, Åbo, Greifswald and Lund (20).

The surgeons' interests were looked after by the Surgical Society, which in a way was a counterpart to the Collegium Medicum, with which it carried on a more and more unequal struggle for power. It was charged with the task of seeing that the Stockholm barber-surgeons employed good journeymen and apprentices so that the needs of field surgeons in the Swedish army and navy could be met. The training provided for the barber-surgeons was long organized on purely practical lines and consequently very defective in other aspects. How a young barber-surgeon's apprenticeship was spent appears from a letter from the Collegium Medicum to the Government Board in 1747:

The apprenticeship is mandatory for three years but is commonly extended to four, five and six, according to contract and what funds the apprentice possesses. When the years of apprenticeship are over, the apprentice receives the official printed certificate, whether he be skillful or unskillful. The first few years are mostly spent doing small tasks and waiting at table, both for the household and for the journeymen, until he gradually becomes accustomed to wielding the razor, opening veins, applying plasters and at most bandaging a wound or a fracture, and he may, in addition now and then be permitted to see a few operations performed by his master. He must learn the rest by reading and practice and the copying of certain manuscripts in which he must demonstrate extraordinary diligence and inclination, although often after he has completed his apprenticeship and become a journeyman.

In 1689 the Surgical Society of Stockholm became the only authority to hold surgical examinations in Sweden. Those who were particularly interested in the profession had access to the same medical training as the medical students, and many barber-surgeon apprentices "travelled in their art", like journeymen in other crafts, mainly in Germany but later also in France where surgery was on a much more scientific level (24, 30).

The training which the Collegium Medicum carried on was concerned chiefly with anatomy. Despite a proposal by four members of the Collegium for real scientific activity by that organization, it came to nothing and even the anatomical demonstrations were rare. During this period a number of doctors gave private lessons in anatomy and some public anatomical demonstrations were held in Stockholm. In 1716 a professorship of anatomy was established in Stockholm and given to Magnus Bromelius (1679-1731), with Salomon Schützer as his assistant as *anatomiae et chirurgiae operator publicus*. Bromelius received no salary and anatomical instruction again stagnated after he relinquished his appointment in 1720.

Sweden's greatest name in natural science is undoubtedly that of Carl von Linné, or Linnaeus (1707-1778). He practiced medicine in Stockholm until he was appointed professor of medicine at Uppsala in 1741, where he was very much esteemed as a teacher in the fields of botany, *materia medica*, dietetics and natural history. His lectures on dietetics, what we now call hygiene, were received with very great interest and brought together audiences numbering several hundreds (47).

No clinical instruction took place in Stockholm during the first half of the eighteenth century, despite some admonitory examples from abroad. In Uppsala such instruction had been fairly successfully established, where Lars Roberg (1664-1742), the learned and eccentric professor of medicine—who wrote some interesting advice to the young medical students (22)—in

1717 founded a *nosocomium academicum* in a little house close by Riddartorget. Here clinical instruction was given for a short period, but Roberg was unsuitable as the head of a clinic and the *nosocomium* fell into decay. It was re-established in 1741 by Nils Rosén von Rosenstein (1706-1773), the leading medical man of the period in Sweden (27).

Thus medical students and young physicians had still to a large extent to make their clinical observations, as in Hoffvenius's time, by following the senior physicians in their professional practice. Probably they also had opportunities of studying at the Danvik Hospital, the Orphan Asylum and the Johannis Cottage Hospital in Stockholm, to which the students of Uppsala paid visits. A proposal put forward in 1748 for practical medical instruction in connection with the work of the parish doctors (the present district medical officers) in Stockholm led to nothing.

The Serafimer Hospital and Medical Training in Stockholm During the Eighteenth Century

In 1561 Vilhelmus Lemnius (1520/1530-1573) (12), Erik XIV's Dutch physician-in-ordinary, produced a plan for the organization of the medical system, including a teaching hospital (20, 31). However, this plan was not to be realized until a much later date. In the reign of Johan III, brother and successor to Erik XIV, stress was laid on the importance of the municipalities taking over the care of the sick and poor. A third brother, who in turn succeeded to Johan III under the name of Karl IX, determined that the sick should be cared for in the small cottage hospitals, which had existed in Sweden under the name of "hospital" since the Middle Ages, and the poor, in the community centers. In 1624 Gustav Adolf II, known in history as the King of Sweden during the Thirty Years' War, directed that a hospital should be established in every province, and his daughter and successor, Queen Christina, promulgated a law on social welfare that became a standard for similar regulations in the Swedish church law of 1686, in respect to hospitals.

The small country hospitals remained, as did much of the old system, and the district medical officers who were coming into existence were entrusted with the charge of the hospitals, confirmed in the first instructions to district medical officers in 1744.

Interest in social-welfare questions increased in Sweden, as in other countries during the eighteenth century, but, curiously enough, Stockholm still had no hospital for acute cases at the beginning of the century. Thus it was that Nils Boy (1683-1739), a municipal medical officer, presented a memorandum to the 1731 and 1734 Riksdag sessions on the establishment of a hospital in Stockholm for the sick poor, which would also serve as a teaching hospital. It was likewise realized as important that this hospital serve the patients from the Stockholm garrison.

In 1783 a state grant from the customs duty on tobacco became available for the project as well as certain other tax receipts and voluntary contributions with which Gripenhielm House, a private property on Kungsholmen in the western part of Stockholm, was purchased. As soon as it was barely adequately converted, the new hospital opened its doors on 30 October 1752 with eight beds, as the national hospital for Sweden and Finland which then formed a united kingdom. Abraham Bäck (1713-1795) was appointed senior physician and Olof af Acrel (1717-1806) (28), senior surgeon.

At first the teaching at the hospital, which had no connection with that given at the universities or by the Surgical Society, was not compulsory except for those who intended to seek public employment as physicians or surgeons. In the regulations issued in 1755 by the hospital's governors—two Knights of the Order of the Seraphim, since the hospital was under the control of that Order (founded in 1748)—there were directions for medical instruction which seem somewhat modern. Originally the doors had been open, even the doors of the operating theatre, to all physicians and surgeons, but gradually this was found to be impractical and it was decided that not more than four of the student physicians should be admitted at one time. They were obliged to take part in the work of the senior physician and the senior surgeon and were themselves responsible for two or three patients, whom they were required, under supervision, "to treat, bandage and nurse, as occasion requires". The senior physician was obliged to "give one lecture each week to the young physicians and surgeons on the patients, their diseases and their present condition, and on the remedies which have been adopted for their maintenance." The students were also required to be present at operations and deliveries. The rules in the regulations of 1788 were similar but also provided that newly enrolled surgeon-apprentices could also receive instruction at the hospital.

During the first few years the young physicians serving at the hospital were subjected to examinations, both during the actual period of service and afterwards, before certificates could be issued testifying to the completion of their service. The student physicians were also given opportunities to carry out a number of operations under the direction of their supervisor.

In the medical department the senior physicians, who changed fairly frequently, did not seem to have exercised a similar influence on the medical training.

The professorships of medicine which came into existence in Stockholm in the middle of the eighteenth century (25) were of the greatest importance for future developments. They emanated from the scientific lectures that were given on the initiative of some of the members of the Surgical Society, among them Acrel. These lectures began in 1746. They dealt with different parts of surgery and were held on premises rented in the city. In 1747 the four lecturers presented a petition to the King-in-council for permission to carry out public surgical operations in the city's anatomical theatre both for their own practice and for training others. The petition was granted on 10

March 1749, the same day that Abraham Bäck was appointed professor of anatomy, a post which he soon relinquished, however. A tug-of-war now began between the Collegium Medicum and the Society in regard to teaching.

In a Royal letter to the Collegium Medicum, 10 August 1752, orders were given for the establishment of a professorship and a prosectorship in anatomy and surgery, primarily for the training of field surgeons. At the same time it was decided that all journeyman surgeons, regimental surgeons and surgeons in the navy should produce a certificate from the professor that they had a satisfactory knowledge of anatomy and surgery before undergoing examinations. The first holder of this professorship was Roland Martin (1726-1788) who was not appointed until 31 January 1756. When Anders Johan Hagströmer (1753-1830) was appointed to this professorship in 1793, a surgeon became professor of anatomy and surgery. The lectures were attended by surgeon apprentices and by some medical students as well.

Training in midwifery had been carried on in Sweden ever since the days of Johan von Hoorn (1662-1724) who, after studies at Paris and elsewhere, began to give public lectures in Stockholm on both anatomy and obstetrics. In 1697 he published, for the use of midwives, a book entitled *The Well-Trained Swedish Midwife*, and in 1715 a continuation entitled *Siphra and Pua*, after two midwives mentioned in the Bible (4). In 1723 he was officially appointed a teacher of midwifery. One of his successors, Eric Elff (1718-1761), also succeeded Abraham Bäck as senior physician at the Serafimer Hospital in 1753. There Elff set up the first obstetrical clinic in Sweden designed for the training of physicians. By a Royal letter of 1757 "the tutor and director of midwifery" was enjoined to urge, through the agency of the Collegium Medicum, all physicians to attend the instruction in midwifery before they were promoted to medical officer. In 1761 David Schulz von Schulzenheim (1732-1823) was appointed Elff's successor with the duty of lecturing to young physicians and surgeons and to prospective midwives, who were to be present at the hospital deliveries and also at serious cases in the professor's private practice. However, Schulzenheim found that the obstetrical work at the hospital was insignificant, and so he founded a separate maternity hospital, first in his own house and later, in 1776 in Rödbodtorget in Stockholm.

Botanical gardens and chemical laboratories had been established at the medical schools a little earlier. Urban Hiärne (1641-1724), one of the four notable physicians of that day with Rudbeck, Hoffvenius and von Hoorn, who contributed to the first period of expansion in Swedish medicine, established both a botanical garden and a chemical laboratory in Stockholm (35). In the middle of the eighteenth century sporadic lectures on botany were given in Stockholm and a botanical garden was also established at the Serafimer Hospital. In 1761 a further professorship of natural history and pharmacy was created and given to Peter Jonas Bergius (1730-1790). The audiences at his lectures included medical students, surgeons and students of pharmacy. Despite the fact that the title of the professorship was later altered to include medicine, no clinical instruction in medicine took place.

The three new professorships were loosely associated; they were all dependent on the Collegium Medicum and had no right to examine. However, they were to constitute the origin of the Karolinska Institute.

Partly because of the hindrance presented by the compulsory preparatory studies in theology and philosophy, the period of medical studies in the middle of the eighteenth century was still very long. At this time the purely medical examinations included a *disputatio exercitii gratia,* an *examen theoreticum* (after which the student became *medicinae candidatus*), and *examen practicum* (which entitled him to the degree of licentiate of medicine), and a *disputatio pro gradu.*

At the same time the surgeon's studies (24) comprised three years for an apprentice and four years for a journeyman or, after 1771, *studiosus chirurgiae.* After this the surgeon who was summoned to take up some vacant post as field surgeon (only in 1792 was this restricted to surgeons qualified for the post) could undergo the examination for master of surgery before the Surgical Society in the presence of, *inter alios,* the president of the Collegium Medicum who examined the candidate in physiology and *chirurgica medica.* The requirements of this examination were extensive and were supplemented in the 1755 regulations by a practical test in which the candidate was required to carry out all manual operations and motions that occurred in surgery ("these may be more carefully executed on dead bodies, when the hospital's organization affords opportunity in the future"). However, the operation test was not carried out at the hospital but in the anatomical theatre in the city. At this period the bath attendants, with their journeymen and apprentices, also took part in nursing the sick, despite their slight or complete lack of knowledge of medicine, and a number of apprentice and journeymen bath attendants were also seen to be present at the lectures. Thus, surgery and medicine were quite separate but, when the ignorance of surgery among the physicians was found to be too manifest, a Royal letter of 1752 enjoined the medical students to acquire at least a minimum knowledge of surgery. This was further emphasized in a letter from the Collegium Medicum in 1755.

The common training in surgery and obstetrics given to physicians and surgeons since 1753 at the Serafimer Hospital and later also at the Maternity Hospital seems to a large extent to have contributed to the union of the two professions. More and more students were trained as medico-surgeons, a development of which Acrel, as a pioneer in the field, was a zealous advocate. However, it was not until 1797 that the formal union took place through a Royal letter by which the Surgical Society was dissolved and the Collegium Medicum alone took charge of both medicine and surgery in Sweden.

Through this new arrangement it became the duty of the faculties of medicine to train the physicians, while the Collegium Medicum attended to the training of the surgeons. According to the above-mentioned Royal letter, the surgeons were to serve for six months at the hospital in Stockholm. After passing the examination of master of surgery, the student received a diploma as *chirurgiae magister*. The surgeons had now been promoted from the status of craftsmen to that of scientific practitioners. Owing to the altered regulations, licentiates and doctors of medicine, by serving six months at the hospital and passing the third part of the examination for *chirurgiae magister* (the operation examination, as it was called), could become *chirurgiae magister* and thus have an opportunity of obtaining a post as surgeon.

Clinical Training in the Nineteenth Century, Especially at the Karolinska Institute

At the beginning of the nineteenth century the training at Serafimer Hospital, which had played such a prominent part in the training of medical students in Sweden, seems not to have been so efficient as it had been in Acrel's day. For this reason Sven Anders Hedin (1750-1821) proposed to the Collegium Medicum that real clinical training be introduced. The administrative board at the hospital expressed approval and the King-in-Council decided, in a Royal letter of 1802, that clinical lectures be given twice each week by the senior physician and senior surgeon and that the student physicians serving at the hospital receive graded marks for diligence and skill.

At this period an important new medical institution came into existence as a consequence of the serious defects revealed by the Swedish military medical services both in 1788 and in the Finnish war against Russia (1808-1809). On the initiative of Erik Carl Trafvenfelt (1774-1835), the Riksdag decided to establish an Institute for the training of army surgeons. This was founded by a Royal letter in 1810 (33). According to an ordinance of the same year concerning the Institute, which later received the official name of Karolinska Medico-Surgical Institute, there were to be among the professors "one for *medicina theoretica* and one for practical medical science, to which two professors the respective parts of theoretical and practical surgery will also belong." Moreover, in a Royal letter of 1813, by which the old Collegium Medicum was replaced by a Royal Health Collegium, it was declared that "the Hospital of the Order of the Seraphim" would be available to the students of the Institute, who were also to be entitled to make use of the clinical lectures given there. The surgical examinations were not transferred to the professors, though the members of the Collegium had the right to take part and the latter institution to issue diplomas.

The Serafimer Hospital (21) still occupied its exceptional position. Curiously enough, both the senior physician and the senior surgeon, with the later approval of the King-in-Council, opposed the idea of the Hospital being connected as closely as possible with the Karolinska Institute. The Institute was, of course, originally a school for surgeons, but it had a good many scholarships at its disposal and a number of its students had opportunities to secure salaried posts as assistant physicians at the Stockholm Garrison Hospital, sometimes without any clinical training in medicine. This was certainly one of the reasons why it now became more common for medical students to transfer their studies not only to the faculties of medicine but also to the Institute. As a further curiosity it may be mentioned that the shortage of doctors led to some interest in medicine on the part of the clergy. Even some theological students held scholarships and were students at the Institute, where they passed the *chirurgiae magister* examination.

In 1812 the King-in-council decided that the *chirurgiae magister* examination should not be compulsory for students who wished to proceed to the doctorate of medicine. But no one was allowed to be nominated for a district medical officer, or similar post, unless he held both degrees.

The 1815 ordinance for the Medico-Surgical Institute (21) required the *Professor medicinae et chirurgiae theoreticae* to lecture on physiology, theoretical medicine and surgery for four hours each week during the term. He was also to instruct in pathological anatomy and oversee preparation of pathological and anatomical specimens. The *Professor medicinae et chirurgiae practicae* was to instruct in pathology and *therapia specialis* and "the principles of surgical bandaging and operations". The last instruction was to be given in the form of operations on cadavers in the anatomical theatre. The professor's lecture program included the principles of "diet and the method of establishing and carrying on medical care in the army and navy in war and peace".

For admission as a student to the Institute, it was required that the applicant pass a medico-philosophical examination at one of the universities or an entrance examination in which special importance was attached to knowledge of languages (Latin, German and French), mathematics and physics. Three medico-surgical examinations, which were held in public, were to be passed at the Institute.

The first examination included natural history, chemistry, anatomy and physiology. For admission to the second examination, that for the "candidate in surgery", comprising theoretical and practical medicine, surgery, obstetrics and medico-legal knowledge, a certificate was required that the applicant had served for a certain period at the Garrison Hospital and the medical department of the Serafimer Hospital. In order to be able to sit for the third examination, in the principles of surgical operations and bandaging and obstetrical operations, the applicant had to have a certificate of six months' service at the General Maternity Hospital. However, licentiates and doctors of medicine who wished to gain the degree of *chirurgiae magister* needed only to carry out the service and undergo the testing required for the third examination. This regulation placed the masters of surgery on an equal footing with the doctors of medicine and both could receive diplomas from the Health Collegium as registered physicians.

In 1816 the Karolinska Institute was moved to a site on Kungsholmen, opposite the Serafimer Hospital, and housed in premises which had formerly been used as a glassworks. It remained there until 1946 when it was removed to its present quarters in the Norrbacka area of Solna, a town just outside Stockholm.

According to tradition, the examinations taken at the Institute were preceded by oral examinations. In 1817 the Health Collegium decided that those who underwent the operation examination should first be examined orally as to their knowledge of cutting for hernia, applying ligatures, trepanning and amputation, that is, knowledge that was of special value to army doctors. In the future the oral examination was to be concerned with operations connected with the examination, which the examinee selected by drawing lots "and which he should also perform".

In 1822 it was decided that no one should be allowed to enroll at the Institute who had not passed "at least the examination for *medico-philosophiae candidatus* at either of the universities." For admission to the third examination certain periods of service, amounting to at least a year, were required, distributed between medical, surgical, and obstetrical clinics and "visits to the General Orphanage at the times when the physician there is making his rounds and when inoculation against smallpox is being carried out". Thus, the third examination had been changed from a pure operations examination to a real master's degree. In 1828 the Health Collegium went the whole length and divided the third examination into an operations examination, which was taken at the Institute, and an examination for the master's degree, which was taken at the Collegium.

Owing to the similar amount of preliminary knowledge required and the expansion of the third examination to form a real master's degree, the examination of 1822 placed all the students at the Institute on an equal footing with those in the faculties of medicine. This led to a very prolonged and heated dispute between the Institute and the faculties of medicine, especially that of Uppsala, carried on primarily by Jöns Jacob Berzelius (1779-1848) (17), professor of medicine and pharmacy, and later by Anders Retzius (1796-1860), professor of anatomy, on behalf of the Institute, and by Israel Hwasser (1790-1860), on behalf of the faculty. This dispute also became political, since the university was supported by the King and the government, and the more democratic Institute, if the word "democratic" may be used without misinterpretation, by the Riksdag.

In 1825 a training committee was appointed. Its work was to result in a real union between the Institute and the Serafimer Hospital for clinical training. According to a Royal letter of 1835 to the Health Collegium, there were to be at the Institute, besides the other professors, a professor of medicine and a professor of surgery, each with an annual salary of 800 *riksdalers;* the former was also to be senior physician and the latter senior surgeon at the Serafimer Hospital. As a result Carl Johan Ekströmer (1793-1860) (5) was appointed to the professorship of surgery. He was the first university-trained student at the Karolinska Institute and had been appointed to the post of senior surgeon at the Hospital at the age of 28. Though Acrel had formed an extremely important school of surgery at the Serafimer Hospital, it was through Ekströmer that a surgical clinic in the modern sense came into existence. He was an unusually energetic and capable man, with great practical, surgical and administrative experience, and in 1849 became director-general of the Health Collegium.

A medical clinic was still needed, however. The teaching carried on at the Garrison Hospital seems scarcely to have justified the description of "clinical". A doctor who carried out his clinical service in the medical department of Serafimer Hospital in the 1830's described it as follows:

The duties consisted in accompanying the senior physician on his morning round, in which he examined the patient very briefly, gave the name of the disease and prescribed the medicine. It was very exceptional for him to say anything in explanation of the disease and its symptoms or to give the reasons for prescribing the medicine. The students' greatest exertion was to copy any prescription formulae which were prescribed in order to make a *vademecum practicum* from them.

A modern system of medical and clinical training was introduced by Magnus Huss (1807-1890) (13), who in the late summer of 1838 returned from an 18 months' tour of study abroad. It had seemed to him important to acquire some knowledge of auscultation and percussion, which at that time were practiced only by a few physicians in Sweden.

Huss returned to Stockholm rich in knowledge but poor in purse and succeeded in getting a post in the medical department of Serafimer Hospital where the same year he was given charge of the clinical instruction of the medical students, and in 1840 became senior physician. In the autumn term of 1840 Huss became associate professor at the Karolinska Institute, and as head of a clinic, it fell to him to hold perhaps the most responsible post in Swedish medicine. He may be said to have been the creator of the first Swedish clinic devoted to internal medicine. His program for medical training contained three main points: correct diagnosis, based on physical examination; simple treatment suitable to the diagnosis; and careful comparison between the symptoms of the disease and the postmortem changes. He gave clinical instruction partly at the bedside on four days each week and partly as formal lectures twice each week. The medical students also were now allowed to make examinations of the patients, to start journals and to follow the patients' progress under the direction of the instructor. The instruction was given in a separate part of the medical department called *Clinicum*. It was not until 1846 that the union between the Karolinska Institute and the Serafimer Hospital became complete; in that year Huss was also appointed a professor of medicine at the Institute.

The reports of the Karolinska Institute (21) provide us with information about the instruction in the medical and surgical clinics at that period. Thus, if we select the year 1851, Huss's program during the spring term was "the acute and chronic chest diseases and typhus fever" and during the autumn term "the diseases of blood circulation and the respiratory organs." The professor also directed the "pathological sections" in the department and "interpreted the phenomena connected with them."

The number of medical students serving at the Hospital was twenty during the spring term and seventeen during the autumn. The assistant physician in the clinical department, who at this period was Per Henrik Malmsten (1811-1883) (29), gave lectures and demonstrations on dermatology "as well as the chronic ailments of the spinal cord, the stomach and the kidneys." He also had to direct and demonstrate pathological sections.

What was the corresponding program in the surgical clinic? There the professor was Carl Gustaf Santesson *the elder* (1819-1886) (21). During the spring term he spoke on "hernia, *anus praeternaturalis*, and *fistula stercorea*", the operations connected with these complaints, the diseases of the rectum, and congenital malformations in the genito-urinary organs. During the autumn term he dealt with the diseases of the genito-urinary organs, the operations related to them, and with the ligation of arteries. During both terms the program included the "diagnosis and treatment of the patients admitted to the surgical clinic and their external complaints, and also demonstrations and examinations of malignant growths." The number of medical students was as great as in the medical clinic.

The duties of the medical students give us a good idea of the detailed arrangement of the teaching. The student, who was either a "candidate of medicine" or had passed the first part of the examination for the degree of *chirurgiae magister*, presented himself each morning to take part in the senior physician's or the deputy senior physician's rounds. At the beginning of his service the student accompanied the assistant senior physician on his rounds in the non-clinical wards, also accompanied the assistant physician on his round, and carried out ordinary venesection, postmortem examinations, etc. In the following week he went over to the clinical wards where he performed the same duties. He also had certain patients allotted to him for the recording of anamneses, the keeping of journals, etc. During his period of service, he was also to take his turn on the daily duty roster at the hospital. In the absence of the assistant surgeon, assistant physician, clerk, the student-physician took their places and received, allocated beds and gave initial treatment to the patients. This also applied to acute cases admitted in the absence of the physicians. The student also assisted when necessary, in night supervision of patients recently operated on. Thus the daily duty roster, with it multifarious tasks, was, even at that period, a valuable complement to the student's training. He was also to attend the evening rounds "as often as possible" and to "try to secure from the ward's clerk the information, particularly information on the methods of physical examination, which there was not time to get on the morning rounds."

During the auscultation period, which, like the service period proper, lasted four months, the student was enjoined to take part in the rounds at least three times each week and to examine and treat by himself three especially allocated cases, on which he was to start journal records and write case reports. It was only after these journal records had been handed over to the professor that the student received a graded mark for his service. This was entered in the student's service record and reported to the Health Collegium.

There were similar instructions for the student surgeons. The student presented himself each morning at a definite time in the operating theatre, equipped with scissors, tweezers, probe and lancet, to take part in the rounds. Each student was responsible for taking over, in his turn, the supervision and care of the incoming patients allocated to him. Records were taken under the supervision of the senior surgeon or his deputy, who also appointed examiners in suitable cases. The student kept a careful case record, possibly combined with a postmortem record, of a patient selected by the senior surgeon. He had also to carry out bloodletting and "such operations as may be contemplated and which the senior surgeon entrusts to him to carry out under his own supervision or that of the surgeon concerned (the assistant surgeon or the clerk)". As in the medical department, the student had to carry out postmortem examinations and watch over the patients who had recently been operated on. In the surgical department also there was a rule that the students must themselves examine, treat and give complete journal descriptions of three cases, together with case reports. One of these was to be an ophthalmological case, since ophthalmology was still part of surgery. There was also a daily duty roster in the surgical department and graded reports were issued to the students.

The Serafimer Hospital and the Karolinska Institute played a leading part in Swedish medical training, and many of the developments in medical training in Sweden are reflected particularly in the history of the Karolinska Institute and Stockholm generally.

DEVELOPMENTS DURING THE LAST HUNDRED YEARS

The conflict between the Karolinska Institute and the universities continued during the 1840's, 1850's and 1860's, when the Institute gradually succeeded in establishing professorships in additional subjects, such as pediatrics (1845) and pathological anatomy (1854). When the statutes of Uppsala (1655) and those of Lund (1668) were revised in 1852 (1), a year of clinical service in Stockholm was prescribed before the student could take the licentiate of medicine degree in the Faculties of Medicine. This, of course, gave rise to renewed controversy. In order to gain the doctorate of medicine, it was now necessary to write an essay in Latin on a given subject and undergo examination in certain subjects that were of importance for medical students (physics, chemistry, botany, zoology and Latin) in the faculty of arts and sciences, pass the "candidate of medicine" examination, perform a certain amount of clinical service, write a thesis for the licentiate of medicine degree, undergo the examination for this degree, and finally publish and defend a thesis for the doctorate. A Royal letter of 22 April 1853 prescribed that those who entered for the licentiate of medicine examination should produce a certificate of service "as assistant physician . . . in the Stockholm clinics for at least a year." The professors at Karolinska Institute drew up a proposal concerning the distribution of this service, but the university faculties protested, and in 1859 a Royal Commission was appointed to prepare proposals for reorganizing medical training. It was not until 1861 that the King-in-Council came to a decision that had been awaited with great excitement. This decision amounted to saying that the faculties alone should be entrusted with the "candidate of medicine" examination, while the licentiate could be taken either in the faculties or at the Institute. It was a very important decision for Swedish medical education, since it simultaneously abolished the *chirurgiae magister* and the master's degree. The licentiate entitled the candidate to hold most of the medical posts in the country. By statutes promulgated on the same day the Karolinska Institute was released from its previous dependence on the Health Collegium and became almost comparable to a medical faculty. A new kind of post (*docentur* or assistant professorship) was also established at the Institute.

It may now be asked what categories of students there were at the Karolinska Institute and consequently at Serafimer Hospital at this period. The students at the Institute were primarily licentiates and doctors of medicine who wished to become *chirurgiae magister*. In 1851 the Institute had 110 students and in 1860-1861, 133. A number of Finns also studied in Stockholm. The training given at the Institute seems to have been of a high standard at this period, as is clear from official acknowledgements from abroad. In 1853 it applied for and received the right to be included in the list of medical schools approved by the Royal College of Surgeons of England. Attendance at these medical schools entitled the student to apply for the MRCS and FRCS. It was natural that the spokesmen of the Karolinska Institute should continue the struggle—which had political overtones—to make the Institute into a complete faculty with the right to hold examinations also for the "candidate of medicine" degree. This right was assigned by a Royal letter in 1873, operative in 1874. It may also be added that since 1870 women have been entitled to study medicine in Sweden.

On 13 November 1874 the Karolinska Institute was also granted the right to hold disputations for doctoral degrees, but it was not until 1906 that it received the right to confer this degree (21).

A fresh change was introduced into medical studies by the examination regulations issued in 1907. For the licentiate of medicine six months' preparatory service at the medical and surgical clinics of either of the universities was required, as well as four months' service at the medical clinic of the Institute, a further two months' service as assistant physician at the Institute or at either of the universities, two months' service at the neurological clinic of the Institute, four months' service at the surgical clinic of the Institute and a month's service as assistant surgeon at either of the universities or at the Institute, two months' service at the ophthalmological clinic of the Institute or the corresponding clinics of either of the Universities, and four months' service in pathological anatomy at the Institute. As a curiosity it may be mentioned that in these examination regulations the history of medicine disappeared as a compulsory subject. The traditional medico-philosophical examination, which had a history reaching back to 1655 was abandoned in 1908, when it was prescribed that prospective students at the university needed only produce a matriculation certificate with certain qualifications.

However, it was only when the Institute had conquered the last bastion defended by the university faculty—the right to give the preparatory instruction lasting six months—that the old conflict finally reached a conclusion. The examination requirements in medicine, surgery and pathological anatomy at the Institute were specially revised in certain provisional regulations concerning the licentiate of medicine announced in Royal letters of 1923, 1925, 1927, 1930, and 1940.

The present organization of basic medical training in Sweden is regulated by a statute which came into force in 1964 and replaced a statute of 1955 (32). The object of medical education is to give prospective physicians the necessary scientific training and a theoretical and clinical basis for further training during the practice of their profession in different spheres of medical work. This training is given in the form of lectures, conferences, seminar exercises, clinical rounds, demonstrations, laboratory work and individual tuition. The "candidate of medicine" degree is reckoned to take at most two academic sessions of ten months each and licentiate of medicine, a further four and a half academic sessions at most. In each examination subject the student must give proof, by undergoing examinations—which may be either oral or written or both—that he has acquired knowledge and proficiency in accordance with the requirements given in the syllabus.

The "candidate of medicine" degree includes anatomy and histology, medical statistics and medical genetics, general and medical chemistry, medical physics and physiology, and psychology. The licentiate of medicine degree is divided into three parts. During the first part, the propedeutic year, the student reads pathology, bacteriology, and pharmacology and takes general courses in medicine and surgery, a propedeutic course in social medicine, a course in methods of clinical investigation, a clinical laboratory course, a propedeutic course in radiology and a demonstration course in nursing technique and physiotherapy. The second part comprises medicine and surgery and also courses in clinical chemistry and clinical physiology, clinical bacteriology, X-ray diagnosis and radiotherapy (including tumor diagnosis). In addition, there are optional combinations of courses in dermato-venereology, clinical epidemiology, phthisiology, hygiene, military medicine, ophthalmology, oto-rhinolaryngology, obstetrics and gynecology, neurology and psychiatry, pediatrics, social medicine and forensic medicine. The course in child psychiatry is taken at the same time as the course in pediatrics. The third part includes four months' service as an assistant in an optional subject. This is followed by examination and registration.

The Karolinska Institute and its departments had to fight hard and energetically for equality of rights with the university faculties and their institutions. But when the victory was won, Stockholm was able to show that it had a very well-equipped school of medicine, with clinics of international standard, a school which was also to be entrusted with the task of awarding the Nobel Prize for medicine (18, 19, 32).

The greatly increased demand for medical treatment and the organization of important projects in public health and medical care in Sweden in the last few decades have resulted in a need for additional training centers, besides the Karolinska Institute and the Faculties of Medicine at Uppsala and Lund (31).

Ever since 1912 an agitation had been going on in Stockholm for the establishment of a new government teaching hospital. This agitation did not reach a successful conclusion until the 1927 Riksdag session, when it was decided to establish the Karolinska Hospital at Norrbacka. This has become the Karolinska Institute's second teaching hospital with about 1800 beds. The Institute of Radiology (Radiumhemmet), founded in 1910, the King Gustav V Research Institute, which came into existence in 1948, are also associated with the Karolinska Hospital. A year ago, clinical instruction in medicine and surgery was likewise introduced at the Söder Hospital, which was founded in 1943 and is one of the largest hospitals in Stockholm.

But let us return to the eighteenth century. In 1765 a Royal letter gave permission for the collection of funds for the building of hospitals in other places than Stockholm (23). Like Uppsala, Lund had plans early in the century for establishing a *nosocomium academicum*, with its value for teaching purposes. For financial reasons, however, the Chancellor of the university recommended instead the establishment of a country hospital in Lund. Since the university planned to use it for medical training, it was willing to contribute to the cost, and its proposal to offer one of its properties known as Munck House for the purpose was presented at a meeting at Malmö. In 1767 the hospital was established in Munck House, at first with two beds. Today Lund has a modern hospital which is also a teaching hospital.

In spite of the great developments in Lund, its clinics became overcrowded in the 1940's, which led to the idea of transferring some of the university's teaching to Malmö. Indeed, since 1908 such teaching had already been carried on at the Malmö Hospital for Infectious Diseases. In 1947 the Riksdag decided that the medical and surgical departments and the department of gynecology at the hospital in Malmö should be used for research and teaching and that the senior physicians should be appointed as professors in the Faculty of Medicine at Lund.

In order to meet the increasing need of doctors, a new medical school was founded in 1949 in Gothenburg and in 1954 was incorporated as the Faculty of Medicine in the newly established University of Gothenburg which had long had a school of humanistic studies. The Sahlgren Hospital with 2400 beds was taken into use as a teaching hospital, and with its multiplicity of departments it was well equipped for the purpose. Originally founded on the fortune of Anton Niclas Sahlgren, a director of the Swedish East India Company, and opened with 24 beds, as early as 1789 it had been determined that the hospital should be not only a charitable institution but also a teaching establishment.

For a long time the most northerly university in the country was Uppsala. However, in 1957 the Riksdag decided that propedeutic and clinical instruction should also be organized at Umea in the county of Västerbotten, where there was already a dental school and an institute of forensic medicine and where a scientific library was under construction. Clinical instruction began there in 1959, and Umea now has a complete Faculty of Medicine.

In 1958, Arthur Engel (1900-   ), recently retired as Director-General of the National Board of Health, brought forward his plan for regional medical care (43). By this Sweden was divided into seven regions, with a main "regional" hospital in which all the specialities are represented. It was natural to use these regional hospitals, with their large capacities, as regards both numbers of patients and staff for medical instruction. It is planned to locate a sixth medical school in Linköping, which possesses one of the regional hospitals and where some postgraduate training is already provided.

A list of Swedish doctors published in 1805 revealed the medical profession to consist of 281 persons, of whom 129 were doctors of medicine who did not hold the degree of *chirurgiae magister* and ninety-nine were only *chirurgiae magister*. Only fifty-two held both the medical and surgical degrees. By 1850, of 463 doctors, 396 were fully trained and only five had solely the surgical degree (2).

As medicine developed the need of specialities made itself felt during the later decades of the nineteenth century, especially in institutional medical care. However, even in 1900 there were separate hospital departments for only ten specialities comprising altogether thirty-four special departments. In 1940 the number of special departments was 332, with a corresponding increase in the number of senior physicians and in 1960 the number had increased to 778 (2).

Similar developments took place in non-institutional care, in which the number of specialists is estimated at half the number of doctors actively engaged in this form of medical care.

Up to now further training has been given chiefly through the physicians concerned taking periods of service as assistant physician at hospitals (2) where the senior physicians are specially qualified in particular fields. In recent years further training courses have been organized for medical officers in the armed forces, for county and local medical officers.

No sharp distinction has hitherto been drawn between further training and postgraduate training. Both these forms were previously left entirely to the physicians themselves, but in recent years they have gained the attention of the government and have been the subject of a commission of inquiry.

The development of the different specialities in Sweden represents a growth from the parent disciplines of medicine and surgery (23). As a result of the great importance of the treatment of syphilis at the time when hospitals began to be organized on a large scale, syphilology was the first speciality to break away from internal medicine. The special hospitals for the treatment of syphilitic patients, of which the first opened in Vadstena in 1795, were called *kurhus*, and the physicians who worked there, *kurhusläkare*. However, these hospitals were few in number and were as a rule replaced by special wards in the general hospitals. Instruction in venereal diseases was not given until 1858 in Stockholm, where an associate professorship in syphilology was established at the Karolinska Institute three years later. It is true that after the end of the nineteenth century the professor of syphilology also gave instruction in dermatology, but this subject was not formally included in his instruction until 1927.

After various discussions, and with the aid of funds from a bequest, an independent professorship in neurology was established at Karolinska Institute in 1887.

In the Swedish Public Health Act of 1874 it was prescribed that cases of infectious diseases should be treated at special hospitals or sickrooms. However, special instruction in infectious diseases was not given until 1911. In the instructional sphere developments were relatively slow and it was not until 1965 that the first professorship in clinical epidemiology was established, in Stockholm.

Tuberculosis was widely endemic in Sweden, especially during the last century. When a large national donation was collected on the occasion of the jubilee of Oscar II's reign in 1897, the King gave the entire sum for the erection of national sanatoria. Although a considerable number of sanatoria have come into being since then, the present trend is to locate the thoracic clinics, with their medical and surgical treatment units, in large hospitals, especially regional hospitals.

Rheumatology has always been counted as part of internal medicine. The initiative in providing specialist treatment for rheumatic patients was taken in 1914 by the National Pensions Board, which is now part of the National

Social Insurance Board. The Pensions Board set up independent establishments for patients suffering from disabling diseases, especially the rheumatic diseases.

The chronically sick were originally excluded from receipt of treatment at ordinary hospitals. In 1927 the Riksdag decided to give government grants for the organization of nursing homes for the chronically sick. These nursing homes, which are now often connected with hospitals for acute diseases, are called "long-term treatment clinics" and the treatment of the patients is almost a speciality in itself, with many points of contact with geriatrics.

Although occupational medicine is a speciality in many countries, it is not yet so in Sweden, although in the last few years clinics for occupational diseases have been organized.

There has been lively discussion in Sweden over the question of the division of internal medicine into independent specialities or the retention of sub-specialities within the framework of the department of internal medicine, each sub-speciality being represented by an assistant senior physician. In many cases the latter course has been adopted; this applies to cardiology, allergology, endocrinology, hematology, and internal renal diseases, although independent clinics have been established in isolated cases. However, there are in any event opportunities for specialist training in these special clinical fields, and developments seem to be moving in the direction of independent clinical units.

As an independent speciality, pediatrics has had a long history in Sweden, even since the days of Nils Rosén von Rosenstein, who was the foremost children's physician in Europe in the mid-eighteenth century. The first Swedish professorship in pediatrics was established in 1845 at the General Orphanage in Stockholm.

As has already been indicated, obstetrics developed at a very early stage, along with surgery. Gynecology was detached as a separate speciality and was first combined with obstetrics in 1861 in Stockholm. However, later development was slow, and at first independent maternity hospitals or maternity wards at the general hospitals were established in preference to departments representing this whole field. In recent times, however, conditions have changed radically, so that in 1960 there were forty departments of gynecology but only three independent maternity hospitals—which were also combined with departments of gynecology.

Pediatric surgery early became an independent speciality, when the Crown Princess Louisa Children's Hospital in Stockholm was divided in 1885 into a medical and a surgical unit, headed respectively by a senior physician and a senior surgeon.

Orthopedics developed at the institutions for the crippled and disabled, which originally came into being on private initiative and had not only medical but also social functions. The first orthopedic clinic in Sweden was started in 1882 at the Princess Eugenia Institution and the first professorship in the subject was established in 1913 at the Karolinska Institute. The tendency has been to transform orthopedics into surgery of the extremities.

The ear, nose and throat diseases were developed as an independent speciality relatively late. The first clinic came into existence at Sabbatsberg Hospital in 1910 and the first professorship at the Karolinska Institute in 1912.

Several surgical specialities have been elaborated in the present century. Neurosurgery has been represented by a professorship at the Karolinska Institute since 1935. The actual work was carried out at Serafimer Hospital until recently. There has been a professorship of plastic surgery at Uppsala since 1960. Thoracic surgery has also been represented by a professorship at the Karolinska Institute since 1948. Urology, for which a senior physician post was established at the Karolinska Hospital in 1958, gained its first professorship in 1967. The surgery of the hand is not represented by a professorship, but the first senior physician post in this field came into existence at Gothenburg in 1957.

Anesthesiology is a speciality of relatively recent date. The first holder of the Swedish professorship in this field entered on his duties in 1964 but had already been working as senior physician at the Karolinska Hospital since 1945 and had helped with the training of the majority of senior anesthetists in Sweden.

The Institute of Radiology in Stockholm was founded in 1910 and was Sweden's first independent radiotherapy clinic. It came into existence on private initiative when a collection of five million kronor was made in 1928 in honor of King Gustav V's 70th birthday.

X-ray diagnosis is perhaps the most important auxiliary speciality in the art of medicine. An institute of X-ray diagnosis was started as early as 1903 at Serafimer Hospital and nowadays there are departments of X-ray diagnosis in all Swedish hospitals.

As aids to therapy, physiotherapy and massage have a long history. There are special departments of physiotherapy at certain Swedish hospitals and, in addition, physiotherapists at all hospitals. The last few years have seen the emergence of special rehabilitation clinics or departments connected with hospitals, as prescribed by the National Board of Health in 1954.

The special hospital departments include laboratories of pathological anatomy, clinical chemistry, clinical physiology and clinical bacteriology, hormone laboratories and blood-donor centers, all of which, like the special departments just enumerated, are headed by specially trained physicians. Naturally, there are also theoretical institutions in anatomy, physiology, chemistry, pharmacology, etc., which train not only medical students but also theoreticians and research workers, some of whom later on transfer to contiguous specialities in clinical medicine.

The care of the mentally ill, which nowadays is being increasingly integrated into the ordinary units for the care of somatic cases, contrary to the previous policy which was to erect separate mental hospitals—which, however, still exist in many places, has a history stretching back to 1817 when the hospitals for the mentally sick and the hospitals for the physically ill were separated. In the act of 1858 it was prescribed that at each mental hospital there was to be a senior physician and at least one assistant physician. In the same year the first professorship of psychiatry was established at Uppsala. Thus the care of the mentally ill became an officially recognized speciality in Sweden.

Over a century ago instruction in the history of medicine, coupled with pathology and pathological anatomy, was given at the Karolinska Institute. In the examination regulations of 1907 this subject disappeared from the curriculum but has been restored in recent years, though in the form of voluntary courses. Since 1958 government grants have been made for this voluntary instruction, which is now given in all Swedish faculties of medicine in the form of ten lectures in all. At Stockholm these lectures are now included in the regular courses in the subjects of medicine and surgery. There is no professorship in this field, but the history of learning has two professorships in the Faculties of Arts and Sciences at Uppsala and Gothenburg.

The number of senior physician posts in institutional medical care has been doubled in the last twenty years and, despite the shortage of physicians, no great difficulties have been observed in meeting the need of senior and assistant physicians in institutional medical care. It has been more difficult to fill certain posts in non-institutional medical care, and most difficult of all to fill the post of district medical officer.

Finally, it may be mentioned that the travels of individual physicians for the purpose of study, especially to the United States, have been of the greatest importance in the training of Swedish specialists. The high standard of Swedish medical education can probably be described as the product of the well thought out and exacting training of students elected by "blocked" intake to the faculties of medicine, and, in addition, of further training which is supported by the individual physician and to a certain extent by the government, in which travels abroad must certainly be considered to play a very important part at the senior level.

Now, what of the developments in the Scandinavian countries bordering on Sweden? Until 1814 Norway was united with Denmark and after that date with Sweden, up to and including 1905. Finland was a part of Sweden until 1809 and after that a Russian grand duchy, until it became independent in 1919. Iceland was an independent state in the tenth and eleventh centuries and was later united with Norway and Denmark, until it became completely independent in 1943. We may therefore say that, up to the beginning of the nineteenth century only two countries were of importance in this connection, Sweden and Denmark.

Denmark

The University of Copenhagen (9, 38, 39, 40), which was originally intended to have four faculties, theology, law, medicine and philosophy, came into existence in 1479. However, it was closed down in 1530 by reason of the disturbances of the Reformation period, and at that time medical instruction does not seem to have been given, although a Faculty of Medicine existed, at least nominally.

Christian III reopened the university on 9 September 1537, and at the solemn ceremony in the Frue Kirke presented the Vice-Chancellor's insignia to the professor of medicine, Christiern Thorckelsen Morsing (1485-1560), who had previously been professor of philosophy at the university but had transferred to the Faculty of Medicine, a not uncommon translation at that period. On 10 November medical teaching began. According to the university statutes, two professors of medicine were to be appointed and, besides teaching, were to practice as physicians at the Court, the University, and in the town and country. Thus the Faculty of Medicine which had now come into being was the first in the Scandinavian countries.

As in Sweden, the medical lectures were based, in accordance with scholastic principles, on the Moslem writers Rhazes and Avicenna, and on the writers of antiquity such as Hippocrates and Galen (40). Physics, on Aristotelian principles, and mathematics, according to Euclid and Ptolemy, were also included in the program. Each professor gave in addition an annual oration in honor of the study of medicine and took his turn in compiling an almanac.

Thus the beginning was modest and traditional, but the Copenhagen Faculty, which more often than not was on a level with other European faculties of medicine, was to be improved substantially in the seventeenth century, primarily through the efforts of the famous Bartholin family, which also helped to insure that the statutory regulation regarding the prominence of anatomical teaching was really adhered to.

As has already been mentioned, the first professor of medicine was Christiern Morsing who had studied abroad. Like many others, he became a teacher in several faculties of the university and had originally professed mathematics. Thomas Fincke, who was at first professor of mathematics, then professor of rhetoric, was finally professor of medicine, and Caspar Bartholin the elder, professor of rhetoric (1611), professor of medicine (1613), ended his career as professor of theology (1624).

Hans Frandsen (Johannes Franciscus Ripensis), who translated several Greek authors into Latin and was a well-known Danish writer, succeeded Morsing, and in turn Frandsen was succeeded by Anders Christensen (1551-1606) who is said to have dissected a cadaver in the Frue Kirke, an act which aroused general horror.

Thereafter arose the golden age of medicine in Denmark, an age which was to a large extent dominated by the Bartholin family which provided no less than ten professors of medicine and an equal number of professors of other faculties, making the family a parallel to the Gregory and Monro families in Scotland and the Darwin and Huxley families in England (42).

Caspar Bartholin the elder (1585-1629) studied medicine in England and Italy and, on the basis of his own dissections, produced a textbook of anatomy entitled *Anatomicae Institutiones Corporis Humani Utriusque Sexus* (1611), the same year in which he was appointed professor at Copenhagen. His book appeared in 30 editions, in Latin, French, English, German, Dutch, and Italian, and became a standard work in Europe for no less than three generations of medical students.

When Bartholin left the Faculty of Medicine, he was succeeded by Oluf Worm (1588-1654) who, in his efforts to promote the study of anatomy, received valuable help from Henning Arnisaeus, the royal physician-in-ordinary, and likewise a good anatomist. In 1628 he donated a skeleton to the university, a donation which formed the foundation of the collection of anatomical teaching material. Worm founded the Museum Wormianum, containing in the style of the time, preparations from the three realms of nature and various curiosities.

Through the agency of his father-in-law Jacob Fabricius, who became physician-in-ordinary to the Danish King in 1637, Simon Paulli (1603-1680), a professor at Rostock, was summoned to Denmark and appointed professor of anatomy, surgery, and botany. He made important contributions to contemporary knowledge, particularly in anatomy and botany, and in 1644 he also succeeded—with the aid of the King and Fabricius—in establishing an anatomical theatre in Copenhagen, the Domus Anatomica, built on the model of the anatomical theatre at Leyden.

After the death of Christian IV, his patron, Paulli retired in favor of Thomas Bartholin (1616-1680), who had recently returned to Denmark after ten year's travel abroad and entered upon a short but intense period of activity. In 1641 Bartholin published an enlarged edition of his father's *Institutiones anatomicae*. He gathered young men of talent around himself but for reasons of health retired at the age of forty to a farm near Holbaek, from which, in his capacity as Dean, he directed the affairs of the Faculty of Medicine. He is also remembered as the originator of the first Danish pharmcopeia and the founder of the first Danish medical journal, *Acta medica et philosophica Hafniensis* (1673). Of Thomas Bartholin's five sons, the eldest, Caspar Bartholin *the younger* (1655-1738), after whom the *ductus Bartholianus* was named, became professor of medicine in 1680.

The foremost member of the Bartholin family, however, was Thomas (49), whose most famous student was undoubtedly Nicolaus Steno or Niels Stensen (1636-1686), noted for his discovery of the *ductus Stenonianus* and his important studies of muscles, glands, and lacrymal apparatus, the ovarian follicles, and what were later known as Peyer's plaques. The study of anatomy was of great importance in the Copehagen Faculty of Medicine.

Medical training was given on the same general lines as in Sweden, and the rivalry between physicians and surgeons continued during the eighteenth century. Although the surgeons were forbidden to treat internal diseases, it was they on whom the common man had to rely in those days when physicians were few, especially during periods of plague when they often fled. The surgeons were trained like ordinary craftsmen and the Faculty showed no sign of wishing to take any interest in them. Although a certain amount of training had been given since 1698, it was of considerable importance that a school of surgery came into existence in 1736, a school which half a century later was transformed into the Royal Academy of Surgery. Students who passed through this school were allowed to practice both surgery and medicine. Thus, there were two categories of physicians, those who had been trained at the university and those who had passed through the Academy. This system remained in force down to 1842, when the Academy was closed and both surgical and medical studies were carried on at the university. It is worth noting that a surgeon, Johannes de Buchwald (1658-1738), not only took the doctorate of medicine but also became professor of medicine.

The violent conflagration in Copenhagen in 1728, which destroyed the Domus Anatomica—it was twelve years before a new anatomical theatre was built—and the university buildings, was not of such great consequence as may be thought, since the professors gave their lectures in their own homes. There were, moreover, no great difficulties because the number of students was from three to five and scarcely a physician a year was produced until 1740.

A great improvement in the conditions of medical training came about in the middle of the eighteenth century. This was the result partly of Linnaeus's work in encouraging the study of the natural sciences and partly of the summoning of the energetic and versatile naturalist Christian Gottlieb Kratzenstein (46), who was a physicist and chemist but also taught medicine. He became *professor medicinae designatus*, that is, he received the promise of the first vacant post, a form of professorship that gave its holder social status but did not yield him any money. This form of professorship was abolished in 1788, and instead assistant professorships were introduced —also unpaid. The third factor in the improvement in training was the coming into being of the school of surgery.

A number of important scientists passed through the school of surgery and some of them were also attached to the Faculty. The first of these was the prosector Georg Heuermann, who gave lectures which were published in the form of a physiological work in four volumes, and a work on the principles of surgical operations in three volumes. He came from Holstein and had written these works in German. When he applied for a professorship, he was rejected on the ground that he could not speak Latin satisfactorily.

At this time the first Danish hospital came into existence, the King Frederik Hospital founded in 1757 by Frederik V. At this hospital, as well as at the maternity hospital known as Fødselsstiftelsen, one of the physicians was Matthias Saxtorph (1822-1900), principally remembered as an accoucheur who also taught anatomy.

During the first part of the nineteenth century pathological anatomy became increasingly important, statistical methods were introduced into medicine and stethoscopy began to be practiced in the clinic. Like Huss in Sweden, Oluf Lundt Bang (1788-1877) worked to introduce the methods of physical diagnosis into Denmark after his travels abroad, during which he visited Laennec in Paris. Bang gave lectures on stethoscopy in the Danish Royal Society of Medicine, but it was primarily his colleague Seligmann Meyer Trier (1800-1863) who popularized the method (8).

The disputes regarding qualifications between physicians and surgeons led to the foundation in 1740 of the Collegium Medicum, the first Danish body for medical administration, in which physicians, surgeons and pharmacists were all members. The amalgamation of the medical and surgical schools was of great importance for medical education and the development of medicine as a whole in Denmark. This amalgamation took place on 1 January 1842. The two different degrees were amalgamated in 1838 after a century of disputes on this question. The examination questions, and the answers, were no longer couched in Latin but in Danish (40). The medical journal *Bibliotek for Laeger*, which was begun in 1809 and is still published, was of no small benefit in the further training of Danish physicians. It published not only original articles but also practical articles in the form of brief notes and reviews of important foreign works. As regards the development of the medical specialities at this period, it may be mentioned that physiology became an independent subject. Experimental physiology was introduced into Denmark by Peter Ludwig Panum (1820-1885) who, after studies with Virchow and Claude Bernard, became a professor first at Kiel and in 1864 at Copenhagen, a professorship which he accepted only on condition that an institute of physiology be erected.

Human anatomy, taught until 1836 as a subsidiary subject, was now for the first time not represented at all in the Faculty of Medicine but only at the Academy of Surgery. Pathological anatomy, combined with general pathology, gained recognition in 1844, and pharmacology, forensic medicine and hygiene, which had been taught in combination sometimes with one subject and sometimes with another until 1832, received separate representation in 1868.

In the sphere of practical medicine—in this case, obstetrics—in 1761 the Faculty was afforded an opportunity to have the medical students trained in a hospital department. However, Johannes de Buchwald, who in 1717 became professor of anatomy, surgery and botany at the university, was the first to give lectures in obstetrics in Denmark. Practical surgical training under the auspices of the Faculty came into existence after 1842, when the Academy of Surgery was incorporated into the Faculty. Regular clinical instruction in internal medicine was not achieved until 1870, when the relations between the Faculty and the hospitals were regularized. Previously individual members of the Faculty who had appointments as senior physicians in hospital departments had occasionally offered the students some clinical instruction.

Foreign advances in medicine reached Denmark at approximately the same time as they reached Sweden; for example, the news of ether anesthesia reached the Scandinavian countries a few months after the original demonstration in Boston, and that of lithotrity, the great development in surgery of the urinary tract, soon after the operation was first performed by Civiale in 1824. Parenthetically it may be added that a Dane, Christian Fenger (1840-1902), who took his degree at the University of Copenhagen in 1867, specialized in pathological anatomy which he introduced as a new subject in America in his capacity as prosector at Cook County Hospital in Chicago. He was also one of the pioneers of antisepsis and as a surgeon had a great influence on American surgeons such as Ochsner, Murphy and the brothers Mayo.

Among the outstanding representatives of internal medicine in Denmark at the end of the nineteenth and early part of the present century was Knud Helge Faber (1862-1956) (6, 7), who was particularly interested in diseases of the gastrointestinal tract. Among the surgeons may be mentioned Oscar Thorvald Bloch (1847-1926), who was an excellent teacher of this subject and was especially interested in both the surgery of the extremities and that of the abdomen. Following his studies under Lister, he introduced asepsis at the King Frederik Hospital.

When the need for further university places became urgent in Denmark, another university was founded in 1928 at Aarhus. After a period of development, the University of Aarhus was able, as from 1955, to produce physicians under similar conditions to those at the parent University of Copenhagen. In 1966 a third university was founded at Odense.

The current medical curriculum in Denmark was laid down in 1967. Three years are devoted to preclinical studies and three and a half years to clinical studies. The medical degrees are called "Part 1" and "Part 2". Part 1 (the preclinical studies) is divided into "Introduction" (chemistry, physics, anatomy, genetics, medical statistics and medical psychology), Part A (anatomy and biochemistry) and Part B (physiology, biophysics and anatomy). All these parts are followed by separate examinations.

The training for the Part 2 degree (the clinical studies) is divided into three periods. Period I is initiated by a propedeutic course in medicine and surgery followed by courses in epidemiology, physiotherapy, microbiology, pharmacology, pathological anatomy, hygiene, preventive medicine and hereditary pathology. During the second period the student must take courses, followed by periods of service, in medicine and surgery, laboratory work, psychiatry, pediatrics, obstetrics and gynecology, neurology and dermato-venereology. The student must during the third period participate in conferences and lectures discussing questions from various parts of the whole of medicine and take courses in medicine and surgery, psychiatry, pediatrics, obstetrics and gynecology, pathological anatomy, dermato-venereology, otorhinolaryngology, neuromedicine, neurosurgery, x-ray diagnosis, radiotherapy, forensic medicine and social medicine. The single examination has been replaced by separate examinations (3).

Finland

In 1640 a university was established at Åbo (now Turku) (41), but the Faculty of Medicine had only one medical post. The doctor invited to fill this post was Olof Regelius, who had been trained at Uppsala but at the time of the invitation was studying in Holland. He never entered upon the duties of his professorship, which instead went to Eric Achrelius (1604-1670), who came to Åbo personally and offered his services as a teacher of *ars medica*. In 1641 Achrelius, who had studied at Uppsala but had not taken the doctorate, became the first professor of medicine at Åbo. He seems to have been an enthusiastic and diligent man, but in the end he became disillusioned, partly on account of the few students and partly on account of their poverty. Poverty also explains why Achrelius never got his dissertations printed. Here it may be added by way of explanation, that until the beginning of the nineteenth century the professors in Finland, as in Sweden in early times, wrote the academic dissertations and the respondents (candidates) were responsible for printing and defending them.

The Swede Elias Tillandz (1640-1693), who had studied at Åbo, Uppsala and Leyden, receiving training under the clinician Sylvius, introduced the dissection of human subjects into anatomical instruction. He had a large medical practice, in the course of which he probably took his students with him, like Hoffvenius at Uppsala. In this way he introduced clinical instruction into Finland.

The little university at Åbo, a typical provincial university, was not much influenced by the great developments in medicine which characterized the latter part of the seventeenth century in the rest of the northern countries. The military operations had a very inhibiting effect on medical development in Finland. The three professors who followed immediately after Tillandz had all studied in the Netherlands; they had short periods of tenure but introduced the teachings of the Dutch school into Finland in their clinical and practical work.

Herman Diedrich Spöring (1710-1747), who became professor of medicine at Åbo in 1728, made important contributions to anatomical instruction and pathological anatomy and worked for the introduction of variolation. Both variolation and vaccination attracted attention at an early date in Finland—this was perhaps the most distinguished contribution to medicine made by the University of Åbo.

During the latter half of the eighteenth century, Åbo was the most northerly university in the world. An anatomical theatre was constructed and opened in 1763 and the following year a chemical laboratory was installed. Johan Gadolin (1760-1852), a chemist, saw to it that the students carried out laboratory training there, at a time when the German and French universities still had no such facilities. Åbo also had a botanical garden from 1758.

Finland secured its first hospital in 1759 when Åbo County Hospital was founded. The senior physician was Johan Johansson Haartman (1725-1788). Like the Serafimer Hospital, that of Åbo was intended from the beginning to be a teaching hospital. Though not connected with the university, it became, on a modest scale, a center of clinical instruction, first under the direction of Haartman, who was appointed assistant professor of medicine at the university. His duties included "illustrating cases of disease at the hospital whenever there is an opportunity". He compiled, on the model of Sauvage and Linnaeus, a nosological system entitled *Sciagraphia morborum* which played a great part in medical studies before pathological anatomy provided a more reliable basis for the classification of diseases.

Around 1800 medical science at Åbo had a brief period of prosperity, until a new war—the Finnish war of 1808-1809, which was to separate Finland from Sweden—broke out and put a stop to developments. At this period Gabriel Erik Haartman (1757-1815), a distant relation to Johan Haartman, was professor of medicine. He had acquired substantial training in surgery and obstetrics at Stockholm and Uppsala, the first physician in Finland to do so. In 1782 he was appointed municipal officer of Åbo and in 1784 was made professor of anatomy, surgery and obstetrics. After the death of his elder relation he became professor of practical medicine in 1789 and in 1794 took over the medical supervision of the county hospital. After having been appointed in 1811 president of the Åbo Collegium Medicum, which had been established after the separation from Sweden, he shortly afterwards abandoned all medical work and became head of the Finance Department.

Haartman took a modern view of medical education. He considered that physiology should be based on animal experiments and that the physiological observations should be compared with the bodily functions in the healthy and the sick. Instruction in pathology should be combined with clinical observations and, in his opinion, well-equipped hospitals with opportunities for real clinical tuition were necessary for teaching practical medicine.

The methods of physical diagnosis, including percussion and auscultation, were early demonstrated to the Swedish Medical Society and thus became known to the Finnish physicians who had studied in Sweden. However, these methods were not used for some considerable time in the clinical instruction at Åbo. Even at the beginning of the nineteenth century the medical students had to accompany their teacher on his visits to the sick, though a certain amount of instruction was given in the hospital.

The clinical institute planned after the war of 1808-1809 and intended "for very necessary practice in the practical parts of medicine" was never brought into use; it was destroyed in the great fire in Åbo on 4 September 1827.

The last professor of medicine at Åbo was Israel Hwasser (1790-1860) (16), who received the professorship in 1817 in succession to Haartman. However, he never gave any clinical instruction, restricting himself to purely theoretical platform lectures tinged with his views on natural philosophy. In 1829 he moved to Uppsala.

After the great fire at Åbo the university was transferred to Helsingfors (now Helsinki) where a new clinical institute was erected and where Matthias Kalm (1793-1833), who had previously been assistant professor of surgery, succeeded Hwasser as professor of theoretical and practical medicine. The Clinical Institute was opened in 1833 with sixty beds, half of which were intended for patients with internal complaints. The principal teacher was Sten Edvard Sjöman (1805-1843), who became particularly well known for having introduced the new methods of percussion and auscultation into Finland. He did not receive the professorship but was given the title of professor in 1842. At that period Finland was a Russian grand duchy and the Tsar Nicholas I had found the Institute *ganz infam schlecht* on his visit to Helsingfors in June 1833. In consequence Sjöman was sent to St. Petersburg (Leningrad) to study its hospitals.

Immanuel Ilmoni (1797-1856), who took over the professorship from Sjöman, had originally thought of devoting himself to anatomy and had studied the technique of dissection and chemistry in Germany. But during his travels (1828-1830) he had also devoted some time to hospital studies in Germany, France and England. Although he did not think that he had employed his time on his travels in a really energetic manner—particularly as regarded the topical subject of physical diagnosis—in the period 1834-1844 he prepared a course of lectures in clinical propedeutics which was of fundamental importance for propedeutic instruction in the Faculty of Medicine. He was aware at an early date of the psychosomatic problems in a way which seems rather modern. He was also of great importance as a clinical teacher generally, though in some respects—for example, febrile diseases—he was still under the influence of old-fashioned ideas. Ilmoni was a skillful anatomist and laid great stress on postmortem examinations, which were performed at the Clinic in the presence of the medical students. He was also well known because of his great work on the history of diseases in the northern countries (14).

In 1848 a new Clinical Institute was opened in Helsingfors with thirty beds. This institute was of great importance over a long period. The breakthrough of the new influences in all spheres of medicine took place in Finland in the years 1848-1849, and at the end of the century all the current methods of investigation in internal medicine were being applied.

Ilmoni's successor, Knut Felix von Willebrand (1814-1893), played a great part in the introduction of new methods and ways of thinking during his time as professor, which extended over two decades (1856-1874). His professorship came to be called the chair of general pathology and clinical medicine. A separate professorship of pathological anatomy was established in 1857 and was held by Otto Edvard August Hjelt (1823-1913), who was also well known as a medical historian (10).

A certain tendency to specialization began to make itself felt. Thus in 1857 obstetrics and pediatrics were set apart as a joint subject of instruction, a professorship of ophthalmology was established in 1871, and an associate professorship of syphilology and cutaneous diseases came into being in 1874.

In Finland the new ideas in the art of medicine made their way, along with physiological medicine and bacteriology, during the later decades of the nineteenth century. The foremost clinician in Finland at this time was Johan Wilhelm Runeberg (1843-1918), the successor of Willebrand as professor of clinical medicine.

It would be going too far to sketch the history of medical education in Finland during the present century when, despite revolutions and wars, the banner of science has been held high. Nowadays the medical students receive instruction in a system very similar to that of Sweden in the University of Helsinki, the Finnish-speaking University of Turku founded in 1920, with a Faculty of Medicine in 1943, and the University of Oulu (Uleåborg) opened in 1959, with a Faculty of Medicine in 1962.

The present Finnish medical curriculum comprises the examinations for the "candidate of medicine" degree and the licentiate of medicine, which can be taken in the faculties of Helsinki, Turku and Oulu, the licentiate degree requiring two and a half years of preclinical studies, three years of clinical studies, and ten months of compulsory service as intern. This last service includes time in a department of internal medicine, an outpatient department and a department of surgery as well as optional service at different clinics (37, 48).

Norway

In 1526, at a diet held in Odense, King Frederik I sought to provide the united kingdoms of Denmark and Norway with a regular medical service (44). Both his proposals and those of his successor miscarried, but in the reign of Frederik II (1559-1588) medicine was finally given consideration. During the seventeenth century ordinances appeared with the aim of regulating the work of pharmacists, physicians and surgeons, particularly with regard to the prevalent epidemics. But the ordinance issued by Christian V in 1672 was the first of any great importance and remained in force, as far as Norway was concerned, until the passing of the Health Law of 1860 and, as far as Denmark was concerned, down to 1906. Under this ordinance the midwives were included in the medical service in addition to physicians and pharmacists. However, much of what was prescribed in the ordinances remained a dead letter and it may be remarked without exaggeration that even at the beginning of the nineteenth century Norway was far behind its Scandinavian neighbors in regard to medical institutions and hospitals.

As in the other northern countries, the division of the medical profession into physicians and surgeons was still in force at the beginning of the nineteenth century. The surgeons were trained at the Academy of Surgery in Copenhagen.

In 1811 Norway secured a university of its own, Universitas Regia Fredericiana, at Christiania (now Oslo) and, despite some doubt in certain quarters, it was decided to establish a Faculty of Medicine with three professors: Michael Skjelderup (1769-1852), from the University of Copenhagen, Nils Berner Sørenssen (1774-1857), and Magnus Andreas Thulstrup (1769-1844). These men had an enormous amount of work to do. Thulstrup was said to have had more to do than any other doctor in the country. He was professor of surgery and obstetrics, surgeon-general of the armed forces and a member of various commissions. The three professors' teaching programs were voluminous, including the medical encyclopedia, the history and literature of medicine, anatomy and physiology, dietetics, pathology and therapy, *materia medica,* pharmacy, surgery and obstetrics, forensic medicine and hygiene. But the number of students was small at that period—there were only three.

The leading man in the Faculty was Skjelderup, who on 18 August 1814 began medical training with lectures on the medical encyclopedia and a survey of the history of medicine. The following year he also began teaching anatomy, and the anatomical theatre that was established became the university's first scientific institute.

Skjelderup's successor was Joachim Andreas Voss (1815-1897), who was regarded for many years as the foremost physician in Norway. Jacob Munch Heiberg (1843-1888), professor for ten years, worked intensively to secure better premises and instructional material and also urged that women should be admitted to medical training.

Gustav Adolf Guldberg (1854-1908) was appointed professor in 1888 and in the same year a demonstrator—who later became the prosector in histology—was appointed. Kristian Emil Schreiner was professor of anatomy from 1908 and Otto Louis Mohr from 1919.

Ten years after the inauguration of the Faculty, a fourth professorship was established in pharmacology and hygiene, given to Frederik Holst (1791-1871), who was particularly interested in the care of the mentally ill. He also helped to complete the Norwegian pharmacopeia (1854).

The different specialities developed rapidly at the Faculty of Medicine in

Christiania. Physiology had already been taught by Michael Skjelderup, and his successor Christian Peter Bianco Boeck (1798-1877), who was professor of physiology, comparative anatomy and veterinary medicine, founded both the zootomic museum and the physiological collection.

The first representative of the subject of physiological chemistry was Jacob Worm-Müller (1834-1889) who during his five-year residence abroad had been greatly influenced by the experimental physiology of Johannes Müller. In 1873 he was made an associate professor and in 1877 professor, and in the same year he founded the laboratory of physiological chemistry.

Pathological anatomy secured independent representation when in 1866 Emanuel Fredrik Hagbarth Winge (1827-1894) was appointed professor of general pathology and pathological anatomy. Hygiene, which in 1824 was combined in a joint professorship with pharmacology, became a separate subject of instruction when, in 1893, Axel Holst (1860-1931) became professor of hygiene and bacteriology.

Ole Rømer Aagaard Sandberg (1811-1883) was the actual founder of psychiatric training in Norway, though considerable contributions were made by his predecessors, for example, Herman Wedel Major. In 1854 Sandberg became the superintendent of the newly established Gaustad Asylum and took up clinical instruction in psychiatry. However, a professorship in the subject did not come into being until 1915; the first holder was Ragnar Vogt (1870-1940). The clinical course at Gaustad Asylum became compulsory in 1908 and psychiatry became a subject of examination in 1914. The establishment of the psychiatric clinic at Vestre Aker in 1926 provided better facilities for the professor's work.

The 16 October 1826 is rightly considered a red-letter day in the history of Norwegian medicine. It was on that day that the newly erected National Hospital (Rikshospitalet) was opened, the hospital at which all Norwegian physicians were to receive their clinical training for over a century. As early as 1815, however, Sørenssen and Thulstrup had taken over the treatment of patients admitted to the civic hospital in Christiania, certain military patients and the patients at Akershus County Hospital. The new National Hospital had 100 beds and very soon the former civil hospital was taken over as a branch. The first senior physicians at the old National Hospital, which was not completed until 1842 and was replaced in 1883 by the new National Hospital, were Sørenssen and Thulstrup, and the rest of the medical staff consisted of two assistant physicians and four "candidates of medicine". At first there was only one department of internal medicine, headed by Sørenssen, but a division into two departments took place in 1868.

Developments in internal medicine in Norway followed the general international currents, and experimental pathology and bacteriology played a great part. Physical diagnosis was introduced in 1833 by Andreas Christian Conradi (1809-1868), who later became professor.

Neurology was detached from internal medicine and became a special subject in 1895 when Christopher Blom Leegaard (1851-1921) was appointed professor of this subject. Lectures on the cutaneous and venereal diseases were originally given within the framework of internal medicine, but when a senior physician was appointed to direct the department of venereal diseases at the National Hospital in 1841, he was made responsible for giving clinical lectures. In 1852 these diseases became a subject of examination; in the preceding year Carl Wilhelm Boeck (1808-1875) had become professor of physiology, cutaneous and venereal diseases.

Pediatrics became a separate speciality when Axel Theodor Johannessen (1849-1926) became assistant professor of this subject in 1891 and professor in 1895.

The first really important surgeon in Norway was Christian August Egeberg (1809-1874), who was also one of the most sympathetic personalities in Norwegian medicine. One of his successors was Julius Nicolaysen (1831-1909) who was a pioneer of many operations in Norway. Alexander Ludvig Normann Malthe (1845-1928), who was both a surgeon and a medical practitioner, had no academic position but was an excellent surgeon and also became well known for his donations, *inter alia,* to the Medical Society in Christiania. Like the Norwegian Medical Association (1886), the Society was of great importance for the further training of Norwegian physicians.

Ophthalmology was separated from surgery and the first professor in the separate subject was Christen Heiberg (1799-1872), who was at the same time a senior physician at the National Hospital. However, it was not until 1897 that a special department of eye, quite separate from the department of surgery, was organized at the hospital. Otorhinolaryngology, in which instruction was given by the professors of surgery, secured an independent professorship in 1895, given to Vilhelm Kristian Uchermann (1852-1929).

Obstetrics and gynecology were originally included in Thulstrup's area, but in 1850 Frans Christian Faye (1806-1890) became professor of obstetrics, gynecology and pediatrics.

During the first few years of the university's existence—until the revision of the statutes in 1845—it was possible for non-academic persons to take an easy, limited, master's degree in medicine. One examined in this way was called *examinatus medicinae* and the examination itself was called the "Norwegian degree examination".

When the need for additional places for training medical students became pressing, a Faculty of Medicine was established in 1946 at the University founded in Bergen.

The training for the master's degree in medicine, which may be taken at the Universities of Oslo and Bergen, covers a period of six years. The period of preclinical study is two and a half years and the period of clinical study, three and a half years. At Oslo the preclinical studies are carried out in accordance with a training plan laid down in 1950 and the clinical studies according to a more modern plan of 1966. After the final examinations, a year and a half of *turnus* service—the compulsory postgraduate period of hospital service—is required before the prospective physician gains the right to practice his profession independently (37).

Iceland

As is already clear from the foregoing account, the physicians who worked in Iceland were for centuries trained in other northern countries, primarily in Denmark and Norway. The University of Iceland in Reykjavik, founded in 1911, is the only one in this country of 185,000 inhabitants. It has a Faculty of Medicine which, according to its statutes is to be an institute of scientific research and, additionally, to train medical officers and general practitioners. Annually from twenty-five to forty students enroll, and in the last few years an average of sixteen students have taken their final examinations after an average period of about seven years' study. Restricted financial conditions have meant that the Faculty's funds have been used mainly for the training of medical officers.

The period of study is divided into three years' training in chemistry, biochemistry, physiology and anatomy, one and a half years' study of pathology and pharmacology, and two to two and a half years devoted to clinical subjects. The small size of the Faculty, which has about ten professors and twenty assistant professors and lecturers, has the advantage of promoting close contact between teachers and students.

It should be clear from this survey that the conditions of medical education in the Scandinavian countries have through the centuries been relatively similar. Sweden and Denmark guided the rest since they had experience in this field derived from Germany and France, and later also from the Anglo-Saxon countries. Nevertheless, each Scandinavian country has to some extent adapted the system of medical education to its own local conditions while, as of the present, seeking ways and means to provide for sufficient uniformity in medical training to allow for a free interchange of physicians.

BIBLIOGRAPHY

1. ANNERSTEDT, C., *Uppsala Universitets Historia*, Vol. 2, pts. 1-2. Uppsala, 1908-1909.
2. BERGSTRAND, H. Läkarekåren och provinsialläkareväsendet, *Medicinalväsendet i Sverige 1813-1962*, ed. W. KOCK. Stockholm, 1963, pp. 107-155.
3. BRØCHNER-MORTENSEN, K., Den nye studieordning, *Ugeskr. f. Laeger*, 1968, 130: 21-24.
4. DJURBERG, V., *Läkaren Johan von Hoorn*. Uppsala, 1942.
5. EKSTRÖMER, C. J., *Kirurgminnen från Karl Johanstiden*, ed. W. KOCK. Stockholm, 1964.
6. FABER, K., *Mit universitetsliv*. Copenhagen, 1943.
7. ———, *Personlige erindringer*. Copenhagen, 1949.
8. GOTFREDSEN, E., Medicine, in *Denmark; an official handbook*. Copenhagen, 1964, pp. 613-618.

9. ———, *Medicinens historie*, 2nd ed. Copenhagen, 1964.
10. HJELT, O. E. A., *Svenska och Finska medicinalverkets historia 1663-1812*. Helsingfors, 1891-1893. 3 vols.
11. HULT, O. T., Petrus Hoffvenius. Den svenska medicinens fader. *Nord. Med.*, 1941, 10: 1431-1432.
12. ———, *Vilhelmus Lemnius och Benedictus Olai. Ett bidrag till Svensk läkarhistoria under Vasa-tiden*. Stockholm, 1918.
13. HUSS, M., *Några skizzer och tidsbilder från min lefnad*. Stockholm, 1891.
14. ILMONI, I., *Bidrag till Nordens sjukdomshistoria 1-3*. Helsingfors, 1846-1853.
15. INGERSLEV, V., *Danmarks laeger og laegevaesen* I-II. Copenhagen, 1873.
16. JOHNSSON (SOININEN), G., *Om sjukdomsorsakerna enligt Israel Hwasser*. Uppsala, 1932.
17. JORPES, E., *Jöns Jacob Berzelius*. Stockholm, 1960.
18. Karolinska Institutet, *The Royal Medical School in Stockholm*. Uppsala, 1960.
19. *Karolinska Mediko-Kirurgiska Institutets historia 1910-1960*, ed. G. LILJESTRAND. Vol. I, pt. 1, III, pt. 1, Uppsala, 1960.
20. KOCK, W., Anatomisk forskning och undervisning i Norden under 1600-talet. *Med.-hist. Arsb.*, 1966.
21. ———, *Kungl. Serafimerlasarettet 1752-1952; en studie i Svensk sjukvårdshistoria*. Jönköping, 1952.
22. ———, Lars Robergs råd till den medicinska ungdomen, *Svenska Läkartidn.*, 1953, 50: 1035-1044.
23. ———, Lasaretten och den slutna kroppsjukvården, *medicinalväsendet i Sverige 1813-1962*, ed. W. KOCK. Stockholm, 1963, pp. 158-239.
24. ———, Läkare—och kirurgyrkets förening i Sverige, *Svenska Läkartidn.*, 1942, 39: 2681-2708.
25. ———, Medicinens historia i läkarutibildningen, *Svenska Läkartidn.*, 1953, 50: 1152-1159.
26. ———, *Medicinhistoriens grunddrag*. Stockholm, 1955.
27. ———, Nils Rosén von Rosenstein, in *Swedish Men of Science*. Stockholm, 1952, pp. 74-80.
28. ———, *Olof af Acrel*. Stockholm, 1967.
29. ———, Pehr Henrik Malmsten—vår mest kände invärtesläkare för 100 år sedan, *Nord. Med.*, 1960, 64: 1573-1579.
30. ———, Som chirurgiska societetens resestipendiat i 1700-talets Europa, *Svenska Läkartidn.*, 1941, 38: 2502-2533.
31. ———, Universitetssjukhusens uppkomst i Sverige, *Svenska Läkartidn.*, 1959, 56: 3282-3294.
32. *Läkarnas Grundutbildning och Vidareutbildning, I: Grundutbildning*, Betänkande av sakkunniga tillkallade av medicinalstyrelsen och universitetskanslers-ämbetet, S O U, 1967: 51.
33. LENNMALM, F., *Karolinska Mediko-Kirurgiska Institutets historia*, I-II, III. Stockholm, 1910.
34. LINDROTH, S., Medicinen i Sverige under stormaktstiden, *Ciba-Jn.*, 1950, 2: 716-747.
35. ———, *Sjukvård och läkare i forna tiders Stockholm, 1252-1952*. Stockholm, 1951, pp. 41-65.

36. ———, Urban Hiärne och laboratorium chymicum, *Lychnos*, 1946-1947, pp. 51-116.
37. NORDISKA KULTURKOMMISSIONEN. *Att Studera i Norden*. Stockholm 1967.
38. NORRIE, G., *Af Medicinsk Facultets Historie, 1750-1841*. Copenhagen, 1934-1939. 3 vols.
39. ———, Om medicinsk fakultet gennem 400 aar, *Bibltk. Laeger*, 1938, 130: 1-19.
40. PANUM, P. L., Vort medicinske fakultets oprindelse og barndom. *Festskrifter Udgivne af det Laegevidenskabelige Fakultet ved Kjøbenhavns Universitet i Anledning af Universitetets Firehundredaarsfest Juni 1879*. Copenhagen, 1879, pp. 1-103.
41. PERRET, L., *Inremedicinsk diagnostik och sjukdomsbeskrivning under 1800-talet; en medicinhistorisk studie med Särskild Hänsyn till förhållandena i Finland*. Helsingfors, 1955.
42. PORTER, I. H., The Bartholins, a seventeenth century family study, *Med.-Hist. Årsb.*, 1964.
43. Regionsjukvårdsutredningen, *S O U*, 1958: 26.
44. REICHBORN-KJENNERUD, I., GRØN, F., and KOBRO, I., *Medicinens historie i Norge*. Oslo, 1936.
45. SACKLÉN, J. F., *Sveriges läkarhistoria, ifrån Konung Gustaf I:s till närvarande tid*. Nyköping, 1822-1853.
46. SNORRASON, E., *Kratzenstein*. Copenhagen, 1967.
47. STRANDELL, B., Läkaren och medicine professorn Carl von Linné. *Svenska Linnésällsk. Årsskr.*, 1961, 44: 1-21.
48. TEIR, H., *Läkarutbildningen i Norden. Rapport från Konferensen i Helsingfors 3-5 Februari 1964 om Grund-, Efter- och Specialistutbildningen*. Helsingfors, 1965.
49. TIGERSTEDT, R., *Dokument ur medicinens historia under sextonde och sjuttonde århundradet*. Stockholm, 1921.

EASTERN EUROPE AND THE FAR EAST

THE HISTORY OF MEDICAL EDUCATION IN RUSSIA*

MIRKO GRMEK
Centre National de la Recherche Scientifique
Paris, France

For many centuries, only physicians coming to Russia from abroad had a professional education obtained through academic study. These trained physicians were few and practiced their art only in the service of the court or for the nobility. In consequence large sections of the population were cared for by monks who were experienced in the art of healing, by other laymen (most often old women) who knew the medicinal virtues of plants and also employed magical procedures, and finally by lay-surgeons called *kostoprav* recognized for their skill in handling wounds. In general their education was obtained by transmission of empirical knowledge through oral tradition and by individual apprenticeship.

During their sojourn in Russia the foreign physicians taught the art of healing to the young men of that country. Let us cite as an example, the certificate made out in Moscow by the physician Marcus Gladebach: "Peter Grigorev", declared the German doctor in this document, "wished to learn the profession of physician from me and for that reason he entered my establishment in service and apprenticeship . . . I hereby give Peter Grigorev leave to go, certifying that he is expert in our practice of medicine . . ." This diploma dates from 1672, but we can take it as an example of a custom which was certainly more ancient. Once in a great while, a Russian subject went abroad for his medical studies, especially to Padua, Leyden, Halle, Bologna, or Cracow.

At the beginning of the seventeenth century there were about twenty physicians in all the territory of Russia who had been trained in Western medicine. Between 1690 and 1730 historians have been able to count 125-150 physicians. Chistovich gives a list of 510 doctors of medicine for the eighteenth century.

This data has only approximate value, but it sufficiently illustrates on the one hand the rate of growth and on the other the severe lack of competent sanitary cadres, particularly notable during epidemics and military campaigns. It was for that reason that in 1654, provoked mainly by the Russo-

---
* Translated from the French by Patricia L. Rather.

Polish War, Czar Alexei Mikhailovich (1645-1676) decided to create a school of medicine which would have as its objective the formation of surgical personnel for the army. The realization of this project was put in the hands of the *Aptekarski prikaz* (Chancellery of Apothecaries). Thus the *Maison Médicale* of the Czar was designated, an institution founded toward the end of the sixteenth century which became an administrative organ and occupied itself with all affairs concerning public health.

In August 1654 thirty persons, either soldiers of the Imperial militia (*streltsi*) or sons of soldiers, were assembled in the building of the *Aptekarski prikaz* in Moscow as pupils of the new School of Medicine. Their education began with theoretical studies: medical botany, pharmacognosy with elements of pharmacology, anatomy (taught by means of a skeleton and anatomical plates, that is, without dissection), and principles of the physiological functions of the human body. The library of the *Aptekarski prikaz* possessed many manuscripts, including Russian translations of various medical books. The students could also work in the pharmacy and thus acquire a knowledge of medicines and their preparation. All of this occurred during the first year of study. The second and third years were devoted to the study of the "signs of sicknesses" and the technique of medical examination. The fourth year was supposed to be devoted chiefly to the study of surgical operations (amputations, treatment of wounds, etc.) and the techniques of bandaging. In view of linguistic difficulties and the desire to obtain rapid results, the teaching was done without extensive formal lecturing and, for the most part, merely by the students giving assistance to the imperial physicians at their daily work. Each student was assigned to a special master whom he was supposed to follow and aid in his treatment of patients.

Historical sources do not let us establish with certainty if the School of the *Aptekarski prikaz* was truly a permanent institution or if, on the contrary, it was only a matter of a series of *ad hoc* courses serving only for one or two graduating classes. Whatever the case may have been, the School no longer existed at the beginning of the reign of Peter the Great.

A thorough training for physicians was not possible without the creation of a stable and ordered hospital base. The first real Russian hospital in the modern sense of the word was opened in Moscow in 1707, as a consequence of a ukase of Peter the Great, dated the 25 May 1706. In founding the hospital the Czar was inspired by the example of Greenwich Hospital, which he had visited during his sojourn in England. From the day of its foundation, this hospital was conceived of as a medical and surgical school primarily serving the needs of the army.

The foundation of the hospital and the Medical-Surgical School of Moscow were part of the vast program of reform by means of which Peter the Great (1672-1721) wished to give some Western aspects to his country and to augment its culture and its economic and military power. He initiated the period of enlightened despotism, so characteristic of eighteenth century feudal society in Eastern and Central Europe.

Peter the Great confided the organization and the direction of his hospital school to Nicolaus Bidloo, the son of the famous Dutch anatomist Govard Bidloo, and a follower of Boerhaave of Leyden. Nicolaus Bidloo, who had come to Russia in 1703 as the personal physician of the Czar, accepted the difficult task of organizing the medical education of the country with youthful ardor and enthusiasm.

During the first period of its existence, from 1707 until the death of Bidloo in 1735, the School of Moscow counted upon the following teaching personnel: a doctor of medicine, who was the director of the School and gave the principal lectures (N. Bidloo), a surgeon (*lekar*), who filled the position of tutor-coach and had the title of *chirurgus-informator* (H. Roepken, and then M. Klanke), an apothecary who taught botany and pharmacy (Ch. Eichler, later J. Maak), a surgeon's-aide and an apothecary's-aide.

According to the law, the School was to consist of fifty students. This number was not immediately attained. Until 1712 Bidloo was not able to find more than thirty-three young men, who were capable and willing to follow the lectures. In fact the greatest problem at the beginnings of the school was the lack of candidates who had the necessary qualifications. However, owing to the embryonic state of primary and secondary education in Muscovy, the only qualifications demanded were youth, a certain natural agility of intellect and a knowledge of Latin. The last represented the major obstacle, but it could not be done away with because there were no medical manuals in Russian and the teachers themselves could not make use of the national tongue without great difficulties.

The students of Bidloo's school had the advantage of being fed, lodged and clothed by the State, and this in the hospital itself. The buildings included dormitories for the patients and for the students, living quarters for the surgeons and a few small rooms where the classes were held. There was also an anatomical theater, a pharmacy with a laboratory and a garden of medicinal plants.

In Bidloo's school the work-day began with patient rounds, but before the visits, the professor explained the cases and indicated the treatment. In the beginning the hospital contained a fairly small number of sick persons, scarcely forty, among whom were many stricken by chronic illnesses, which certainly did not favor clinical teaching. During the eighteenth century the capacity of the Moscow hospital was increased to 700 beds.

After the rounds there were lectures called *collegia*: the professor dictated and the students took careful notes. Books were too precious to be given directly to the students, who had therefore to make up their own manuals from the professor's dictation.

For anatomical instruction Bidloo used the works of Blasius and Blancardus; later he had recourse to the manual of Heister and the German edition of Winslow's book. He had, moreover, brought his father's beautiful anatomical atlas to Moscow. The instruction was regularly accompanied by dissections, at which the Czar himself was sometimes present in person.

The greatest importance was attached to surgical operations. The lectures, given by Bidloo himself, followed the procedure of Rau and later that of Lorenz Heister. The hospital's surgeon demonstrated the techniques of operation on dead bodies and also held a practical course in bandaging and in the treatment of wounds.

The students were divided into three classes, according to the length of their studies, age and ability. There were examinations three times a year. About five or six years were necessary to finish the studies, but certain less talented students remained in the school up to ten or twelve years.

The first graduation took place in 1713. After a public examination the students received their diplomas, written in Latin and signed by the director of the establishment, conferring upon them the title of *subchirurgus* (*podlekar*). They then had still another practical term as physicians in a regiment, after which they were awarded the title of *chirurgus* (*lekar*).

In 1721 Czar Peter informed his agents in Western Europe that he wanted to attract some distinguished savants to Russia. Leibniz and Christian Wolff had recommended to him the creation of a university, but the monarch decided upon an institution which would be at the same time a scientific society, a *lycée* and a school for advanced study. Thus the Academy of Sciences (*Akademiya nauk*) was founded in 1724. The academicians were required to give regular lectures at the School of the Academy. Among the chairs, we note that of anatomy and physiology; from 1725 onwards it was filled by the famous physician and mathematician Daniel Bernoulli (1700-1782). In 1725, for example, this professor gave public lectures (four hours per week) on "the mathematical principles necessary to medical theory and to the study of physiology". It is a curious fact that the great mathematician Euler came to Petersburg (officially at least) as professor of physiology. He quickly ceded that place to J. Weitbreicht (1702-1747), renowned anatomist and the author of valuable studies on syndesmology. Afterwards several illustrious physicians belonged to the Petersburg Academy (K. F. Wolff, J. G. Duverney, A. Protasov, P. A. Zagorski, etc.), but although their works of research deserve high praise, still that institution never really played the role of a Faculty of Medicine. The university division of the Academy was officially abolished in 1765.

Since it had become the seat of army headquarters as well as of the central administration and of public health, Petersburg had to have hospitals. A hospital for the infantry was opened in 1715. At the same time the Admiralty founded two hospitals: in Petersburg and in Kronstadt. In January 1733 these three hospitals organized a medical-surgical teaching program modelled on the example of the school of Bidloo. At first the two hospitals in Petersburg each had twenty students and ten apprentice-surgeons, and the one at Kronstadt had eight students and four apprentices. The teaching body of each hospital was composed of one doctor of medicine, one surgeon-in-chief, two surgeons, and a pharmacist.

After the death of Bidloo, the School of Moscow underwent a grave crisis. A remodeling of its organization and a revision of the curriculum were imposed. Thanks to the archiater Johann Bernhard Fischer (1685-1772), medical education in Russia was regulated and standardized. According to the "General Rules for the Hospitals," conceived by Fischer and promulgated by the Empress Anne, 24 December 1735, four hospitals were to serve for the education of surgeons.

They were put under the supervision of the archiater and of the medical chancellery (previously the Moscow School and Hospital were supervised by the St. Synod, and the Petersburg medical institutions by military authorities). A unique academic curriculum, fairly rigid in fact, was to be adopted in all the establishments. Although commanding an enlarged theoretical basis, the instruction remained tied to the routine of the hospital. The academic year was divided into three parts: *collegium anatomicum, collegium chirurgicum* and *collegium pharmaceuticum*. Each *collegium* lasted four months. We may note in passing that the old Russian hospital schools ignored the idea of academic vacations. The lectures were in Latin and, as in the time of Bidloo, consisted above all of dictation.

An anatomical atlas, *Syllabus seu index omnium partium corporis humani figuris illustratus, in usum chirurgiae studiosorum qui in nosocomis Petropolitanis aluntur*, edited by the Academy of Sciences (1744), was of great service to the students. It consisted of twenty-six illustrations, drawn and engraved by Martin Schein (1712-1762), "Master Designer of Hospitals", a student of the Petersburg School and later chief-physician of that institution. Made in accordance with direct investigations on cadavers the Atlas was really the result of team work under the direction of J. F. Schreiber (1705-1760), professor of anatomy and surgery at Petersburg.

In the Hospital Schools, dissection was done with much enthusiasm. It was chiefly practiced by one of the hospital surgeons, who had the title of *operator* or of *chirurgus anatomicus*. The new curriculum of studies insisted on the importance of autopsy. It is necessary to observe here that Fischer was one of the pioneers of anatomical pathology and the author of a basic work on aspects of the pathologic anatomy of senescence. The bodies of patients who died in the hospital were generally submitted to anatomic examination. This also served to teach students and the surgeon's-aides the principles of legal medicine.

Surgery was also taught by the director of the School or, as in Moscow after 1754, by a secondary doctor, who had the title *doctor legens sive docens*. He had to follow the Heister manual faithfully. All of the operations then known were reviewed and the technical details were demonstrated to the students either in operating on patients or on cadavers. The lessons in surgery embraced general pathology and the treatment of some internal diseases. For each hospitalized patient a *historia morbi* was established according to the precepts of Boerhaave.

In 1763 a course in obstetrics was introduced. In the second half of the eighteenth century, Russia already had several schools for midwives (the first was founded in 1757 by the Empress), but this art was taught to men in exceptional cases only.

The last third of the academic year (which began and finished in September) was devoted to pharmacy. For botany the professor (who was the chief pharmacist of the hospital) used Tournefort's manual; for chemistry and *materia medica,* the books of Boerhaave; for prescriptions and preparation of medicines, the work of Gaub and the pharmacopeia of Leyden.

At the end of each four months' period the students had to pass an examination; at the end of their studies (that is, after an academic term of five to seven years) they were subject to a public examination.

Between 1735 and 1786 several professors of the Hospital Schools distinguished themselves. In Moscow first of all was Laurentius Blumentrost *junior* (1692-1755), the son of a famous Muscovite practioner. He was of German parentage, but since he was born in Moscow, he knew the Russian language and the customs of the country. After having studied medicine in Halle, Oxford, and Leyden, Blumentrost returned to Russia, accompanied Peter the Great during several of his voyages (it was he, for example, who bought the collection of Ruysch), held the position of first president of the Academy of Sciences in Petersburg, and in 1738 was named chief physician of the Hospital School of Moscow. He taught there with devotion and discernment until the year of his death.

Another person of above average ability was Konstantin J. Shchepin (1728-1770). With a degree in medicine from Leyden, he perfected his surgical knowledge in the hospitals of Paris and London before becoming *doctor legens* in Moscow in 1762. The vigor and originality of his lectures were immediately noticed. He awoke interest in Russian mineral waters and baths. Transferred to Petersburg in 1764, he undertook the first course of anatomy and surgery in the Russian language at the Hospital School of that city. The absence of scientific terminology was very inconvenient and Shchepin remarked that it was easier to give ten lectures in Latin than a single one in Russian.

In 1782 the famous physician and epidemiologist, D. Samoilovich (1746-1805), wrote the *Discourse for Auditors of the Hospital Schools of the Russian Empire.* He specified the goal and the principles of medical education. Samoilovich drew inspiration from the example of French schools. A reform in Russian education was then imminent. It was carefully prepared. The government had ordered two physicians, Terekhovski and Shumlyanski, to visit and study the organization of various medical schools in Europe and to make a report on the subject (1785).

Finally, by the law of the 15 July 1786, the hospital services were separated from medical instruction. The effective forces of the military and naval schools were combined and the new law gave the name *Mediko-khi-*

rurgicheskoe-uchilishche (Medical-Surgical School) to the three institutions resulting from the former Hospital Schools, located in Moscow, Petersburg, and Kronstadt. Each school had four ordinary chairs, and the title of professor was officially given to the holders of these chairs. For example, in Moscow the first professors were Y. Rinder (anatomy, physiology and surgery), H. Frese (general pathology and therapeutics) and F. Stefan (botany, chemistry and *materia medica*); the fourth chair (obstetrics) at first remained vacant.

The essential characteristic of the reform of the academic curriculum consisted of the abandonment of clinical empiricism and a more scientific orientation of the studies. Although the professors saw themselves removed from a part of their responsibilities in the hospital routine, the greater number of students (fifty per school) and the enlargement of the curriculum rendered the appointment of assistant professors indispensable.

The division of courses into three classes was maintained. With the exception of the Atlas of Schreiber and Schein and the *Pharmacopeia Rossica* (1778 and 1782), the manuals still came from abroad, (Jacquin for chemistry, Plenck for anatomy, Hoffmann for pathology and therapy, Blumenbach for physiology, and Callisen for surgery).

On the 12 February 1799 the *Collegium Medicum* submitted the plan of Czar Paul for the complete reform of medical education; the School of Kronstadt went out of existence and the two Schools of Moscow and Petersburg changed their names to *Mediko-khirurgicheskaya Akademiya* (Medical-Surgical Academy). It must be noted that the title of "Academy" for the School of Petersburg had already appeared on the 18 December 1798 in a ukase ordering the construction of a new building for that institution, which was to serve as the cradle of the most important accomplishment of Russian medicine in the nineteenth century.

In contrast to the first Hospital Schools, the two Medical-Surgical Academies were impregnated with a nationalistic spirit, especially after the Napoleonic Wars. In Moscow four of the eight professors in practice in 1800 had German names, whereas in Petersburg there was only a single German (the obstetrician and pharmacologist, Johann Ringebroig) at the side of seven Russian professors (Zagorski, Bush, Sobolevski, Severgin, Petrov, Bazilevich and Sapolovich). We may note in passing that the professors were relatively well paid (1200 rubles per year) and that they were encouraged to write their own manuals; for each approved academic manual, the author received a prize equal to his yearly salary.

Toward the end of the eighteenth century, instruction in Russian came to the fore. In order to satisfy the need for physicians in the Baltic provinces Catherine II founded, in 1783, the Medical-Surgical Institute attached to the secret hospital for syphilitics at the Kalinkin Bridge in Petersburg: *Kalinkinskii Mediko-khirurgicheskii Institut*. The instruction was carried out in German. The majority of the students, notably the thirty imperial-scholarship holders, were descendants of foreigners who had settled in Russia.

In the course of the eighteenth century at least two other hospital schools existed on the territory of the Empire: one attached to the military hospital of Elisabethgrad (which functioned from 1788 to 1796), the other in the miners' hospital in Barnaul (Altai).

One can roughly estimate that until the last ten years of the eighteenth century the hospital schools trained at least 1500 physicians (Moscow 700-800, Petersburg and Kronstadt 500-600, and the others about 300). Along with the physicians and surgeons educated abroad there were enough for the most urgent health needs of the army and navy. Let us not speak of the needs of the civilian population, because the lack of hospital personnel over this great expanse of territory was to remain an unresolved problem until the twentieth century. One can understand the necessity for training physicians more expert in practice than instructed in theory.

The first Russian university, that of Moscow, was founded on 24 January 1755, as the result of the vigorous and far-seeing initiative of the writer, naturalist and chemist, Mikhael V. Lomonosov (1711-1765). According to the initial plan it was to have twelve professors distributed within three Faculties: Philosophy (six), Law (three), and Medicine (three). The guiding idea of Lomonosov was the creation of an institution of a high scientific level which would not merely blindly transmit knowledge. Practice ought to result from precise theoretical knowledge and rigorous logical analysis. According to Lomonosov the study of medicine at the university level should not be limited to learning by rote symptoms, recommended therapeutic procedures, etc.; on the contrary it should be essential to acquire a knowledge of the fundamental structures and the laws of living beings, sick and well, and to adapt the mental discipline which permits the correct application of generalized knowledge to a concrete case.

Lomonosov's pedagogical recommendations remained a dead letter for a long time. They were to influence Russian medicine in the nineteenth century, but they could not be applied in the initial stages of the University of Moscow. The medical department of that University was at first an extremely modest one. Nine students enrolled for the lectures in medicine. Their instruction was supervised by a single man, Johann Christian Kerstens (1713-1802), who in fact taught physics, chemistry, natural history and *materia medica*. The Faculty of Medicine did not actually start until 1764. At that time Johann Freidrich Erasmus, a physician from Strasburg (d. 1777), gave lessons in anatomy, followed by dissections and surgical training. In Moscow he published a series of anatomical drawings for the use of the students (1767).

During this period, the central figure of the Faculty of Medicine was Semion G. Zybelin (1735-1802), reformer of Russian clinical medicine. He was responsible for a penetrating analysis of the role of the external milieu in the genesis of ailments. A talented orator and connoisseur of the national literature, Zybelin inaugurated the medical program of the University in the Russian language. Another professor, P. D. Veniaminov, deserves to be mentioned for his activities in public hygiene and the education of the youth. Yet it is also necessary to indicate the weak point of the Faculty of Medicine of Moscow: the absence of a university clinic and of laboratories and, as a consequence, the too theoretical and bookish training. The Faculty performed its pedagogical task in a manner opposite to that employed by the Medical-Surgical School. The synthesis of the two procedures was achieved only in the institutions of the second half of the nineteenth century, in particular in the Medical-Surgical Academy of Petersburg.

Until 1791 the University of Moscow did not have the right to confer the title of doctor of medicine, but only that of *lekar*.

Among the first doctors of the University of Moscow the most important was Foma J. Barssuk-Moysseiev (1768-1811) who, on 29 March 1794, publicly defended a thesis on the physiology of respiration (*Dissertatio Medico-Physica de Respiratione*, Moscow, 1794). After graduation, he was to become in his turn professor of physiology and of therapy at the Faculty of Medicine.

It was at the beginning of the nineteenth century, during the reign of Alexander I (1801-1825), that higher education in the Russian Empire underwent a remarkable expansion. The University of Moscow was enlarged, and the number of students soon passed 500. The law of 5 November 1804 gave a certain autonomy to its teaching staff. The University had four Faculties, among which the Faculty of Medicine had an honorable place (six ordinary chairs). Theoretically, the length of studies was not long: three years. Nevertheless, the professors were more and more demanding and at least four or five years were actually necessary to finish one's studies in medicine.

Through territorial conquests the Russian Empire had won, in the course of the eighteenth century, two university cities: first Dorpat, which was taken from the Swedes, and then Wilno, which fell to Russia on the occasion of the division of Poland. Dorpat (in Estonian, Tartu) had an ancient University, founded by Gustav Adolphus in 1632. However, the medical training had no real impetus unitl the reorganization of that institution and the foundation of a Russian University (1802). This new university did not use Russian as the official language of instruction. Although insisting on a political orientation favorable to imperial centralism, the government allowed in this special case, the use of a non-Russian language. Instead of choosing the idiom of the Estonian people, preference was given to German, that is to say, to the language which was habitually used by the aristocracy of that country.

In Wilno (today Vilnius) the capital of the grand Duchy of Lithuania, the Jesuits had opened an Academy (1578) which, thanks to the Pope and the Polish king, had all the privileges of a university. The Faculty of Medicine was founded on paper in 1641 and in fact after 1780, when the Rector,

Marcin Poczobut, brought in French professors (Reignier, Briotet, Bisio). The Russians enlarged the medical program, especially after having given this institution the rank and title of University (1802).

New universities were founded in Kazan (1804) and at Kharkov (1805). The Pedagogical Institute of Petersburg also became a University (1819).

All of the territory of the Russian Empire was divided into six academic departments, each having as its administrative center a university (Moscow, Petersburg, Kazan, Kharkov, Dorpat, Wilno).

From the date of its foundation, the University of Kharkov (Ukraine) had a Faculty of Medicine, and its scientific influence was very considerable. At Kazan the Faculty of Medicine was founded with a certain delay on 14 May 1814. The University rapidly became the cultural center of the vast area around the Volga and the Ural Mountains. But it should be mentioned that the doors of the University were closed for a long time to Tatars, Bashkirs and other non-Russian inhabitants of these regions.

If the government was little worried about nationalist recriminations, it was uneasy about the growing number of non-aristocratic students, the undeniable growth of liberal ideas and the profession of a certain "materialistic thought" which seemed to be more and more widespread in academic milieus. It would be unjust to deny to the first part of the reign of Alexander I a desire for the modernization of the country and even a kind of sympathy for the new intellectuals. Since the State was engaged in wars, first against the Turks and then against the Swedes and the French, it knew how to appreciate its physicians and technicians. But this situation did not last very long. The Czar's inspectors denounced the Universities, and in particular the Faculties of Medicine, as the seat of an especially dangerous ideology. Thus, for example, in the official report on the situation at the University of Kazan in 1818, it was above all the question of the dangers to which the teachings of science expose the true faith. The imperial envoy had even found it advisable to hinder dissection and to close the anatomical museum, because, he wrote, "it is loathsome and impious to anatomize the human body, which is made in the image of God."

However, the professors of medicine in Russia at that time seem to have been much further from militant materialism than those past inspectors, as certain Soviet historians of today would have us believe. The training in medicine in the Russian schools during the first third of the nineteenth century followed more the example of German romanticism than that of the new anatomo-clinical orientation of the School of Paris.

After the revolt of the Decembrists and the ascent of Nicolas I (1825-1855) to the throne, the czarist administration became even more distrustful, and the moral pressure on the universities grew. An autocrat of conservative ideas, the Czar sometimes personally intervened in university affairs and insisted forcefully on order and uniformity, hierarchical obedience on the military model, respect for religion and a pragmatic orientation for medical, technical and even legal instruction.

After the Polish insurrection the University at Wilno was done away with (1832), and only the Faculty of Medicine persisted for a while in the form of a Medical-Surgical Academy.

In 1834 the University of Saint Vladimir at Kiev was created, but this was only to replace the University of Wilno. The new University was at first deprived of its medical program, and only in 1841, that is to say, after the suppression of the Medical-Surgical Academy of Wilno, did it see itself endowed with an Institute of Medicine, the nucleus of the famous school of the future.

Vast regions, for example, Siberia and the Caucasus, still remained without any institution of higher education, and the students who came to university cities from afar were very numerous. Their movement was facilitated by the construction of railways. At Dorpat the students from outside the region were at first no more than fifteen per cent of the total, but by the middle of the century more than one third of those registered at the university did come from outside the Baltic region. They came from the Ukraine, from central Russia, and even from the Caucasus. The renowned surgeon Nikolai I. Pirogov (1810-1881) completed his medical studies at Dorpat. He wrote memoirs which give testimony of the old patterns of medical education in Russia.

Among the students at Dorpat, we may note K. E. von Baer, Inozemtsev, Filomafitski and Ovsyannikov. A knowledgeable and extraordinary master, Filip V. Ovsyannikov (1827-1926), went on to found the laboratory of experimental physiology at Kazan before becoming professor at Petersburg. The University of Dorpat was a brilliant center of pharmacological research and teaching, especially at the time of the German professor Rudolph Buchheim (1820-1879) and his pupil Oswald Schmiedeberg (1838-1921) of Courland.

In 1804 the famous Johann Peter Frank (1745-1821) accepted a call to take the clinical chair at Wilno, while his son Joseph was given that of pathology. A little later, however, Johann Peter left for Petersburg, where he was given the post of first physician to the Emperor and the chair of practical medicine at the Medical-Surgical Academy, while Joseph Frank (1771-1842) succeeded his father in the clinical chair at Wilno. The good sense and the immense experience of Johann Peter Frank, who was later to become rector of the Academy, contributed much to improve the teaching of medicine in Russia. Under his direction, the Academy became an educational establishment of the first order; courses in Latin and German were established, new chairs were created and special clinics organized and opened. Peter A. Zagorski (1764-1846) gathered a group of anatomists around him and laid the basis of a movement which was to be called "the Russian school of anatomy." D. M. Vellanski, professor of physiology in Petersburg, was a romantic, who drew inspiration from Schelling and Mesmer, denied the value of experimentation in physiology and opposed vivisection. His biological philosophy and methodological attitudes were criticized by the new generation of physicians at Petersburg and even in his own time by professors at Dorpat and Moscow, notably by Alexei M. Filomafitski (1807-1849). This last, the author of a good manual of physiology (1836), may be considered as the founder of experimental physiology in Moscow.

The work of Frank in Petersburg was continued by an excellent Scottish doctor, Sir James Wyllie, who directed the Medical-Surgical Academy from 1808 to 1838. According to a statement of Granville, who visited that institution in 1826, it was functioning in an exemplary manner. The complete cycle of studies took four years. In the first year medical propedeutics was taught, veterinary surgery, natural history, mineralogy, zoology, natural philosophy, anatomy, mathematics, Latin and German. The second year was devoted to the study of physiology, pathology, botany and chemistry, as well as to dissection. In the third year the pharmaceutical art, general therapy, clinical medicine and surgery were taught. Finally, in the fourth year the following courses were included: *materia medica,* "medical police", (i.e. hygiene) and ophthalmology.

Among the new courses we may note the history of medicine. This discipline was first taught (1825) by N. D. Lebedev (b. 1799) at the University of Moscow. In Petersburg it became obligatory, but in spite of the existence of good historians of medicine in that city the lectures were given in turn by physiologists and clinicians. The lectures were inspired by the manuals of Sprengel and Renouard.

After 1811, the Academy published its own scientific review. We may note that after 1828 Professor Prohor Charukovski (1790-1842) introduced the stethoscope of Laennec in his demonstrations of clinical surgery.

In 1840 the Academy of Petersburg was enriched again by a chair of "clinical hospital therapy," which illustrates well the concern for medical education directed toward practice and toward direct experience at the patient's bedside.

At the University of Moscow, the teaching of clinical medicine was primarily linked with the activity of Matvei Y. Mudrov (1776-1831), the successor of Zybelin in the chair of therapeutics. Mudrov complained about his former professors at the Academy of Moscow and reproached them for the inadequacy of anatomical dissections and clinical demonstrations and insisted on improvement of the practical work. An excellent practitioner and at the same time a convinced humanist, Mudrov drew the attention of his pupils not to the disease, but rather to the sick man. He may be regarded as one of the pioneers of psychosomatic medicine.

The teaching of surgery was dominated by a great concern with anatomical precision and rational justification of every operative intervention. If Richter at Moscow, toward the end of the eighteenth century gave fine lectures, his successors surpassed him in practical demonstrations. In 1807-1808 appeared the first complete manual of surgery in Russian, written by J. F. Bush, a professor in Petersburg. In the former capital the professor of clinical medicine, Efrem O. Mukhin (1768-1850), a teacher of Pirogov, published a manual of orthopedics and traumatology (1806), as well as a course of anatomy (1815).

In mid-century two great surgeons stand out clearly on the Russian medical scene. These two indisputable masters were Nikolai I. Pirogov (1810-1881), a professor at Petersburg since 1841, and Fedor J. Inozemtsev (1802-1869), holder of the chair of surgery and therapeutics in Moscow. Their interests were encyclopedic, their personalities fascinating. Pirogov served as the example for many generations of Russian physicians.

It is difficult to make an honest and impartial judgment of the pedagogical and scientific value of the training dispensed in the Russian faculties in the course of the first half of the nineteenth century. Suffering from the chicanery of a police force, which in the name of orthodox religion and *raison d'état,* impeded liberty of speech and research, the professors had every reason to stay with the mediocre and to keep their original ideas to themselves. If one refers to the academic manuals and theses, the training was correct and often little more. The tendency was toward conservatism and eclecticism. Of course there were some exceptions, some original research, some non-conformist professors, and this in particular at the Academy of Petersburg. Although under military dictatorship, the latter paradoxically, enjoyed greater liberty, at least as far as medical research was concerned.

During the first half of the nineteenth century, Russian medical literature developed considerably. Several hundred works were published. For the most part they were sharply oriented toward problems of practical medicine. If, however, it could be said that Russia had its own medical bibliographical resources, the same could not be said with respect to the equipment of the hospital laboratories. With the exception of the University of Petersburg, the university clinics were badly organized and did not have at their disposal sufficient material.

In the course of the second half of the nineteenth century, Russia underwent a transformation. The development of industry, the Crimean War, the cholera epidemics and other events forced Alexander II (emperor from 1855 to 1881) to undertake great administrative, social and even cultural reforms. The abolition of serfdom was without doubt the decisive turning point (1861). In the domain in which we are here interested two events assume great importance: the promulgation of new university statutes (1863) and the establishment of a medical-social system, the *Zemstvos* (1864).

The law of 1863 gave the Russian universities "autonomy". This meant that the assembly of professors could elect the rector of the university and the deans of the faculties of their own choice, and had the right to establish a curriculum and to make decisions in a large group of personnel and administrative questions concerning the professors and the students.

In order to encourage young scientists who found themselves faced by the choice of a career of teaching or one of research the title *privat-docent* was henceforth awarded and in accordance with the scheme then in force in the German universities. For it is necessary to recognize that from the middle of the century to the First World War, Russian university life and practical medicine were profoundly influenced by the example of Germany.

The medical services of the *Zemstvo* (a kind of local administrative unit) represented an original attempt to get a system of public health on its feet which favored the rural population. Let us recall here only that this new organization required a considerable number of physicians. It was necessary to train them in a more ample and rapid way. The Medical Faculties of Moscow, Kharkov, Kazan, Kiev and others accepted more students than before, but this could only be a long term solution to the problem. It was thus decided officially to employ practitioners without academic training. This class of sanitary personnel had the name *feldsher* (derived from the German *Feldscherer*). Their training, a working one, was by apprenticeship. In order to regulate and assure a certain level of professional competence to this new kind of health officers the *Zemstvos* opened about twenty schools for *feldshers* between 1867 and 1876. The course lasted for two years and the training was oriented toward the demand of practice. Although this procedure might seem debatable, it allowed the public health services to operate small hospitals and rural health centers. From 1870 to 1910 the number of doctors and *feldshers*, salaried and working in establishments of *Zemstvos*, rose from 610 to 3100. The number of free practitioners was even higher but this regular rise was revealed as insufficient in the face of demographic growth.

The Empire was prey to serious political tensions. How could it be otherwise, considering the out-of-date officialdom, the rise of nationalism, the situation of a peasantry freed but very poor and scarcely assured of the minimum for survival, and the birth of an industrial proletariat which was clashing with the autocratic and aristocratic state. After the assassination of Alexander II (1881) the police harshly expressed its distrust of the less conformist students. Once again discriminatory measures were applied to direct the recruiting of intellectuals. The *Statute for Russian Universities*, 23 August 1884, suppressed the "university autonomy". The size of the student body continued to grow. It was 3659 in 1855 and 16,357 in 1900. These figures include the students of all the faculties; nearly one third of that total were medical students.

New universities were founded. First of all, that of Tomsk (1888). It was to serve for the immense territory of Siberia. At the beginning, the University of Tomsk had only a single faculty, that of medicine. Another Faculty of Medicine was opened at the University of Odessa (1896).

There was also a Russian university on Polish territory. After the Congress of Vienna, Warsaw found itself under Russian domination. This part of Poland, at least at the beginning, had a certain autonomy which permitted the creation of the University of Alexander in 1817, Polish in language, in the nationality of the professors and the students, as well as in political orientation. Its Faculty of Medicine showed the first signs of prosperity when, as a consequence of the insurrection of November 1830, the University was closed and dissolved. In 1859 the Russian authorities opened a Medical-Surgical Academy in Warsaw, which was replaced in 1862 by one Main School with four Faculties. More than 250 students enrolled immediately in the first year in the Faculty of Medicine. Although the more distinguished professors were of Polish nationality (for example, the physiologist J. Mianowski and the physician T. Chalubinski), the teaching was carried out in Russian. After 1869 the school had the title of Imperial University. Its curriculum was uniform with that of all other universities of the Empire. At once the center of scientific education and the instrument of national oppression, the University of Warsaw played an ambiguous historical role. In 1915 it withdrew to Russian territory and gave birth to the new University of Rostov on the Don.

Toward the end of the nineteenth century, the Russian Empire had eight faculties of medicine (Moscow, Dorpat, Kharkov, Kazan, Kiev, Tomsk, Odessa and Warsaw). It is necessary to add here a faculty which was situated in the autonomous territory of Finland. In fact, at Helsingfors (today Helsinki) there was a former Swedish university with its own School of Medicine (founded in 1640). The Russians governed it from 1809 to 1917.

All of the faculties at this time depended on the Ministry of Public Instruction. Their structure, curriculum, details of teaching and examinations, as well as the hierarchical structure of the teaching staff were uniform.

The Imperial Academy of Petersburg held a special place. It officially became a military institution under the jurisdiction of the Ministry of War. After 1881 it was called *Imperatorskaya Voenno-meditsinskaya Akademiya* (Imperial Military Medical Academy). It must be remarked that this was a privileged school, generously endowed and accepting only the elite of the Russian youth.The military budget permitted for the Academy's expenditures greatly surpassed the small means at the disposition of the Ministry of Public Instruction for all other medical schools. In 1850, the Academy had twenty-seven professors (sixteen ordinary and eleven assistants); in 1909 it would have thirty-three professors and 189 assistants.

The foreign physicians who visited the Military Academy toward the end of the nineteenth century agreed in considering it a model establishment with respect to medical progress.

Besides the departments of clinical medicine, which are very numerous and comprise special clinical instruction in gynecology, obstetrics, ophthalmology, laryngology, etc., this school has a service of medical jurisprudence to which are assigned all simulated diseases, and in its amphitheatre are performed all autopsies pertaining to legal medicine. In proximity to this academy there is an anatomical and physiological institute where these two branches are taught, as well as the auxiliary sciences. This Imperial Academy has for its object not only to qualify students to practice medicine, but also to provide the army with military physicians. All the professors and officers of this Academy, including the students, wear a special robe, and all make their visits in state (Dujardin-Beaumetz, 1888).

An orientation toward the so-called fundamental sciences, the development of specialized clinics, the introduction of scientific research as a part of professional education, the creation of laboratories and the affirmation of an original national medical literature were some of the things which characterized Russian medical training between 1850 and 1917. As far as the backdrop of ideology was concerned, the influence of naturalism and the materialistic philosophy of the revolutionary-democrats such as Chernyshevski, Dobrolyubov, Pisarev and others was felt. In medicine's own domain, the new orientation of Russian science was linked to the activities of Botkin and Sechenov. They elaborated neural theory and gave physiopathological foundation to the practical medicine taught in Russian schools. The clinic could no longer do without the help of laboratories.

Sergei P. Botkin (1832-1889) studied medicine at Moscow in the middle of the nineteenth century (graduation in 1852). Being very dissatisfied with the bookish character of medical education and with the authoritarian transmission of knowledge, he decided to remedy it by offering a method of teaching aimed more at developing the critical sense of the students than at inculcating them with ready-made ideas of older experts. In the inaugural lecture of his course in therapeutics at the Academy of Petersburg in 1862, Botkin revealed his scientific program and pedagogical ideas. He revolted against localism in pathology, insisted on the importance of physiological research for medical theory and practice, underlined the social aspects of disease and revealed the insufficiency of the written word when faced with an extremely complex nosological, anthropological and social reality. He was interested in prophylaxis and considered the sick man rather than the disease.

Pavlov began his scientific research in the laboratory of Botkin's clinic. In general, the students of Botkin became the occupants of a series of chairs in the specialities.

To illustrate the growing specialization of teaching we may note the chairs created, or rather detached from a vaster discipline, at the Academy of Petersburg after the middle of the century: pathological anatomy (1859), physics (1860), botany (1860), practical anatomy (1860), diseases of the eyes (1860), psychiatry (1860), zoology and comparative anatomy (1862), hygiene (1865), operative surgery (1865), pediatrics (1865), experimental pathology (1878), ear-nose-throat (1889).

It is to a schoolmate of Botkin, Gregorii A. Zakharin (1828-1897, graduated in Moscow in 1852) that praise is due for reorganizing clinical teaching at Moscow and for overseeing the first steps of the new specialities. His students learned above all carefully to observe, listen and understand their patients. A remarkable observer and practitioner, he disputed the validity of that bookish approach which had been so displeasing to his own generation of Muscovite students. Among the professors of therapeutics toward 1900, we may point out A. A. Ostroumov (1844-1908) in Moscow and V. P. Obraztsov (1851-1921), cardiologist and chief of the school in Kiev.

In the history of Russian surgery the greatest master after Pirogov was Nicolai V. Sklifosovsky (1863-1904), head of the Surgical Clinic of Petersburg. It was he who introduced antiseptics and modern abdominal surgery into Russia. As in the case of nearly all the other professors mentioned in that period, for Sklifosovsky medical research, work at the hospital, field experience and teaching were combined into a harmonious unity in which each part reinforced the others.

The clinicians themselves did the autopsies, but toward the middle of the century this became the task of specialists in pathology. The clinic could not do without the regular service of pathological anatomy. In 1849 the first chair of pathological anatomy and physiopathology was founded in Moscow. Its head, Professor A. J. Polunin (1820-1888), insisted on the influence of the nervous system in pathological processes. Thus he searched for a way of explanation different from the humoral concepts of Rokitansky and from Virchow's localism. Cellular pathology became rapidly known in Russia. Professor M. M. Rudnev (1837-1878), founder of the chair of pathological morphology in Petersburg (1859), introduced the microscope into the teaching of pathology.

At the pedagogical level no one had an influence comparable to that of Ivan M. Sechenov (1829-1905), creator of the Russian school of neurophysiology. He taught at Petersburg, Moscow and Odessa. He was suspected by the government and pursued by the police for his materialistic ideas but was loved by his students. Sechenov launched a scientific movement which was maintained and reinforced by the physiologists Tarkhanov and Wedenski, by the pharmacologist Kravkov and especially by the pathologist V. V. Pashutin (1845-1901). In 1878 this last separated the chair of experimental pathology from that of general pathology. The physio-pathological orientation of research and teaching characterized a whole period of the Academy of Petersburg, and found its best expression in the work of Ivan P. Pavlov (1849-1936), a first hand student of Botkin and spiritual heir of Sechenov.

From the second half of the nineteenth century on a series of excellent medical manuals aimed at Russian students came into existence (Botkin, Sechemov, Pashutin, Erismann, etc.).

In order to enroll at a faculty of medicine, it was necessary to have proof of one's graduation from a *gymnasium,* or analogous school. A knowledge of the Latin language was considered indispensible. The courses were given in Russian, however, with the exception of Dorpat (German) and Helsingfors (Swedish). All of the faculties had a *numerus clausus:* the number of newly enrolled students was not to surpass a certain figure, established in advance by the Ministry. There were also special restrictions for the admission of Jewish students or Polish nationals. The school most sought after was clearly the Academy of Petersburg, but it took only a relatively small number of students (125-150 per year). The enrollment costs were modest, and although the students had to pay fees directly to the professors under whom they took courses, the annual expenditure remained comparatively low: 60 to 90 roubles (about 32-50 dollars at the time). However, the good students were often able to obtain grants. The majority of the medical students were poor and lived in extreme simplicity. Beginning in the second year of their studies the students of the Academy of Petersburg received a stipend if they were willing to promise to serve as physicians in the army for four years.

For a long time the Russian universities were unwilling to accept women students. Women were of course excluded from the teaching body. However, a large number of young women eagerly thronged to the study of medicine. At first they studied abroad, notably in Switzerland. Between 1872 and 1882 medical courses for women were given at Petersburg, and by 1888 out of about 15,000 physicians in Russia there were some 750 women. The suppression of these courses by Alexander III provoked sharp criticism. Finally a *Zhenskii meditsinskii institut* (Institute of Medicine for Women) was opened in Petersburg (1897). In the administrative division of Advanced Courses for Women in Moscow a department of medicine was organized in 1906. It is interesting to note that this medical department became, after the Revolution, the Faculty of Medicine of the Second University of Moscow.

In the universities and at the Military Academy medical studies lasted for five years. Strict attendance of courses was obligatory. As Louria pointed out in 1909, medical education in Russia:

... has always been supervised and controlled by the Government. Indeed, the Minister of Public Instruction has full charge of all schools, from the primary grades to the university. This insures a uniformity of educational practice throughout the Empire, and the curricula are identical in the University of Dorpat, the nearest neighbor of the illustrious models of German teaching, and in the University of Tomsk, thousands of miles away in Siberia. There are no private institutions of medical teaching supported by the endowments of kind but frequently interested benefactors.

The program of study was divided into two parts: preparatory (two years) and clinical (three years). The first year was devoted to the basic sciences (then called "accessory sciences"): botany, zoology, physics and chemistry. During the second year the students studied the pre-clinical sciences: anatomy, histology and physiology. Starting in the third year they were affiliated with the so-called propedeutical clinics. The students took master courses given by professors and in addition worked in clinics under the direction of assistants. Their clinical work was carefully supervised and they were allowed no responsibility. It was only during the last year that they had free access to the patients and enjoyed a limited independence in their work. During the three clinical years the students had to follow instruction in about twenty specialities. Each discipline taught had an examination to be passed. The program of study precisely prescribed the names of the courses and their chronological order, and the students were allowed absolutely no choice. The system was authoritarian and paternalistic.

The studies were concluded by a state examination and were sanctioned by the diploma of *lekar* (physician). This degree conferred the right of professional practice and corresponded exactly to the degree of doctor of medicine in the German and French universities. The degree of *doktor* (doctor) was a more advanced one and not obtained without at least two years of complementary studies. The number of physicians who obtained the title of *doktor* was about ten per cent.

In order to win the title of *doktor* of medicine the candidate, who was already necessarily in the possession of the diploma of *lekar,* had to work two years in a scientific establishment, pass a new set of examinations and publicly defend a thesis based on an original piece of work.

In 1885 a Clinical Institute, a genuine school for improvement, was opened in Petersburg. It was destined to complete the training of physicians preparing for their doctorate. The Institute gave free courses. It had not only clinics but well equipped laboratories in addition, notably an important department of bacteriology. The Institute was retained by the Soviet government as a center of post-graduate study for physicians.

In Petersburg in 1890 Russia's most advanced school of medical science was founded, the *Institut eksperimentalnoi mediciny* (Institute of Experimental Medicine). It was at once a center of investigative science and instruction, and was destined to train researchers of the highest quality. The Institute had an anti-rabies station, a laboratory for the study of histology, biochemistry, experimental pharmacology, etc. The most famous was without doubt the physiology laboratory organized and directed by Pavlov.

At the moment of the collapse of the czarist regime, there were sixteen schools on Russian territory which served for the education and advanced training of physicians. This figure is obtained by adding up the institutions mentioned here plus the other schools of medicine admitting women only and the Faculty of Iekaterinoslav (founded in 1916) today Dniepropetrovski). All of the schools together trained about 1500 physicians each year.

We can only briefly examine the development of medical education in the USSR.

During the first years of the Soviet regime, the program of studies underwent slight modifications only. In 1918 the *Narkomzdrav*, the committee performing the office of the Ministry of Public Health, took up the question of a radical reform in medical instruction. D. K. Zabolotny suggested a substantial strengthening of the chairs of hygiene, epidemiology, bacteriology and social medicine, in conformity with the prophylactic orientation which it was desired to impose on physicians. An abridged course (three years) was planned for the training of epidemiologists.

Three major ideas concerning medical education governed the first decisions of the Soviet administration: the desire for a marxist-leninist ideological alignment, the fear of epidemics in a period already difficult both socially and economically, and finally the urgent need for a greater number of physicians. The authorities hurried to create new medical schools, especially in regions hitherto deprived: Tbilisi in Georgia (1918), Astrakhan (1918), Voronezh (1918), Tashkent in the Uzbekistan (1919), Smolensk (1920), Omsk (1920), Krasnodar (1920), Minsk in Bielorussia (1921), and Erivan in Armenia (1922).

In the years 1920-1922 began a period which was characterized by a series of experiments in the program of studies, in the details of the instruction and in the duties imposed on the students. How could the need to train rapidly a sufficient number of practitioners be reconciled with the need to maintain a certain minimum standard in their education? How could one adapt the old university structure to the new ideological demands? Contrary to that which went on in certain other domains of scientific and cultural life, the procedure here was marked by a gradual change with much prudence and thought. The first steps toward a new form of medical education were linked with the name of Nikolai Semashko (1874-1949), people's commissar of health and the first Russian professor of social hygiene.

The main structure of the program, modeled on the old German system, was not touched. The preparatory part remained separated from the clinical part of instruction, the students came into contact with sick persons only much later and their clinical exercises were strictly supervised, and major theoretical lectures and laboratory work remained the cornerstones of instruction. However, the time devoted to practical exercises in the theoretical and clinical sciences was considerably augmented. New disciplines were included in the program: general biology, physical and colloid chemistry, infectious diseases, industrial hygiene and social hygiene, as well as political subjects.

The teaching body did not undergo significant reshuffling. The professors who desired to collaborate with the new regime were not touched. It was only later, during the Stalinist years, that strong political and ideological pressures were put on the scientist.

At the student level, radical changes were brought about by the October Revolution. The doors of the university were opened to all. This was so in theory at least. In practice it meant that the old discrimination on the basis of race, religion and sex was effectively suppressed, but that between the nobility and the bourgeoisie on the one hand and the proletariat and the peasantry on the other was inverted. In order to favor young people from the masses, the administration reduced admission requirements. A knowledge of Latin was no longer required. Proof of graduation from a high school was replaced by a wide variety of recommendations emanating from diverse scholarly or political organizations. The length of study was shortened by one year, but each academic year was increased to thirty-six weeks of work (instead of the former thirty). In addition, the students had to complete a practical course of two months in the summer. It was almost a return to the old custom of education without vacation. But was this, too, not in order to meet the same demands?

What the Soviet regime fully inherited from ancient times was a paternalistic attitude, rigidity of curriculum and uniformity of all schools. The student was always guided; he had neither the care nor the pleasure of choice. We may note also that Russia had never had medical schools that were founded, financed or directed by private persons or groups. The nationalization of medical education was something that had been carried out a long time before the Revolution.

The crucial problem posed to the Soviet authorities, as before to Peter the Great, was without doubt that of pre-medical education. It could not be resolved without the creation of a vast network of primary and secondary schools. While waiting for this development the temporary solution was to bring the faculties into action with evening courses for workers.

On the eve of the First World War Russia had about 20,000 physicians. In 1928 they numbered 62,000. But there is no doubt that this growth was accomplished at the price of a general decline in the scientific level of studies.

In 1930 the whole system of higher education was radically reorganized. The specialized advanced schools (including medical instruction) passed from the administration of Commissariats of Public Instruction to that of the Commissariats of the corresponding techniques. In our domain this meant that the Faculties of Medicine were separated from the Universities, reorganized into independent units and placed under the responsibility of the Commissariats (today the Ministries) of Public Health. The Faculties of Medicine were transformed into Institutes of Medicine.

This was also the incentive for the creation of nineteen new schools during 1930-1934. For the most part they were in regions very distant from existing centers (for example, Alma Ata, Samarkand, Irkutsk, Arkhangelsk, Ashkhabad).

In 1934 the organization of the Institutes of Medicine received in general outline, the forms which they have conserved until this day. Each Institute included three Faculties: (a) medical prophylaxis and general medicine (for the training of physicians for hospitals, polyclinics and rural services); (b) sanitary hygiene (for the training of epidemiologists and health inspectors); (c) pediatrics (training specialists in diseases of infants). The Soviet Union aimed at the parallel formation of three categories of physicians: general practitioners (called "therapists"), "specialists in hygiene" and "pediatricians".

In Leningrad, the Institute of Pediatrics, an autonomous establishment, was founded in 1935 with the aim of education for maternal and infant protection. The number of medical schools continued to grow. In 1960 the figure of seventy-seven schools was reached. The number of students varied from school to school from 250 (Irkutsk) to 6000 (the First Institute of Moscow), and most often ranged between 1000 and 3000. The number of students graduating from each school per year was from seventy to 1200, but most often from 200 to 500. In 1960, for example, 26,452 students received the diploma of medicine throughout the Soviet Union.

The number of physicians increased to 105,600 (in 1937), to 130,400 (in 1940) and to 364,300 (in 1959).

The length of studies was once more set at five years (1934) and even prolonged to six years (after 1945). Medical education is free, without tuition fees.

From 1930 on the Institutes of Medicine in the Soviet Union followed programs of study which were no longer copied from a Western model. Sigerist has given an excellent presentation of the evolution of medical training in Soviet schools from 1930 to 1945. He explained the reasoning behind the choices made by the Russians and evaluated their efficacy. It will suffice for us to sum up the resulting program from recent official documents.

For the "therapists" the curriculum includes the subjects which follow: During the first and the second years biology, physics, chemistry, anatomy, histology, physiology, biochemistry, microbiology, parasitology, the principles of marxism-leninism, Latin and one modern language and lastly, physical culture are studied. The special aim is to teach the student scientific method. The instruction is accompanied by practical work in the laboratory. The third year is devoted to pathological anatomy, physiopathology, pharmacology, an introduction to medicine, surgery and several clinical disciplines.

During the fourth and fifth years, the students continue their clinical studies (general medicine, surgery, obstetrics, pediatrics, etc.). An important place is given to hygiene. Special attention is accorded to independant work by the young student. The last year is dedicated to supervised practical work in various disciplines.

The "specialist in hygiene" and the "pediatrician" follow a similar program, but a more important place is given to the disciplines of particular interest and pertinence in their speciality.

Medical instruction is closely tied in with scientific research. The professors divide their time between both tasks. The students are encouraged to participate in working seminars and discussion groups which aid them in original research.

In the Soviet Union the completed study of medicine leads to the diploma of *vrach* (physician). As before, the doctorate is reserved for physicians who choose a scientific career and complete additional studies. In effect the title of *vrach* corresponds to that of *docteur en médecine* in France or to the M.D. in America, while the Russian doctorate approaches the competitive examination known as the *agrégation* in France.

Although a genuine comparison would be difficult if not impossible, the general impression of Western authors is that the *vrach*, on the whole, receives an education slightly inferior to that of the Western physician, but on the other hand holders of the Soviet doctorate in medicine have received a scientific training of the very first order.

SELECTED BIBLIOGRAPHY

I. Slavonic

BAGALEI, D. I., and SKVORTSOV, I. P., *Meditsinskii Fakultet Kharkovskogo Universiteta za Pervye Sto Let Ego sushchestvovaniya (1805-1905)*. Kharkov, 1906.

BIELINSKI, J., *Uniwersytet Wilenski*. Cracow, 1900.

BOGOYAVLENSKI, N. A., K 300-letiyu otkrytiya pervoj lekarskoj shkoly v Rossii (1654 g.), *Feldsher i akush.*, 1955, no. 1, pp. 40-45.

———, Studencheskii rukopisnyi uchebnik anatomii cheloveka Moskovskoi meditsinskoi shkoli Nikolaya Bidlo. *Arkh. Anat.*, 1960, **40**(9): 108-111.

BUTYAGIN, A. S., and SALTANOV, J. A., *Universitetskoe Obrazovanie v SSSR*. Moscow, 1957.

CHISTOVICH, Y., *Istoriya Pervyh Meditsinskih Shkol v Rossii*. St. Petersburg, 1883.

CVETAEV, D. V., *Mediki v Moskovskoi Rusi i Pervyj Russkij Doktor (P. V. Posnikov)*. Warsaw, 1896.

EGOROV, Y. N., Reaktsionnaya politika tsarizma v voprosah universitetskogo obrazovaniya v 30-50 godah XIX veka. *Istor. nauki*, 1960, no. 3.

GERTSENSTEIN, G. M., Ocherk istorii meditsinskago obrazovaniya v Rossii v proshlom stoljetii. *Vrach*, 1883, **4**: 412, 429, 444, 460, 476, 493, 509, 524.

GETSOV, G. B., *Proizvodstvennaya Praktika Studentov Meditsinskih Institutov*. Moscow, 1949.

GRMEK, M. D., Shkole meditsinske, *Med. Enciklopedija* (Zagreb), 1964, **9**: 419-423.

Istoriya Imp. Voenno-Meditsinskoi (b. Mediko-Khirurgicheskoi) Akademii za sto let (1798-1898). St. Petersburg, 1898.

KANEVSKI, L. O., LOTOVA, E. J., and IDELCHIK, H. J., *Osnovnye Cherty Razvitiya Meditsiny v Rossii v Period Kapitalizma*. Moscow, 1956.

KAVETSKI, R. E., and BALITSKI, K. P., *U Istokov Otechestvennoj Meditsiny*. Kiev, 1954.

KHAMITOV, H. S., 150 let vysshego meditsinskogo obrazovaniya v g. Kazani. *Kazanskii Med. Zh.*, 1964, no. 5, pp. 9-19.

LAKHTIN, M. Y., Gde priobretali svoi znaniya russkie vrachi v Moskovskom gosudarstve. *Vestn. Vospit.*, 1904, **3**(9): 71-77.

LEBEDYANSKAYA, A., Pervaya voenno-meditsinskaya shkola v Rossii. *Voenno-Istor. Zh.*, 1941, no. 3, pp. 97-101.

LYUBIMENKO, I. I., Prosveshchenie i nauka pri Petre I i osnovanie Akademii nauk. *Izvest. Akad. Nauk, Ser. Istor.*, 1945, **2**, no. 3.

MATYUSHENKOV, I., *Tsil i Zadachi Universitetskago Meditsinskago Obrazovaniya*. Moscow, 1876.

MULTANOVSKI, M. P., *Istoriya Meditsiny*. Moscow, 1967.

Myasnikov, A. L., *Russkie Terapeuticheskie Shkoli.* Moscow, 1951.

Nazarenko, I. I., Iz istorii prisuzhdeniya pervyh doktorskih stepenej v Rossii. *Sov. Zdravookhr.,* 1951, no. 5, pp. 55-60.

Ocherki po Istorii I Moskovskogo Ordena Lenina Meditsinskogo Instituta imeni Sechenova. Moscow, 1959.

Ovcharov, V. K., *Razvitie Vysshego Meditsinskogo Obrazovaniya v SSSR i Rol v nem 2. Moskovskogo gos. Universiteta, 1918-1930.* Moscow, 1955.

Palkin, B. N., *Russkie Gospitalnye Shkoly XVIII Veka i ikh Vospitanniki.* Moscow, 1959.

Pekarski, P. P., *Istoriya Imp. Akademii Nauk v Peterburge.* St. Petersburg, 1870-1873.

Petrov, B. D., Rol Moskovskogo Universiteta v razvitti meditsiny. *Terap. Arkhiv,* 1956, pp. 75-81.

———, Rol nauchnykh shkol v razvitii meditsiny. *Sov Zdravookhr.,* 1964, no. 10, pp. 62-69.

Petrov, P. N., Pervoe meditsinskoe uchilishche v Rossii (shkola N. Bidlo pri Moskovskom gospitale). *Vrach,* 1887, no. 46, p. 901.

Popov, M. F., *Kratkii Istoricheskii ocherk Tomskogo Universiteta za Pervye 25 let ego Sushchestvovaniya (1888-1913).* Tomsk, 1913.

Rodov, Y. I., Nikolai A. Semashko kak pedagog i dejatel meditsinskogo obrazovaniya v SSSR. *Sov. Zdravookhr.,* 1950, no. 4, pp. 33-37.

Rossijski, D. M., *200 let Meditsinskago Fakulteta Moskovskogo gos. Universiteta.* Moscow, 1955.

Smirnov, S. A., Mediko-khirurgicheskaya akademiya v Varshave. *Mosk. Med. Gazeta,* 1859, no. 41, pp. 331-332.

125 let Kievskogo Meditsinskogo Instituta. Kiev, 1966.

175 let Pervogo Moskovskogo gos. Meditsinskogo Instituta. Moscow, 1940.

Tikhomirov, M. N. et al., *Istoriya Moskovskogo Universiteta 1755-1919.* Moscow, 1955.

Tikotin, M. A., *Zagorski i Pervaya Russkaya Anatomicheskaya Shkola.* Moscow, 1950.

Tomilin, S. A. Pervaya meditsinskaya shkola v Rossii. *Vrach. Delo,* 1930, no. 7, pp. 481-484.

Vladimirskij-Budanov, M. F. *Istoriya Universiteta sv. Vladimira,* Kiev, 1884.

Zmeev, L. F., Pervyj vrachebnyj diplom vydannyj v Rossii. *Russ. Arkhiv,* 1887, no. 6, pp. 254-255.

———, *Biloe Vrachebnoj Rossii.* St. Petersburg, 1890.

II. English, French and German

Chervyakov, A., and Endukidze, A., The training of physicians in Russia; decree of the Central Executive Committee of U.S.S.R. *J. Am. Med. Ass.,* 1934, 103: 1798.

Dujardin-Beaumetz, De l'enseignement médical et de la pratique médicale en Russie. *Gaz. Hebd. Méd. Chir.,* 2 ser., 1888, 25: 809-811.

Forrest, W. P. Medical education in Ukraine. *Lancet,* 1948, 2: 579-582.

Fuster, Organisation des études médicales en Russie. *Montpell. Méd.,* 1904, 19: 455-463.

GANTT, W. H., A review of medical education in Soviet Russia. *Brit. Med. J.*, 1924, **1**: 1055-1058.
GARRISON, F. H., Russian medicine under the Old Regime. *Bull. N.Y. Acad. Med.*, 1931, **7**: 693-734.
HEINE, M., *Medicinisch-historisches aus Russland*. St. Petersburg, 1851.
HUARD, P., and WONG, M., Structure de la médecine russe au siècle des lumières. *Concours Méd.*, 1965, **87**: 125-134.
Imperial Military Medical School at St. Petersburg. *J. Ass. Milit. Surg. U.S.*, 1904, **15**: 427-429.
KURASHOV, S. V., Medical education in the Soviet Union. *Brit. Med. J.*, 1954, **2**: 510-512.
LOURIA, L., Medical education in Russia. *Med. Rec., N.Y.*, 1909, **75**: 58-61.
MARCOU, L'Allemagne médicale en Russie. *Archs. Gén. Méd.*, 1903, **80**: 1825-1828.
Medizinische Facultät der Universität Dorpat in den Jahren 1802-1870. *Dorpat Med. Z.*, 1870-1871, **1**: 351-357.
OSSIPOW, E., POPOW, I., and KOURKIN, P., *La médecine de zemstvo en Russie*. Moscow, 1900.
RICHTER, W. M., *Geschichte der Medicin in Russland*. Moscow, 1813-1817.
RUBAKIN, A., *La réforme de l'enseignement de la médecine dans l'URSS*. Nancy-Paris, 1931.
SEMASCHKO, N., Das medizinische Studium und die ärztliche Fortbildung. *Dtsch. med. Wschr.*, 1924, **49**: 1587-1588.
SHIMKIN, M. B., The new Soviet curriculum in medicine. *Amer. Rev. Sov. Med.*, 1947, **4**: 271-274.
SIGERIST, H. E., *Medicine and health in the Soviet Union*. New York, 1947.
VUCINICH, A., *Science in Russian culture; a history to 1860*. Stanford, 1963.
World Directory of Medical Schools. Geneva, 1964.
ZEISS, H., Johann Peter Franks Tätigkeit in St. Petersburg. *Klin. Wschr.*, 1933, **12**: 353-356.

MEDICAL EDUCATION IN INDIA SINCE ANCIENT TIMES

NANDKUMAR H. KESWANI
All-India Institute of Medical Sciences
New Delhi, India

The status of medical education in a particular country, in a given period, cannot be divorced from the quality of the general education and the level of creative and analytical spirit of the people of that country. Since remote antiquity India is known to have enjoyed a civilization, of which medicine and other natural sciences formed an essential part. The ancient Indian civilization is almost synonymous with the Hindu civilization, since originally the term "Hindu" had a territorial but not a credal significance. The Persians with their dialectal difficulties with the sound "sa", christened the ancient people living in the region of the great river "Sindhu" as the "Hindu", and the later Western invaders of Persia, particularly the Greeks, who had similar dialectal difficulties with the sound "ha", in turn labelled these people as the "Indu", and hence the name "Indus" for the river "Sindhu". The first reference to the word "Hindu" in ancient Indian literature appears in the *Padma Purāna* (Lakshmi Pathi), and was applied to the aboriginal tribes, savage and half-civilized people, the cultured Dravidians and the Vedic Aryans, worshipping different gods and practising different rites (*Kūrma Purāna*)—a living symbol of unity in diversity. This is the reason that even up to this day, to many uninformed persons Hinduism has seemed to be a museum of multitudinous gods, a medley of beliefs, a mumbo-jumbo of rites—a mere name without matter!

Hinduism has never been a canonized and codified creed, bound by mechanical adherence to authority, but a way of life with many variegated hues, enabling every component community to retain its past traditions and preserve its interests and individuality. It has been a way of conduct rather than a form of creed, like the universal religion of Spinoza in which justice and charity have the force of law and ordinance. It has always subordinated authority to experience, to foster and promote the spiritual life in accordance with the nature and laws of the world of reality; and, therefore, on no occasion in the long history of Hindu civilization has there been a warfare between science and theology. Hinduism, even today, continues to be "a movement, not a position; a process, not a result; a growing tradition, not a fixed revelation" (S. Radhakrishnan).

Since earliest times the Hindus have strived to achieve four supreme ends of life—righteousness (*dharma*), wealth (*artha*), cultural and artistic values (*kāma*), and spiritual freedom (*mōksha*); and the fundamental principles governing the ideals, practices and conduct of the Hindus to achieve these supreme ends were welded into a comprehensive thought, the so-called Hindu region or *Dharma*. It is this Hindu *Dharma*, rather than any political or economic factor, that has, in the course of the long history of the Hindus, moulded and fashioned the ancient Indian civilization. The ancient Indian literature bears "an exclusively religious stamp; even those latest productions of the Vedic age which cannot be called directly religious, are yet meant to further religious ends. This is, indeed, implied by the term 'Vedic'. For *Veda* primarily signifying "knowledge" (from *vid*, to know), designates "sacred lore" as a branch of literature. Besides this general sense, the word has also the restricted meaning of 'sacred book'" (Macdonell). This unique tendency of Hindu thought, for these reasons, is more manifest in the sphere of education than in any other activity, since through the ages learning in India has been pursued as a part of, and for the sake of *Dharma*.

The life of an individual, according to the Hindu *Dharma*, has a four-fold plan. The first period of his life is that of *brahamacharya*, for the training and discipline of the body and mind, attained through proper education; the second stage is that of *grahastha*, to work for the world as a householder; the third stage is that of *vānaprastha*, when the responsibilities of the home are given up to retreat from the social and familial bonds into the solitude of the forest to meditate on more metaphysical matters; and, the last is the stage of *sannyāsa*, or the period of renunciation and expectant awaiting of spiritual freedom to gain the spirit of equanimity and blissful salvation (*mōksha*).

Thus being a part of one's religious duty, education was necessarily individual, in which the individual was the main concern and center. Unlike the modern day institutions or abstraction called schools, in which a "class" of invisible, intangible and sometimes intractable material of different minds and moral conditions is educated *en masse*, the ancient Indian ideal of education was to treat it as a sacred and a secret contract between the *brahmachārin* (pupil) and the *gurū* (teacher), whose personal and intimate contact with and constant vigil over the pupil brought out, in full, the latent capacities and potentialities in the pupil's personality. This intimate relationship between the teacher and the pupil was initiated through a ceremony called the *Upanayana* (leading to; meaning, taking a student to a teacher in order to hand him over to the latter for his education). At this ceremony "by laying his right hand on [the pupil], the teacher becomes pregnant [with him]" and "he who enters on a term of studentship becomes an embryo," the teacher impregnates the pupil with his spirit and delivers the pupil in a new birth (*Shatapatha Brāhamana*). The *Manu Smrti*, in the same vein, emphasizes that all men are born unregenerate (*shūdra*) by the first or physical birth, but become regenerate (*dvija*, twice-born or born afresh) by the second or spiritual birth after education. After the initiation, the student lived with his teacher as a member of his family and was treated by the teacher in every way as his own son. This living relationship between the student and the teacher, wherein the pupil belonged to the teacher, was the ideal of ancient Indian education.

While the primary and secondary education was individual in character with a residential basis, where in his solitary hermitage a teacher gathered around himself only a few pupils, whose education he could conveniently manage, higher education and advanced training in the humanities, sciences, and technical and professional fields were imparted at special centers for higher studies, very much akin to the universities of today (Mookerji).

The earliest Hindu scripture, the *Rig Veda* mentions *Sanghās,* or the assemblies of seers, gathering together to debate and discuss the philosophy and language of the *Vedās*. The *Brāhmanās* and the *Upanishads,* which explain the sacred significance of the sacrificial ceremonies and the mystical and philosophical speculations on the nature of things, respectively, speak of periodic conferences of the *Rishīs* (sages) at the courts of ancient kings, convened for the advancement of knowledge through the process of free discussion on the new philosophical theories propounded by some of the *Rishīs*. The new theories were incorporated into the texts only after they had been accepted by the whole *Parishad,* or the academy of the learned people, to which repaired not only the local sages but also the *Upanishad charakas,* the scholars who wandered through the country in the quest of truth and learning and, in turn, often became the source of enlightenment to these congregations. The women of Vedic India did not lag behind the men either, and one hears in the *Vedās* and the *Upanishads* of Gārgī Vāchaknavī, a learned lady holding her own in the debates against the men-philosophers at the court of King Janaka of Videha. The other familiar female figures met with in these scriptures are Vishwavārā, Apālā, Vadavā, Prātitheyī, and Sulabhā Maitreyī. The learned kings, besides Janaka of Videha, who promoted the cause of higher learning in Vedic times, were Ajātshatru of Kāshī, Pravāhan Jaivali of Panchāla, and Ashvapati of Kekaya. Ashvapati claimed that "In my kingdom there is no thief, no miser, no drunkard, no man who neglected his duties, no ignorant person, no adulterer, much less an adulteress" (*Chāndōgya Upanishad*).

Buddhism, which was built on the basis of monastic brotherhood (*Sangha*) of its monks (*Bhikshūs*), later expanded the small residential schools of the Brahmanical system into large residential establishments called the *Vihārās,* or the monasteries, which functioned mainly as centers for higher learning and anticipated the modern universities. Some of these notable *Vihārās* were situated in Takshashilā, where now only the ruins of Taxila stand; in Sāranāth and Kashī (Benares) on the banks of the Holy Ganges, which even today is reminiscent of the Vedic age; in Nālandā, where the vast expanse of the ruins still reflects the glory of that once world-renowned university; and in Vikramshilā, Vallabhī, Mithilā, and Nādia.

Ancient Indian history, before the advent of the Muslims, can be broadly divided into four periods, which also accommodate the conflicting view of the Indian and Western scholars regarding the chronology of the ancient Hindu texts (Altekar). The first period, from prehistoric times to 1000 B.C., can be conveniently called the Vedic age since most of the Vedic literature was composed during this period. The second, Upanishadic Age, extends from 1000 B.C. to 200 B.C., when the *Upanishads,* the *Sūtrās,* the Buddhist canons, and the Epics were composed. This can also be called the Age of the Nandās and the Mauryās who were the powerful rulers during the later part of this period. The third period, extending from 200 B.C. to 500 A.D., can be conveniently described as the Age of the Dharmashāstrās, or the Age of the Shungās, the Sātavāhanās, the Vākātākās, and the Guptās, the leading dynasties of the time. The fourth period, from 500 A.D. to 1200 A.D., was the Puranic Age when the *Purānās* and the *Nibandhās* (digests) guided the Hindu society during the rule of eminent kings like Harsha (c. 606-647) and Bhōja (840-890).

Although there is no direct evidence to show that the art of writing was known in the early Vedic period (2500 B.C.), one can infer from the inscribed seals found among the ruins of the Indus Valley Civilization (fourth millennium B.C.) that the so-called Aryans, even though ignorant of this art, must have learnt it from the Indus Valley people, as there is clear cultural evidence to show that the art of writing was known during the later Vedic period (Ojha).

Primary education during the Vedic period must have been given in the family itself, probably by a family priest who had the charge of educating children from the age of five, in the elements of arithmetic, grammar, phonology and metrics. The children probably had to recite and memorize the sacred hymns, which must have been composed for this specific purpose, like the nursery rhymes of modern times. During the Upanishadic and Dharmashāstra periods, when the art of writing was well known, the three R's forming an integral part of primary education, were assigned a definite place in the educational system by the exaltation of its commencement at the age of five or six, into a religious ritual known as *Aksharasvīkarana* or *Vidyārambha Sanskāra*. On an auspicious day the child started the ritual by worshipping Saraswatī, the goddess of learning, the tutelary deities of the family, and the primary teacher. At the end of the ceremony, the teacher and the other invited Brāhmanās were given suitable presents, and finally, the child was handed over to the care of the teacher. The students, it is recorded, were first taught to write all the alphabet on grains of rice with a specially prepared gold or silver pen (Altekar). Later on, in the absence of paper, the children of the rich used to write on wooden boards with some kind of color, and those of the poorer class learnt by writing with their fingers or pointed sticks on the ground covered with dust or sand. When the entire lesson was mastered by the student, the teacher would write on a palmleaf with an iron stylus and hand over the leaf to the pupil for tracing on the same leaf with charcoal ink. Still later the students learnt to write on plantain leaves first, and then on palm leaves. Arithmetic and elementary grammar were taught through metrical texts, which rendered memorizing easy. This system of teaching persisted in India almost down to the end of the eighteenth century.

Hiuen Tsāng (or Yuān Chwāng), the famous Chinese pilgrim who travelled for sixteen years (620-645) through Central Asia and India, gives a detailed account of the content of the Indian education of those days (Samuel Beal), which is also corroborated by another Chinese traveller I-tsing, who studied at Nālandā for a period of ten years (675-685) and made a "successful study in medical science, but did not follow it up and specialize in it because he was out on a different mission" (Takakusu). According to their accounts, primary education began at the age of six years, with study of the first book of reading called *Siddhirastu* (or *Siddham*), which gave forty-nine letters of the alphabet and 10,000 syllables arranged in 300 *shlōkās* (stanzas or hymns). The primer was completed in six months, and the student was introduced to the second book of reading, the *Sūtra* (formulae) of Pāninī, containing 1000 *shlōkās* which were to be learnt by the eighth year. This was followed by study of a book on *Dhātu* (roots) and another on three *Khilās* (supplements), which were begun at the age of ten and were to be mastered after three years' diligent study. When the student was fifteen years old he began the study of the commentary on Pāninī's *Sūtra* by Jayāditya, called *Kāshikāvritti*, which took him years to understand.

According to Hiuen Tsāng, after the mastery of *Siddham* the child was also introduced, at the age of seven, to the great *Shāstrās* (sacred texts or injunctions) of five *Vidyās* (learning), which were compulsory in the curriculum. These five subjects were *Shabdavidyā* (grammar and lexicography), *Shilpasthānavidyā* (arts), *Chikitsāvidyā* (medicine), *Hetuvidyā* (logic), and *Ādhyātmavidyā* (science of universal soul, or philosophy). Only on completion of this course was a student considered fit for advanced studies in any special field. I-tsing has also given some details of the medical syllabus, and mentions the eight traditional branches of Āyurveda, the elements of which were to be studied by all the students, including those intended for monkhood since, as he puts it, "Is it not a sad thing that sickness prevents the pursuit of one's duty and vocation? Is it not beneficial if people can benefit others as well as themselves, by the study of medicine?"

That the elementary knowledge of medicine was considered essential for all students shows the importance attached to the medical science by the ancient Indians, who by their religious tenets believed that the purity of the heart and the mind could not be achieved in an unsound and unclean body, and therefore rigorously enforced the cleanliness and preservation of sound health in every religious ceremony. The later *Dharmashāstrās* are full of injunctions regarding purity, ablutions, diet, regulations, behavior, and mental and physical discipline. It is no wonder, then, that medical science became the most popular and a highly developed science of the Hindu civilization, and prompted Charaka to remark that Āyurveda is a science "which teaches mankind what constitutes their good in both the worlds" (*Sūtrasthānam:* I.43).

The profession of the physician is as old as the history of mankind itself. The excavations at Mohanjo-Daro in Sind and other sites of the Indus Valley Civilization, have not yet revealed any direct evidence to show what the status of medicine was in India of that remote period, although the public health facilities as found in the ruins, far surpass those found in the ruins of other ancient civilizations. One interesting engraving on a steatite seal found in Mohanjo-Daro shows a divinity closely resembling Shiva Pashupatí —the Lord of animals, who, in the *Purānās,* is considered to be the first propounder of Āyurveda, and is often quoted as an authority on the medical subjects in the later treatises (Mukhopadhyaya). Shiva, however, does not figure by name in the *Rig Veda,* but in the *Yajur Veda* it is the name given to Rudra, a powerful Rig Vedic god of medicine. Some scholars believe that this god was one of the many elements of the Indus Valley culture taken over and assimilated into the Aryan culture.

The main source of Aryan culture and its medicine was the four *Vedās* which, according to the Hindu belief, were originally revealed by Brahmā, the Creator, to the sages some 6000 years before the Christian era; but Western scholars consider it improbable that the *Rig Veda* was compiled before the second millennium B.C. The *Rig Veda* is considered as the original source of Hindu medicine. It is a collection of hymns in the praise of deities, many of whom are extolled for their medical and surgical skills. The *Sāma Veda* and the *Yajur Veda* are closely related to the *Rig Veda* and contain many hymns borrowed from the latter. The fourth, *Atharva Veda,* and its *Kaushika Sūtra,* are replete with prayers, incantations, charms and spells to protect people against all kinds of disease and disaster. But, side by side with the magico-religious formulae, empirical and rational measures are also given in these texts. The word Āyurveda (from *Āyuh,* life; and *veda,* knowledge), the "Knowledge of life" or the knowledge by which the nature of life is understood and thus life prolonged, does not figure in the *Vedās,* but the lay-physician (*bhishak*), in addition to the priest-physician (*bhishagatharvan*), is mentioned in all the *Vedās*. The *Rig Veda* declares: "Various are our acts, [various] are the occupations of men; the carpenter desires timber, the physician disease, the *Brāhman* a worshipper. . . . I am the singer, papa is the physician, mama throws the corn upon the grinding stones" (IX.7.9.1-3). In the *Atharva Veda,* "four hundred leeches are in this, yea, a thousand healing herbs" (II.9.3.), refers to an amulet which possesses the healing power of a hundred physicians and a thousand medicinal herbs;

and, again the poet continues, "Let him who made it also heal, he truly is the deftest leech. Pure, with a leech he verily shall give thee medicines that heal" (II.9.5).

The traditional Vedic medicine must have flourished for centuries, handed down orally from master to pupil. It was at a much later stage that the traditional Hindu medicine was named Āyurveda, and Brahmā, the fountainhead of all knowledge, naturally became the first composer of the *Āyur Veda*. During the Upanishadic period when rational medicine really began, the *Āyur Veda* became intimately associated with the *Atharva Veda* as its *Upānga* (secondary part), and was sometimes called the *Upaveda* (secondary *Veda*) of the *Rig Veda*. Unfortunately the *Āyur Veda* is no more available in its original form, but the Ayurvedic tradition is revealed by the *Samhitās* (compendiums) of Bhela, Charaka and Sushruta. "The first has reached us in a single and incomplete manuscript. The other two are not available in their original form but only as they were revised by later authors. None of these three texts represents a first effort at systemic description of medical science; on the contrary, all three suppose an already established tradition. They limit themselves merely to collecting the facts of the above mentioned tradition and teaching it" (Filliozat).

Brahmā, who is said to have been the "First teacher of the Universe" according to Hindu mythology, originally composed the *Āyur Veda* in 100,000 hymns divided into 1000 chapters; but realizing that it would be beyond the comprehension of the mere man, he abridged the *Veda* and divided it into eight parts, with *Kāya Chikitsā* (medicine) and *Shalya Tantram* (surgery) as the main subjects.

According to Sushruta, the original *Āyur Veda,* in addition to *Kāya Chikitsā* (medicine) and *Shalya Tantram* (surgery), contained chapters on *Shālākya Tantram* (diseases of the region above the clavicles, *i.e.* eye, ear, nose and throat diseases), *Bhūtavidyā* (demonology), *Kaumārbhritya* (pediatrics), *Agad Tantram* (toxicology), *Rasayana Tantram* (science of rejuvenation), and *Vājīkarana Tantram* (science of aphrodisiacs).

Having propounded the science of healing, Brahmā propagated this knowledge through Daksha Prajāpati, who taught the science to the legendary Ashwinī Kumāras—the celestial physicians to the gods. These Ashwinīs have also been called Dasra and Nāsatya in the *Vedās*, and it is interesting to note that the names of Nāsatya, Indra, Varuna and Mitra—the Vedic gods, also appear in the documents found in the excavations at Boghaz Koyi in Cappadocia in northwest Mesopotamia. It is therefore believed that the Mitanian Kings used to worship the Vedic gods as early as 1600 B.C. (Winkler). The Ashwinīs imparted the science of medicine to Indra—the Chief of the gods, who, in turn, is said to have transmitted this knowledge to its mortal protagonists.

According to the Ātreya School of Medicine, represented by Charaka, the first mortal who received the science was Bharadwāja. Bharadwāja repaired to the court of Indra, on being delegated by the congress of the *Rishis,* who had assembled in the Himalayas to deliberate upon and find out ways and means to make man happier and healthier, and decided to appeal to Indra to impart to them the knowledge of Āyurveda for the redemption of suffering humanity. This, incidentally, is the first recorded medical conference held on the soil of India, in which a galaxy of medical sages participated. Bharadwāja imparted the science of medicine to Ātreya and other great sages, of whom Agnivesha seems to have been the first to compose a book on Āyurveda. This text was later reconstituted by Charaka, who claims to have reproduced the actual words of Ātreya Punarvasu.

But according to the Dhanwantari School of Surgery, represented by the celebrated surgeon Sushruta, Indra favored Dhanwantari with the entire knowledge of Āyurveda. And, Dhanwantari "who warded off death, disease and decay from the celestials", appeared in the form of Divōdāsa, King of Kashī, and on being approached by a group of sages who were moved by human suffering, agreed to admit them to his hermitage and gave them the science of healing. The sages selected Sushruta to be their spokesman and leader, who is supposed to have recorded the very words of Dhanwantari. In the South, however, the tradition credits Rishī Agastya with the dissemination of the knowledge of Āyurveda during the ancient times.

From this survey of the origins of Āyurveda from remote antiquity to the rational period of ancient Indian history, it will be seen that the only texts on the science of medicine which are extant in a complete form today, even though redacted and revised by later authors, and used for many centuries as the text books in India, are the *Samhitās,* of Sushruta and Charaka, which contain all information regarding the state of medical education in ancient times. These two *Samhitās,* on the whole, have similar contents, identical or analogous divisions, and corresponding theoretical and practical data, except for the fact that the *Sushruta Samhitā* is richer in the field of surgery. Charaka in his writing has the combined role of moralist, philosopher and above all, physician; whereas, Sushruta has tried to cast off whatever shackles of priestly domination remained at his time, and created an atmosphere of independent thinking and investigation, which later characterized Greek medicine. Both presented an abundance of material in an extremely condensed form, and professed a rational approach. The magic and the *mantrās* (holy incantations) were brought into play by them only in cases of delirium, in demonic possession, sometimes in diseases of children, and in the ceremonies connected with birth. The state of health and disease was explained on the basis of the interplay of the constituent elements of the body, the general and alimentary regimen, and the influences of time and season.

To assign any definite date to the early medical authors and teachers such as Bharadwāja, King Divōdāsa, Ātreya and Sushruta, in the absence of any direct chronological evidence in texts, is not an easy task; but various scholars, on the basis of content, form and language of at least the earlier parts of their extant texts, have, with every caution of long marginal variation, assigned to them the period about the beginning of the first millennium B.C. This was a period of great ferment and productivity of Hindu thought, when most of the earlier *Upanishads,* the six *Vedāngās* (appendages of the *Vedās*), and the *Sūtrās* dealing with the various branches of humanities, were composed. The *Shad Darshana* (six systems of philosophy) were also fully developed during this period, and Kapilā's *Sānkhya,* Patanjali's *Yōga,* Gautamā's *Nyāya,* Kanādā's *Vaisheshika,* Jaimini's *Pūrvamīmansā,* and Vyāsā's *Uttarmīmansā* had made their appearance. Great progress was also made in the field of the exact sciences such as mathematics (including arithmetic, algebra and geometry), astronomy and physics. The greatest work of this period, which also throws some light on the medical knowledge and practice of those times, was the *Ashtādhyāyī* of *Pānini* (750 B.C.–R. G. Bhandarkar), the most authoritative work on Sanskrit grammar, wherein the three humors of the body (*Vāta, Pitta* and *Shleshman*) are mentioned, and their imbalance is stated to be the cause of various diseases defined in this grammar. With such a highly developed state of the Hindu sciences and philosophy during the eighth and seventh centuries B.C., the medical sciences, as manifested in the *Samhitās* of Sushruta and Charaka, may well be regarded as belonging originally to this age.

The principles and practices governing general education were also applicable to vocational education. The same individual, intimate relationship between the pupil (*shishya*) and the teacher (*guru*) living together as members of one family, was also operative in regard to medical education. The *Upanayana* (initiation) ceremony which was primarily intended for Vedic education, and was performed at the time of commencement of primary education by the students of the first three classes, i.e., *Brāhmana* (priestly class), *Kshatriya* (martial class), and *Vaishya* (agricultural and commercial class), in accordance with the rules of their order, was performed again by all classes of students, including *Shūdrās* (serving class), for admission to the medical course. Sushruta (*Sūtrasthānam:* II) in his discourse on formal initiation of a pupil (*shishyopanayanīyam*) enjoins that a *Kshatriya* or a *Vaishya* physician can also become a teacher for the students of his own caste. It is probable that even *Brāhmana* pupils may have been initiated by non-*Brāhmana* medical teachers. Sushruta declares that even a *Shūdra* student of good character and parentage could be initiated into the mysteries of Āyurveda, but by omitting the *mantrās* required to be recited by the students of the other three classes on such occasion. As a matter of fact, the *Kshatriya* and *Shūdra* surgeons, by virtue of their environment and tradition, must have been better adepts in the use of the knife than their *Brāhmana* and *Vaishya* counterparts. But, irrespective of the class to which the candidate belonged, the admission to the medical course was restricted to those students only, who were "of tender years; born of a good family; possessed of a desire to learn, strength, energy of action, contentment, character, self-control, a good retentive memory, intellect, courage, purity of body and mind, and a simple and clear comprehension; and commanded a clear insight into the things studied;" and were also "found to have been further graced with the necessary qualifications of thin lips, tongue and teeth; a straight nose; large, honest and intelligent eyes; with a benign contour of the mouth; and, with a contented frame of mind, pleasant in speech and dealings, and usually painstaking in efforts."

The Ayurvedic *Upanayana* ceremony was performed on an auspicious day. Prayers were offered, first by the teacher and then by the pupil, to the deities associated with Āyurveda—Brahmā, Prajāpati, Ashwinīs and Indra, and to the ancient *Rishīs*—Dhanwantari, Bharadwāja and Ātreya. Then both teacher and pupil circumambulated the fire on the altar, and in the presence of sacred fire the *guru* charged the student as follows:

Thou shalt renounce all evil desires, anger, greed, passion, pride, egotism, envy, harshness, meanness, untruth, indolence, and other qualities that bring infamy upon oneself. Thou shalt clip thy nails and hair close, wear brown garment, and dedicate thyself to the observance of truth, celibacy and salutation to the elders. Devoting thyself, at my bidding to movement, laying thyself down, being seated, taking thy meal and study, thou shalt be engaged in doing whatever is good and pleasing to me. If thou shouldst behave otherwise, sin will befall thee, thy learning will go fruitless, and will attain no popularity. If I do not treat thee properly despite thy proper observance of these behests, may sin befall me, and my learning will go fruitless. The *Brāhmanās,* the preceptor, the poor, the friendly, the travellers, the lowly, the good and the destitute—these thou shalt treat when they come to thee, like thine own kith and kin, and relieve their ailments with thy medications (without charging for it any remuneration whatever). Thus behaving, good will befall thee. Thou shalt not treat a hunter, a bird-catcher, an outcaste and a person doing sinful acts. Thus thy learning will attain popularity and will gain for thee friends, fame, righteousness, wealth and fulfilment (*Sushruta Samhitā, Sūtrasthānam:* II.5).

Similar oaths of initiation, which have much in common with the famous Hippocratic Oath, are found in the *Charaka Samhitā,* in the *Kāshyapa Samhitā,* a text on pediatrics, and even in the *Hastyāyurveda* of Pālakāpya, a veterinary text dealing with the treatment of the diseases of the elephant. Since the physician was expected to deal with patients of both sexes, and of all sorts and conditions, it was imperative that he be of sound and healthy body, observe rules of hygiene and avoid all kinds of defilement, infection and contamination, and be a man of strict morals.

The admission of a student to the medical course was conditional, and he was on probation for a period of six months, according to Sushruta, when he was under observation regarding his fitness for a medical career.

As in the Vedic education, the pupil learnt by first memorizing the texts and then concentrating his mind; the student had to "go over the aphorisms in order, repeating them over and over again all the while *understanding their import fully,* in order to correct his own faults of reading and to recognize the measure of those in the reading of others. In this manner, at noon, in the afternoon, and in the night, ever vigilant" the student applied himself to study (*Charaka Samhitā, Vimānasthānam:* VIII.9). The teacher laid special emphasis on a student's proper understanding of the texts that he was to memorize; and Sushruta considers a pupil foolish "who has gone through a large number of books without gaining any real insight into the knowledge propounded therein", and compares him with "an ass laden with logs of sandalwood, that labors under the weight which it carries without being able to appreciate its value" (*Sūtrasthānam:* IV.3).

It was impressed upon the student that he acquire full knowledge through three different vehicles: authoritative texts and instruction (*āptōpadesha*), direct observation from practical experience (*pratyaksha*), and last but not least, the inferential method (*anumāna*). The training comprised logical elaboration, and further supplementation by the intelligent teacher or student, based on speculative or imaginative corollaries drawn from the main data of the theoretical texts, as well as the practical experience acquired in the past. Like the modern-day aphorism that "to study medicine in the wards alone is to sail on an uncharted sea, but to study from the books alone is not to sail at all", Sushruta compares the one who knows only one branch of his art, with a bird having one wing; and admonishes that "he who is versed only in books will be alarmed and confused, like a coward on a battlefield, when confronted with active disease; he who rashly engages in practice without previous study of written science is entitled to no respect from the best of mankind, and merits punishment from the king; but he who combines reading with experience proceeds safely and surely like a chariot on two wheels" (*Sūtrasthānam:* III.18).

The selection of the textbooks was of great importance, and meticulous care was given to the sanction and authorization of the texts. Charaka advised the students that there are many treatises on medicine current in the world. From these he should choose that treatise which has obtained great popularity and is approved by wise men; which is comprehensive in scope and held in high esteem by those who are worthy of credence, suitable alike for understanding by the three grades of students, *i.e.*, very intelligent, average, and slow; free from faults of repetition, revealed by a sage, arranged in well-formed aphorisms with well-authenticated commentary and summary, free from vulgar usage and difficult words, rich in synonyms, with words of traditionally accepted sense; and concerned mainly with the true nature of things, relevant to the theme, orderly in its arrangement of topics, elucidating, and enriched with definitions and illustration. "Such a treatise", he declared, "is like the unclouded sun, dispelling darkness and illuminating everything" (*Vimānasthānam:* VIII.6). The texts (*Samhitās*) were therefore prepared in such a way that they served as complete works of reference, and at the same time gave impetus for further research and progress by showing the line of research to the highly intellectual students. Hence the ancient texts like the *Ātreya Samhitā*, the *Sushruta Samhitā* and the *Kāshyapa Samhitā* were redacted and rewritten; the old theories were examined in the light of new experience, and the texts were brought up to date. The two Vāgbhattās of Sind (? 2nd century B.C., ? 7th century A.D.) who sensed the setback in the progressive spirit of Āyurveda, composed two treatises—the *Ashtāngahrdaya* and the *Ashtānga Sangraha*, which while presenting a summary of the *Charaka* and the *Sushruta*, with gleanings from other medical texts known in their time, brought the subject up to date. They introduced new drugs, and made valuable modifications and additions in surgery, despite strong opposition from the orthodox schools of medicine; thus earning for themselves a permanent place in the history of Ayurvedic literature by being ranked, later on, as equal to Charaka and Sushruta. These three medical teachers are known, even today, as the *Vrddha Trayī* or the "Triad of Ancients". The redactors of the *Sushruta,* the *Charaka* and the *Vāgbhattās*, although believing in the sanctity of the basic principles enunciated in these texts, were nevertheless alert to progressive additions to the texts as and when required according to time and place. They were also ever ready to assimilate useful things from whatever source available. Charaka states in unequivocal terms: "The entire world is the teacher to the intelligent, and foe to the unintelligent. Hence knowing this well, thou shouldst listen and act according to the words of instruction even from an unfriendly person, when they are worthy and such as bring fame and long life to you, and are capable of giving strength and prosperity" (*Vimānasthānam:* VIII.14). These same principles guided the authors in their compilations during still later days of specialization, and although each book was written on a special branch of medicine, the basic knowledge of all the other branches was incorporated in a concise form in each book.

The Ayurvedic texts lay down what the qualities of a real teacher should be, since not every person could become a teacher. He was required to be an ideal to the pupil, and an undenying source of knowledge and inspiration to him. Hence Charaka requires of a teacher to be "one who is thoroughly versed both in theory and practice; who is skillful, upright, pure, deft of hand, well equipped, possessed of all faculties; who is conversant with human nature and the line of treatment, and possesses special insight into the science; who is free from self-conceit, envy and irritability; who is endowed with fortitude, is affectionate towards his pupils, proficient in reading, and skillful in exposition. The teacher endowed with such qualities equips quickly the good disciple with all the qualities of a physician, just as the rain clouds at the proper season endow the fertile field with the best of crops" (*Vimānasthānam:* VIII.4).

In his daily conduct, the student was required to observe strict rules as enunciated in the Ayurvedic *Upanayana*. He was required to abstain from his studies on the holidays; and was forbidden from resort to studies when in hunger, thirst, disease or indisposition, and when the weather was inclement or the day was inauspicious.

The exact duration of the medical course is not mentioned in any of the classical texts. Apparently the student was permitted to leave school and practise only after he was considered proficient by his teacher, which in the case of a good student took from six to seven years. In the case of Jīvaka, the famous physician to the Buddha (563-477 B.C.) and the physician-in-chief to King Bimbisāra of Magadha (543-491 B.C.), it is related in the Tibetan texts that he was permitted to go home very reluctantly by his teacher Bhikshu Ātreya at Takshashilā, even after he had spent seven years in the study of medicine. After completion of the medical course, the student (*Brahmachārī*) was known as *Adhyayanāntagah* (one who has completed his studies), and was permitted to enter the second stage of life—the *grahastha*. Before his departure the student paid a token of gratitude in the form of the fees for the whole course to his teacher, and went through the final ceremony akin to the modern convocation. There was, however, a class of students who pursued their studies all through their lives and took a vow to remain *Naishthika Brahmachārī* (life-long scholars). There were also those, and they exist always and everywhere, who went from one teacher to the other, and never stayed long enough with any teacher or institution to be of any good to themselves or to the others, and were known as *Tirthakākās*, meaning the "wandering crows".

The students were required to go through both practical and theoretical examination at the time of admission, as well as at the final stage of their career, and the tests were not very easy. At the entrance examination hardly twenty to thirty per cent were successful, and the doors of the universities were zealously guarded by experts called the *Dwāra Pandits* (the scholars guarding the doors of entry). At the final examination, Charaka insisted that a medical student should be examined by another physician in each of the eight classical divisions of Āyurveda, in each "system and its interpretation, the main sections of the system and their interpretation, the chapters [in each section] and their interpretation, and the questions and their interpretation; and thus being examined, he should give his answers, leaving out nothing, by verbatim quotations, by explanations of the quotations, and by further elucidation of the difficult parts of the explanations" (*Sūtrathānam:* XXX.30).

The practical examination was also very comprehensive, and at times tested the knowledge of a candidate not only in the medical sceinces but also in allied disciplines. An interesting anecdote is told of Jīvaka, who before being declared successful was given a final test by his teacher, who asked him to "Take the spade and seek round about Takshashilā a *yōjana* (approximately eight miles) on every side, and whatever plant you see which is not medicinal, bring it to me." Jīvaka studied the plants of the entire region specified by his teacher, but could not discover a single one which was devoid of curative properties and useless to human beings. The teacher was satisfied with the answer, and permitted his pupil to go out and practise.

The graduates were encouraged to attend conferences, and travel to different centers of learning to hold discussions (*Sambhāshā*) with other medical men; since, as Charaka advises, "the discussions increase their zeal for knowledge (*Samharsha*), clarify knowledge, increase eloquence, bring renown, remove doubts about learning previously acquired, and strengthen the convictions. In the course of these discussions, many new things may be learnt, and often, out of zeal an opponent will disclose the most cherished teachings of his own teachers" (*Vimānasthānam:* VIII.15-18).

The convocation ceremony was a solemn occasion, at which an oath, reminding the graduate of the vow he had taken at the initiation ceremony, was administered to the candidate who was on the threshold of a new life and was about to embark on a medical career. It was called *Samāvartanam*, meaning thereby the returning home of the student after completing his course at the teacher's residence. The oath as enunciated by the ancients, enjoined on the graduate that he must have completed the course of theoretical texts, and must have fully understood their interpretation. He must have witnessed the performance of the surgical operations. He must be neat in appearance, and put on clean, white clothes and shoes. His dress must not be foppish. His mind must be pure and good, and his speech non-violent. He must be honest in his dealings with the patients, and guarded in his pronouncements; for, to the sick man a physician is like a father, to the man in health like a friend, and after the sickness has passed and the health restored he is like a preserver. The details of these commandments to the fresh graduates are eloquently described by Charaka. After the ritual offering of the prayers to the medical deities and the teacher, the student is charged in the following words:

Acting at my behest, thou shalt conduct thyself for the achievement of the teacher's purpose alone to the best of thy abilities. If thou desireth success, wealth and fame as a physician, and heaven after death, thou shalt pray for the welfare of all creatures beginning with the cows and *Brāhmanās*. Day and night, however thou mayst be engaged, thou shalt endeavour for the relief of the patient with all thy heart and soul. Thou shalt not desert or injure thy patient even for the sake of thy life or thy living. Thou shalt not commit adultery even in thought. Even so, thou shalt not covet others' possessions. Thou shalt be modest in thy attire and appearance. Thou shouldst not be a drunkard or a sinful man, nor shouldst thou associate with the abettors of crimes. Thou shouldst speak words that are gentle, be pure, righteous, worthy, true, wholesome and moderate. Thy behaviour must be in consideration of the time and the place, and heedful of the past experience. Thou shalt act always with a view to acquisition of knowledge and the fullness of equipment. No persons who are hated by the king, or who are hated by the public, or who are haters of the public, shall receive treatment. Similarly, those that are very unnatural, wicked and miserable in character and conduct, those who have not vindicated their honour, and those that are on the point of death, and similarly women who are unattended by their husbands or guardians, shall not receive treatment. No offering of meat by a woman without the behest of her husband or guardian shall be accepted by thee. While entering the patient's house thou shalt be accompanied by a man who is known to the patient and who has his permission to enter, and thou shalt be well-clad and bent of head, self-possessed, and conduct thyself after repeated consideration. Thou shalt thus properly make the entry. Having entered, thy speech, mind, intellect and senses shall be entirely devoted to no other thought than that of being helpful to the patient, and of things concerning him only. The peculiar customs of the patient's household shall not be made public. Even knowing that the patient's life span has come to its close, it shall not be mentioned by thee where if so done, it would cause shock to the patient or to the others. Though possessed of knowledge, one should not boast very much of one's knowledge. Most people are offended by the boastfulness of even those who are otherwise good and authoritative. There is no limit to Āyurveda. So thou shouldst apply thyself to it with diligence. This is how thou shouldst act. Again, thou shouldst learn the skill of practice from another without carping. The entire world is the teacher to the intelligent and the foe to the unintelligent. Hence knowing this well, thou shouldst listen and act according to the words of instruction from even an unfriendly person, when they are worthy and such as bring fame and long life to you, and are capable of giving you strength and prosperity (*Vimānasthānam:* VIII.13).

After termination of his studies and his practical training, when the student was permitted by his teacher to start his professional career, Sushruta (*Sūtrathānam:* X.2-3) demanded of the future physician that he also secure the permission of the King for the practice of medicine. This license from the State was considered necessary because otherwise the quacks would force their existence on the gullible masses, and prove to be a public calamity. Charaka denounced these quacks in no uncertain terms, and condemned "those who putting on the garb of a physician, thus gull their patients just as the bird-catchers in the forest gull the birds in nets by camouflaging themselves; such outcasts from the science of healing, both theoretical and practical, of time and of measure, are to be shunned, for they are the messengers of death on earth. The discriminating patient should avoid these unlettered laureates, who put on the airs of physicians for the sake of a living; they are serpents that have gorged on air" (*Sūtrasthānam:* XXIX. 8, 10-12). He further warned the public that, "One may survive a fall of a thunderbolt on one's head, but one cannot be expected to escape the fatal effects of medicine prescribed by an ignorant physician" (*Sūtrasthānam:* I. 128-132); and, advised them that they should constantly proffer salutations to those "who are learned in science, skilfull, pure, expert in performance, practised of hands, and self-controlled" (*Sūtrasthānam:* XXIX.13). Thus it would appear that in ancient India utmost care was taken to safeguard the welfare of the people from the unauthorized and ignorant exploiters.

The content of the curriculum for the training of a physician was comprehensive. The entire subject of medicine was divided into eight main fields, and the science of Āyurveda was known as *Ashtānga Āyurveda*. The eight branches were (1) *Kāyachiktsā* (medicine, which included principles of physiology and pathology), (2) *Shalyachikitsā* (surgery including anatomy), (3) *Shālākya Chikitsā* (eye, ear, nose and throat diseases), (4) *Kaumārabhrtya* (pediatrics including obstetrics and embryology), (5) *Bhūta Vidyā* (demonology, which included psychotherapy and analysis of dreams, etc.), (6) *Agada Tantra* (toxicology), (7) *Rasāyana* (rejuvenation, with reference to geriatrics), and (8) *Vājīkarana* (virilification). Apart from these fields of medical science, a medical student was expected to know ten arts, that were considered indispensable in the preparation and application of his curative measures, and in his clinical and experimental work. These were: distillation, operative skill, cooking, horticulture, metallurgy, sugar manufacture, pharmacy, analysis and separation of minerals, compounding of metals, and preparation of alkalis.

Although the curriculum in general had a clinical bias, and the basic medical sciences were not taught as distinct disciplines, proper emphasis on the teaching of anatomy, physiology, pathology, microbiology and pharmacology was laid during the teaching of relevant clinical subjects. For example, anatomy was taught as part of surgery, embryology was included in pediatrics and obstetrics, and the teaching of physiology and pathology was interwoven with the teaching of all the clinical disciplines, particularly internal medicine.

Sushruta laid great stress on the direct and personal observation of anatomy of the body by a surgeon, and gave details regarding dissection (*Avagharshana*). He stated: "Any one desirous of acquiring a thorough knowledge of anatomy should prepare a dead body and carefully observe [by dissecting it] and examine its different parts. For, a thorough knowledge can only be acquired by comparing the accounts given in the *Shāstrās* (books on the subject) by direct personal observation." "A dead body selected for this purpose should not be wanting in any of its parts, should not be of a person who has lived up to a hundred years [i.e., too old], or of one who died from any protracted disease or of poison. The excrementa should first be removed from the entrails, and the body left to decompose in the water of a solitary and still pool, securely placed in a cage [so that it may not be eaten away by the fish, nor drift away], after it has been covered entirely with the outer sheaths of the *Munja* grass, *Kusha* grass, hemp or with a rope, etc. After seven days the body should be thoroughly decomposed, when the observer should slowly scrape off the decomposed skin, etc., with a whisk made out of grass roots, hair, *Kusha* blade or with a strip of split bamboo, and carefully observe with his own eyes all the different organs, external and internal, beginning with the skin as described before" (*Shārīrasthānam:* V.49-56). Charaka also emphasized that "The physician who knows the anatomical enumeration of the body together with a description of all its different members is never a victim to the confusion arising from the ignorance thereof" (*Shārīrasthānam:* VII.19). It would appear from the methods employed for dissection during ancient times, that the detailed and minute anatomy of most of the organs was not known, and probably was not needed. However, the topographic description of the various organs, bones, joints, muscles, ligaments, vessels and nerves was given with a fair amount of accuracy. As the method of dissection was not perfect enough to provide the precise knowledge of regional anatomy, so essential to a surgeon, this difficulty was surmounted by designating vital points (*Marmās*) on the surface of the body, and the physician was warned to approach these parts with extreme caution (*Sushruta Samhitā, Shārīrasthānam;* VI; Keswani, 1967).

During later times, the discontinuance of dissection, due to the prevalence of stricter notions of ceremonial purity, when touching a corpse became a taboo, proved fatal to the progress of surgery. Accordingly, even the profession of the physician, who during the earlier period of Hindu history was deified, began to be held in low esteem, and a physician was considered as a defiler of the company at the dinner table (*Manu Smrti:* III.152).

The science of embryology was a subject of considerable speculation and controversy in the various schools of philosophy and medicine, and one whole *Upanishad* was devoted to speculations regarding the formation and development of the human embryo (Keswani, 1962, 1965). This *Garbhōpanishad* (a brief treatise on the embryo) is considered to be of greater antiquity than the rest of the *Atharvan Upanishads*—the class to which it belongs (Keswani, 1965). There are ample references to the development of the human embryo in the *Vedās* themselves, and it is surprising to read in them and in the various medical texts, the ideas of the ancient Hindus regarding biological evolution, reproduction, generation, preformation and spontaneous regeneration, which at a first glance appear as if borrowed from modern day texts on embryology.

The Ayurvedic theory of *tridōsha* or *tridhātu* (three elements), namely *Vāyu* (air), *Pitta* (bile), and *Kapha* (phlegm), on which the whole of the physiology and pathology of the ancient Hindus was built, has been misunderstood by many scholars to mean literally the elements of "air, bile and phlegm". The ancients used these terms in a very broad sense, with variable meanings depending on the context in which they were used. The term "*Vāyu*", for example, comprehends all phenomena of motion that come under the function of life, cell development in general, and the central nervous system in particular. "*Pitta*" does not mean essentially bile alone, but signifies the functions of metabolism and thermogenesis, which include digestion, formation of the blood and various secretions and excretions, which are either the means or the end of tissue combustion. The term "*Kapha*" does not mean mere phlegm, but is used primarily to imply functions of cooling and preservation—thermotaxis or heat regulation, and secondarily the production of various protective fluids, like mucus, synovial fluid, etc. The imbalance of these elements (*dōshās*) beyond normal variations, was considered to cause bodily dysfunction and disturbance.

The body was considered by the Indian medical writers as a conglomeration of the modified *Pancha Mahābhūta* (five elements), consisting of *prithvī* (earth), *ap* (water), *tejas* (fire), *vāyu* (air), and *ākāsha* (ether). These combined to uphold the body by means of seven *dhātūs* (basic tissues), which were: *rasa* (plasma), *rakta* (blood), *māmsa* (flesh), *medas* (fat), *asthi* (bone), *majjā* (marrow), and *shukra* (semen). By decrease and increase in the basic elements of the body, the deteriorating changes occurred. *Rasa*, for example, if decreased caused heart disease, trembling, a feeling of emptiness and thirst; and, if increased, caused nausea and salivation. By decrease in the secretion of *shukra*, pain was caused in the penis and the testes, impotency, slowness in the emission of semen, and the ejaculate was mixed with blood. The quintessence of all the seven elements was called *ōjas* (vitality) or *bala* (power), which governed the functions of the external and internal organs of the body. *Ōjas* was oily, white, cold, soft, etc., and pervaded the whole body. It was demolished by injuries, grief, exhaustion, hunger, etc. There were three stages in the derangement of the *ōjas*—the worst of them led to death. The functions of the liver, spleen, heart and brain were explained on the basis of the external flow or the exchange of these tissue elements in the body. The heart was considered as the chief receptacle of the three most important fluids of the body, *rasa* (plasma), *rakta* (blood) and *ōjas* (vitality). According to Sen, Charaka (*Sūtrasthānam*: XXX.8-10) stated that "From the great center [the heart] emanate vessels carrying blood into all parts of the body—an element that nourishes the life of all animals, and without which life would be extinct. It is that element which goes to nourish the fetus in the uterus, and which flowing into its body returns to the mother's heart." The character of the blood and its functions, the respiration, the metabolism, and the activity of the central nervous system, in normal as well as in abnormal and unhealthy condition, were well discussed.

The pathogenesis of all diseases was considered to be a result of the derangement of the *tridōsha*, and from this viewpoint Sushruta divided all diseases into 1120 types, whereas Charaka considered them as innumerable. The disease so caused was called a *nija* disease, as compared to the conditions caused by the external factors, which were called *āgantuka* and included mental afflictions. Another classification based on the etiological factors, divided diseases into seven categories: *ādibalapravrtta* or the hereditary conditions caused by diseased germ cells; *janmabalapravrtta* or congenital diseases; *dōshabalapravrtta* or diseases due to the *dōshās*; *sanghātabalapravrtta* or injuries and wounds; *kālabalapravrtta* or seasonal diseases caused by changes in the weather; *daivabalapravrtta* due to the divine will; and *svabhāvabalapravrtta* or natural, like hunger (*Sushruta Samhitā, Sūtrasthānam*: XXIV. 4-11).

In the field of *materia medica* and pharmacy, the properties of drugs and foods were investigated. The drugs were described by a terminology which, when transliterated, does not fail, in many instances, to give a correct insight into their therapeutic uses. The drugs of the Hindu *materia medica* found a prominent place in Hippocratic and Muslim medicine. The Ayurvedic works of the later period, also incorporated in their own *materia medica* some of the foreign drugs that were found by them to have valuable therapeutic properties, *viz.* rhubarb, opium, jamaica, sarsa, etc. (*Bhāva Prakāsha*).

The diagnosis, it was enjoined, should be made by the five senses supplemented by interrogation. Diagnosis was built on cause (*nidāna*), premonitory indications (*pūrva-rūpa*), symptoms (*rūpa*), therapeutic tests (*upashaya*), and the natural history of the development of the disease (*samprāpti*). Sushruta declared that "The physician (*bhishak*), the drug (*dravya*), the attendants or the nursing personnel (*upasthātā*), and the patient (*rōgī*) are the four pillars on which rests the success of the therapy." And the physician, with a profound understanding of the science of medicine was considered the first and the strongest pillar of therapy (*Sushruta Samhitā, Sūtrasthānam:* XXXIV.7; *Charaka Samhitā, Sūtrasthānam:* IX.3). The different methods of treatment, based on the diagnosis of the case, were outlined, and the drugs were classified into seventy-five types according to their therapeutic effects. The whole doctrine of medical care was considered to consist of the following factors, which were to be kept in mind in achieving successful results: the organism (*sharīra*), its maintenance (*vritti*), the cause of disease (*hetu*), nature of disease (*vyādhi*), action or treatment (*karma*), effect or result (*kārya*), time (*kāla*), the agent or the physician (*kartr*), the means and instruments (*karana*), and, the final decision on the line of treatment (*vidhi vinihchaya*).

The microbial origin of the disease, and the infective nature of some of them, such as erruptive fevers, leprosy, smallpox, tuberculosis, etc., were clearly indicated. Sushruta stated that "All forms of leprosy and some other skin conditions are due not only to the derangement of the *vāyu, pitta* and *kapha,* but are also of parasitic origin. . . . Various skin diseases, leprosy, fever, tuberculosis, ophthalmia, and epidemic diseases borne by air and water, are usually capable of transmission from one man to another" (*Nidānasthānam:* V.4-8,26). In *Uttara Tantra* (LIV. 8-11), he further stated that "There are various fine organisms which circulate in the blood and are invisible to the naked eye; usually these look like round bodies of copper colour and are without legs. They give rise to the various forms of skin disease", etc. (Sen).

In surgery, the progress made by the ancient Hindus was phenomenal, indeed, and justified the remarks of Sir William Hunter (1718-1783) "that Hindu medicine was an independent development. Arab medicine was founded on translations from Sanskrit treatises, and in turn, European medicine down to the seventh century, was based on the Latin version of the

Arabian translation" (Gordon). The training of the surgeon was thorough and methodical, and the principles of surgery were based on well-established principles. The students were instructed to practise surgery on various objects, such as cucumbers, gourds, melons, all types of phantoms, and stretched skins of animals with hair still on them. They practised puncturing on distended bags, probing on the openings in worm-eaten wood or lotus roots, extracting teeth on dead animals, suturing on thick cloth or soft leather, and bandaging on the limbs of manikins.

Sushruta classified surgical operations into eight categories: excision (*bhedana*), incision (*chhedana*), scarification (*lekhana*), puncturing (*vedhana*), probing (*eshana*), extraction (*āharna*), evacuation and drainage (*visrāvana*), and suturing (*sīvana*). A variety of instruments and appliances was included in the armamentarium of a surgeon. Sushruta grouped them under two types: *yantra* or blunt instruments numbering 101, and *shastra* or sharp instruments numbering twenty; but the physician was impressed that he must consider the hand as the first, the best, and the most important of them all, as it was with its assistance that all surgical operations were performed. It was enjoined that the instruments should be made out of the best steel, should be well-shaped, with sharp flawless edges, and should be kept in a handsome portable case, with a separate compartment for each instrument. Fourteen types of bandages (*bandha* or *patta*), with their specific uses were described. The surgical operations on all parts of the body have been given in detail, such as laparotomy, craniotomy, caesarian section, forceps application, plastic repair of the torn ear lobule, cheiloplasty, rhinoplasty, excision of cataract, tonsillectomy, excision of laryngeal polyps, excision of anal fistulae, repair of hernias and proplapse of rectum, lithotomy, amputation of bones, reduction of dislocations, and many neural surgical procedures. Preoperative preparation of skin, with some medicaments, and dressing of wounds with medicated oils was routinely done. Hemostasis was achieved by the use of ice, caustics or actual cautery. Venesection and cupping were in general use, and poultices and fomentations formed a routine part of the treatment. Anesthetics were not unknown either. Medicated wines were liberally used pre- and post-operatively to assuage pain. Sushruta distinctly taught that wine should be used before operation to produce insensibility to pain (*Sūtrasthānam*: XVII.13); and Charaka advised the use of wine to alleviate the pain of operations (*Chikitsāsthānam*: XXIV.64). A drug called *Sammōhinī* (producer of unconsciousness) was administered before major operations, and another drug, *Sanjīvanī* (the restorer of life), was employed to resuscitate the unconscious after operation or shock (Keswani, 1961, 1967).

The science of obstetrics and gynecology progressed well in spite of the fact that it was not considered a distinct division of the original Āyurveda. Menstrual disturbances, diseases of the female genital tract, and their treatment were specified. The clinical course and the various stages of labour, the management of puerperium, miscarriage (*makkalla*) and abortion of various types, and difficult labor (*mūdha garbha*) were discussed in detail. The different malpositions of the fetus at birth were well understood, and the different methods of treatment by version, embryotomy, caesarian section, etc., were described.

The general care of infants, the diseases peculiar to children, including nine types of conditions causing convulsion, were described. Since the suddenness with which severe illness appeared and disappeared in children could not be explained easily, and since the innocent little children were easily vulnerable to the evil influences of the demons, children's diseases were particularly attributed to demoniac influences. Jambha and Naigamesha were the two demons described in the *Atharva Veda* (VII.10), as making children sick. Sushruta mentioned nine *grahās* (evil spirits) which torment the children. The voluntary offenses of the mother or the wet-nurse were blamed for the attack by these *grahās*, which could be driven out by medicines, baths, fumigation and special sacrifices to propitiate the individual *grahā* (*Uttara-Tantra:* XXVII-XXXVII,LX).

The diseases of the head (*shirōrōga*), nervous diseases (*vātavyādhi*), and other mental afflictions were well classified, and their treatment was detailed. The various nervous disorders mentioned were: *ākshepaka* (convulsions), *apatantraka* (apoplectic convulsions), *dāruna apatānaka* (hysteric fits), *dandāpatānaka* (stick-like stiffness of the body), *dhanuhstambha* (bow-like body or tetanus), *bahirāyāma* (dorsal bending), *abhyantarāyāma* (ventral bow-like bending of the body), *pakshavadha* (hemiplegia), *sarvāngarōga* (total paralysis of the body), *ardita* (facial paralysis), *hanugraha* (lock-jaw), *manyāstambha* (stiff neck), *jihvāstambha* (paralysis of the tongue), *grdhrasī* (sciatica), *kalāyakhanja* (St. Vitus' Dance), *vepathu* (paralysis agitans), etc. Fainting (*mūrchhā*), giddiness (*bhramarōga*), and apoplexy (*sanyāsa*) were treated with cooling agents and stimulants. Four kinds of epilepsy (*apasmāra*) were distinguished and the treatment outlined. It was enjoined that once the attack was over, one should not rebuke the patient, but should attempt to cheer his soul. Six kinds of madness (*unmāda*) were differentiated. The worst forms were attributed to demoniac influences and called *bhutōnmāda* (possession). Along with the violent remedies suggested it was also recommended that the possessed one should be cheered with friendly talk. Sushruta devoted one complete chapter (*Sūtrasthānam:* XXIX) to the analysis of dreams, and he believed that the favorable or unfavorable termination of a disease could be predicted from the messengers, omens and dreams.

Dietetics and hygiene formed as much a part of the religious ritual as of the medical practice of the ancient Hindus. Therefore the dietetic regimen and the hygienic principles were almost mandatory. A sensible physician was advised always and in all diseases to take into consideration the regulation of diet and then turn to the treatment of the disease with drugs.

The science of medicine received the greatest support and stimulus during the Buddhist period, since the Buddha himself was a votary of medicine and regularly attended on the sick in his camp. And when the Buddha declared that "This is the noble truth of suffering, that to be born is to suffer, to die is to suffer, and to fall sick is to suffer" (Keswani, 1961), his followers considered tending the sick as one of their religious obligations. But the practice of surgery received a great setback since during its later period Buddhism put a stop to animal sacrifice and prohibited dissection. Yet Jīvaka, the personal physician to the Buddha, practised surgery with success and is known to have performed cranial surgery on many occasions.

Ī-tsing in his works quotes from a *sūtra* on medicine, which according to him was preached by the Buddha himself. The old Buddhist text of *Mahāvagga*, belonging to the Mahāyāna sect (4th century B.C.), deals with the practice of medicine as prevalent in those days. It was in no way different from the Ayurvedic medicine of the period, and speaks of *tridōsha,* eye ointments, nasal remedies, various drugs of the Ayurvedic *materia medica,* including asafoetida, fomentations, maggots in the head, and even laparotomy. Charaka is believed by some scholars to have been the court-physician of the Buddhist king Kanishka (100 A.D.), whose wife he attended as obstetrician during a difficult delivery. Nāgārjuna (? 200 B.C.), the famous physician and alchemist who revised the *Sushruta Samhitā,* was a Buddhist. The *Samhitās* of the Vāgbhattās show traces of Buddhistic tendencies. The famous Bower Manuscript, which was found in a Buddhist stupa in Kashgar and was probably written in Indian Guptā script by a travelling Hindu about 450 A.D., has several references to the Buddha as Bhāgava and Tathāgata. The later Mahāyāna Buddhists even raised one of the Bōddhisattva Mahāsattvās to the status of a god of medicine, the Bhaishajyaguru, who in China and Tibet was known as Bh-guru, and in Japan as Yakusha Niyorai.

Thus the ancient Āyurveda flourished in the Buddhist *vihārās* (monasteries), and they became the hub of medical education also. The most noted of these monasteries were situated in Takshashilā, Nālandā, Vikramashilā, Sāranāth, Ōdāntapurī, Vallabhī, Jagadala, Mithilā, and Nādia; and Hiuen Tsāng and Ī-tsing have given excellent accounts of them. After the death of the Buddha, his son Rāhul, and later his famous disciple Emperor *Ashōka* (Acc. 273 B.C.-232 B.C.) established charitable hospitals for animals and men. In the fourth century, another Buddhist king Buddhadāsa of Ceylon, who was himself a physician and wrote a medical text *Saratthasamgaha,* maintained physicians for his troops as well as for his elephants and horses, and erected hospitals all over his country (*Mahāvamsō*).

With the spread of Buddhism to the Asian countries, the science of Āyurveda also travelled far and wide, and found a congenial soil in those countries where even to this day the ancient Āyurveda, in the garb of indigenous medicine, is being taught and practised in the native schools of medicine. More than two thousand years ago, the Hindu culture and civilization reached Indonesia, and Āyurveda became the medicine of those islands. Bali has preserved a rich Ayurvedic theory and practice, and more than two hundred and fifty medical texts on palmleaf manuscripts, like *Usada* (from Sanskrit *aushadha*) and *Trinadī* (the three channels), are known to exist in the land. The medicine of Tibet, untouched by wars until the Chinese occupation, is based on that of Āyurveda, too. Several Sanskrit texts that were translated into Tibetan, including the *Rgyud bzhi* (Four Tantras) of which the original Sanskrit is now lost, have preserved the rich heritage of Indian medicine in Tibet. The Chakpori Medical College at Lhasa was the oldest institution of its kind. From Tibet, Āyurveda travelled to the Khalkhas of Inner Mongolia, and the distant Buryats of north-east Siberia; and all the Buddhist monasteries in those areas became great centers of medical learning. The Medical College of the Kumbum Monastery in Sinkiang attained special renown, as it was the birth place of Tson-kha-pa, the founder of the yellow sect, and has in its possession a vast array of charts illustrating the ancient medical texts. Another famous Mongolian Medical College was at the Yung Ho Kung Monastery, a monument of a great era in Peking, where once the Mongol students from Urga, Kiakhta and Kobdo, and Buryats from the Baikal Lake, the Kalmuks from the Volga River, the Manchus from Tsitsihar, the Tanguts from Koko-Nor, and the Tibetans from Lhasa, repaired for medical training. One of the famous physicians from these Mongolian medical schools, who was known for his Āyurvedic practice and his great success in this therapy during recent times in Leningrad, was the Siberian N. N. Badmaev, who counted among his patients some of the prominent communist leaders like Bukharin and Rykov, the famous writer Alexei Tolstoy, and even Joseph Stalin (Lokesh Chandra).

During the later part of the Puranic period, in the eighth century, the texts of Charaka and Sushruta were revised by Mādhava or Mādhavakara, son of Indukara, who made a further advance in the enumeration and description of the diseases, and discussed them in seventy-nine *Nidānās* under their causes, symptoms and complications. He devoted a special chapter to smallpox (*masūrikā*), which was classified as a minor disease by the earlier writers. This text of Mādhava called *Nidāna* and known chiefly for its section on pathology, remained as a standard textbook for a long time to come and is quoted profusely by the sixteenth century author Bhāva Mishra in his *Bhāva Prakāsha*; it was also translated into Arabic under the various titles: *Nidān*, *Badān* and *Yedān*. The style of presentation, which was a great improvement on the texts of Charaka and Sushruta, shows that it must have been composed for the guidance of students and practitioners of medicine, and that during the Puranic period, before the Arabs conquered Sind and Northern India, the training of physicians had been very systematic and thorough.

From ancient times until about the fifteenth century, the Muslims served as an important medium through which the products of India, China and Persia passed to the Western world. During the spread of Buddhism, and along with it that of Āyurveda, into northern Asia, scholars travelled through Persia and other Muslim countries, which thus had close contacts with India even before the rise of Islam. Many words of Indian origin like *Kāfōōr* (*Karpūra* in Sanskrit, or camphor) and *Sandal* (*Chandan* in Sanskrit) appear in the Qur'an; and, there is even a tradition current among the Muslims that Adam descended from Paradise onto the Indian soil and received his first revelation in India. Some Indians are said to have been seen in the company of the Prophet Muhammad himself (Siddiqi); and it is reported that during the Prophet's time there existed at Senaa in Southern Arabia a famous medical school, the principal of which, Hārit Ben Kaldah, had acquired his knowledge in India (Lassen). It is also recorded that many Hindus were settled in Mesopotamia during the early part of the Umayyad regime; and when Sind was conquered by the Arabs in 636-637 A.D. and annexed to the Umayyad Empire, many Arabs settled there. The superiority of the Hindu civilization that they found in India astonished them, and the sublimity of the Hindu philosophical ideas and the richness and versatility of their intellect came as a revelation. This started a greater influx of Indian men, materials and manuscripts in science and in medicine into the Muslim world. The court at Baghdad extended its patronage to Hindu scholarship, and the second Abbāsid Khalīfā, al-Mansūr (753-774 A.D.), received emissaries from Sind, one of which included Hindu pandits who presented him with two Indian books on astronomy, the *Brahma Siddhānta* and the *Khandakhādyaka,* which were translated into Arabic by al-Fazārī and perhaps also by Yākūb ibn Tarīq. Another influx of the Hindu learning into Arabia took place during the reign of Khalīfā Hārūn-al-Rashīd (786-814 A.D.) under the influence of the Barmecides, who had attained the highest position in the Abbāsid court. These Barmecides, who were the descendants of the high priest (*Pramukh*-Arabicized into *Barmak*) of the Buddhist temple at Balkh, and therefore had special interest in Indian sciences, encouraged the translation of Sanskrit medical texts into Arabic (al-Bīrūnī, 973-1048 A.D.), and appointed many Indian physicians to run the Royal Barmecides Hospital in Baghdad. The first of these was Manka (Mānikya), a successful physician and a philosopher of saintly character, with a profound knowledge of the Persian and Sanskrit languages. The second was Ibn Dhan (a descendant of Dhana, Dhanwantari or Dhanapati), who was appointed the director of the Royal Hospital by Yahyā, the Barmecide. Both these scholars translated several Indian medical texts into Persian or Arabic. Another famous Hindu practitioner in Baghdad during the reign of Hārūn was Sāleh bin Bhela (Arabicized name of Sāli, the son or descendant of Bhela) who seems to have cured Ibrāhīm, a cousin of Hārūn, after the Khalīfā's personal physician Gabriel, who was an expert in Greek medicine, declared Ibrāhīm dead (Siddiqi).

Among the various Sanskrit texts that were rendered into Arabic during the Abbāsid Caliphate, the following medical works in Arabic were commonly used by the Muslims until the end of the ninth century:

1. The *Charaka Samhitā* is said to have been translated into Pahlawī by Manka, and later into Arabic by Abdullā bin Alī and was popularly known as "*Sharak*".

2. The *Sushruta Samhitā* was translated by Manka and was known to the Muslims as "*Sasrad*" or "*Susrud*".

3. *Ashtāngahrdaya* was rendered into Arabic by Ibn Dhan and was known as "*Astankar*", "*Astagar*" or "*Asankar*".

4. *Nidāna of Mādhavakara* was known in its Arabic rendering as "*Nidān*" or "*Badān*".

5. *Siddhayōga* was translated by Ibn Dhan and was known as "*Sindhastāq*" or "*Sindhashān*".

6. A treatise on women's diseases, written by an Indian woman whose name appears as Rūsā in the Arabic version, was translated into Arabic in the eighth century (Nadvi).

Abū Sahl Alī bin Rabban-at-Tabarī gives at the end of his book *Firdausu'l Hikmat* (Paradise of wisdom, 850 A.D.) a short account of the whole system of Indian medicine on the basis of the *Charaka,* the *Sushruta,* the *Ashtāngahrdaya* and the *Nidāna;* and Abū Bakr Muhammad bin Zakriyyā-ar-Rāzī (Rhazes, 850-923 A.D.), who was familiar with the works of Alī b. Rabban quoted the above mentioned medical works along with many other Indian texts in his *al-Hāwī (Liber Continens).* The later Muslim medical writers made reference only to the *Charaka* and the *Sushruta,* as they were largely influenced by Greek medical theories. But the Persian books produced in India appear to have been largely influenced by the Indian system of medicine; and in one of the extant tenth century Persian works on pharmacology by Abū Mansūr Muwāffaq, who made a journey to India in search of Hindu *materia medica,* abundant Indian material is included, and many Indian works are cited which, unfortunately, are not known to exist today (Jolly).

The early Muslim invaders who came to India returned to their homelands with all the plunder, after ransacking the rich religious institutions that were also the centers of Hindu culture and education. But the later Muslim invaders from the twelfth century onwards came to stay, and continued their scornful attitude to everything that smacked of Hindu civilization. They brought with them their own physicians and their own medicines, little realizing the fact that the medicine of these conquerors was what they had originally imbibed centuries earlier from the conquered. Therefore during the early Muslim period the indigenous schools of Hindu medicine languished, and due to lack of patronage from the rulers, and the devastation caused to the medical and education institutions by the successive onslaughts of the earlier Muslim invaders, the cause of Hindu medicine suffered an irreparable and permanent blow.

Muslim royalty recognized only their own *hakeems* (physicians) who were imported from the schools of Tabrīz, Mashahad, Gīlān, and Shīrāz in Persia, where Greek thought ruled supreme; and therefore Greek (Yūnānī or Ūnānī) medicine made further and further inroads into the indigenous medicine of India. These royal *hakeems* attracted more and more students not only from India but also from various places in Persia. However, the system of medicine that they followed was neither Indian nor Persian, but a hybrid Muslim version of Greek medicine. There were no established schools of Tibbī medicine—the hybrid Muslim system of medicine as it came to be known later on. Each well-known *hakeem* had a few students under his preceptorship to be trained in the Tibbī and Yūnānī systems of medicine. For want of State patronage, the collective and organized system of teaching of Āyurveda also receded into the background and was replaced by the preceptor system. Gradually the conquerors realized the value of the ancient Hindu system of medicine, and as an indirect gesture of recognition of its supremacy over their own hybrid system, they allowed their *hakeems* to translate the Indian medical texts into Persian. The first such work in Persian to be published in India in 680 *Hijrī* (1281), was entitled *Tibbe-Fīrēzeshāhī*, which was probably later dedicated to Sultān Allāudīn Khiljī (1288-1296) who was also known as *Fīrōzeshāh* (Zillur Rahman). Some scholars, however, believe that this work was written later in the days of Fīrōzeshāh Tughlak (1351-1388).

It seems that during the early period of Muslim rule, medicine was mainly transmitted from the teacher to the student, who invariably happened to be the son of the teacher, and there is recorded evidence of many such families of *hakeems* whose successors even to this day are engaged in Tibbī practice. One such genealogical tree of a family of *hakeems* is available in the *Imtihān-ul-albā-le-quāfat-il-at-tibā* which traces the generations of teachers from the fifteenth century to the present:

Hakeem (Hk.) Masīhā Khān, who was physician to the Shah of Iran and was appointed as *Hakeem Bashi*, trained his son
Hk. Arshād Khān, who came from Persia in the days of Bābar the Great, trained his son
Hk. Shāh Mohd. Khān, who trained his son
Hk. Dōst Mohd. Khān, who trained his son
Hk. Habībullāh Khān, who trained his son
Hk. Abdul Salām Khān, who trained his son
Hk. Baqā Khān Āazam, who trained his son
Hk. Ismāīl Khān, known as Baqā Khān Ashghar, who wrote *Qarābadīn-e-Baqāi* and trained his son
Hk. Isahāq Khān, who wrote *Ghāyat-ul-fahūm* and trained his son
Hk. Zakāulā Khān, who wrote a pharmacopeia called *Qarābadīn-e-Zakāī*, trained his pupil
Hk. Ahsānulā Khān, who was Prime Minister to Emperor Bahādur Shāh Zafar of Delhi (acc. 1837, deposed 1857), trained his Hindu pupil
Hk. Baldev Sahāi of Meeruth, who trained his pupil

Hk. Syed Karam Hussain, who trained his two sons

Hk. Syed Atīqul-Qādir and *Hk.* Syed Fazlūr Rahmān, of whom the latter trained his son

Hk. Syed Zillur Rahmān, who is a keen student of the history of medicine and publishes a *Tibbī* journal, *al-Hikmat*, from Delhi.

During the days of Akbar the Great (1555-1605), when the arts and sciences received the greatest fillip due to the personal patronage of the Emperor himself, many students came to India to study medicine under the famous personal physicians to the Emperor, *Hk.* Fatehullāh Khān and *Hk.* Alī Gīlānī. *Hk.* Fatehullāh Khān wrote a commentary in Persian on the "*al-Quānoon*" (the Canon) of Ibn Sennā (Avicenna); and Alī Gīlānī was the teacher of *Hk.* Sadrā, son of a famous Persian *Hakeem*, Fakhruddīn Khān. It was during this period that Bhāva Mishra wrote his universally esteemed *Bhāva Prakāsha* (1558-1559) which mentions syphilis for the first time, since the Portuguese had already brought this scourge to India in 1535. Comparable to the *Bhāva Prakāsha* in its extensive character, size and content is the text of *Āyurvedasaukhya* which formed a part of *Tōdarānanda* (1589), inspired by the famous Hindu minister of Akbar, Rājā Tōdar Mall. The *Phiranga rōga* (syphilis) and its treatment with *chōbachīnī*, opium (*aphīma, ahiphena*), and quicksilver (*pārada*) are mentioned in these texts.

Although there were attempts at the revival of the ancient system of Indian medicine during Akbar's time, and the Yūnānī and Tibbī systems were being amalgamated with it, no advances were made in the field of medicine similar to what was happening then in the West. The only significant achievement of this period of Indian medicine was the translation of most of the Arabic texts into Persian, the Court language of the period. By the time of Emperor Ālamgīr's rule (1658-1707) all the Arabic texts that were used by the Yūnānī and Tibbī schools in those days, were also available in the Persian language. Akbar Arzānī, the personal physician to the Emperor Ālamgīr, was himself responsible for the translation of many of those texts, which came to be known as "*al-Arzānī*" meaning "making it cheap", and are used in these schools even to this date (Zillur Rahman). It was during the latter part of the nineteenth century that the famous translator *Hakeem* Ghulām Hussain Qantoorī of Oudh translated into Urdū most of these Persian texts, including a five volume edition of *al-Quānoon*, which were published by the Naval Kishore Press of Lucknow.

The first Tibbī school where regular, organised and collective training was imparted in Tibbī medicine, was started in Delhi in 1882 by Hāziqūl-Mulk *Hk.* Abdul Majīd Khān, Khāndān-e-Sharīf, and was known as Madarsāh Tibbia. By the end of the nineteenth century another school, Takmī-ul-Tib was started by *Hk.* Abdul Azīz, Khāndān-e-Azīzī in Lucknow. These two schools were pioneers in the teaching of Tibbī medicine, and were followed by many more schools of very mediocre standard all over

India. To stop further diffusion of such low-standard schools, the Government laid down very strict rules governing their recognition. This curbed their mushroom growth in the country.

The so-called modern medicine was introduced into India by the Portuguese in the sixteenth century, when Albuquerque (d. 1515) conquered Goa in 1510, and founded the Royal Hospital which was later handed over to the Jesuits in 1591. The Jesuits managed this institution remarkably well and in 1703 introduced a rudimentary form of medical training at the Hospital, with Cipriano Valdares as its master. By 1842 this was converted into the School of Medicine and Surgery (Neelameghan). Although the Portuguese first brought modern medicine to India, it was the French and the British who later established and consolidated the modern medical system in India.

The British built Fort St. George in Madras in 1640, and among the employees of the East India Company there were physicians who, like the ordinary servants of the Company, were more concerned with commerce and colonization than with the practice of medicine. In 1772 the Company founded the Madras General Hospital, and their physicians hired some local people as their assistants and gave them elementary training in medicine. The city of Calcutta was founded by the British in 1690, and as the activities of the East India Company expanded, they paid more attention to the health of their civilian and military personnel. In 1740, the Medical Department of the Company was created, which was comprised of the British military surgeons and their local assistants who, except for some practical experience at the Military Station Hospitals, had no other qualification in medicine. These Indian physicians in the Company's service later came to be known as "Native Dressers" in Madras, and as "Black Doctors" in Bengal (Reddy). By 1767, one such local assistant, called the "Native Doctor", was attached to each battalion of the Sepoys. The "Native Doctors" commenced their career as compounders (pharmacy assistants) and had to submit to an examination before they were entitled to a higher rank and pay.

In Bengal the Court of Directors of the East India Company, impressed with the antiquity of the Sanskrit language and the excellent treatises in Āyurvedic medicine available in the language, set about reestablishing the Āyurvedic schools in Bengal and advocated incorporation into their curriculum of the modern advances in medical sciences. In October 1824 a medical school was started in Calcutta with Surgeon Bretton as its Superintendent. Bretton succeeded Surgeon James Jamieson who was named the first Superintendent of the proposed medical school by the General Order of 21 June 1822, but he died in June 1823 before the opening of the school. In 1827 Dr. John Tytler, a versatile oriental scholar, started a series of lectures on mathematics and anatomy according to the Western system at the Calcutta Sanskrit College. In the same year more teachers were appointed to lecture in Sanskrit to the Āyurvedic students, and to teach them the works of Charaka,

Sushruta and others. Pandit Madhusūdan Gupta, who later became famous for being the first Indian in modern times to dissect a cadaver, was one of the students of this school, the Native Medical Institution, where medicine was taught according to both the Āyurvedic and Western systems. Simultaneously classes in Urdū were held at the Calcutta Madarsāh where the Tibbī students were taught the works of the Muslim physicians, in addition to various other treatises on medicine in Urdū. Bretton died in November 1830, and John Tytler became the Superintendent of the school in addition to his teaching duties. Tytler "translated many European texts into Indian languages for the teaching of the students, and maintained that the education of Indian pupils should be given in their own mother-tongue, if possible, and that the English language should not be thrust on such a rich medium of instruction as the Sanskrit language" (Sarbadhikari, 1961).

In 1833 Lord Bentinck, the Governor-General of India, ordered revision of the medical education, and on 28 January 1835, his Lordship in Council issued an Order abolishing the Native Medical Institution in Calcutta, which was the first teaching institution of its kind in British India. Along with this the medical classes conducted at the Calcutta Sanskrit College and at the Calcutta Madarsāh were also discontinued. The Governor-General's Order of 1835 also decreed that a new medical college should be established for the instruction of a certain number of Indian youths in the various branches of medical science, in strict accordance with the mode adopted in Europe; and also provided that the college should be open for admission to all classes of people between the ages of fourteen and twenty years, without distinction of caste or creed; and that the instruction should be given through the medium of English, which had become the Court language in 1835, after Macaulay's celebrated "Minute on Education" which came down with a "vehement and shallow attack on Sanskrit literature", and declared that "the content of higher education should be Western learning, including science, and that the language of instruction should be English" (Spear). The period of study was fixed at four years, and at the end, the final examination was to be conducted under the supervision of the Council of Education. It was decided that the students who were successful in the final examination should receive the Certificate of Qualification to practise medicine and surgery; and if employed by the Government would receive a salary of thirty rupees, and be designated "Native Doctors".

The first batch of fifty students was admitted on 20 February 1835, and the New Medical College, Calcutta, as it was then called, was started in the premises of old Petty Court Jail. The books and other equipment of the abolished Native Medical Institution were made over to the New Medical College, and Pandit Madhusūdan Gupta, who was then the Vaidya Professor at the Native Medical Institution, was transferred, along with his two assistants at the Sanskrit College, to the New Medical College. Assistant Surgeon M. J. Bramley was appointed Superintendent, and Assistant Surgeon H. H.

Goodeve as his only assistant, to teach full-time at the College. Later during the same year, Bramley was designated the Principal, and Goodeve became the professor of medicine and anatomy, and a new appointee, Assistant Surgeon William B. O'Shaughnessy, was added to the staff as the professor of chemistry and *materia medica*. When Bramley suddenly died at the age of thirty-four, in 1837, the staff situation at the College was reviewed and a few more teachers were appointed to the staff. Dr. C. C. Egerton was named to the chair of surgery, Dr. J. McCosh to that of clinical medicine, and Dr. N. Wallich, superintendent of the botanical gardens, to the chair of botany, and Richard O'Shaughnessy was appointed demonstrator in anatomy and assistant to the Lecturer in Chemistry. They were all asked to function as the Council for the management of the affairs of the College, and a well-known local philanthropist, David Hare was nominated as the Treasurer and Secretary to this Council.

The students started their work with no museum of anatomic preparations, no library, and no hospital. Besides, the teachers were confronted with the task of combating the strong national prejudice against dissecting cadavers or handling anatomic material. To overcome this, parts of the human body were first introduced in illustration at the time of teaching, and these were gradually replaced by the organs from domestic animals, wooden models and metallic representations. It took Goodeve six months before he could place a cadaver on the table, which created much interest and excitement among his students, who probably "went through the same kind of mental stress that a Westerner might experience who contemplated a change of religion which would involve social outlawry" (Spear). But soon the students became familiar with the sight of a cadaver from daily repetition, and on 10 January 1836, Pandit Madhusūdan Gupta, accompanied by his pupils, Umā Charan Set, Dwārkā Nāth Gupta, Rāj Kristō Dey and one other whose name is not recorded, joined Goodeve in an outhouse of the College building, and began dissection on a cadaver. Rāj Kristō Dey is believed to have been the first Hindu student who actually dissected the human body on that day (Sarbadhikari, 1962). That day will be marked in the annals of the history of Western medical education in India, when Indians rose superior to the prejudices of their traditional education and flung open the gates of modern medical science to their countrymen. Two years later Goodeve, who had carefully watched the reaction of the students toward dissection, was pleased to address his new batch of students thus: "in less than two years from the foundation of the College, practical anatomy has completely become a portion of the necessary studies of the Hindu medical students as amongst their brethren in Europe and America. The practice of dissection has since advanced so rapidly amongst us that the magnificent room, recently erected, in which upwards of 500 bodies were dissected and operated upon in the course of last year, has already become too small for our purpose; we have been compelled to construct an adjoining shed for the convenience of the class, now amounting to upwards of 250 youths of all nations, colours, religions and castes, co-mingling together in this good work as freely and amicably as the more homogeneous frequenters of an European school. I know nowhere a more striking example of the powerful influence of science, in promoting liberality and good feeling amongst her votaries, than in this very interesting example" (Sarbadhikari, 1962).

In Madras a similar beginning was made when the Madras Medical School was opened on the last day of June 1835 for training subordinates for the medical services in the army. The first batch of eleven "locals" and ten "Eurasions" were admitted for a four-year course at this institution; and after qualifying, the "Eurasians" were employed as "apothecaries" and the "locals" were designated as the "dressers" in the army. In 1838 the School was opened to civilians, but the prejudices against Western medicine ran so high at that time that for eight years no civilian came forward to register. When some of the civilians finally joined, it was decided to conduct three different courses, one for physicians, to extend over a period of five years, another for apothecaries, to run for a period of over four years, and a third for training the dressers for over three years. The medical students after qualifying were awarded a diploma of "Graduate in Medicine, Midwifery and Chirurgery"— G.M.M.C. In 1850 the School was raised to the status of a College, and when the University of Madras was founded in 1857, the College was affiliated to it. It is of great interest to record that for the first time in the history of modern medical education in India, women students were admitted to the Madras Medical College in 1875, when they were debarred from the other medical colleges in India, and even in Britain their admission to the medical colleges was being much debated and discouraged (Reddy; Bannerjee).

The first examination at the Calcutta Medical College was held on 30 October 1838, after the students had completed a course in anatomy, physiology, chemistry, natural philosophy, *materia medica*, general and medical botany, physic and surgery, for a period of three and a half years. The examination lasted for seven days, and only four out of eleven students were successful. It is interesting to note the conclusions of the examiners at the end of the final examination:

The ordeal through which these young men have passed, is one of no common kind, and affords a very gratifying measure of capacity and acquirements. The result is such as to satisfy us with their average knowledge of a solid and well grounded character. We have unanimously come to the decision of granting them letters of testimonial, and we consider them competent to practice medicine and surgery. We beg to recommend them accordingly to the liberal consideration of the Government, as the first Hindus who, rising superior to the trammels of prejudice and obstacles of no ordinary character, have distinguished themselves by attaining to complete Medical Education upon enlightened principles. . . . We further recommend that at the end of five years, they should undergo another examination and that for the purpose of frequent and habitual reference they

be supplied, before quitting the College, with the following works:
Phillips' Translation of the London Pharmacopoeia.
Thomson's Elements of Materia Medica and Therapeutics.
Dr. O'Shaughnessy's Manual of Chemistry.
Cloquet's Anatomy by Knox.
Sir C. Bell's Institute of Surgery (just published).
Dr. Geo. Gregory's Elements of Medicine.
Twining on the Diseases of Bengal.
Cooper on Dislocations and Fractures, etc.
Clarke's Commentaries on the Diseases of Children.

(Sarbadhikari, 1962)

The Calcutta Medical College was donated its present site by Motilal Seal in 1835, and the foundation stone of its Hospital was laid on 28 January 1835, and the building completed in 1848. The sanctioned bed strength of the Hospital is 1400, and besides the departments of medicine, surgery, ophthalmology and emergency, special departments of chest diseases, dentistry, otorhinolaryngology, dermatology, venereology, neurology, psychiatry, orthopedics, radiology and cardiology were created. In 1857, when the University of Calcutta was created, the College was affiliated to it.

With all this happening in Madras and Calcutta, it was not long before a medical college was founded in Bombay. The foundation stone of the Grant Medical College was laid by the Lord Bishop of Calcutta on 30 March 1843, and the College was formally opened by the then Governor of Bombay, Sir George Arthur, on 3 November 1845 (Bannerjee). In 1857 the College was affiliated to the University of Bombay, which had just been founded. More and more medical schools were opened, and by 1900 there were seventeen such schools and colleges where training in modern medicine was being imparted. In 1916 the Lady Hardinge Medical College, exclusively for the training of female physicians, was opened in New Delhi, and it remains the only institution of its kind in India.

In 1947, when India achieved its independence, there were only fifteen medical colleges with an annual admission of about 1000 students, and today there are more than ninety medical colleges with an annual admission of about 11,000 students. The diploma-granting medical schools have been gradually upgraded to degree-granting colleges, which are affiliated to various universities. Today in more than sixty medical colleges post-graduate training is given, and special institutes for postgraduate training and research in medical sciences are being established throughout the country. To begin with, the All-India Institute of Medical Sciences was established in New Delhi in 1956 as an autonomous institution of national importance under an act of Parliament, since it was realized by the authorities that institutions, like individuals, can develop in stature and fulfill themselves only in an atmosphere of freedom and responsibility. The most important goals of this Institute, in undergraduate medical education, are to evolve patterns of education that will facilitate the practice of scientific medicine of the highest quality under the conditions existing in India, and to prepare the students to meet the needs of the immediate future. In the field of postgraduate education, the most essential function of the Institute is to provide opportunities for training teachers for the medical colleges in India in an atmosphere of research and enquiry. This All-India Institute of Medical Sciences has already, in little more than a decade, attained great eminence in setting new patterns of medical education. The Health Survey and Development Committee, popularly known as the Bhore Committee, in 1946 recommended establishment of six more such Institutes of Postgraduate Medical Education and Research. Three Institutes at Calcutta, Chandigarh and Pondicherry are now in the process of development, and more are planned. The continuing education of the physician has been receiving great attention from the specialist organizations as well as medical associations in India, as in the other advanced countries. Two colleges of general practitioners are already functioning in New Delhi and Hyderabad as "University without walls" for continuing medical education, to keep the general physician and the specialist abreast of the latest advances in the medical sciences, and thus to improve the standard of medical care in the country.

As early as 1843, the Royal College of Surgeons of England recognized the three medical colleges in Calcutta, Madras and Bombay, and the medical degrees awarded by their respective universities were registered in the Register of the General Medical Council of the United Kingdom. Thus, until 1933, when the Medical Council of India was constituted, medical education developed under the direct, general supervision of the General Medical Council, and the standard of medical education in India reached the minimum level obtained in Great Britain during that period. The Medical Council of India was vested with powers similar to those enjoyed by the General Medical Council in the United Kingdom and had Provincial Councils under its jurisdiction which maintained Provincial Registers and handled matters of discipline. The chief responsibility of the Medical Council of India was to maintain uniform minimum standards of university medical education in India, and to further the recognition abroad of the degrees awarded by the Indian universities. The watchword of the Medical Council was "efficiency at home, and honour abroad". The Council was empowered to recommend to the Government of India recognition of various medical institutions, whether under control of a university or of any other autonomous body which, in its opinion, conferred medical degrees or diplomas of a standard considered satisfactory by the Council. The Council was authorized to call for any information from a medical institution, and to send its inspectors or visitors to the institution to report on the manner in which it was run, on the quality of the training, and on the standard of examinations conducted by the affiliated university. On the basis of such reports the Council evaluated the nature and quality of training given at a particular medical center, and accordingly sent its recommendations to the Government regarding the advisability of recognizing the medical qualifications awarded by the university concerned. Soon after the transfer of power, the Parliament of India enacted further legislation which created Nursing, Dental and Pharmacy Councils on the same pattern.

With amazing advances in biology, physics and chemistry, and the increased need for specialization in medical sciences during recent times, the Government of India felt that premedical and medical education needed critical evaluation and revision, in keeping with advances elsewhere and suitable for the particular requirements of the country. With this view in mind, the Government organized the first Medical Education Conference in the autumn of 1955 which in particular considered the proceedings of the World Medical Education Conference with special reference to the country's needs, and recommended major reforms in medical education in India. The Conference conducted exhaustive discussions on premedical education; the selection of students and their entrance requirements; the content of medical courses and the curricular hours; the technics of assessment and examination; the establishment of full-time teaching departments; the creation of departments of social and preventive medicine, psychological medicine and statistics; the training and selection of medical teachers, the conditions of their service, etc. Many of the recommendations of this Conference have been generally accepted by the Government but unfortunately, are far from being implemented. A direct outcome of the deliberations of this Conference, however, was that fresh legislation on medical education was enacted by Parliament in 1956, which repealed the Indian Medical Council Act, 1933, and gave more powers to the Medical Council with regard to the maintenance of an All-India Register, undergraduate and postgraduate medical education, etc. This Act of Parliament, 1956, was further amended by the Indian Medical Council (Amendment) Act, 1964, which vested the Council with powers to prevent quackery, etc.

The Second Medical Education Conference was convened in 1959, but most of its recommendations were not implemented either by the State Governments or by the Universities. In the same year the Government of India set up a Health Survey and Planning Committee to formulate health programs for the country in the Third and subsequent Five Year Plans. The Committee made a panoramic survey of the whole field of professional education of physicians, nurses, paramedical and other personnel, and made comprehensive recommendations for undergraduate and postgraduate medical education, which are much more far-reaching than the famous Flexner Committee and Goodenough Committee Reports. "If only these recommendations are implemented in their totality, the standard of medical education in this country will be second to none in the world" (Rao). The Committee, among other things, laid special emphasis on the quality of medical education, while suggesting the establishment of one medical college for every five million of the population; on the introduction of new technics of teaching and assessment; on the qualifications and training of teachers in medical colleges; and on the conversion of the internship to that of a compulsory housemanship. This Committee also pointed out the ways and means of meeting the great shortage of teachers and medical research workers, and recommended the inter-institutional relationship between the Indian medical centers and those in the United Kingdom, Canada and the United States of America. The last recommendation has already been implemented and a beginning has been made by having collaborative exchange between the Universities of Baroda and Edinburgh, and between the John Hopkins University and the Calcutta University.

The success of any medical educational system rests on its four pillars—the student, the curriculum, the teacher, and research. The teacher, although very much neglected by the powers that be, is the main actor in this drama of medical education. Realizing this fact, the medical teachers and educationists in India have created a forum for furthering the cause of medical education, and in 1961 formed the Indian Association for the Advancement of Medical Education, which meets every year to discuss one particular aspect of medical education. Their sixth annual conference in 1966 discussed Medicine and Society as a prelude to the Third World Medical Education Conference which met in November 1966 in New Delhi, with the main theme of the Conference: Medical Education: factor in socio-economic development.

During the past fifty centuries of its unrecorded and recorded history, India has always been conscious of the fact that "The health education of today shall determine the patterns of health care of tomorrow" (Keswani). But in a country as vast as India, with its ever-increasing population and a limited range of resources available to tackle its health problems, the solution is formidable, indeed. With the population explosion all the efforts of the country seem neither fast enough nor soon enough. In a country where today at least 400,000 physicians are needed, there are hardly 90,000 to combat disease and save life, and where more than 10,000 medical teachers are required, most of the medical colleges are working with only forty-five per cent of their proper teaching personnel. Yet medical education is growing with incredible speed, and India is training physicians with the requisite knowledge and basic skills, to develop in them essential habits and attitudes, to make them understand professional and ethical principles, and to inculcate in them the capacity to analyze local problems and contribute towards their solution, and thus play a vital role in organizing community resources for the solution of the medical problems of the country.

BIBLIOGRAPHY

ALTEKAR, A. S., *Education in ancient India*. Varanasi, 1965.
Atharva Veda, trans. into English by RALPH T. H. GRIFFITH. Benares, 1916. 3 vols.

BANNERJEE, J. N., History of medical teaching in India from the preliterary period up to modern times. *India J. Med. Educ.*, 1966, **5** (2): 76-90.
BEAL, S., *Chinese accounts of India*, trans. from the Chinese of Hiuen Tsiang. Calcutta, 1957-1958. 4 vols.
Bhela Samhitā, ed. by ASUTOSH MOOKERJEE. *J. Dep. Lett. Calcutta Univ.*, 1921, 6.
Bower manuscript. Facsimile leaves, Nagari transcript, Romanised transliteration and English translation, with notes, by A. F. R. HOERNLE. Calcutta, 1893-1912. Pts. 1-2, Archaeological Survey of India, New Imperial Series, Vol. 22.
Charaka Sahmitā, ed. by SHREE GULABKUNVERBA with trans. into Hindi, Gujarati and English. Jamnagar, 1949. 6 vols.
―――, ed. by PANDIT HARIDUTT SHASTRI AYURVEDACHARY. Lahore, 1940 (in Hindi)
FILLIOZAT, J., *The classical doctrine of Indian medicine*. Delhi, 1964.
GORDON, B. L., *Medicine throughout antiquity*. Philadelphia, 1949.
HIUEN TSANG, see BEAL.
I-TSING, see TAKAKUSU.
JOLLY, J., *Indian medicine*. Poona, 1951.
Kashvapa Samhitā, Hindi trans. by SRI SATYAPAL BHISAGACARYA. Varanasi, 1953.
Kaushika Sutra, The Kaucika Sutra of the Atharva-veda, ed. by MAURICE BLOOMFIELD. *J. Am. Orient. Soc.*, 1890, 14.
―――― *Atharvavediva Koushik Sutra*, trans. by THAKUR UDAYA NARAIN SINGHA. Madhurapur, Dist. Muzaffarpur, 1942.
KESWANI, N. H., Anaesthesia and analgesia among the ancients. *Indian J. Anaesth.*, 1961, **9**: 231-242.
―――, Evolution of Surgery, *Medicine Surg.*, Baroda, 1961, **1** (8-9): 9-20.
―――, The concepts of generation, reproduction, evolution and human development as found in the writings of India (Hindu) scholars during the early period (up to 1200 A.D.) of Indian history. *Bull. Nat. Inst. Sci. India*, 1962, No. 21, pp. 206-225.
―――, Garbha Upanishad, a brief Sanskrit treatise on ancient Indian embryology. *Clio Med.*, 1965, **1**: 65-74.
―――, Ancient Hindu orthopaedic surgery. *Indian J. Orthop.*, 1967, **1**: 76-94.
―――, Modern medicine in Indian traditional setting—a new brew in an old vat. *Medicine and Culture*. London, 1969, pp. 677-688.
LAKSHMI PATHI, *A textbook of Ayurveda*. Vol. 1, sect. 1, Jamnagar, 1944.
LASSEN, C., *Alterthumskunde*. Leipzig, 1843-1872. 4 vols.
LOKESH, CHANDRA, Personal communication.
MACDONNELL, A., *A history of Sanskrit literature*. Delhi, 1958.
MADHAVAKARA, *Madhava Nidanam*, ed. by BHISHAGRATNA PANDIT SHREE BRAHMA SHANKAR SHASTRI. Benares, 1954.
Mahavagga, in Vinaya texts, trans. T. W. RHYS DAVIDS and HERMANN OLDENBERG. Oxford, 1881. Sacred Books of the East Series. Vols. 13, 17, ed. M. MÜLLER.
Mahavamso, in Roman characters, with trans. and introd. essay by GEORGE TURNOUR. Cotta (Ceylon), 1837.
Manu Smrti, The laws of Manu, trans. by G. BUHLER. Oxford, 1886. Sacred Books of the East Series, Vol. 25, ed. M. MÜLLER.

Mookerji, R. K., *Ancient Indian education*. Delhi, 1960.
Mukhopadhyaya, G. N., *History of Indian medicine*. Calcutta, 1923-1929. 3 vols.
Nadvi, *Arab aur Bharat Ke Sanbandha*. Allahabad, n.d. (in Hindi).
Neelameghan, A., The Royal Hospital at Goa as described in some seventeenth century travel accounts. *Indian J. Hist. Med.*, 1961, 6: 2.
Pālākapya, *Hastyāyurveda*, ed. Mahadeva Chimanaji Apte. Poona, 1894. Anandashrama Sanskrit Series, Vol. 26.
Panini, Sutra text with German translation, ed. by Bohtlingk. Leipzig, 1887.
———, *Kāshikāvritti* (Benares Commentary) ed. Jayaditya and Vamana. 2 ed., Benares, 1898.
Purana, *Kurma*, trans. by Panchanana Tarkaratna. Calcutta, 1332 B. S. (in Bengali).
———, Bombay, 1905 (in Sanskrit).
———, *Padma*. Poona, 1893. Anandashrama Sanskrit Series.
Puschmann, T., *A history of medical education*. New York, 1966.
Radhakrishnan, S., *The Hindu view of life*. London, 1949.
Rao, K. N., Medical education. Dr. Sir Lakshmanaswami Mudaliar Endowment Lectures in Medicine, University of Madras, 1965-1966. Pt. 3, March 1966.
Reddy, D. V. S., *The beginnings of modern medicine in Madras*. Calcutta, 1947.
Rig Veda Samhitā, trans. by H. H. Wilson. Poona, 1925-1928. 6 vols.
Sachau, E. C., *Alberuni's India*, trans. into English. Delhi, 1964.
Sāma Veda, The hymns, trans. by R. T. H. Griffith. Benares, 1926.
Sarbadhikari, K. C., Western medical education in India during the early days of British occupation. *Indian J. Med. Educ.*, 1961, 1: 27-32.
———, The early days of the first medical college in India. *Indian J. Med. Educ.*, 1962, 1: 203-298.
Sen, G. N., *Hindu medicine. An address on Ayurveda delivered at the Foundation Ceremony of Benares Hindu University in 1916*.
Shatapatha Brāhmana, trans. by Julius Eggeling. Oxford, 1881-1900. Sacred Books of the East Series, Vols. 21, 26, 41, 43, ed. M. Müller.
Siddiqi, M. Z., *Studies in Arabic and Persian medical literature*. Calcutta, 1959.
Spear, P., *India, a modern history*. Ann Arbor, Michigan, 1961.
Srī Bhāva Mishra, *Bhāva Prakāsha*, ed. by Bhishagratna Pandit Shree Brahama Shankara Misra. Varanasi, 1961 (in Hindi).
Sushruta Samhitā, trans. and ed. by Kaviraj Kunjalal Bhishagratna. 2 ed. Varanasi, 1963. 3 vols.
Susrutasamhitā, ed. by Vaidya Jadāvji Trikamji Acharya and Narayan Ram Acharya. Bombay, 1938 (in Sanskrit).
Takakusu, I. T., *I-tsing: a record of the Buddhist religion as practised in India and Malay Archipelago (671-695 A.D.)*. Oxford, 1896.
Upanishads, trans. by Swami Mikhilananda. London, 1951-1959. 4 vols.
Vāghbhatta, *Ashtāngahrdaya Samhitā. A compendium of the Hindu system of medicine with the commentary of Arundatta*, revised and collated by Anna M. Kunte. Bombay, 1880. 2 vols.
———, *Astāngahrdaya Samhitā: ein altindisches Lehrbuch der Heilkunde*, aus dem Sanskrit ins Deutsche übertragen, mit Einleitung, Anmerkungen und Indices, von Luise Hilgenberg und Willibald Kirfel. Leiden, 1937.

Yajur Veda, the Texts of White, trans. by R. T. H. GRIFFITH. Benares, 1957.
Yajur Veda, Krishna, The Veda of the Black Yajus School, trans. by A. B. KEITH. Cambridge, Mass., 1914.
ZILLUR RAHMAN, HAKIM SYED, Personal communication.

MEDICAL EDUCATION IN SOUTH-EAST ASIA (EXCLUDING JAPAN)

PIERRE HUARD
Université de Paris
Paris, France

Using the cultural aspect as a basis of classification, the following three groups may be considered: (a) Countries with a Chinese cultural background: China, Formosa, Hong-Kong, Vietnam; (b) Hindu countries: Malaysia, Thailand, Burma; or partly Hindu: Indonesia with its large Muslim population; (c) Christian countries: Philippine Islands. In each of these groups special attention has been given to recent developments of medical education, but it will be of some interest to give some attention to earlier times in some of them, specifically China and Indo-China.*

Ancient Period

CHINA

The teaching of medicine in China is attested as early as the fourth century B.C., that is, the first Chinese medical teachers were contemporaries of Hippocrates and his successors. Bridgman has dedicated an interesting study to these ancient Chinese medical schools which produced physicians refusing any connection with witchcraft. At the time of Charlemagne in France, the *Tang* Dynasty (598-907) created the *tai-yi-chu* (great medical service), one of the most ancient examples of medical instruction supervised by the state. This institution was highly favored under the *Sung* (960-1270). We owe to it the diffusion of the first printed medical books. Different medical specialities were taught: internal medicine, *materia medica*, ophthalmology, charms and magic, forensic medicine, dietetics, sexual hygiene, pediatrics, gynecology and dermatology. Acupuncture and moxas were demonstrated on the occasion of numerous "practical exercises" with the help of the famous "Man of bronze" (1026); dolls made from wood and cardboard served the same purpose. Education included the learning of medical ethics of the highest character. The *tai-yi-chu* was maintained with a few modifi-

---

* The author is extremely grateful for help provided by the World Health Organization of Geneva; by the Embassies of China, Viêtnam, and the Kingdom of Cambodia in Paris; by the Deans of the Faculties of Medicine of Saïgon and of Phnom-Penh; by Professors Pham biêu Tâm and Le Xuân Chât of Saïgon, and by Mr. Nguyên Huu Dung.

cations under the *Yuan* Mongol Dynasty (1270-1368) and the *Ming* Dynasty (1368-1644) but declined under the *Ch'in* (1644-1911).

In the nineteenth century there existed in China only a small Imperial College, very inferior qualitatively and quantitatively to the analogous institution in Japan. Its role was limited to the preparation of physicians attached to the Imperial Palace. In the remainder of the country the decline of traditional medicine continued during the whole nineteenth century and the first part of the twentieth, that is, until the accession to power of the present government. The value placed upon all popular lore has forced the state to a rehabilitation of classic medicine and a reorganization of its teaching. Native Chinese medicine now possesses its own institutes and hospitals, but it is practiced in close relationship with modern Western medicine. The Western physicians must know the traditional medicine in the same way that the Chinese physicians must know Western medical ideas.

Western medicine was first brought to China by the Jesuit scholars of Peking, but they compiled books only for the use of the Emperor, and Western medicine was only slightly diffused outside the "Forbidden City". The *Manchu Anatomy* prepared in 1723 by R. P. Parennin did not make an impression similar to that provoked in Japan by the *Kaitai-shinsho*. The publication of the latter work in 1744 was a noteworthy event from all points of view: medical, scientific and cultural.

The story of the penetration and teaching of Western medicine into China has been repeatedly studied, in particular by Wong and Wu and by the present author. Some of this teaching took place at the Christian missions, chiefly Anglo-Saxon, and many Chinese scholars were also trained abroad. Western schools were opened in Peking (*Tong Wen Kuan*), then in Canton (1870) and Tien-Tsin (1881). The first Chinese surgeon with Western training was Kwan A-To (1818-1874). The first physician to take a foreign medical degree was Wonk Cheuk-Hing (1828-1878) who became a doctor of Edinburgh (1853). The first woman physician took her bachelor's degree in 1885. At present every Chinese student following the modern curriculum receives at the same time some traditional medical instruction in acupuncture and moxas, massage, respiratory techniques and physical culture. This traditional medicine partly compensates for the lack of technical and pharmaceutical equipment and fits better to the private means of the average Chinese. It is a chief object of the present cultural revolution to use more subjective, psychic and moral factors as compensation for the technological gap.

Ex-French Indo-China

It is necessary to distinguish an eastern Chinese group, the Vietnamese, who had their own chronology and dynastic annals, from a western Hindu group, Laos, and Cambodia, without chronology and almost without historical past.

Vietnam, Chinese Colony (3rd century B.C.-A.D. 968)

Vietnam passed through several cultural stages. As a Chinese colony it was the *Kiao-Tche* to which the Buddhist missionaries came probably bringing there, as elsewhere, some rudiments of Indian medicine. Chinese was used as the political and cultural language, but the importation of scientific books and Chinese medical plants was forbidden. The only possibility was for the Vietnamese to go to China to learn from Chinese teachers. A few examples of this have been published by Maurice Durand and by the present author.

Vietnam, Satellite of China (969-1884)

As a satellite of China, Vietnam was known as Annam. The teaching system was still the Chinese system that did not provide for the free spread of knowledge but rather the inculcation of a particular kind of education considered in a double sense as education and religion. "Education is the blood and veins of the Empire. The law is a temporary curb. Only education can bind forever", said the emperor K'ang Hsi.

In a country in which the urban population as late as 1939 represented only three and a half per cent of the nation, it was a political necessity to control the mass of "Black Heads"—that is to say, the multitude—by a philosophical bureaucracy of Confucian scholars who were the result of triennial competitions which, suppressed in China in 1905, were to exist in Vietnam until 1919. The competitive system, adopted in Hanoï in 1075, took more than seven centuries to implant itself in the South. But it succeeded in doing so and played a more important part in Vietnam than in China. The Emperor considered himself as the "First Scholar", and the prestige of this title was such that, as the ancient French monarchy had suppressed the title of "High Constable", so the *Nguyên* dynasty suppressed the too powerful title of "First Doctor of the First Degree". In this atmosphere the satellization favored China but not Vietnam. In the fifteenth century, during six years (1407-1413), the Ming troops occupied North Vietnam, carried off scientific works of importance and sent the chief intellectuals of the country to China. The xylography known to the *Sung* as early as the tenth century made its effective appearance in Vietnam only in the seventeenth century.

Under these conditions the training of physicians was very difficult. It was carried out in the following manner:

1. Every scholar who had access to the classic Chinese literature could become a physician either by chance or by filial piety to look after his aged parents.

2. Every member or friend of a medical family could become a physician by apprenticeship. Physicians were grouped in "corporations" which contributed also to the spread of medicine by preparing students and, at certain dates, celebrating the protective spirits, that is, the great Chinese-Viet-

namese physicians of former times. This homage took place in the temple of medicine that continued to exist in Hanoï unil 1954.

3. There were also private schools, analogous to Chinese "studios" where a scholar lectured, embracing at the same time the purely professional, the literary and the philosophical aspects of medicine.

4. Finally, as in China, there existed a Health Service of the Imperial Palace, with a hierarchy possessing the degrees of the mandarinate and recruiting "pupils" by quadrennial competition, to whom a formal teaching was given. It existed even in the eighteenth century and was renewed in 1850.

In fact, two different medical systems coexisted in Vietnam. The one was southern and consisted essentially of popular non-written recipes representing Vietnamese oral tradition. The other, called "northern medicine", was a learned system, recorded in books in Chinese characters and rooted in the classic Chinese medicine. It was the only system taught in a regular way.

The results were far from negligible. Such scholars as Tuê tinh and Lan-Ông held an important place in the classic Chinese medicine, and gave proof in many instances of a critical mind. In their opinion, the Chinese medicine could not be admitted to a tropical country, such as Vietnam, without undergoing necessary adaptations.

From the fifteenth century, the Court of Annam, following the example of the Court of China, welcomed Western physicians who were often French missionaries. Mention may also be made of the British surgeon Duff (1747-1824), who operated with success upon King Vo Vuong for an anal fistula, and the American surgeon George Finlayson who came with the Crawford mission in 1781. Although they gained some Vietnamese desciples, they had no influence upon the medical professional as a whole.

Vietnam, French Colony

It seems that it was not so much repression as it was the retrograde system of education that was the "keystone" of an Empire governed by an Emperor—who was less powerful than the President of the United States—and a small number of officers. To interfere with this system was to run the risk of causing dangerous cracks in an already rotten structure. Thus the problems of scientific education in general and of medical training in particular were not considered in technical and cultural terms but according to a political and emotional situation. In addition, the hostility towards France lasted from 1858 to 1883 and created resistant military and ideological movements causing a situation unfavorable to reform of the educational system along European lines.

If officers and Court were as a whole hostile to the creation of scientific and technical schools with the help of foreign professors, it was desired by the reformist movement which sought the teaching of Western languages and sciences and the creation of scholarships for study overseas.

This attitude of intellectual opposition, associated with armed resistance, was common to all clandestine nationalist movements. Thus, if at the beginning of the colonial period native scholars accused French teaching of destroying the traditional culture, reformists no longer used this reproach, because their enthusiasm had been aroused by the Japanese victory over the Russians. Moreover, the rising generation began to accuse the French administration of not opening the doors of French and Indo-Chinese universities widely enough to the Vietnamese.

The first school for superior training to be founded in Vietnam was a school of medicine. The idea dated from 1886, and the model was the school of medicine of Pondichéry, opened in 1857. After long discussions as to location, whether Saïgon or Hanoï, it was opened in the North in 1902. The famous A. Yersin, who isolated the plague bacillus (Hong-Kong, 1894), was the first director. The school was created by Hoang Cao Khai (d. 1933), not in Hanoï but two miles outside in the village of Thai-Hà-Âp, in the province of Hadong. A few years later the first school teaching the Latin alphabet was opened by the viceroy of Tonkin in his own home.

Hoang Cao Khai was convinced that the fanaticism and conservative patriotism of scholars was not clear-sighted and that intellectual emancipation, together with scientific and technical studies constituted the apprenticeship of freedom leading to national independence. Thus it was not by chance that the new school of medicine was built on his lands. It was completed by a hospital of forty beds for medical instruction and a library. However, the location was unfavorable. The school, outside the capital, was without a good water supply and was in an area of endemic paludism. It was finally transferred to Hanoï where it has remained. The new building was opened 1 March 1905 and admitted twenty-nine students paying eight piastres a month. Recruitment had been made through competitive examinations in Tonkin (where fifteen candidates were selected from 121), but without such distinction in Annam, in Cochin-China and in Cambodia. Until the month of June a pre-medical course was offered based on natural history, anatomy and physiology, arithmetic, geometry, French, history and geography. Examinations reduced the students from twenty-nine to fifteen, who began their medical studies in October.

What was their mental attitude? Between the creation of the school and the end of the first graduation there occurred the Russo-Japanese war (1905) and the nationalistic pro-Japanese movement, which affected the thinking of the students. It is not known whether or not they were deliberately reformists like the revolutionaries who took refuge in Japan, Pham-Boï-Châu (1867-1940) and Prince Cuong-Dê (1882-1950), or whether they were attached to the still powerful mandarinate. The triennial competitive examinations, suppressed in China in 1905, were to last in Vietnam until 1919. With the purpose of putting on a "face of importance" they tried not to break with the system of traditional values and requested their assimilation with the laureates of the literary competitive examinations. Thus they asked on

25 March 1902, for permission to wear a distinctive badge, an ivory one analogous to a mandarin badge, the *The' Bai* on which seven Chinese characters were engraved.

The Improvement Council of the school of medicine welcomed this petition, and the badge, allowed from the start, was officially adopted in April 1905. But the prestige of the mandarinate fell out of favor with the youth and the badge was quickly forgotten (Nguyên Xuân Nguyên).

The initial program of the school proposed by Yersin was free of any chauvinism or paternalism. He wanted his school to be an institution of scientific research with laboratories open to the students and scientists of every nationality. These possibilities were carefully examined in 1902, but Yersin, appointed to a new office, left his post in July 1904.

The first auxiliary physicians to be graduated were sent into the provinces in 1907 where, although they were at first received with distrust, they became rapidly appreciated by the administration and the population.

The brightest student was Nguyên Van Thinh (1888-1946), a southerner. He profited by the First World's War to continue his studies in France, where he gained his French baccalaureate and the P.C.B. in 1909-1910, entered at the Faculty of Medicine of Paris in 1912, worked as an extern fellow of Paris hospitals in 1914-1915, and as an intern in 1919. He was a student of Professors Achard, Lannelongue, Blanchard, Nicolas and Loeper. As Mesnil's disciple at the Pasteur Institute in 1921 he sustained a thesis on beri-beri. Going back to Saïgon-Cholon, Thinh practiced medicine and became President of Council of the first Republic of Cochin-China, but was unable to overcome the Vietnamese nationalism opposed to his French medical culture. So like many other statesmen, philosophers and disciples of Confucius, he killed himself.

From 1902 the school of medicine trained health officers ("Indo-Chinese physicians") destined to serve in Vietnam, Cambodia and Laos. In 1936 it began to train doctors of medicine. In 1941 it was transformed into a mixed medical and pharmaceutical faculty with special departments for medicine, pharmacy, dentistry and midwifery. In 1947 it opened an annex in Saïgon later transformed into a faculty.

Thanks to the institutions of the colonial regime, the first patriots' program was realized even if in a different manner than they had hoped. Gifted Vietnamese students were able to be on equal terms with European students; so they broke the climate of dependence and prestige which, more than force, was the support of colonialism. The idea of subordination, founded on values in which personal merit did not find a place, appeared insupportable to liberated Vietnamese. The conquest of political emancipation was the consequence of this intellectual emancipation. Much more than the Confucian moral, the study of sciences and techniques was the apprenticeship of freedom. Nationalism united with science became gradually a very explosive feeling. For the majority of medical students it constituted an exacting and obeyed religion. The revolution of 1946 gave impressive evidence of it.

Modern Period

CONTINENTAL CHINA

In 1916 there were in China twenty-six medical schools (thirteen of them of the missionary type) where 1940 students were registered. By 1935 these figures grew to thirty-three faculties of medicine, 20,799 beds distributed in 500 hospitals, 3528 students, 14,000 to 20,000 practitioners trained in the West (*Si-yi*) of whom only 7000 were registered (Hughes). These figures increased until about 1940 when the Japanese invasion, the transference of the northern universities to the south, and the civil war made the running of expensive organizations difficult, when the supply of medicines, instruments and various installations posed insoluble problems.

Many witnesses allow us to form some idea of the medical distress of this period, in the course of which Norman Bethune and Dwarkanath Kotais were distinguished, finally dying victims of their devotion to the Chinese people. By the time of the liberation (1949) out of 600 million Chinese, millions had died from malnutrition, tuberculosis, cholera, schistosomiasis, typhoid fever, dysentery, and the huge incidence of parasitic diseases.

There existed no more than eighty-seven hospitals with 6400 beds at their disposal. There were at Peking only 1100 beds for 3,000,000 inhabitants, and at Shanghai only 1700 for 7,000,000 inhabitants. Nearly 1000 new hospital units were built or put back into working order.

The number of physicians of Western training fluctuated between 20,000 and 35,000, most of them qualified or specialized abroad. That is to say, one modern physician for 25,000 inhabitants at a time when in the highly civilized countries one physician was scarcely sufficient for 500 to 700 inhabitants; 20,000 physicians, that is, only twice the total number of Belgian physicians, served a population larger than the combined populations of the United States, U.S.S.R., and Europe.

When peace was restored with the 1948 liberation, the People's Republic permitted the reorganization of training centers and of important hospitals where none of the more recent aspects of modern medicine was neglected. In 1950 the National Congress of Health Technicians examined the duration of medical studies which included a short course (two-and-a-half years) and a long course (eight years). It recommended that health doctors be trained in four years, and specialists (surgery, pediatrics, obstetrics and gynecology) also in four years.

Actually a new program in 1959 anticipated much longer courses of six years, with the mastery of one foreign language, and of eight years with knowledge of two foreign languages (English and Russian) taught by language teachers at the medical school itself. The students admitted to these courses are selected by a very strict method, their training being undertaken by members of the Academy of Medical Sciences. The best students go on to do research work at the laboratories of the Academy.

The increase in the numbers of physicians of Western training has been as follows:

1945	1950	1952	1954	1955	1957	1964
34,600	41,400	51,736	63,046	70,000	73,573	100,000

The total number of Chinese physicians has risen to 120,000, that is, one for 5000 to 6000 inhabitants. In this number are included the "health officers", i.e., 500,000 traditional physicians concentrated especially in the southern provinces, then distributed throughout the whole territory. Because of this, 1000 inhabitants are served by one physician.

For the university year 1957-1958 there were 7974 teachers, of whom 497 were professors, 535, assistant professors, 2296, lecturers, and 4656, assistants.

There were thirty-eight institutes or schools of medicine plus two of pharmacy and four of traditional medicine. The medical and pharmaceutical schools had 49,107 students, of whom 19,747 were women and 1909 were students from national minorities. The women students represented, according to the schools, 30-40 per cent of the actual total. The most important center of education is the Peking University of Medical Sciences. It developed from 403 students and 134 beds in 1949 to 3700 students and 2308 beds in 1965. It has a faculty of medicine, a faculty of pharmacy, a faculty of hygiene, and a faculty of basic sciences (physics, chemistry and biology). From 1937 to 1949 it granted 1069 medical degrees. In the period 1949-1965 this number increased to 5700. More than seventy such establishments exist in the provinces.

The Academy of Medical Sciences, founded in 1957, is made up of several sections: physiology, pathology, histology, biochemistry, pharmacy (synthetic and physical chemistry). It is forming several new research institutes devoted to children's diseases, obstetrics, nutritional diseases, epidemiology, hematology, and medical biology.

The provinces have not been neglected. The animal biology laboratory of Hangtcheous (Tchekiang) was built in the year of the visit of the French medical expedition (Professors J. Bernard, Matey, Denoix, cf. *Le Monde,* 9 November 1958). Subsequently the Institute of Endocrinology and the Institute of Parasitology in Shanghai were established.

Some 500,000 physicians, surgeons, dentists, nurses and others have been qualified since 1949, to increase the ranks of the public health service, already making a total of one million medical workers. In each large town there are now well-equipped modern hospitals, and each region possesses one or two hospitals. The people's communes each have one health center and most of the production units have their own clinics. The largest factories and mines also benefit from fairly large hospital centers. Since the liberation 120,000 students have left the medical institutes in the country. Moreover, some 330,000 young health technicians have completed three or four years of lectures in the 200 undergraduate medical schools in the country.

At present each province or each self-governing region, with the exception of Tibet, possesses one or several medical institutes, some of which cater to specialists. The first institute for medical specialists for miners has been established in Tangchan, the center of the famous coal-mining region of Kaillan. In these institutes the length of the courses varies from five to six years. Each is attached to one or several hospitals where the students carry out their studies. Nowadays, 90,000 students are educated, that is eight times the number of students registered in all the Chinese medical institutes in 1947, the record year before the liberation.

Since 1958 much stress has been laid on the combination of theoretical and practical studies, which must be developed equally. Besides this, the students are compelled to do ten weeks of manual labor, whether it be at the university or in the urban or rural communities. The reason for this is to exorcise the ghost of the scholar of former times, whose inordinately long fingernails were indicative of contempt for any manual activity. The goal is to make the student graft onto himself a second identity, that of the manual worker which he had never been and which he must become. The standard is, in fact, that of the manual worker, and the cultured manual worker is only an ideal. Indeed, these arrangements do not help higher education and research work. They have not been well received in all parts of the country. Generally speaking, the desired end is to form a "skilled Red" personnel, that is to say, politically developed and professionally competent.

Throughout the extent of the Chinese territory the number of qualified physicians has increased as follows:

1954	1958	1961	1963	1964
4527	5393	19,000	25,000	90,000

Finally, a great effort has been made to educate the rural masses in hygiene in order to obtain their effective cooperation in the field of preventive medicine.

Traditional medicine has found new favor, in so far as medicine of the people is taught in about twenty universities and two research institutes, one of which is the Research Institute of Peking founded in 1955. It is also taught in many specialist hospitals. Not only classical Chinese medicine has been restored to a position of honor, but also the medical inheritance of the different racial minorities (Manchurians, Mongolians, Tibetans, etc.). In many centers of training and treatment the Westernized physicians (*Si-yi*) and the traditionalist physicians (*Tchong-yi*) cooperate in the closest way.

The Islands of South-East Asia

Formosa (Taiwan)

Population: 10,612,000
Physicians: 6901
Number per physician: 1500
Medical Schools: 4

Medical Center of National Defense, founded in 1902 (six-year course)
Medical School, University of Tainan, founded in 1945 (seven-year course)
Medical School of Kaoshiung, founded in 1954 (six-year course)
Medical School of Taipei, founded in 1960 (six-year course)
Number of medical students: 2597

Curriculum: Six years of primary education, six years of secondary education, one or two years of premedical studies, two years of basic sciences, two years of clinical sciences and one year of residence in a hospital.

Degrees awarded: Bachelor of Medicine, Bachelor of Medical Science, with supplementary years in a research institute in order to become a Master of Medicine.

Tainan is also an active center of traditional Chinese medicine, but we lack information about this.

Hong Kong (1960)

Population: 3,075,000
Physicians: 1001
Number per physician: 3100
Schools or Faculties of medicine: 1
Number of students: 395
Curriculum: see Burma
Degrees awarded: see Burma

The Philippines (1960)

Population: 27,792,000
Practicing physicians: 3949
Schools or faculties: 7

The Faculty of Medicine, St. Thomas' University, founded in 1871
The College of Medicine, University of the Philippines, founded in 1907
The College of Medicine of the South-West, Cebu, founded in 1947
The College of Medicine, Central University of Manila, founded in 1947
The Institute of Medicine, University of the Far East, founded in 1952
The College of Medicine of the East, Quezón, founded in 1956

The College of Medicine, Cebu Institute of Technology, founded in 1957
Number of students: 11,967
Curriculum:
Degree awarded: Doctor of Medicine

The Rockefeller Foundation of Manila is a research establishment which teaches postgraduate public health for physicians, health officers, nurses, and laboratory assistants.

Indonesia (1960)

Population: 93,506,000
Doctors: 1938
Number per physician: 75,000
Schools or Faculties of Medicine: 6
  The Faculty of Medicine, University of Indonesia, at Djakkarta
  The Faculty of Medicine, University of Gadjah Mada, founded in 1946 at Djakkarta
  The Faculty of Medicine, Airlangga University, founded in 1948 at Sourabuya
  The Faculty of Medicine, University of Northern Sumatra, founded in 1952 at Medan
  The Faculty of Medicine, Anadalos University, founded in 1955 at Padang
  The Faculty of Medicine, Hasan Udin University, founded in 1956 at Macassar
Number of Students: 5201

Curriculum: At the time of the independence, everyone was admitted to the university, and the results were disastrous. Actually the standard of teaching and education of the population has become considerably higher, a method of selection has been introduced and the results are satisfying.

The curriculum is modelled on that of the Netherlands, with a premedical course (first year), a preclinical course (second and third years), a theoretical and clinical period (fourth and fifth years): a practical clinical period (sixth and seventh years) approved by the *Candidaat examen,* the *Doctoraal examen,* and the *Arts examen.* It has been somewhat modified because of an association between the Medical School of the University of California and the University of Indonesia. For ten years it has shown excellent results, but it has taken all the tenacity of the teachers to overcome the doubts of the local governments and the American administration, and the changes which have been brought about in the supervisory departments (F.S. Smith). Nearly thirty per cent of qualified physicians are retained in the faculties to form a national framework for the students, and to fill the vacancies (there were 160 teaching posts vacant in 1964). The rest of the qualified physicians do a course of at least two years in the rural sectors which are all too often left to the djami (traditional physicians).

English-Speaking Indo-China

Malaya (1960)

Population:
Doctors:
Number per physician:
Schools or Faculties of Medicine: 2
  King Edward VII College of Medicine, founded in 1905 in Singapore
  Medical School of Kuala Lumpur, founded in 1963
  Curriculum: See Burma
  Degrees awarded: See Burma

Thailand (1960)

Population: 26,258,000
Doctors: 3402
Number per physician: 7700
Schools or Faculties of Medicine: 3
  Faculty of Medicine, University of Medical Sciences of Bangkok
  Faculty of Medicine, University of Medical Sciences of Chiengmai
  Siriray Medical College, University of Medical Sciences of Thonbun, founded in 1880
Number of Students: 1268
Curriculum: See Burma
Degrees awarded: See Burma

Burma (1960)

Population: 20,662,200
Doctors: 1962
Number per physician: 11,000
Schools or Faculties: 2
  Faculty of Mandalay
  Faculty of Rangoon
Number of Students: 1349
Curriculum: Secondary studies, two premedical years, five medical years and two months as a house surgeon.
Degrees awarded: Bachelor of Medicine and Bachelor of Surgery (M.B., B.S.), recognized by the General Medical Council of the United Kingdom.

Indo-China—Formerly French Colonial Territories

As mentioned earlier, under the colonial system a medical school of Indo-China was founded in Hanoï in 1902 by Yersin. It instructed health officers (called Indo-Chinese doctors) who were to serve in Vietnam, Cambodia, and Laos. In 1936 it began to train doctors of medicine. In 1941, it was changed into a joint faculty of Medicine and Pharmacy, comprising a medical section, a pharmaceutical section, a dental section, and a midwifery section. In 1947, it had a branch at Saïgon, itself changed into a faculty.

In 1954, the record office and most of the personnel of the Faculty of Hanoï were evacuated to Saigon. From this time on, the teaching of medicine, previously French and standardized, progressively broke up into four national centers which were developed separately in North Vietnam, South Vietnam, Cambodia, and Laos.

The Democratic Republic of Vietnam (North Vietnam)

Number of inhabitants: 20,000,000
Number of medical doctors: 8800?
Traditional doctors: 16,000
Assistant doctors: 2349
Faculties of Medicine: 1
Schools for assistant doctors: 20

The Democratic Republic of Vietnam gives a striking example of adaptation in the education of physicians in a state of chronic war and with much reduced foreign aid. Besides this, there was a lack of teachers, a lack of material and technical bases, and a lack of a national scientific language. Under these conditions, it was necessary to change completely and integrate medical teaching into a vast strategic health scheme. According to this, whatever his standard, the physician becomes a medical worker, a revolutionary officer, specialized in medical work, whose job is to preserve the health of his fellow men. Besides a professional and technical training, he must have a political and ideological education which occupies eight to twelve per cent of the total number of lecture hours. In order to meet quickly the enormous requirements of the higher health staff, the academic system was inadequate. Doctors of medicine, it was admitted, could only form the technical and scientific peak of a completely nationalized pyramid of medicine and health. This pyramid covered the whole country, spreading from the capital to the main provincial and regional towns, and from there to the remotest communities and hamlets. At the base of the pyramid, the collectivized network includes the assistant country doctors, the kindergarten teachers, the nurses, the country midwives and the hygiene propagandists. Above them come the auxiliary staff of the state system (midwives and nurses), who have above them the secondary staff (medical auxiliaries and laboratory technicians), and higher staff (doctors of medicine). Vertical channels allow more gifted medical workers to change rank. The best nurses and midwives are changed into medical auxiliaries in the provincial schools, founded in 1960. The best medical auxiliaries are made into doctors of medicine by the faculty of medicine after a shorter program of study than that of the ordinary students (four years).

Furthermore, it was essential not to discard the traditional practitioners likely to become the higher state staff, but following China's example, "to integrate them into socialist medicine in a united medical system, which would prevent and treat illnesses in an original way, unique to Vietnam, thus building up a specifically national medicine."

They have gone even further in developing "a sense of respect towards traditional medicine whose heritage must be continued and expanded, not being content with exploiting a few remedies." Medicine is, in fact, a creation of the rural masses which must be continued and built up by the scientists to raise it to a higher standard. The medical schools also include, at all levels, a course in traditional medicine. An institute of traditional medicine, with clinical departments and laboratories, works in collaboration with the faculty of medicine.

The teaching syllabus for traditional medicine includes: the study of its basic principles and of their application to daily practice; preventive medicine and therapeutic medicine. It is completed by reports on acupuncture, Vietnamese medical topics, effective popular remedies, and medical history. It comprises fifty theory lectures at the faculty, seventy-eight in the schools for medical auxiliaries, without counting the hospital residential period, the study of drug determination, and the examination of plants in the botanical gardens. Thus traditional medicine is considered to be a specialist course, just as important as the others.

The teaching of traditional medicine is closely associated with that of modern medicine, with the following guiding principles:

1. Priority is given to preventive medicine. "A cured patient is a focus of contamination which has been supressed and will become, if he is well trained, a good propagandist for preventive measures." "Attack disease at its source before it breaks out." Hygiene takes up ten per cent of the total number of lecture hours.

2. Teaching is essentially practical and centered on Vietnamese pathology, which is different from that of Western countries and even of some others. It is not particularly important for the auxiliary or secondary staff. Hospitals and training centers work in conjunction with one another everywhere.

3. Manual work is not neglected. The students must remember, that they are workers like other people and must help in the upkeep of the faculties, schools and hospitals.

4. The academic training is a complete syllabus of political and social activities outside the academic year, forcing the students to integrate with the common people, to take part in their work and to contribute to their health education.

5. Each higher member of staff must know one foreign language; each

scientific worker or member of the teaching staff must know at least two.

6. The technical and moral standard of all medical health workers is tending to be constantly raised under the strict control of the state.

The higher medical studies take six years, after which the young doctors (sixty per cent of whom are sons of workers and peasants) do a period of three years as a practicing physician, at the end of which four options are open to them:

- a. To continue to practice while attending post-university refresher courses.
- b. To attend a specialist course which lasts two to three years and results in a higher qualification
- c. To attend the special preparatory training leading to a professorship or to scientific research.
- d. Courses abroad are arranged.

The results of this effort, which is as vast as it is original, have been the following: There is one medical auxiliary for 1000 inhabitants on the plains, and one for 500 in the mountains. After ten years of independence, up to 1964, there were fifteen times more doctors of medicine than in 1954. Health standards of the population have improved to conditions which are quite remarkable. The technical level of the teachers from the Faculty of Medicine at Hanoï is vouched for by their scientific publications, which have gained international recognition.

South Vietnam

There are five universities—Hue, Dalat, Can Tho, and Saïgon. A state university coexists with Catholic universities and a Buddhist university (Van Hanh). Only Hue and the State University of Saïgon offer a medical training.

The faculty of medicine of Hue began to operate in 1961. It has just awarded its first diplomas (the syllabus is not the same as that of the Faculty of Medicine of Saïgon, but consists of six years and one preparatory year). The total number of medical students at Hue was 218 (for the year 1965-1966), with sixteen male students and 202 female. (In general the male students who want to enter medicine go to Saïgon, a better equipped faculty with teachers of higher intellectual powers, more university qualifications and training). However, some teachers from Saïgon go to teach at Hue. Hue has both a school of midwives, which has trained fifty-eight practitioners, and a school of health technicians which has trained 186 health inspectors, 124 of whom are women and sixty-two of whom are men.

The state university of Saïgon originally comprised only one joint faculty of medicine and pharmacy. The Faculty of Pharmacy was founded in 1962 and the Stomatology Department in 1964.

In the three branches (medicine, pharmacy and dentistry) there is first a preparatory year. Then:

Medicine: 6 academic years
Pharmacy: 4 academic years
Stomatology: 5 academic years

From 1965, in order to be admitted into the preparatory year, a competitive examination in which only full bachelors can be registered has been necessary.

The number of students admitted varies each year, but averages 200 for Medicine, 200 for Pharmacy, 100 for Stomatology.

Concerning student numbers, the total figures are listed in Table 4. For the following years I have been able to obtain more detailed information

TABLE 4
NUMBER OF STUDENTS ADMITTED, 1954–1956

Year	Medicine	Pharmacy	Ch. Dentistry
1954–55	362	211	21
1955–56	458	225	35

(Table 5). But the information concerning 1966-1967 and 1967-1968 is not yet registered and, moreover, incomplete. Table 6 shows the number of medical and allied medical degrees awarded by the State University of Saïgon.

The Faculty of Medicine of Saïgon was transferred to the government of the Republic of Vietnam in 1955. It has had to face up to numerous problems concerning teaching accomodation, professional structure, syllabus of studies, and scientific terminology. These have been studied in depth by Professor Nguyen Huu, who has drawn up a statement in the thesis of his pupil, Mrs. Nguyen Van Nha (Saïgon, 1962), and also by the Dean of the Faculty, Phạm Biêu Tâm.

1. The question of notoriously inadequate premises has actually been mainly resolved, thanks to the cooperation of American aid, of which the first schemes date back to 1956.

However, the general dearth in personnel, materials, and financial resources, and also the administration's and the public's lack of understanding, delayed the inauguration of the first part of the Medical Education Center until 1965, the date of its inauguration at Cholon. The second part will include a teaching hospital with 500 beds and instruction in all the clinical disciplines.

2. The teaching staff could be entirely Vietnamese if all the competent Vietnamese living abroad returned to their own country. They do not do this because of the political instability, the insecurity, the difficulty of obtaining exit visas, and the low rate of pay. The readjustment of these factors necessitates lengthy delays. The collaboration of foreign teachers, whether

TABLE 5
Number of Students Admitted, 1956-1967

Academic Year	Medical Students Male & Female	Pharmacy		Dentistry	
		Male	Female	Male	Female
1956–57	544	141	116	28	12
1957–58	607	147	153	73	23
1958–59	730	134	198	85	25
1959–60	859	243	264	87	38
1960–61	920	391	432	126	37
1961–62	1017	640	621	149	30
1962–63	1136	978	834	80	18
1963–64	1160	1062	1045	71	18
1964–65	1194	1473	1434	69	21
1965–66	1152	1355	1362	70	24
1966–67	1157				

French, American or whatever, can only, in fact, be temporary. Therefore it is important to increase the hard core of actual teachers (seventy), whose competence has, in many cases, been tested by the American Universities and the competitive examination for admission to the teaching staff of the French faculties of medicine.

The recruitment of assistants (forty) is considerably slowed down by the permanent state of war which calls all the students who have finished their sixth year to active service, except for a small contingent selected by the competitive examination. This proportion is much too small to assure a good teaching profession and the training of future teachers. This involves many students being sent abroad where some succeed in finding a more agreeable post than in Vietnam and are lost to their country.

3. Availing itself of the privilege of new countries which can venture to

TABLE 6
Degrees Awarded by the State University of Saïgon

Academic Year	Medical Doctorate Male & Female	1st Diploma in Pharmacy		Doctorate of Stomatology	
		Male	Female	Male	Female
1956–57	13	28	29	2	0
1957–58	29	29	20	2	0
1958–59	45	13	20	2	1
1959–60	45	9	12	5	2
1960–61	69	6	14	8	7
1961–62	66	16	35	12	10
1962–63	169	19	39	22	4
1963–64	112	39	24	12	2
1964–65	85	64	32	13	3
1965–66	110	107	27	8	2
1966–67	139				

free themselves from the sterile past, and choosing from what seems most advantageous in the reforms proposed by different countries, South Vietnam has arrived at a very fascinating and flexible curriculum, which can be modified as required and which we shall quickly summarize:

a. The premedical stage, which is included to develop the taste for observation, common sense, and critical ability, and to increase the manual competency. Its syllabus is a reorganized P.C.B., supplemented by a course in biostatistics, social sciences, and by two foreign languages, French and English.

b. The stage of basic or doctoral training (undergraduate) is preceded by a selective screening, which, from 1962, has eliminated students who are obviously unsuited.

During the course of the year some examinations precede the transitional examinations associated with the system of "credits". The course of study lasts six years; four spent in theoretical lectures and practical work, and two in internship (the second year of which must be spent in the provinces). The end in view is to give a training which the student does not undergo passively, but is already well grounded, to acquire his own experience which will serve him throughout his professional life. There is no preliminary stage. From the first year, contact is made with disease. During these two years, the methodology of all the medical disciplines is taught, as well as the common syndromes and minor medico-surgical operations. During the third and fourth years the student performs the function of an extern, and passes through all the specialist sections in two-month stages. In the two final years, the student, relieved of all study courses, devotes himself entirely to the hospital. Formerly he had to perform the duties of the external student. Only the *internat des hôpitaux* remains in force. The thesis is replaced by a short treatise devoted preferably to a subject of local significance.

c. The stage of specialist study strengthens the position of the student in a general or specialist medico-surgical practice, or even in research or in the teaching profession. The internal medical stage in hospitals, and the monitory stage of the faculty, are the two competitive examinations which lead to a career in teaching where the nominations are made on the standard reached.

d. The post university stage is by necessity as long as the professional stage. It has already shown excellent results in orthopedic surgery.

4. Scientific terminology is a huge problem which confronts the world (in order to impose a universal nomenclature) and also the nation (in order to translate the universal nomenclature into the different languages). Anatomical nomenclature is already advanced and is present in three languages, Vietnamese, Latin, and French, according to the standards laid down by the P.N.A. (1955).

The end pursued by the university of Vietnam is to arrive quickly, but under good conditions, at a training given wholly in the national language.

However, even when this condition is fulfilled, foreign books will continue to be in the majority in the libraries, and that is why every doctor must know at least two foreign languages.

Meanwhile, French, which was the preferred language of the secondary and higher education, is seeing its influence decrease and it is possible that it will subsequently be eliminated in favor of English. An English section was, moreover, founded in the faculty in 1964. It is developing by stages at a yearly rate parallel with the French section.

Outside the training given in Vietnam a record must be kept of the students registered in foreign Faculties of Medicine. Their importance is indicated in Table 7. I believe that no precise significance can be drawn from these figures. There are many different reasons for going abroad—fictional and actual, serious and undeclared. The year 1965, particularly, was rife with political upheavals. That explains the considerable drop in numbers of students going abroad for their studies.

In principle, only those students interested in subjects not taught in Vietnam are awarded grants or are authorized to go at their own expense. For this reason, there is no grant awarded for medical, pharmaceutical or dental studies abroad. Those who want to study these three subjects abroad register in another school (preferably a technical school), so that they may leave first of all at the risk of changing their subject later. This only concerns those who go at their own expense. The grant-aided students are not allowed to change their subjects.

The students who want to go away at their own expense must be full bachelors and be less than twenty-one years of age. But this age has just been reduced to eighteen. Will there then be changes concerning studies pursued abroad? It is probable, and from then on few students will be able to go away.

If the faculties of medicine are linked up with the Ministry of National Education, by contrast, the following establishments are dependent on the Ministry of Health.

NURSING SCHOOLS. A national school of nurses was formed in Vietnam at

TABLE 7

STUDENTS REGISTERED IN FOREIGN FACULTIES OF MEDICINE

Years	Scholarships		Self-Supporting	
	Male	Female	Male	Female
1962	162	62	278	110
1963	125	64	177	17
1964	97	48	118	4
1965	36	24	15	?
1966	61	?	289	108

Saïgon in 1952. It was abolished in 1961 to be replaced by six auxiliary schools of health, opened in Saïgon, Hue, Can-Tho, Long-Xuyen, Da-Nang (Tourane) and Ban-Mê-Thoût (Darlac). These schools were, in their turn, abolished in 1967.

The average number of nurses and qualified assistants each year is fifty. Length of studies is one year. The condition for admission is the Certificate of Primary Studies.

TECHNICAL MEDICAL OFFICERS. These are the medical auxiliaries, destined to be sent into the districts (of each of the forty-three provinces of Vietnam), 350 in all. The school of Medical Technical Officers, formed in 1955, was abolished in 1960. Length of studies is three years. Admission is determined by competitive examination among candidates who have first class certificates of Secondary Education (which corresponds to the junior secondary certificate). There have been four years of intake in all (64 + 86 + 43 + 45), having produced 238 officers.

The section of Technical Medical Officers has also been abolished, and replaced by the Health Technicians section.

HEALTH TECHNICIANS (HEALTH TECHNICIAN SCHOOL OF SAÏGON AND HUE). The two schools for technicians (Hue and Saïgon) were formed in 1960. The length of studies and the conditions of admission are the same as for the technical medical officers. The syllabus is slightly different and is especially concerned with medical care (nursing). Two modern languages are taught— English and French. In short, the Health Technicians are nurses. That is why the nursing schools (of too low a standard) have been closed down. The number of qualified Health Technicians is 469.

SCHOOLS OF MIDWIFERY (SAÏGON AND HUE). The midwifery school was dependent on the Faculty of Medicine until 1957. Since then it has come under the Ministry of Health. Length of the courses and conditions for admission are the same as for the Health Technicians. Each year these two schools turn out eight midwives. The number of midwives trained, up to now, is 800. As for the Technicians, the language used is Vietnamese, but there are lectures in English and French.

With approximately ten hospitals, and with nursing homes which revolve around it, the Faculty at Saïgon is an extremely important hospital network. It publishes the *Acta Medica Vietnamica* and takes a very important part in the editing of the *Bulletin of the Institute of Cancer of Vietnam*. The Saïgon Faculty has become known throughout the world from its students' theses and the publications of its professors; it constitutes one of the best medical training centers in South-East Asia. It is in contact with the Faculty of Medicine in Paris, and, for the last few years, has had strong connections with the University of Missouri and of Michigan. It also has received assistance from the A.M.A.

Laos

Population: 2,000,000
Physicians: 40
Number per Physician: 45,000
Medical Schools: 1
Number of students:

Curriculum: Four years of studies after the entrance examination which attracts particularly well-trained nurses and midwives, but from which graduates are exempt.

Degrees awarded: Assistant Doctor. It gives entrance, without need of sanction, into the third year of the course in French schools and faculties of medicine, and can thus lead to a French university Doctorate.

At the time of the accession of their country to independence, most of the students of Laotian medicine had stopped attending the Vietnamese medical training centers in order to pursue their studies, either in Phom-Penh, or in France. Once admitted into the schools of the French Health Service of the Army (Lyons) and the Navy (Bordeaux), they have had difficulty in attaining the Doctorate in Medicine. Since few of them are graduates, they cannot gain the university Doctorate. Such were the reasons for the establishment of the Royal School of Medicine at Vientiane, in 1957. This was done with the help of the Mission of Economic and French Technical Aid and the French Cultural Service, the Colombo Plan, and the Asia Foundation, in association with a school of the Military Health Service. The language used in teaching is French.

For want of a sufficient number of graduates, who have been attracted to other careers, it has only trained assistant doctors up till now, the best of whom have been able to continue their studies in France and go on to attain the state Doctorate. The increase in the actual numbers taking secondary education will allow consideration, in the near future, of the formation of a medical school leading up to the state Doctorate.

Cambodia

Population: 4,952,000
Physicians: 183
Number per physician: 27,000
Faculties of medicine: 1
Number of students: 483

Curriculum: Trainee health officers study for four years, differing from medical students whose studies are largely inspired by French standards.

Degrees awarded: Diploma of Health Officer, and Cambodian State Doctorate. These diplomas give entrance, without need of sanction, into the third year of the French faculties and schools of medicine and can thus lead to a French university Doctorate.

At the time of the proclamation of independence, a School for Health Officers was formed in 1946, and then changed into a Royal School of Medicine, Pharmacy, and Allied Medical Sciences (1962). In 1964, it became the Faculty of Medicine and Allied Medical Sciences. It coexists with a military health service school, which is linked to similar French schools and which does not award any degrees. The Faculty is placed under the supervision of the Ministry of National Education, while the previous teaching establishments were dependent on the Ministry of Health. An alliance with the Faculty of Medicine of Paris, drawn up in 1953, has facilitated the transformation of the School of Health Officers into a Medical School, has secured the control of examinations and has permitted French professors to be sent to Cambodia, and Khmer physicians to be sent to Paris, two of whom have been received into the competitive examination for admission into the teaching staff of the French faculties of medicine. The Khmerization of the teaching body is thus initiated in a fitting way, and very probably this evolution will continue. The language used for teaching is French. I have not been able to find out how many students continue their studies abroad.

Effectively, students have had a dual function which has been a measure of discretion and adaptation to the needs of the moment. Gradually, as secondary training gains importance, the number of trainee health officers diminishes, and the number of students increases. Furthermore it is possible for a health officer to take up his studies again and, under certain conditions, to end up with a doctorate.

The total number of students is constantly growing:

1962-63	1963-64	1964-65	1965-66	1966-67
293	299	344	414	483

The School of Medicine of Phom-Penh benefitted at its foundation by subsidies from the French Mission for Economic and Technical Aid, by substantial encouragement from the O.M.S., the Asia Foundation, and from the Colombo Plan.

Conclusions

1. In the whole of South-East Asia, although in different proportions, the effective medical force is still too small, and the output of the schools and faculties of medicine is insufficient. From this results the necessity, first to form, on a parallel with medical doctors, health officers, sanitary inspectors, or nurses, who will assure that some of the doctors' functions are carried out; and secondly, often to utilize doctors of traditionalist training.

Everywhere, in fact, a popular medicine, sometimes complex in structure, but perfectly adapted to the racial groups in which it has evolved, coexists with Western medicine.

2. The teaching profession must take these particulars into account, in according as much importance to preventive medicine as to curative medicine,

in bringing about training centers common to all categories of health personnel; in easing the passage of this personnel to one or other of these categories and in disposing of an extensive educational scheme, ranging from the training of the specialist to the education in hygiene of the rural masses. In Continental China and in North Vietnam, the popular doctors even carry out the functions of an official teaching staff.

3. The vernacular languages are tending, more and more, to become used for teaching, with an important regional time-lag. From this point of view, China, Indonesia, and North Vietnam have a real advance over other countries. Nevertheless, the problem is nowhere entirely resolved, because scientific terminology is not yet at a desired level and because the use of foreign books and the necessity of studies abroad cannot be eliminated. On the other hand, publications in the vernacular would only have a limited audience. Also linguistic teaching, both French and English, is in the syllabus of many faculties and medical schools, and continental China itself continues to publish scientific reviews in English.

4. The administrative structure of a faculty or school of medicine varies. The state establishments are sometimes under the jurisdiction of the Ministry of National Education (Vietnam), and sometimes under that of the Ministry of Health.

There are also schools of a military character, dependent on the Ministry of the Armed Forces (China, Vietnam, Cambodia, Laos).

Private, lay, or religious institutions (St. Thomas of Manila), coexist with the official institutions in the countries whose social structures still remain subdivided. Moreover, their necessity is unavoidable when the official schools lack space. That is why the U.P. University of Manila only admits about 100 students per year. The remainder fall back on the free colleges, which cannot cover their costs, although taking in a great number of students.

5. The student population is particularly dense in countries where a leisured or a middle class exists. By these the doctoral degree is sought after, less for its professional value than as an honorary title and a mark of social prestige. In socialist nations, or in those less socially hierarchical, the medical students are also teachers or young people who need to earn their living. That is why, in Indonesia, they can carry out a part-time job from 2 p.m. onwards. The proportion of women students rises to about fifty per cent in the free colleges of the Philippines, in Indonesia, in Thailand, and in Burma. Their results are often better than those of the male students. Elsewhere, a grant system functions obliging the recipients, once their studies are finished, to remain in civil or military service for a certain number of years.

In most cases students do not find the necessary conditions for a perfect intellectual, moral and physical expansion in their own homes. Halls of residence of university cities with reading rooms and sports grounds are necessary.

6. At the entrance into the faculties or schools of medicine, the idea of

numerus clausus has been adopted by many countries using very different methods. The syllabus of studies leading to the doctorate in medicine is almost the same everywhere, with a special bias towards tropical pathology, dietetics and psychiatry. Almost everywhere the teaching staff is insufficient. The teachers in employment have not an enviable position with regard to facilities; foreign teachers only remain for a limited time, and many future masters sent overseas to complete their studies do not return, contributing to the brain drain.

The emphasis must therefore be placed on local recruitment of a quality necessitating a substantial improvement in socio-political status, and in the scales of pay for teachers, who often lack the junior personnel needed in order that their work and their research may be profitable.

7. External aid will still be necessary for a long time in the countries ravaged by war.

BIBLIOGRAPHY

AHEARN, A. M., Viet-Cong medicine. *Milit. Med.*, 1966, **131**: 219-221.

BRUMPT, L., and MEYNARD, M., *Rapport sur un voyage d'etudes médical dans le Sud-Est Asiatique*, 1955. Unpublished.

CHU TAT DAC, *La formation professionelle et technique du personnel infirmier et auxiliaire au Vietnam de 1945 à 1960*. Saigon, 1961. Thesis.

FABER, KNUD, *Rapport sur les écoles de médecine en Chine*. Geneva, 1931.

HUARD, P., L'introduction de la médecine européenne en Asie Orientale, in *Education*, Saigon, 1948. Reprinted in: *Médecine Tropicale*, Marseille, 1949.

HUARD, P., and WONG, M., Education et développement dans le Sud-Est de l'Asie, in *Colloque de Bruxelles; Institut de Sociologie*, 1966. Contains an important bibliography.

———, *La Médecine des Chinois*. Paris, 1967.

LAIGRET, J., L'Ecole Royale de Médecine de Ventiane. *World Med. J.*, 1964, **11**: 330-331.

MEYNARD, M., L'enseignement de la médecine au Cambodge. *Press Méd.*, 1956 64 (2nd pt.): 1079-1080.

NGUYÊN-HUU, et al., *Danh tu co-thé hoc* [Nomina anatomica]. Saigon, 1963.

NGUYÊN-HUU, and PHEN HUY TRUONG, *Nomina anatomica*. Saigon, 1965.

NGUYÊN VAN NHU, *La médicine Vietnamienne à la croisée des chemins*. Saigon, 1962. Thesis.

PHAM BIÊU TÂM, Medical education in Saigon. *J. Med. Educ.*, 1962, **37**: 956-961.

——— Viet-Nam du Sud. Principaux problèmes et leur solution actuelle. *World Med. J.*, 1964, **11**: 333-334.

PHAM BIÊU TÂM, NGUYEN DUC NGUYEN, *Notes pour le fonctionnement de la nouvelle Faculté de Médecine de l'Université de Saigon*. Saigon, 1965.

Répertoire mondial des écoles de médecine. Organisation Mondiale de la Santé. Geneva, 1964.

SAING-SOPHONN, La Faculté Royale de Médecine du Cambodge. *World Med. J.*, 1964, **11**: 328-329.

SMYTH, F. S., Health and medicine in Indonesia. *J. Med. Educ.*, 1963, **38**: 693-696.

TRÂN MINH TUNG, *Aspects juridiques de l'internat en médecine dans les hôpitaux de Saigon*. Saigon, 1956. Thesis.

THE HISTORY OF MEDICAL EDUCATION IN JAPAN: THE RISE OF WESTERN MEDICAL EDUCATION

JOHN Z. BOWERS
Josiah Macy, Jr. Foundation
New York, New York

When Commodore Matthew Perry arrived in Japan in 1853, an increasing number of Japanese physicians had been teaching and practicing Western medicine for more than seventy-five years. This was in sharp contrast to China where the physicians followed slavishly the traditional medicine and a handful of Western missionaries were the only practitioners of Western medicine. Both countries were essentially closed to the West. The foreign traders in China were restricted to a small island in the Pearl River at Canton. The Dutch, the only Western nation allowed to trade with Japan, were restricted to a small island, Dejima, in Nagasaki Bay. But since 1641 physicians serving in the Dutch company at Dejima had been teaching Western medicine to Japanese interpreters, physicians and students.

Four factors were significant in the rise of Western medicine and medical education in Japan. One was the island culture of Japan and the long tradition of turning outside for new knowledge. This attitude was rooted in the close cultural relationship with China and, during the T'ang Dynasty, 618-906 A.D., there was a massive flow of Chinese culture, including art, language, religion and medicine, into Japan.

A second factor was political. Beginning in 1610, the Japanese were ruled by a strong, unified, central government, and the country was able to react quickly and as a unit to external contacts.

A third factor that influenced the rise of Western medical education in Japan was the nature of its early contacts with the West. The Portuguese were the first Westerners encountered by the Japanese, and a major goal for the Lusitanians was peaceful conversion of the people to Christianity. The amiable nature of the relationship was established by the world's greatest missionary, St. Francis Xavier, who expressed his affection for the Japanese in one of his letters: "These are the best people so far that we have encountered and it seems to me that among unbelievers no people can be found to excel them" (1). When Xavier decided that his mission was not achieving the success that he desired, he obtained permission from Jesuit headquarters at Goa to launch a mission to China. His reasoning was interesting and was based on the history of the relationship between China and Japan. Xavier believed that if he could convert the Chinese to Christianity, the Japanese, who were their cultural heirs, would follow suit. But he died on Sanchan Island off the Chinese coast before he could enter the mainland.

In contrast to the religious theme of their relationships in Japan, the Portuguese were militant to the degree of barbarity in their relationships in China. Fired by a zeal bred by centuries of enmity toward Islam, they unleashed all of their loathing for pagans on the Chinese. They were eager for trade and had no hesitation whatsoever in killing, bombarding and pillaging to gain their ends.

A fourth factor favoring the rise of Western medicine in Japan was the keen interest of the Japanese in medicine, their deep faith in drugs, and a remarkable sensitivity to health which continues today. Probably the rugged climate, the frequent epidemics and repeated widespread famine were several of the factors that fostered these attitudes. During the two centuries of *Sakoku* (closed country), for many of the Japanese the sole reason for outside trade was the import of medicines. Today the sensitivity of the Japanese to health and their zeal for drugs at least equals that of citizens of the United States.

For many centuries Japanese medicine was essentially Chinese medicine which the Japanese called *Kampo* (literally, Han technique). There were Chinese influences in Japanese medicine as early as 200 B.C., but in the seventh century Japanese medicine became Chinese medicine. This coincided with the founding of the T'ang Dynasty which rapidly brought China to the position of the most powerful and the richest country in the world. In 602 A.D. Kwan Roku, who was a Buddhist priest-physician, came to Japan from Korea and taught Chinese medicine to thirty-four young men selected by the Imperial Court. In 608 A.D. twelve Japanese students were sent to China to study medicine.

Influenced in part by the great achievements of the T'ang Dynasty, in 702 A.D. an elaborate system of laws, the Taiho Code, was promulgated in Japan. One section of the Taiho Code called for the establishment of a university (*daigaku*), with a medical school in the capital city, then Nara; and for the establishment of colleges (*kokugaku*), with medical training, in each province. Seven years were established as the required length of training in internal medicine; for surgery and pediatrics, five years; and for ophthalmology and otorhinolaryngology, four years. However, because of incessant civil war, the Taiho program in medical education never became effective.

The first contact with the West was in 1542 through shipwrecked Portuguese sailors who were swept ashore in southwest Japan. In the Treaty of Tordesillas, promulgated in 1494, Pope Alexander VI divided the world in half, and Japan fell in the half allocated to Portugal. Francis Xavier arrived in Japan two years later to begin a period that Boxer has appropriately termed the "Christian Century" in Japan (2).

The Jesuits wrote frequent reports on their work and their observations in Japan. Some of the most detailed observations were recorded by Father Luis Frois, and from one of his letters we have a description of the universities in Japan in the sixteenth century:

When the universities of Japan are spoken of, it must not be imagined that they resembled the universities of Europe. Most of the students are *bonzes*, or study to become *bonzes*, and the principal end of their work is to learn the Chinese and Japanese characters. They endeavour also to master the teaching of the different sects (that is, their theology); some little astronomy, some little medicine; but in the method of teaching and learning there is nothing of the strict system which characterizes the schools of Europe. Furthermore, in Japan there is but one single University with a semblance of United Faculties; it is in the region of Bandou, in the place called Axicanga (3).

In another letter Father Frois gave an interesting comparison of diseases and medical practices between East and West:

1. Amongst us, scrofula, pain from stone, gout and bubonic plague are frequent things; all of these diseases are rare in Japan.
2. We use bleeding; the Japanese, "buttons of fire" with herbs.
3. The men amongst us are accustomed ordinarily to bleed from the arms; the Japanese with leeches or with a knife on the forehead, and the horses with a lancet.
4. We use clysters or syringes; they in no case use this remedy.
5. Amongst us the physicians prescribe through pharmacies; the Japanese physicians send the medicines from their houses.
6. Our physicians take the pulse of men and of women first on the right arm, afterwards on the left; the Japanese, for men, first on the left and for women, first on the right.
7. Our physicians look at the urine in order to have more information of the illness; the Japanese in no case look at it.
8. The flesh of Europeans through being delicate, heals very slowly; that of the Japanese, through being robust, heals much better and more quickly from severe wounds, abrasions, infections and accidents.
9. Amongst us wounds are sutured; the Japanese place on them a little adhesive paper.
10. We make all bandages with cloth; the Japanese make them with paper.
11. Amongst us all abscesses are burnt with fire; the Japanese will die before using our harsh surgical remedies.
12. If our sick are fasting, one works to force them to eat; the Japanese consider this cruel and if the sick are fasting they thus let them die.
13. Our sick are in cots or beds with sheets, mattresses and pillows; the Japanese upon a board on the floor with a block of wood and their kimono as cover.
14. In Europe one has hens and chickens as medicine for the sick; the Japanese consider this poisonous and order to give them fish and salted turnip.
15. We extract teeth with pincers, tooth forceps, pliers, etc.; the Japanese with

chisel and mallet, or with a bow and arrow tied to the tooth, or with blacksmith's tongs.
16. Our spices and medicines are ground in a mortar; in Japan they are ground in a boat-shaped dish of copper with a wheel of iron held between the hands.
17. Amongst us pearls are used for personal ornamentation; in Japan they serve for nothing more than to be ground to make medicines.
18. Amongst us, if a physician is not examined, there is a penalty and he cannot practice; in Japan, in order to make a living, whoever wants to can be a physician.
19. Amongst us for a man to become ill with a venereal disease is always a filthy and shameful thing; the Japanese men and women consider this a common thing and take no shame for it (4).

The first person to practice Western medicine in Japan was a Portuguese, Luis d'Almeida, who had been trained in surgery at the famous Todos-os-Santos Hospital established in 1498 in Lisbon (5). During his first eight years in Asia, Almeida sailed as a successful trader between Siam, Malacca, Sanchan Island and Japan. He then decided to use the substantial sums of money that he had accumulated for a nobler cause and affiliated with the Jesuits in Japan as a lay brother. Almeida's fortune was a principal source of support for the Jesuit missions in Japan. He established a few foundling homes and leprosaria, and instructed several Japanese in medicine and surgery. However, in 1582, the Jesuit Visitor-General, Alessandro Valignano, came to Japan and directed that the practice of medicine by the Jesuits should be restricted to the nobility and the samurai. Sensitive to the apprehensions of the military rulers about their presence, the Jesuits played a cool hand and avoided public attention.

The Spanish friars in Manila were eager to extend their missionary programs to Japan and, when Portugal was conquered in 1580, they became increasingly forceful in their demands. Finally, in 1593 Franciscan friars followed by other orders from Manila entered Japan and in contrast to the Jesuits openly proclaimed the virtues of Christianity. They placed greater emphasis on medical alms than the Jesuits and established hospitals in Nagasaki, Kyoto and Tohoku, the last several hundred miles north of Edo (now Tokyo). They instructed their Japanese assistants in the care of the sick poor, and soon there was open hostility between the Jesuits and the Spanish friars.

The missions of the Iberians were doomed to failure. Tokugawa Ieyasu succeeded in obtaining full military control of Japan in 1600 and in 1603 assumed the title of shogun. He thereby separated the power of the national military ruler seated in Edo from that of the mikado who henceforth was powerless in his magnificent palace in beautiful Kyoto.

Ieyasu determined to establish a completely feudal state, and he felt that this would be more readily achieved if he did not have zealous Christian missionaries stirring up his people and bickering with each other. Ieyasu and his successors were also concerned lest Portuguese or Spanish arms might join with rebellious Christian converts to seize the Empire. In 1638, after stern repressions culminating in mass crucifixions, the last Portuguese were expelled from Japan. Yet the shogun wished to maintain a "window looking out on the rest of the world" (6). His choice was limited to a single nation—The Netherlands. The Dutch had established a trading post on Hirado Island, northwest of Nagasaki in 1609, and, learning a lesson from the problems of the Spaniards and Portuguese, had sedulously avoided any signs or activities relating to Christianity. In 1641, three years after the departure of the last Portuguese, the Dutch were moved to Dejima, a small island in Nagasaki Bay, and here for more than two centuries they were the only contact between Japan and the West.

Dejima, encompassing an area of 600 by 120 feet, was connected to the mainland by a single short bridge and closely guarded day and night. The surrounding waters were patrolled, and even the gutters were serpentine to thwart any exchange of materials through those channels. The Netherlanders were permitted to leave the island on occasional passes, primarily to enjoy the pleasures of the nearby geisha quarter.

Each year they were required to make a *hofreis* to Edo to report to the shogun on the state of the world, and especially on the activities of the Portuguese. The military rulers of Japan held an abiding fear of the Portuguese and the possible reintroduction of Christianity. Therefore, they had decreed in 1630 that thirty-two religious and scientific books written in Chinese on Christianity and Western science could not be brought into Japan. This effectively cut the Japanese off from all contacts with books describing advances in Western science, since they could only read Chinese. The ban applied principally to the books of the great Italian Jesuit scholar, Matteo Ricci, who was in China from 1583 to 1610 and served as the astronomer and cartographer for the court in Peking.

The physician at the factory was normally a Dutchman, but there were also physicians and surgeons from Germany, Sweden and Poland. When the captain of the factory made the annual *hofreis* to Edo he was usually accompanied by the physician. Just ten years after the Dutch were "isolated" at Dejima, Caspar Schamberger, a German surgeon at the factory, 1649-1650, lectured on and demonstrated Western medicine and surgery in Edo. He was greeted with such enthusiasm and plied with so many questions that henceforth the physician always accompanied the captain on the *hofreis*.

The first certificate of satisfactory instruction in Western medicine was issued at Dejima twenty-five years after closure:

We, the undersigned, bear witness and attest as the truth that the Japanese named *Chōan*, servant of the lord of Hirado, has studied for a considerable time under the Dutch surgeons and is well instructed (so far as we can tell) in the art of Surgery. He is therefore well acquainted with the potency of Dutch medicines, of which he has given us sufficient practical proofs, and we hereby declare him to be an accomplished practitioner.—*Japan, in the agency of Nagasaki, this 21st January 1665.* [signed] Jacob Gruijs. 1665. Nicolaes de Roy, D. Busch, surgeon on the island of Deshima (7).

In 1690 Engelbert Kaempfer, a brilliant German physician-explorer from Westphalia, who had spent seven years studying the medicine, natural history and culture of Russia, Persia, Malabar and Java, came to Dejima. He gathered a massive documentation on Japan and his book *The History of Japan* stands today as the most complete and widely quoted description of the Empire during the Tokugawa Shogunate (8). Kaempfer's information on Japan was gathered in part through barter—he attributed his success to the fact that he was always prepared to exchange his advice as a physician, his medicines, and his plentiful supply of European liquors for information. Soon after his arrival at Dejima, Kaempfer began to tutor a young Japanese apprentice in physick and surgery and in reading and writing Dutch.

During the seventeenth century a number of schools primarily for Chinese studies were established in the fiefs. In the words of Dore: "The Chinese language was the royal road, and the only road, to all knowledge at the beginning of the period. The classics, the most instructive history, the most refined literature, the most authoritative works on military strategy were all written in Chinese. So, too, were books on medicine, astronomy, mathematics and law" (9). Chinese culture, Confucian teachings, calligraphy, military arts and a smattering of arithmetic were included in the curriculum.

Training in medicine was based exclusively on the study of Chinese traditional medicine, *Chung-i*, with practical instruction through anatomical models and clinical demonstrations. A garden for the cultivation of medicinal plants usually adjoined each of the schools, and the great Chinese botanical atlas, *Pen-t'sao-kang-mu*, was the major text for *materia medica* (10). The students at the Bakufu school in Edo were permitted to attend dismemberments at the execution grounds and also benefited from the fact that there was a hospital for clinical demonstrations attached to the medical school.

Two of the major barriers to the introduction of Western medicine were the lack of a common language and the ban on books. In the second quarter of the eighteenth century the first major breakthrough occurred during the rule of a wise and progressive shogun, Tokugawa Yoshimune. He felt it to be his personal responsibility to assure that his people had a correct calendar since so many events, from planting crops to weddings, were determined by the calendar. Yoshimune recognized that the official calendar was incorrect and his consultant advised him that it could only be revised by reference to Chinese books on Western astronomy that had been banned. Therefore, he decreed that books which did not actually expound Christian doctrine might be imported and studied, and eleven Chinese translations of European astronomical texts were specifically included in the new decree.

This was a first step but a far more significant development occurred in 1739 when Yoshimune ordered a court physician, Noro Genjō, to study Western medicine with the Dutch at Nagasaki, and the court librarian, Aoki Konyō, to study Dutch and compile a dictionary. After two years, Noro Genjō published a brief atlas on Dutch plants of medicinal value. Aoki Konyō, facing a far more difficult task, published a Dutch-Japanese dictionary in 1758.

The publications of Noro and Aoki heralded a mounting Japanese interest in Western knowledge. They were sanctioned by the shogun and this indicated the ruler's support for the study of Western medicine and Dutch. Further, Noro and Aoki were two of the most distinguished scholars in Japan. Admittedly, the books were primitive but the impressive point is that just one century after the country was sealed from the West the shogun and two of the country's leading scholars had acknowledged the importance of Western medicine.

The pivotal event in the rise of Western medical education in Japan was the publication in 1774 of a Japanese translation of a Dutch version of a German anatomical text. It marked a turning point not only for medicine but for the advancement of all aspects of Western knowledge in Japan.

For centuries the Japanese had shared the Chinese attitude of complete opposition to dissection of the human body. Writs were periodically issued to allow the dismemberment of a corpse but these were always performed by an *eta*, a pariah outcaste, who shouted out his findings to conform with the anatomical concepts of Chinese traditional medicine, *Chung-i*. It is true that these concepts included a heart, lungs and abdominal viscera, but there were numerous misconceptions. Seven tubes were believed to emanate from the heart including one to the liver, one to a kidney, and another to the spleen. The heart was connected with the small intestine. The lungs had six "leaves"; the liver four "leaves" on the right and three "leaves" on the left. There were three "warmers": one for the thorax, one for the diaphragm and a third for the abdomen. Situated between the kidneys was a structure called "the gate of life." In some Chinese atlases the heart was shown in the right thorax while in others in the midline; there was no knowledge of the circulation. It is no wonder that the Chinese concluded that Western man must be constructed abnormally when they first viewed Western anatomical plates.

The first significant development in the introduction of dissection and of Western anatomy into Japan occurred on 7 February 1754 when a Kyoto physician, Yamawaki Tōyō, attended a dissection and recorded the first effort at rational observation of human anatomy. It is probable that he had with him a copy of the *Syntagma Anatomicum*, a German anatomical text by Johann Vesling, which had been translated into Dutch by Ger Blaes and published at Leiden in 1661 (11).

Yamawaki Tōyō represents a study in contrasts. On the one hand, he was a leader of the second generation of the *Koho* school of medicine which advocated a return to the ancient teachings of Chinese medicine. Yet he could not accept the anatomical theories of Chinese medicine and persuaded a friend, Kosuki Genteki, physician to a powerful *daimyō,* to let him accompany him to a dismemberment. Yamawaki made a number of sketches at the dismemberment, and these were published in a two-volume atlas, *Zoshi* (1759) which, although it contained inaccuracies, was a landmark. It stimulated two young Edo physicians to undertake a more complete study of Western anatomy and to publish an epochal translation. In the words of one of them, Sugita Gempaku, "Yamawaki showed what was the true method of inquiry" (12).

The two pioneer physicians, Maeno Ryotaku (1723-1803) and Sugita Gempaku (1733-1817) were a perfect team. Sugita was a hard-driving, ambitious, questioning yet pragmatic physician; Maeno, on the other hand, was a scholar who was dedicated exclusively to the acquisition of knowledge. He had entered the field of medicine because it afforded the richest opportunities for the study of all aspects of Western knowledge.

Maeno spent his youth in the service of the lord of Nakatsu in Buzen, an "outer" *daimyō;* Sugita under the lord of Kohama, fief of Wakasa. They moved in due course to Edo; Sugita to study medicine and Maeno to study Dutch with Aoki Konyō. Maeno was determined to master as many languages as possible so that he would be able to read all books about Europe.

In 1767, Maeno took Sugita to visit the Dutch when they came to Edo on the *hofreis*. The same year Sugita copied some of the illustrations from Lorenz Heister's *Chirurgie* which was by all odds the most popular European surgical text of the period (13).

It was about 1770 when Maeno decided to go to Nagasaki for an intensive study of Dutch since both he and Sugita had by now become totally dedicated to the mastery of Western medicine. On the long journey to Nagasaki, Maeno evinced most strikingly his selfless dedication to scholarship when he stopped at Dazaifu and vowed to seek no worldly gain until he had completely mastered Dutch (14). Maeno spent one hundred days studying Dutch with the interpreters at Nagasaki and learned about six or seven hundred Dutch words. He acquired a simple Dutch dictionary and a Dutch translation of *Tafel Anatomia,* a concise German anatomical text and atlas (15).

While Maeno was studying at Nagasaki, Sugita had also obtained access to a copy of the anatomical treatise of Johann Kulmus which was owned by a colleague from Nakatsu, Nakagawa Jun'an. When Sugita compared the anatomical plates in Kulmus with Kampō, he noted a number of differences and sought an opportunity to gather first-hand information by observing a dismemberment. On 4 March 1771 Sugita, Maeno and Nakagawa took the copies of Kulmus to the execution ground in Edo where they observed a dismemberment. They soon realized that the structure and distribution of organs agreed with Kulmus and not with the charts of Kampō. As they left the dismemberment, they determined to translate Kulmus and to reproduce the plates.

The translation was a heroic undertaking; after one year they were able to translate about ten lines a day and the manuscript was rewritten eleven times in four years. Maeno, with his knowledge of six to seven hundred Dutch words, was the real leader of the team. In 1773 he returned to Nagasaki and obtained a translation into Japanese of a Dutch and French dictionary by Peter Marin as well as several Western medical texts.

As they approached the end of their task, Sugita and Maeno became apprehensive that the publication of a reproduction of a Western book might bring confiscation by the *bakufu*. They recalled that in 1767 a small monograph about Holland had met such a fate. Therefore, as a trial balloon, they published in 1773 a preliminary report of just five pages, *Kaitai Yakuzu*, (Short Atlas of New Anatomy). When this met with no untoward reaction, they completed the translation and, as an added safeguard, presented copies to the *bakufu* and to the mikado's court in Kyoto. Again, as a sign of the rapidly rising acceptance of western knowledge, there were no hostile reactions. In August 1774, an epochal and heroic event in the history of medical education was accomplished with the publication of the *Kaitai Shinsho* (New Book of Anatomy), in five volumes with Sugita Gempaku as the senior author.

Maeno remained steadfast to the vow that he had taken at Dazaifu and refused to have his name listed as one of the authors of the translation. There could be no recognition until he had completely mastered Dutch.

The translation was made in classical Chinese which Sugita and Maeno hoped would bring to the attention of physicians and scholars in China the fact that there were progressive scholars among their island neighbors. Yet the use of the classical Chinese language made the *Kaitai Shinsho* of limited value to the average practitioner of medicine in Japan. Further, the reproductions from wood block prints lessened the accuracy desirable in an anatomical atlas.

Professor Charles R. Boxer, to whom I am deeply indebted for scholarly guidance and encouragement in my studies, has pointed out that the *Kaitai Shinsho* includes reproductions from Valverde's *Vivae imagines partium corporis humani aeris formis expressae* (Antwerp, 1566), and from Bidloo's *Ontleding des menschelycken lichaams* (Amsterdam, 1960) (16). And he notes that these works go back to the famous Flemish anatomist, Vesalius.

In the words of the great historian of Japan, Sir George Sansom, the *Kaitai Shinsho* was "the first European work to be printed and published by Japanese in Japan and therefore a landmark" (17). It was the turning point in the rise of Western medicine.

One year after the publication of *Kaitai Shinsho*, a young Swedish botanist-physician, Carl Pieter Thunberg, came to Dejima as the factory physician. Thunberg, a graduate of Uppsala, was a favorite disciple of Linnaeus and was persuaded by friends of Linnaeus in Amsterdam to collect botanical specimens in Japan. After a period in Paris, Thunberg sailed for the Cape Colony for the sole purpose of learning Dutch so that his work in Japan would be meaningful. But influenced by the opportunities for pioneer botanical studies at the Cape, he remained for three years and became the first authority on Cape botany (18).

Thunberg, in the mold of Engelbert Kaempfer, was a scientific explorer, a physician and a botanist. He was repeatedly impressed with the interest of the Japanese in the study of Western medicine and botany. In his diaries he recorded that at Edo he was surrounded day and night by enthusiastic and intelligent students whose incessant questions ranged over medicine, natural history, botany and rural economy. The nature of their questions indicated to Thunberg that they had acquired a reasonable knowledge of Western medicine and natural history. Two of the young court physicians, Katsuragawa Hoshū and Nakagawa Junan, came daily to study botany and medicine and carried as their texts Johnston's *Historia Naturalis*, Heister's *Chirurgie*, and Dodoens' *Herbal*.

When Thunberg returned to Uppsala he was able to carry on an active correspondence and exchange of specimens with Katsuragawa and Nakagawa. Their letters were written in Dutch and show that they had sufficient command of the language to communicate effectively. Thirty-two letters from his young Japanese students to Thunberg are in the University Archives at Uppsala, and the ease with which scientific materials flowed between Japan and Sweden is clearly shown in one of the letters:

Mijn Heer Carel Pieter Thunberg
    ik Bedank U Ed. Onderwijst kruijd en aptheek kunsten welk weetenschappen zijn voor leeden jaar, ook bedank ik ben ontfangen naar docter J. Hoffman U. Ed. gezonde drie pees boeken. Met kijn heer diensaar Sijemon zende ik nu aan Ued 100 maar andere eenige zaad von planten en weenige gedroogde bladen ik zal volgende zenden aankomende jaar indien voor U. Ed. Japaesche boeken schrijt op briefge wat U Ed. gelieven zal. ik verzoek twee groot woorden boeken namentlijk voor hollandse en latijnische van Pr. Marin, ook nieuwe verbeterde aptheek boeken. Deselfs prijzen voor U Ed. besoekt't scrijt op briefe
        Heer Dr. Hossie bedankt aan den Mijn Heer geteuijgenise
        Uw Ed. Dienaar
        Na. Zjunnan
        (Nakagawa Junnun) circa 1780

My dear Carel Pieter Thunberg
    I thank you who teaches herb and apothecary arts which sciences are from last year, also I thank you since I have received three copies of books you sent for Dr. J. Hoffman. With my lord servant Sijemon I will now send you 100 various seeds of plants and a few dried leaves. Following this I shall send you next year Japanese books if you write a letter saying what you want. I request two large dictionaries namely for Dutch and for Latin by Prof. Marin and also new improved apothecary books. I request you to write the prices of the above in a letter.

> Dr. Hossie thanks my Lords for his testimonies
> Your servant
> Na. Zjunnan
> Edo, 11 March 1777 (Nakagawa Junan) circa 1780

Another evidence of the penetration of Western medicine into Japan during the last quarter of the eighteenth century may be found in the records of Dejima's greatest *opperhoofd,* Isaac Titsingh (1745-1812). By Japanese decree, the *opperhoofd* was not allowed to remain at Dejima for longer than one year, but he could return to the post at the end of the year's absence. Titsingh served at Dejima between 1779 and 1785: "During my residence in Japan, several persons of quality at Yedo, Miyako, and Osaka applied themselves assiduously to the acquisition of our language and the reading of our books" (19). He was especially impressed with the progress in the study of Dutch by Thunberg's favorite pupils: Katsuragawa and Nakagawa. They corresponded with him as they had with Thunberg, and Titsingh returned their letters with corrections of their Dutch. The Governor directed that this correspondence should pass unopened as a sign of his support for Western learning.

In the library of the famed Kyoto National University, the letters written to Isaac Titsingh by his Japanese correspondents have been deposited and a reproduction of sections of two of these clearly demonstrates the facility which they had acquired in writing Dutch (Figures 29-32).

The publication of *Kaitai Shinsho* opened an era designated as *Rangaku* (Dutch study) in which there was a rapid proliferation of the study of Dutch and Western medicine. The center for the pursuit of Western learning now shifted from Nagasaki to Edo where zealous students of Western medicine, natural history and Dutch flourished. The leader of *Rangaku,* Ōtsuki Gentaku, was the son of a Sendai physician who studied under Maeno Ryōtaku and in 1785 went to Nagāsaki for further Dutch study (20). He returned to Edo the following year and opened a school, Shirandō, for the study of Western medicine and Dutch. This was the first school to teach Western medicine in Japan and the total number of students trained in Western medicine by Ōtsuki has been estimated to have been as high as ninety. One of Ōtsuki's students, Hashimoto Sokichi, at the turn of the century opened Shikandō in Osaka, the first school to teach Western medicine in the Kansai area.

In addition to his outstanding ability as a teacher, Ōtsuki Gentaku was as well a prolific author and translator. In 1788, he published *Rangaku Kaitei* (Ladder to the Dutch Studies), which was the first book dealing exclusively with a European language written in Japanese by a Japanese and published in Japan.

The translation of Western medical texts from Dutch sources was a principal activity of the Rangakusha. Professor Ranzaburō Ōtori lists seventeen books on basic medical sciences and thirty books on clinical medicine translated between 1772 and 1866 (21). Some of the leading texts were: *Medici-*

nae Compendium of Johannes de Gorter; *Chirurgie* of Lorenz Heister; *Enchiridon Medicum* of Christoph W. Hufeland; and *Natuurkundig Handboek voor Leerlingen in den Heelen Geneeskunde,* a Dutch translation of a handbook of physic for medical students by J. N. Isfording.

Thus we see that more than seventy-five years before the opening of Japan there were Japanese scholars teaching Western medicine and translating Western medical texts, and Japanese students and physicians were corresponding on scientific subjects with Western scholars with whom they had studied when the Dutch came to Edo on the *hofreis*.

From time to time in history a concatenation of events brings the right man to the right place at the right time. In this study the shining example of such a confluence occurred on 8 August 1823 when the "Three Sisters" sailed into Nagasaki Bay for the annual trade, carrying as supercargo the new *opperhoofd,* Colonel der Steuler, and the new physician for Dejima. The latter was a twenty-seven year old German from the Bavarian city of Würzburg, Philipp Franz Balthasar von Siebold, descendant of a distinguished medical family. His grandfather, Karl Kaspar von Siebold of Würzburg, was renowned as one of Germany's leading surgeons and a man who had been instrumental in moving surgery from the bloody practices of the barbers to academic rank. His principal interest was in obstetrics and he developed the pelvic curve on the forceps. Philip's father was professor of physiology at Würzburg and died just two years after his son's birth on 16 February 1796.

The medical faculty at Würzburg was based on the famed Julius-Spital, Germany's first teaching hospital (22). A wise and noble Prince Bishop of the Duchy, Julius Echter von Mispelbrunn, had directed the establishment of the hospital for use in the religious wars that pitted Roman Catholics against Protestants, and he dedicated it on 12 March 1576. The Bishop had an eye on the financial stability of the institution and decreed that all citizens of the duchy must donate an annual tithe of their harvests of agricultural products, timber and grapes to support the hospital (23). With such an endowment the Julius-Spital prospered and was the first in Germany to erect an anatomical theatre.

Philipp von Siebold completed his medical studies in 1820 and, as was customary, spent two years in general medical practice to gain practical clinical skills. At the end of this period instead of returning to Würzburg for further studies he obtained an appointment as court physician at The Hague. However, the drowsy court life held no attractions for an adventurer, and on 11 July 1822 he was commissioned with the rank of surgeon-major in the East Indies Army of the Netherlands. He sailed to Batavia where he was selected for the post of surgeon at Dejima with a special charge to promote the scientific and educational image of the Dutch and to gather all available information on every aspect of Japanese culture.

With Raffles' seizure of the fortress at Batavia and Napoleon's conquest of the Low Countries, the status of the Dutch in Japan declined. For several years there was no trade between Java and Nagasaki. The Dutch felt that their status might be restored by a factory physician with the scientific abilities of Thunberg who had deeply impressed the Japanese. It was also clear that the time when Japan would be opened to the West by treaty or by force was approaching and the Dutch wished to continue as the primary Western repository of knowledge of Japan.

On his arrival, Siebold skirted a disaster that could have resulted in his prompt return under guard to Java. In the earlier years the guards at the water gate of Dejima had not been able to differentiate the Dutch language from German but now they could do so. When they interrogated Siebold they recognized that he was not speaking Dutch and barred his entry. However, the *opperhoofd* quickly announced that Siebold was "*Orandayama*", a Dutch mountaineer, and he was allowed to enter.

Siebold found a band of students waiting for the arrival of the new physician, and he immediately began to teach Western medicine and natural history and to collect scientific specimens and information on Japanese culture. The number of students increased rapidly and classes were transferred to the city of Nagasaki to the home of the physician-interpreter, Narabayashi. Thus Siebold became the first foreigner to teach Western medicine systematically on the Japanese mainland. Attracted by his accessibility in Nagasaki, as well as his outstanding abilities as a teacher, eager Japanese students came to Nagasaki from across the empire.

Siebold's teaching program included lectures and demonstrations in medicine and natural history. He relied heavily on practical clinical demonstrations in which he introduced his students to the diagnostic and therapeutic methods of Western medicine such as thoracentesis and paracentesis, percussion, and pelvic examination. He taught the use of the truss in the treatment of hernia, surgical procedures for the repair of harelip and hydrocele and the removal of breast tumors. Siebold performed the first cataract surgery in Japan after experiments on a pig and a fish, and demonstrated the use of belladonna in dilation of the pupil.

Siebold was a uniquely resourceful scholar. The students were not required to pay tuition and in the long-standing Japanese tradition brought many gifts to their teacher. It was not long before the gifts became botanical and zoological specimens as well as Japanese art works; the students soon recognized Siebold's primary interests. As another evidence of his resourcefulness, he required every student to prepare a dissertation in Dutch on an assigned topic. The topics were chosen to bring new information to Siebold's storehouse of knowledge on Japan; as examples: "On the Whales Produced in the Kyūshū Area," "How to Make Salt", "A Method of Cultivating Tea Trees and Making Tea in Japan", "A Study on the Ancient History of Japan", "The Acupuncture Method among the Chinese", and "On the Causes of Measles and Smallpox in Children" (24). With characteristic

Teutonic thoroughness, Siebold read every word of every dissertation, and all mistakes were corrected with a red pencil.

After a year, the flow of physicians and students, the former often accompanied by their patients, was so great that Siebold, with the cooperation of his friend and leading student, Shige Dennoshin, and Narabayashi determined to gain access to property in Nagasaki. Foreigners could not purchase land, but a lovely hillside house and spacious grounds in the Narutaki section of Nagasaki were acquired under title to Shige for Siebold's use. Siebold transferred all of his activities—as a teacher, as a collector and as a physician to the house at Narutaki.

The house soon became his clinic, lecture hall and demonstration room as well as the museum for his collection.

Two years later the house at Narutaki became Siebold's bridal suite. He had been captivated by a beautiful geisha, Kusamoto Taki, whom he had met at the home of one of his patients. A year later a daughter, O-Inc, was born, who subsequently became a midwife and attended the court.

Philipp Siebold made the *hofreis* to Edo in 1826; it was his most memorable opportunity to collect information on Japan. His rigorously scheduled days were frequently interrupted by his former students: "At the crack of dawn, my pupils and other physicians of the neighborhood came with their sick and asked for advice and assistance. The disorders were, as usual, chronic, neglected and unhealable diseases, and the inconvenient consultations cost much time and patience. I did it all for my pupils' sake" (25).

In addition to patients some of the students brought their dissertations to fulfill an agreement with Siebold when they were awarded the diploma at Narutaki.

At Edo, Siebold found the company of the Rangakusha led by Ōtsuki Gentaku more stimulating than the interpreters at Nagasaki. Indeed, Siebold was so carried away with his opportunities to teach and collect in Edo that he overplayed his hand. He presented a map showing the Dutch colonies and a copy of "A Voyage Around the World" by Krusenstern to the court astronomer, Takahashi Sakasaemon. In return Takahashi permitted Siebold to copy a highly guarded prize, the Imperial secret map of Japan, which was of immense value to any potential invader because of its detailed information on Japanese topography, including the shoreline and harbors.

Siebold returned to Nagasaki delighted with his accomplishments on the *hofreis*. In addition to teaching medicine and natural history he studied and sorted out the vast amount of information in his diaries on practically every foot of country that he had traversed on the journey. He planned to return to Europe in 1828, but in the fall of that year the Nagasaki authorities were informed by jealous neo-Confucianists that Siebold possessed a copy of the Imperial secret map. After one year of imprisonment on Dejima he was released to return to the Netherlands—banished from Japan for life.

Siebold bequeathed to Japan a talented group of physicians whom he had trained in Western medicine, as well as young scholars with more catholic scientific interests. Two of his students, Ito Genboku and Totsukā Seikai, were the first physicians trained in Western medicine to be officially designated to serve the shogun's court. This occurred in 1858 and, five years later, Ito Genboku was designated as a director of the new medical school, Igaku-jo in Edo, the forerunner of Tokyo Imperial University Faculty of Medicine. Three of Siebold's students became Japan's leading surgeons; one continued to perform major surgery after one of his arms had been amputated! In other fields, Kozeki Sanei was astronomer for the shogunate and Iwasaki Kanyen was a distinguished botanist.

Siebold's most versatile, most talented—and most turbulent—student has been described as "probably the most accomplished Dutch scholar of his day in Japan" (26). An intensely dedicated and idealistic scholar, Takano Chōei studied medicine, natural history and Dutch with Siebold and then, as physician to the *daimyo* of Hirado Island, translated Western medical and obstetrical texts and wrote a number of treatises including *Comparison of Chinese and Western Anatomies, Autopsy of the Human Body,* and *Essentials of Medical Science.* At the height of the national reactions stirred by the neo-Confucianists against Western learning that had been precipitated by the Siebold case, Takano Chōei fled Hirado and hid at Satsuma fief on Kyūshū. As the Siebold storm abated he returned to Edo where he was welcomed warmly by the Rangakusha and continued the study of Dutch and Western medicine. He also continued to be a prolific writer, and in one of his books boldly advocated that Japanese ports should be open to Western commerce. This was too much for the *bakufu* and he was imprisoned for five years. He escaped, translated a book on military strategy from Dutch into Japanese, and the quality of the translation was so good that the police who believed him dead realized that Takano Choei was alive. He was arrested and died ignominiously in Edo on 30 October 1850. As a final flourish he endeavored to commit *seppuku* but missed the vital spot and bled to death in the mud.

Philipp von Siebold continued to lead a life of productive scholarship on his return to Europe. He settled at Leyden where he sponsored the founding of the national ethnographic museum and his hundreds of items from Japan were the major collection in the museum. Near Leyden he developed a large botanical garden in which he cultivated many Japanese botanical specimens including the magnolia, peony and lily. To the garden came many foreign dignitaries from The Hague. The first was the Tzarevitch Alexander, who later became Alexander II of Imperial Russia and played a significant role in Siebold's later career.

One of Siebold's major occupations was the classification of his massive Japanese collections and the preparation of books based on his Japanese materials. His personal observations and activities were published in *Nippon,* the first volume of which appeared in 1832 (27).

In the same year the first volume of *Fauna Japonica* (28) appeared and, between 1835 and 1841, *Flora Japonica* (29).

Siebold's life never lacked for adventure. Even though he had been expelled from Japan, his thoughts constantly turned back to the Empire. He considered the opening of the country under the most favorable and peaceful circumstances for the Japanese to be his personal responsibility. In 1852, the U. S. Departments of State and of the Navy determined to launch an expedition to open Japan. They naturally turned to The Hague for maps, navigational charts and other information on Japan. The first packet of materials included copies of the maps from Siebold's *Nippon* (30).

News of the proposed Perry expedition reached Siebold and he informed the U. S. Chargé d'Affaires, George Folsom, that he was prepared to join the expedition. Folsom reported to Washington: "I am authorized to make a tender of his services to you. He consents to this with the hope that, if his offer should be accepted, he may be able to contribute to the success of an enterprise which he regards as one of the most important that can employ the energy of a great nation" (31).

Meanwhile, the Russians had decided to send an expedition to Japan, and in January 1853 Siebold received an invitation from the Russian Foreign Minister, Count Nesselrode, to advise the Tsar's court on Oriental problems. Siebold, his vanity touched, accepted the invitation to St. Petersburg not knowing that by so doing he would destroy any possibility of becoming a member of the Perry expedition.

According to the narrative of the expedition, Perry refused Siebold: "on personal grounds since from information received from abroad he suspected Von Siebold of being a Russian spy and he knew that he had been banished" (32). Yet it might have been calamitous if Siebold and Perry had sailed together; they were too much alike—imperious, arrogant, demanding and intolerant.

After repeated efforts, Siebold returned to Japan in 1859 with the consent of the government, but his personal victory was doomed to end again in personal tragedy. For a time he taught and practiced medicine at Nagasaki and was summoned to Edo in the role that he had dreamed of—as advisor to the *bakufu* on relations with the Western powers seeking treaties with Japan. But he was caught in the crossfire between the European nations and the Japanese when he tried to perform the impossible task that he had desired as intermediary, and at the insistence of the Dutch Ambassador was forced to leave Japan. Siebold returned to Munich where he established a second exhibition of Japanese culture and where he died on 18 October 1866.

Today at Narutaki, still maintained as a memorial to Siebold, an inscription on the gate reads: "Siebold is the one who deserves the glory for the great achievement of having introduced knowledge to the Japan of today." His tomb in Munich is inscribed with Japanese characters that read: "such a strong bridge". This describes admirably his singular role as the outstanding person in the two-way flow of information between Japan and the West during the two and a half centuries of the Tokugawa shogunate.

Beginning in the last half of the eighteenth century, "private" schools for teaching Western medicine were established in Edo, Osaka and Kyoto. They were "private" in the sense that they were established by a physician and financed from student fees and the physician's practice. Thirteen private schools were established between 1765 and 1850; twelve of them taught Western medicine.

Shōsendō, established in Edo by one of Siebold's students, Itō Genboku, was one of the most popular of the private medical schools. At Shōsendō there were always a few hundred students selected from the various Han, for Ito was one of the three leading teachers of Dutch, and his students came to study Dutch as well as medicine. Ogata Kōan from Osaka, Aoki Kenzo from Choshu, and Oishi Ryohei from Saga were the leading students, whom Itō designated as Snow, Moon and Flower.

The rules and regulations established by Itō were strict and emphasized the importance attached to the study of Dutch:

Juku Rules and Regulations:

1. Reading any books other than Dutch books and their translations is prohibited.
2. Do not drink or chit-chat.
3. Students may go outside five (5) times a month, but are absolutely prohibited from going out more than this. If you have to stay out or are late under any circumstances, bring back with you an explanatory letter with the *In* [seal] of your sponsor.
4. When going to the bath or the *Kamiyui* [men's hairdresser], ask permission by submitting your card and pick it up without delay when you come back.
5. Submit your card to the office by 8 o'clock every morning and pick it up at 10 o'clock every night.

Those who violate these regulations will be ordered 20 days curfew and night duty at the pharmacy room, and those who violate repeatedly will be ordered to leave this Juku.

Entrance Ceremony: [Paid to]

1. Entrance fee Y 50.00 Seisei [Itō Genboku]
   A fan in a box
2. Expense for paper Y 25.00 Okugata [Mrs. Itō]
   handkerchiefs Waka Sensei [Younger
3. Y 12.50 teacher]—Heir-
   presumptive to the Sensei—
4. Y 12.50 Jukuto [Head of the Juku]
5. Y 50.00 Jukuchu [all students]
6. Y 12.50 Boku [Helpers]
At the time when you enter Juku.

Shōsendō Steward (33)

The most outstanding Japanese teacher of Western knowledge during the Tokugawa shogunate period, Ogata Kōan, is revered in Japan today as the perfect exemplar of scholarship, virtue and teacher. After studying medicine, natural history and Dutch at Osaka, Nagasaki and Edo, at the age of twenty-eight Ogata opened a school, Tekiteki-sai Juku, in Osaka. Ogata taught a remarkably dedicated group of brilliant, progressive students in medicine, natural history, chemistry, physics and Dutch. The most illustrious of them was Fukuzawa Yukichi, a leader in the Restoration movement and the founder of one of Japan's most progressive institutions of higher learning, Keio University. In his autobiography, Fukuzawa paints a vivid picture of how Ogata Kōan's students studied around the clock stirred by the example of their beloved teacher (34). Ogata encouraged competition among the students by conducting classes in which they openly interrogated each other; an unusual program for Japan even today. Each student was assigned a single *tatami* mat, 6 feet by 3 feet; it served as his study hall, his locker and his dormitory room. To encourage competition, the location of a student's mat in the hall was determined by class rank. The low-ranking students were required to live by the entry where they might be stepped on during the night, or by the wall where the light was so poor that artificial illumination was necessary even in the daytime. There was only one copy of a Dutch-Japanese dictionary and its importance has been described by Nagayo Sensai:

As more than a hundred students in the entire Juku had to rely on this single set of the dictionary, the room was constantly occupied by the students, who came in and out in turn and used the dictionary, pulling it left and right, front and back, and it was almost impossible even to touch it . . . there were many who went to use it in the middle of the night when there were not many students about in the room. Thus the light in the dictionary room was on throughout the night all year around (34).

The study of Dutch had now become of such importance that feudal lords paid the students a substantial fee to copy pages of the Doeff-Halma dictionary for their use.

Ogata Kōan did not restrict his activities to teaching. He translated several Western medical texts into Japanese. One of these, *Byōgaku Tsūron*, 1847-1849, was the first Japanese text on general pathology and was drawn from abstracts of the writings of European physicians, principally Christian Wilhelm Hufeland. Another translation by Ogata Kōan, *Fushi keikun ikun*, published in 1843, was taken from the section entitled "Praxis" in *Enchiridion medicum* by Hufeland. The Dutch source was entitled *Enchiridion medicum. Handleiding tot de geneeskundige praktijk* by H. H. Hageman, Amsterdam, 1838.

Ogata Kōan was called to Edo in 1862 to take the position of Director of the Institute for Western Medicine. He had suffered from tuberculosis since his youth, and died from a massive pulmonary hemorrhage just one year after moving to Edo.

One of the most exciting and yet little known achievements in the advance of medical education in Japan was the establishment of the first international medical school at Nagasaki in 1857. It was the handiwork of a remarkable pair of young physicians: one a Hollander, Johannes Lydius Catherinus Pompe van Meerdervoort, the other a Japanese, Matsumoto Ryōjun.

In 1853, the shogun asked the Dutch for help in the establishment of a naval military school at Nagasaki. A frigate, the *Soembing*, and a detachment of naval officers to staff the new school sailed to Nagasaki. One year later the Japanese sent a request to The Hague for a second ship and the services of a Dutch physician to teach medicine. The captain of the second detachment, W. J. C. Ridder Huyssen van Kattendyke, chose Pompe van Meerdervoort, a twenty-eight-year-old graduate of the military medical school in Utrecht, to be the medical officer—it was a brilliant selection.

Pompe van Meerdervoort was born in Brugge, 5 May 1829, just one year before the separation of Belgium from the Netherlands. When he completed his medical studies in 1849 he was commissioned as "Army Surgeon, third class for marine duty". He was assigned to the forces in the East Indies, returned to the Netherlands in the summer of 1855, and an entry in his service record, dated 1 February 1857, is for an assignment to the "propellor ship, *Japan*", for passage to the Dutch East Indies (56).

Shortly after his arrival at Nagasaki on 22 September 1857, Pompe was informed by the Netherlands Commissioner, Donker Curtius, that the Japanese government wished to have students instructed in medicine and surgery and that Pompe should make whatever arrangements were necessary. He was also advised that this request was based upon the fact that the shogunate now fully accepted the superiority of Western medicine over *Kampo*.

From the beginning, Pompe van Meerdervoort was determined to develop a program of medical education at Nagasaki patterned after the best in Europe. He never wavered from that determination. With no opportunity to prepare a program, no resources and no Western associates, Pompe could have limited his classes to a few lectures and demonstrations on Western diagnostic and therapeutic methods. Instead he determined from the outset to establish a full five-year program beginning with the premedical sciences. His accomplishment is all the more remarkable because it was made against great difficulties.

Shortly after Pompe's arrival he was joined by a twenty-five-year-old physician to the shogun's court in Edo, Matsumoto Ryōjun. Matsumoto was a practitioner of *Kampō* but had also studied Western medicine with his distinguished father, Satō Taizen. When Matsumoto learned of the plan to teach Western medicine at Nagasaki, he obtained permission from the court to join Pompe both as a student and as an assistant.

The first class was convened on 15 November 1857. Communication was a major problem since the twelve students that Matsumoto had assembled knew very little Dutch and Pompe knew very little Japanese. A complex but effective procedure was developed. Pompe lectured in Dutch with an interpreter beside him who translated orally into Japanese. Matsumoto copied the translation and this translation was then made available to the students. Pompe recognized that the interpreter made many errors in medical terms and arranged for an instructor from the Naval School to teach Dutch to the medical students. He also prepared summaries of his lectures which were translated into Japanese and distributed to the students before the lecture. Two-and-a-half years after the opening of the school, the complicated lecture system which was still in use was described in detail by a visitor from Hong Kong (37).

Despite the language problem, the number of students rose rapidly and, just seven months after the opening of the school, there were twenty-three studying medicine and surgery, and twenty, the natural sciences: physics, chemistry, geology and mineralogy.

The intense dedication of the students to their studies was Pompe's greatest satisfaction. In the classes in physics, their ambition, in Pompe's words, was "unbounded. Seldom do I see anyone of them who is not all attention" (38). A singular accolade was given to Pompe in 1859 when Ogata Kōan sent his son, Ogata Heizo, to study medicine with him at Nagasaki.

The serious lack of teaching materials was a major problem for Pompe and the greatest drawback was his inability to obtain cadavers for dissection. He literally bombarded the officials with requests for a cadaver but, although hundreds of autopsies had been performed since the publication of *Kaitai Shinsho*, the officials were hesitant to permit a foreigner to dissect a body. For two years he relied on plates and *papier mâché* models from Paris for anatomical illustrations. Finally, on 2 September 1859, excited students reported to him in great secrecy that an execution was to take place in six days and that the body would be turned over to him for dissection. Twenty-one medical students and twenty-four physicians attended the dissection which began on 9 September 1859. Pompe dissected until noontime, and then turned the cadaver over to his students. These continued to dissect and study the cadaver into the evening, and when darkness fell their enthusiasm was so great that they went to the residence of the Governor and obtained permission to retain the body and to continue their dissection on the following day.

A second body was turned over to Pompe two months later, and of the sixty medical students and physicians who attended the dissection one was a woman who was keenly interested and assisted Pompe in the dissection; her name was Ine, the daughter of Philipp von Siebold.

From the first year of the medical school's program at Nagasaki, Pompe had campaigned for the construction of a teaching hospital and in June 1858, he prepared a detailed memorandum on the design of such a building.

A copy of this memorandum has been included in the detailed and extremely valuable study of Pompe van Meerdervoort by Professor Numata Jiro of the Department of Historiography at Tokyo National University. Pompe placed great emphasis on the importance of fresh air, a hilltop location, spacious wards and a completely adequate supply of clear water—he even recommended that all windows be kept open twelve months a year and that a large window be cut in the ceiling of each ward. Approval for the hospital was supported by his invaluable service as a physician and public health advisor during a severe cholera epidemic which invaded Nagasaki from China in the summer of 1858. Pompe attributed the epidemic to the arrival of the *U.S.S. Mississippi,* but an investigation of the records of the *Mississippi* and the descriptions of that cruise give no suggestion of cholera. It is probable that it was introduced by one of the numerous unregistered ships that sailed constantly between China and Nagasaki.

Pompe's hospital, Yōjōsho, was opened on 21 September 1861, with one hundred and twenty beds, the flag of the Rising Sun flying from one end of the roof and the Dutch tricolor from the other. Adjoining the hospital was a new medical school building which included a lecture hall, demonstration rooms and a dormitory for the students.

To a backbreaking schedule as the one-man faculty of the medical school, Pompe now added the responsibilities of the chief and only member of the staff of Yōjōsho. He climbed the hill from Dejima to Yōjōsho early each morning to start his ward rounds at eight o'clock and to assign clinical tasks to the students. He lectured from four to six hours each day and supervised the preparation of all medicines used in the hospital. At the end of the day he visited the sick in their homes, accompanied by Matsumoto and several students. The evenings at Dejima were spent in preparing lectures for the following day.

The full five-year curriculum that Pompe developed was exemplary. The students began their studies with the premedical subjects—physics, biology and chemistry. These were followed by instruction in anatomy, histology, physiology, pathology, pharmacology and therapeutics, general medicine, surgery and ophthalmology.

In 1862, Pompe requested orders to return to the Netherlands. His successor, Antonius F. Bauduin, arrived at Nagasaki in September of that year, and Pompe had the deep satisfaction of awarding diplomas to sixty-one students. During his five years in Japan he taught a total of one hundred and fifty students.

It is surprising that despite his dedication as a teacher in Japan Pompe made no effort to continue that role in the Netherlands. Instead he entered the practice of medicine, including obstetrics, in The Hague. He was active in the International Red Cross and served as advisor to Japanese students who came to The Netherlands to study medicine, science and engineering. Pompe van Meerdervoort died in September 1908, aged eighty years, at Bergen op Zoom in The Netherlands.

Today who would dream of sending a twenty-eight year old medical officer with no previous experience as a teacher to establish a medical school. Pompe van Meerdervoort's devotion to excellence, his unique capacity for hard work and his ability as a teacher should bring him a noble memorial in the history of medical education. There is no instance in history in which a medical school was established under such adverse circumstances—but with such outstanding accomplishments.

In 1868 the shogunate collapsed; the Emperor Meiji made a symbolic march from Kyoto to his new capital, Tokyo, and the role of the Emperor was restored to a position of true Imperial function. Japan was rapidly launched on a course to adopt the programs of selected Western nations that seemed best suited to bring her rapidly to a leading position in the world community of nations. In the Charter Oath of Five Articles proclaimed at Kyoto, 6 April 1868, the young Emperor Meiji enunciated the Imperial attitude: "Wisdom and knowledge shall be sought after in all parts of the world to establish firmly the foundations of the empire" (39).

In 1858 a group of physicians had established an Institute for Vaccination in Tokyo; in 1860 the shogunate afforded partial subsidy for the program and the following year the name was changed to Institute for Western Medicine. In 1869 the school was designated as the East College of the University under the administration of Dr. Iwasa Jun and Dr. Sagara Chian.

As the Japanese began to turn their attention toward the determination of those foreign legal, governmental and educational systems that might best be adopted for the Empire, the question arose as to the system of medical education to be adopted. There was just one choice, the German system. In the second quarter of the nineteenth century, splendid research institutes in the basic sciences opened at university centers such as Berlin, Bonn, Leipzig and Königsberg staffed by scholars dedicated to fundamental research. In contrast to the dominance of the clinical departments in the British and French patterns of medical education, the German medical schools focused on histology, embryology, physiology and pathology to place medicine on a firm scientific basis.

Another factor that influenced the Japanese to adopt the German system was the memory of Philipp von Siebold, enshrined as the greatest scholar in the history of Japan.

Moreover, a majority of the medical books that had entered Japan through Dejima were German texts that had been translated into Dutch and this was not lost on the Japanese.

An interesting figure in the decision on medicine was an American missionary of Dutch descent, the Reverend Guido Fridolin Verbeck. Verbeck had opened a mission school in Nagasaki and taught several of the men who were now the leaders in the vigorous move to westernize Japan. When he was asked for his opinion he stated that there was no better medicine in the world than that of Germany.

On the Japanese side, the leading advocate for the German system was

Dr. Sagara Chian. In 1870, he forwarded an official request to the government that German professors be invited to teach at the East College. Soon after, the government announced that Japanese medicine would be based on German medicine and the German Consul, von Brandt, was asked to arrange for professors from German medical schools to come to Tokyo University. One year later the first pair of German teachers arrived and thus began a flow of German professors to Japan and in turn Japanese students to Germany. There is probably no instance in history where a country not under the colonial domination of another adopted an external system of medicine as completely as the Japanese adopted the German system.

The first professors from Germany, Leopold Mueller and Theodore Hoffmann, were both from military medicine. Because of the poor preparation of the students they established a three-year pre-medical program, followed by four years of medical school. In the German tradition, instruction was based on the lecture, the clinical lecture and the demonstration.

In 1875, national regulations were promulgated which required the study of Western medicine for all physicians; *Kampō* was not barred but could only be practiced by a physician trained in Western medicine.

In 1886, a postgraduate program was established which was intended to resemble the demanding German program for the achievement of the title of *Privatdozent*. In Germany, the program which was reserved for young medical scientists pointing towards careers in academic medicine required four years of full-time research culminating in an examination, *Habilitationschrift*, in which the candidate presented and defended his research thesis before the medical faculty. However, in Japan the program rapidly deteriorated so that essentially every medical school graduate enrolled in the program and all who enrolled were in due course awarded the degree of *Igaku Hakase*, Doctor of Medical Science.

A second Imperial Medical School was established at Kyoto in 1897. In the meantime, second-level medical schools had opened, entry to which was directly from middle school without the superior education that students in the Imperial Universities were required to obtain in the eight numbered higher schools. These were called *igaku semmon gakko* and, since entry was directly from middle school, they accepted candidates with far lower educational background than the Imperial Universities.

The total number of Imperial medical schools in Japan increased to seven before World War II, and two additional Imperial schools were established in the Japanese territories of Korea and Formosa.

During World War II, medical education suffered severe dislocations and the number of *semmon gakko* rose to fifty-one.

With the end of the war, the separation of medical schools into Imperial Universities and *semmon gakko* was ended. Twenty-three of the latter were closed and the remainder elevated their programs to meet a single standard applied to all medical schools.

The flow of medical graduates to Germany for post-graduate training before the war has been replaced by a flow to the United States. However, the American impact on Japanese medical schools has been almost exclusively in the research programs and the influence on undergraduate medical education has been minimal.

An unfortunate American "contribution" was the rather precipitous establishment of an internship in 1946. The idea of requiring a year of practical experience in the care of patients after the lecture-oriented medical school program was a good one. However, the essentials of an internship—a meaningful clinical program for inpatient care, supervision, dormitories and a stipend were not included. The internship is the weakest link in medical education in Japan today.

In 1968, there were forty-six medical schools in Japan and no new schools are under development. The standards of medical education are dictated by the Ministry of Education in Tokyo and this, combined with the abiding concern for medical research as the primary goal of a medical school, has resulted in a lack of any significant ferment in medical education. As a result, Japan is probably the only major country which has not undertaken a major reform in medical education since World War II.

Medical education in Japan today continues with the German philosophy and pattern adopted one hundred years ago.

NOTES

1. Sir G. B. SANSOM, *The Western World and Japan.* London, 1950, p. 115.
2. C. R. BOXER, *The Christian century in Japan.* London, 1951.
3. JAMES MURDOCH, *A History of Japan.* Vol. 2, London, 1949, p. 154.
   Axicanga was Ashikaga in northeast Japan, where there was a justly famed school primarily concerned with the study of Chinese and Confucian classics.
4. LUIS FROIS, *Kulturgegensätze Europa-Japan (1585). Kritische Ausgabe.* Tokyo, 1955. Monumenta Nipponica, No. 15.
5. SEBASTIAO COSTA SANTOS, *Oinicio da Escola de Cirugia do Hospital Real de Todos os Santos, 1504-1565, Faculdade de Medicina de Lisboa. Primeiro centenario de Fundacao da Regia Escola De Cirugia De Lisboa. MDCCCXV-MCMXXV.* Lisbon, 1925.
6. E. O. REISCHAUER, *Japan, past and present.* New York, 1946, p. 91.
7. C. R. BOXER, *Jan Compagnie in Japan,* 1600-1850, 2 ed. The Hague, 1950, p. 46.
8. ENGELBERT KAEMPFER, *The history of Japan,* translated by J. G. Scheuchzer. London, 1727, 2 vols. (The printed title-page is preceded by an engraved title-page in Latin. "J.G." appears on the English title-page, "Johannes Casparus Scheuchzer" on the Latin title-page).
9. R. P. DORE, *Education in Tokugawa Japan.* Berkeley and Los Angeles, 1965, p. 136.
10. LI SHIH-CHEN, *Pen-t'sao-kang-mu.* The *Pen-t'sao* as it is usually designated was compiled between 1552-1578 by Li Shih-chen and published in 1596.

11. W. Reinhardt, *Yamawaki Toyo, anatomist, philosopher, physician, poet, scholar, teacher.* Unpublished manuscript.
12. Sugita Gempaku, *Keiei yawa* (Night Story with a Shadow). Edo, 1810.
13. *Chirurgie, in welcher alles was zur Wundarzney gehoeret nach der neussten und besten Art.* Nuremberg, 1718. Sugita used the Dutch translation, *Heelkundige Onderwzingen waar in alles wat ter Helingen Genezing der Uiterlyke Gebreken Behoost.* Amsterdam, 1755.
14. He made this vow at a Temmangu Shrine in Dazaifu, near Fukuoka. The importance of this shrine in education continues today, and families travel to pray for success of a son or daughter in the grueling university entrance examinations.
15. *Anatomische Tabellen,* Danzig, 1722, by Johann Adam Kulmus of Breslau and as *Tabulae Anatomicae,* Amsterdam, 1731. It was referred to in Japan as *Tafel Anatomia.*
16. C. R. Boxer, *Jan Compagnie in Japan, op. cit.*
17. Sir G. B. Sansom, *The Western World and Japan,* 1962, p. 204.
18. Carl Pieter Thunberg, *Travels in Europe, Africa and Asia, made between the years 1770 and 1779,* 3d. ed. London, 1795-1796.
19. I. Titsingh, *Illustrations of Japan,* translated from the French by Frederic Shoberl. London, 1822, p. 182.
20. Sekiuma Fujihiko, *Seiigaku Tōzen Shiwa,* Vol. 2. Tokyo, 1933, pp. 40, 65, 72, 73.
21. Ranzaburō Ōtori, *Monumenta Nipponica,* Vol. 19, nos. 3-4, 1964, p. 20.
22. *Das Juliusspital Würzburg in Vergangenheit und Gegenwart.* Würzburg, 1953.
23. Today the basement corridors are lined with hundreds of barrels of delicious Frankenwein from the grape tithe, and there is a popular Weinstube in one corner of the hospital.
24. Mrs. Atsumi Minami, Oriental Collection Librarian at University of California Medical Center, San Francisco, assisted me in these studies.
25. *Nippon. Archiv zur Beschreibung von Japan und dessen Neben und Schutzländern: Jezo mit den südlichen Kurilen, Krafto, Koorai und den Liukiu-Inseln, nach japanischen und europäischen Schriften und eigenen Beobachtungen bearbeitet.* Ausgegeben unter dem Schutze Seiner Majestät des Königs der Niederlande. Leiden: beim Verfasser; Amsterdam: J. Muller; Leiden: C. C. van der Hoek 1832 [-1858]. 7 Teile mit vielen Abbildungen, Karten und Tabellen, p. 318.
26. Sir G. B. Sansom, *op. cit.,* p. 260.
27. *Nippon.* 1832 [-1858]. See Note 25.
28. P. F. Siebold, C. J. Temminck, H. Schlegel, and W. de Haan, *Fauna Japonica sive descriptio animalium, quae in itinere per Japoniam, jussu et auspiciis superiorum, qui summum in India Batava imperium tenent, suscepto, annis 1823-1830 collegit, notis, observationibus et adumbrationibus illustravit Ph. Fr. de Siebold. Conjunctis studiis C. J. Temminck et H. Schlegel pro vertebratis atque W. de Haan pro invertebratis elaborata. Regis auspiciis edita.* Leiden: beim Verfasser, 1833 [-1850]. 5 vols. I: Crustacea, II: Pisces, III: Mammalia, Reptilia, IV: Aves.
29. P. F. Siebold, *and* J. G. Zaccarini, *Flora Japonica sive plantae, quas in*

imperio Japonico collegit, descripsit, ex parte in ipsis locis pingendas curavit Dr. Ph. Fr. de Siebold. Regis auspiciis edita. Sectio prima continens plantas ornatui vel usui inservientes. Digessit Dr. J. G. Zaccarini, Leiden: beim Verfasser; 1835-1841. *Volumen secundum, ab auctoribus inchoatum relictum ad finem perduxit F. A. Guil. Miquel,* Leiden: 1870.

30. A copy of one of Siebold's maps is included in the official narrative of the expedition: F. L. HAWKS, *Narrative of the Expedition of an American Squadron to the China Seas and Japan,* Washington, 1856.
31. Letter of George Folsom to Daniel Webster, Secretary of State, Legation of the United States, The Hague, 6 August 1852. The National Archives, Washington, D. C. Letter No. 34.
32. F. L. HAWKS, *op. cit.,* p. 79.
33. ITŌ SAKAE, *Shōsendō: Itō Genboku den.* Tokyo, 1916, pp. 105-128. Made available by Mrs. A. Minami.
34. *The Autobiography of Yukichi Fukuzawa.* Revised translation by Eiichi Kiyooka. New York, 1966.
35. NAGAYO SENSAI, *Shōkō Shishi,* ed. YAMAZAKI TASUKU, Tokyo, 1958, (*Nihon Ishi Gakkai,* ed. IGAKU KOTENSHU, V. 2). When the Napoleonic Wars interrupted the trade with Batavia, Hendrik Doeff, the *opperhoofd,* prepared a Dutch-Japanese dictionary. His Dutch reference source was *Woordenboek der Nederduitsche en Fransche taalan,* a Dutch-French dictionary by François Halma.
36. A complete copy of Pompe van Meerdervoort's service record was prepared for me by Major J. Karbaat of the military hospital in Utrecht.
37. GEORGE SMITH, *Ten weeks in Japan.* London, 1861.
38. J. L. C. POMPE VAN MEERDERVOORT, *Vijf Jaren in Japan (1857-1862).* Leiden, 1867-1868, p. 75.
39. YANAGA CHITOSI, *Japan since Perry.* New York, 1949, p. 107.

WESTERN HEMISPHERE

MEDICAL EDUCATION IN IBEROAMERICA

FRANCISCO GUERRA
Wellcome Institute of the History of Medicine
London, England

IBERIAN ROOTS

Dawn of Medical Education in Medieval Iberia

The medical profession in Spain grew out of the Moslem texts and the Jewish practitioners of the medieval period. During the thirteenth century the tradition of outstanding medical teaching at the Moslem schools of Toledo and Córdoba passed over to the schools annexed to the cathedrals in the Christian area, and the medical chairs founded in the *studia* entirely replaced, for all practical purposes, the apprenticeship and hospital training of physicians in the Peninsula.

The short-lived *studium* at Palencia's cathedral (1212) marks the beginning of the Spanish universities. It was followed in 1218 by the establishment of the *studium* of Salamanca by Alfonso IX, King of León, confirmed in 1243 by the royal privileges of Ferdinand III, and again in 1252, and in 1267 by Alfonso X. A Bull of Alexander VI, 25 March 1254, declared the *studium* of Salamanca one of the four *studia generalia* of Christendom, with the same privileges as Bologna, Paris and Oxford. Medicine was taught fairly early at Salamanca, because on the 9 November 1252 Alfonso X endowed two teaching positions for master physicians with 200 maravedís a year. In 1300 the *studium* of Salamanca came under direct jurisdiction of the Pope and was endowed with higher rents by Boniface VIII. In 1416 Pedro de Luna, the Antipope, who was supported by the Spaniards as Benedict XIII, increased the endowment of the Prime and Vespers Chairs of Medicine to 150 and 113 florins a year. The medical curriculum of Salamanca was shaped definitively in 1422 by a Bull of Martin V, granting the Constitutions that established the pattern of medical education in the Spanish universities for the next six centuries.

During the reign of Sancho I of Portugal, Canon Mendo Días went to Paris to study medicine, and on his return in 1212 tried to introduce medical education in Coimbra. This, however, became possible only after 1288 when the Prior of Santa Cruz of Coimbra, under the royal protection of Dionysius, joined with other notables in founding the *studium generale* of Coimbra. In

1290 a Bull of Nicholas IV granted the *studium* the privilege of conferring degrees in grammar, logic, law, canons and medicine, but transferred its site to Lisbon. Another Bull of Clement V restored the *studium* to Coimbra, and the location of the university alternated between Coimbra and Lisbon in the years 1338, 1354, 1377 and 1537, when finally John III ordered its definitive transfer to Coimbra. To the Prime Chair of medicine, endowed by King Dionysius in 1309, John II added the Vespers Chair in 1493.

During the Spanish reconquest several cities of the Peninsula followed the example of Salamanca and eventually obtained Papal Bulls to establish universities and to grant medical diplomas. About 1260 the city council of Valladolid founded a *studium generale* which had its privileges confirmed in 1346 by Clement VI for the study of all arts and sciences, and in 1404 Henry III endowed a chair of medicine with 1500 maravedís. In 1300 Lérida received from James II of Aragon the privilege for a *Universitas studentium*, with constitutions which included provision for medical teaching. In 1354 Huesca received from Peter IV of Aragon the privilege to establish another university, but medical teaching was not introduced until 1461. Through a Bull of Sixtus IV Zaragoza received confirmation of the *studium* of Arts already in existence, and further prerogatives from John II. However, higher education began there after 1542 when the Emperor Charles V confirmed the original privilege at the Courts of Monzón. The council of Valencia founded its *studium* in 1411 and in 1500 a Bull of Alexander VI gave it the right to grant medical degrees. The members of the Barcelona city council tried to establish higher education in 1450, and later that year were supported by privileges from King Alfonso V of Aragon, but medical education had little importance in Barcelona until the end of the eighteenth century.

Aragón, under Alfonso II, was the earliest Spanish realm to make compulsory the examinations of apprentices prior to practice. In 1254, shortly after the reconquest of Seville, Alfonso X asked for residences to be occupied by physicians "from abroad" so that they could carry on teaching. This training by apprenticeship spread to other areas of Spain because the *Fuero Real* of León and Castile, contained regulations for physicians and master of ulcers, amplified in the *Código de las Siete Partidas*, which was written between 1256 and 1265 and became the basis of Spanish law. In Valencia B. Flubia received a royal privilege in 1369 from Peter of Aragon to teach medicine and F. de Suria received a similar privilege in 1392 from John I. During the fifteenth century it was the council of the city of Valencia that appointed *fisichs de autoridat* as medical examiners of apprentices, indicating that, as in Castile, apprenticeship was still the usual method of medical training.

Hospital training in Spain also had an early beginning in the Monastery of Guadalupe in Extremadura, founded in 1337 by Alfonso XI, soon to be made famous by the Infirmary and the Hospital where some of the greatest Spanish and Portuguese physicians received their training up to the seventeenth century. But it must be pointed out that the education of the physician in both Spain and Portugal took place in the university and only the surgeons can trace their training from hospital practice.

The fact that since the thirteenth century the Spanish kings had appointed examiners for medical and surgical apprentices and that these examinations continued for centuries, even after universities gained permission to confer medical degrees, frequently led to a conflict of authority. In Castile John II gave the privilege of medical examiners to his royal physicians, but the Courts of Zamora (1432) and several cities in the Spanish realms issued local medical ordinances for examining and licensing physicians, surgeons, apothecaries and farriers. Eventually, after the union of the kingdoms of Castile and Aragon in 1479 and under the centralizing policies of Ferdinand and Isabella a royal medical board, the *Protomedicato* was established. This was empowered to licence and control all medical matters, including examination of apprentices and medical graduates, a prerogative that was challenged by all the medical schools of Spanish universities, particularly Salamanca, which carried on litigation against the Protomedicate until 1741. A similar situation existed in Portugal because John I had decreed that before being licensed to practice, medical graduates had to pass an examination with the *Fisico-môr*, or Physician-in-chief, and this regulation was in force even for graduates of Coimbra until 1556.

Growth of Medical Schools During the Renaissance

Many universities were founded throughout Spain during the sixteenth century as a result of the competitive efforts of prelates, religious orders and the nobility, eager to obtain from the Papacy the local privilege of providing studies and conferring degrees, the latter formerly a distinction limited to Salamanca. However, the only other Spanish university with cultural momentum comparable to that of Salamanca was Alcalá de Henares, founded by Cardinal F. Ximenez de Cisneros. In 1499 Pope Alexander VI granted Alcalá the privilege of establishing the *Collegium Scholarium* of St. Ildefonso, modeled on the colleges of Salamanca and Valladolid, to teach theology, canon law and arts. Further privileges from Pope Julius II in 1508 established the University of Alcalá which, together with Salamanca and Valladolid, became one of the three great universities of Castile. The constitutions of Alcalá (1510) provided for two medical chairs, and medical teaching began at that date, but the privilege of conferring degrees of Bachelor, Licentiate and Doctor of Medicine was granted by Pope Leo X only in 1514.

The University of Santiago de Compostela, founded by the Bishop in 1501 as a school of humanities, received its privileges in a papal Bull (1504) of Julius II. Its medical teaching began in 1648 with two chairs, Prime and Vespers, which were complemented in 1654 with another of Method, followed in 1751 by those of Anatomy and Surgery. In 1502 Seville received a privilege from the monarchs Ferdinand and Isabella to establish a *studium generale,* confirmed as a university in 1505 by Pope Julius II. In 1508 this Pope granted medical studies to Seville, which by 1572 had three chairs of medicine, Prime, Vespers and Method. Contrary to the leading role of the city in Spanish history, the University of Toledo was of minor cultural significance after the sixteenth century. Founded in 1520 as the College of St. Catherine, established in 1485, the university obtained privileges for medical teaching in 1529 and graduated a small number of physicians up to 1807. In 1526 Granada received the privilege of a university from the Emperor Charles V, and in 1531 Pope Clement VII granted it the right to confer degrees, including those of medicine. In 1776 Granada reformed the curriculum of its medical school that has continued to function until today. The University of Osuna was founded in 1548 by the Duke of Ureña; it received its privileges in 1549 from Pope Paul III and from the beginning had a medical chair endowed with 80 ducats. Despite the proximity of Seville and Cadiz, the University of Osuna enjoyed a good reputation in Andalucia until its closure in 1820. Oñate in the Basque country received the privilege of a university in 1540 by a papal Bull of Paul III, including medical education, but the chairs of medicine were never endowed. Pamplona also had a university privilege from 1608, confirmed in 1623 by Urban VIII, and in 1630 by a royal order of Philip IV, but medical education was not begun until 1757-1780, and again 1817-1840. Sigüenza had a college functioning as a university in 1476, which in 1483 received privileges from Pope Sixtus IV. The chair of medicine and the conferring of medical diplomas was granted in 1552 by Pope Julius III. It is an interesting fact that the University of Sigüenza had a hospital for the teaching of "charity" to surgeons and male nurses from the fifteenth century, but medical education came to an end there in 1807.

The universities created in the kingdom of Aragon, Catalonia and Valencia were the result of the efforts of local authorities and had more popular support than those of royal or ecclesiastical promotion in the old kingdom of Castile, Leon and Andalucia. The University of Valencia, established by the city council, granted degrees from 1500, and from 1585 had one of the most populous medical schools, outstanding in the sixteenth and eighteenth centuries for its anatomical teaching. The University of Barcelona began medical education in 1665, but in 1714 the university was closed by Philip V as a reprisal for the support Catalonia had given to Charles of Austria, and the medical school was transferred to Cervera. However, a minor college for the training of surgeons, which did not confer degrees, was continued in the Hospital of the Holy Cross. In 1821 Barcelona recovered its university from Cervera but lost it again in 1823 for political reasons, finally regaining it, with its medical school, in 1836. After the failure of the Miramar College, founded by Raymund Lull in 1275, Mallorca in 1483 received from Ferdinand and Isabella the privilege of establishing a university; in 1503 it obtained further support from Ferdinand II, from the Emperor Charles V in

1526, from Philip II in 1597, and finally in 1673 from Pope Clement X. In Palma there were chairs of medicine and surgery in 1691, and by 1697 the medical school had five professors and seven graduates, but in 1842 it ceased to function and was incorporated in the medical school of Barcelona. In the kingdom of Aragon the best record in medical education belongs to Zaragoza where, from 1476, the university was closely connected with the city. Despite early privileges from Popes Julius III in 1554 and Paul IV in 1555, the Statutes were not issued until 1583 and medical teaching began only in 1618. In the eighteenth century its medical population compared favorably even with that of Valencia. In 1714 Philip V suppressed all the universities in the Catalan cities that had supported his opponent Charles of Austria and transferred them with their medical schools to the loyal university town of Cervera near Lérida. Cervera then had six medical chairs, and in 1731 its degrees were validated with the same privileges as those of Alcalá, Salamanca, Valladolid and Huesca. The universities of Cervera and Mallorca were in turn incorporated with Barcelona in 1842.

During this period a number of universities of ephemeral existence were established by religious communities. In the kingdom of Valencia the Duke of Gandía, better known as St. Francis of Borja, established the University of Gandía, operated by the Jesuits, which in 1547 received the usual papal privileges from Paul III. Gandía had four chairs of medicine but in 1768 it was closed due to the expulsion of the Jesuits from Spain. The University of Orihuela was founded with papal privilege in 1522 by the Archbishop of Tarragona for Dominican friars; it received further privileges from Pope Pius V in 1568 and 1569, and carried on medical education until 1790.

The origin of the university and medical school of Madrid, the central university, is complex and of late development despite the fact that the city had been the capital of Spain since the sixteenth century. In 1623 it was suggested that the University of Alcalá be transferred to Madrid, but in 1625, in the teeth of opposition from Salamanca and Alcalá, the Society of Jesus took the decisive step of establishing in Madrid the Imperial College of San Isidro with emphasis on physics, mathematics and natural sciences. With the expulsion of the Jesuits in 1767 Charles III reformed the faculty of Imperial College, but it was not until 1836 that the University of Alcalá was definitively transferred to Madrid and the present structure of the University of Madrid began to take shape. The medical school, known as San Carlos, did not have a university foundation but developed from the Royal College of Surgery founded in 1783. Even so, by the middle of the seventeenth century there were twenty medical schools in the thirty-two Spanish universities then established, but only one in Portugal.

Evolution of the Medical Curriculum

Medical education in Spain and Portugal seems to have been based on the same medical doctrines and to have followed practically the same medical texts from the thirteenth to the nineteenth century. In fact, the medical curriculum in Spain leapt without evolution from ancient tradition to modern clinical medicine. This fact cannot be interpreted in simple terms because those six centuries embrace the growth and decline, as well as the golden era, of Spanish and Portuguese medicine.

The Spanish universities approached the study of medicine very much like that of theology and law, and appreciation of this attitude is vital for an understanding of their role in medical education. Medicine was considered to be a body of doctrines established in the texts of accepted authorities—Hippocrates, Galen, Avicenna—which should be lectured and commented on by the teacher and memorized by the student. The education of a physician was not a learning process in which the student moved from structure to function, or from disease to treatment; the order to be followed was immaterial—anatomy was taught after therapeutics—because the aim in education was coverage of the doctrines in the statutory texts, and the subject could be subdivided irrespective of sequence to suit terms, courses, or years of study. Initially, therefore, the title of the Chair was not a matter of subject but of academic hierarchy, and the title began to have real meaning only after the middle of the sixteenth century.

The Spanish medical education born in 1252 at the University of Salamanca had its umbilical cord in the Moslem school, and the initial intercourse between Christian physicians and Moslem scholars led to an early translation of Avicenna's *Canon* into Latin. Writing in 1569 on the cultural background of Salamanca, Chacón stated that from the beginning the Medical Chair was Avicenna's and had not changed since, because his doctrine was more concise and better organized than Galen's. The Statutes of the University of Salamanca, first published in 1561, have paramount importance in the history of Spanish medical education. They were the model for the universities and ruled them for centuries; they specified the compulsory program of medical study at every stage during the academic year. They also explain the dominant role of Avicenna as the text of reference.

1. It is ordered that [the professor] in the Prime Chair of Medicine lecture to the first year students on the first *Fen* of Avicenna. During the first two months before Christmas he must lecture on the first three chapters on the three doctrines, as far as the fourth doctrine and exclusive of *de humoribus*. 2. *Item*, from the beginning of January through February, he must lecture on the chapters of the fourth doctrine; that is, one, *de humoribus*, and the other, *de qualitate generationis eorum*.

In this way the statutes provided for every term of the academic calendar for the first year students, and then set out the program for the subjects to be taught by the Prime Chair to the students of the second, third and fourth years. Subsequently the statutes specify the teaching program of the Vespers Chair.

13. In the Vespers Chair [in the first year] there must be lectures on the first

of [Hippocrates'] *Aphorisms,* until Christmas, and the second until Easter, and the third until St. John ['s day]. 14. In the second year the fourth must be lectured on until Lent, and the fifth until Easter, and the sixth and seventh until St. John ['s day].

Further on the statutes indicate that the ninth book of Rhazes' *ad Almansorem* was lectured on from the Vespers Chair for third year studies, and concludes with the program of Method or Therapeutics.

The statutes of Salamanca are also of great interest for the study of renaissance anatomy and the Vesalian influence in Spain. A full chapter is devoted to the then recently instituted Chair of Anatomy with the program of dissections and the texts to be followed.

1. We establish and order that the Professor of Anatomy perform six complete anatomies from the day of St. Luke until [that of] St. John: one of only the muscles, another of only the veins, another of only the bones, another of only the nerves; two complete of the whole body. And in that period he will perform twelve partial [dissections]: two of the head; two of the eyes; two of the kidneys; two of the heart; two of the muscles and veins of the leg. The six complete [dissections] must be performed in the anatomical building constructed to this end; and the twelve partial [dissections] either in the hospital for study or in the general [hospital] of Medicine, not more than one and a half hours of the time of the anatomy class must be spent on them. That is, complete anatomies must be undertaken after the class of Prime has been left and until the class of Vespers in the evening; in such a way that the lectures from the Prime and Vespers Chairs are never missed. 2. *Item,* that due to the stench of the complete anatomies, they must not be carried beyond two or three days, dealing only with the use and names [of the parts], and indicating precisely where they are dealt with by Galen, and Vesalius, and other selected [authors]. . . . 3. *Item,* that they [the professors] in their Chair from [the day of] St. Luke until the holidays have as salary . . . sixteen thousand maravedís, and for each complete anatomy performed, two thousand maravedís; and for each partial dissection one thousand maravedís. And only those [dissections] proved to have been performed perfectly and fully will be paid for. 4. *Item,* that having been given the royal commission and sufficient powers by the University, the above-mentioned Professor [of Anatomy] should be diligent in obtaining human bodies for the aforesaid dissections; and if they cannot be obtained, that he lecture from his Chair to his class and show it the plates and figures of Vesalius so that what is being lectured about can be understood. And within the year sometimes to have discussions of anatomy at which the Professor must be present.

The constitutions given by Cardinal Cisneros to the University of Alcalá in 1510 (published in 1514), were less explicit than those of Salamanca, though they confirmed the influence of Avicenna's texts and the general pattern of medical training at Spanish universities during the crucial sixteenth century.

Chapter 49 . . . let there be two Chairs of Medicine, held by two physicians of great erudition and tested experience. Each of them must provide two lectures on every working day: one before noon, the other afterwards. In such a way that one of them lectures on the course or *Canon* of Avicenna, which course he is ordered to elucidate completely in two years. The other professor at the same time must lecture on the course of the Art of Hippocrates and Galen, and in the same way he must complete this course in the term of two years. At the end of such courses both professors will exchange subjects in such a way that the one who was in charge of the course on Avicenna will undertake the course of the Art and will continue it during two years, and the other the opposite.

A more detailed program, similar to that of Salamanca and including anatomical teaching, was approved by the faculty of Alcalá in 1561.

The curriculum at Coimbra followed very closely the development of Salamanca and Alcalá, and after the new statutes of 1591 granted by Philip II, then King of Spain and Portugal, the medical studies had the following chairs: Prime at 6 a.m., lecturing on Galen's books *Tekhne [iatrike]* and *De locis affectis* in the first three years; *De morbis et symptomatis* in the fourth; *De differentiis febrium* in the fifth; *De simplicium medicamentis* in the sixth; Vespers at 6 p.m., lecturing on the *Aphorisms* of Hippocrates for the first two years; Book IX of Rhazes' *ad Almansorem* in the third; the *Regimen* of Hippocrates in the fourth; the *Epidemics* and *Prognostics* in the fifth. The third or Avicennan Chair lectured on the *Canon* at 9 a.m. The Chair of None, named for the ninth canonical hour, lectured at 3 p.m. on Galen's works and the surgery of Guy de Chauliac or John of Vigo. There were, furthermore, general dissections three times a year and regional dissections six times a year.

Medical teaching in Spain and Portugal remained for many centuries entirely theoretical. The professor lectured from the official texts required by the statutes of the university, and his reputation depended on the elegance with which he was able to comment on them or refer to ancillary literature. Erudition and rhetoric in the presentation of medical doctrines remained the best assets of professors as well as the elements to be judged during the graduation exercises. After the class it was customary for the teacher *estar al poste*, the expression indicating that he was near the column of the cloister next to the classroom to answer the queries of the students and resolve their differences. Although clinical teaching at the university hospital was practiced, theoretical training was dominant. This partly explains the fact that the medical school of Alcalá prospered and increased considerably in student population during the sixteenth and seventeenth centuries thanks to brilliant teachers, while Salamanca declined despite a good university hospital, St. Mary the White, as well as other hospitals and a much larger city population for private practice. In 1617 Philip III gave detailed instructions on the lectures on medical texts by the professors, the time to be allocated to notes, repetition and questions and the manner of performing tests, that perpetuated awkward pedagogic procedures in the medical schools of Spain and Spanish America.

A second Chair of Medicine was added at Salamanca in 1416, at Coimbra in 1493 and at Valladolid in 1534. The original Chair was then named *Prima*, or Prime, with classes in the morning, while the second Chair was known as *Vísperas*, or Vespers, because the classes were at the sixth hour of the Catholic prayers in the evening. During the sixteenth century the universities of Spain and Portugal progressively increased the number of Chairs. Salamanca created the Third or *Tercera* in 1530, Valladolid in 1592; Coimbra's *Tercia* also created in the sixteenth century was entirely devoted to Avicenna, and in the seventeenth century it became the Chair of Prognosis. Probably the most interesting facet of renaissance medicine in the Spanish universities occurred in the middle of the sixteenth century with the arrival of the anatomical scholars from abroad. In 1551 the faculty of Salamanca established a *cursatoria* or complementary Chair of Anatomy, in 1556 another of Surgery, and in 1573 one of Simples or *Materia Medica*. Valladolid established a Chair of Surgery in 1594. Alcalá followed the same expansion in the curriculum, enlarging its Chairs of Prime and Vespers with another two complementary ones in 1566. During one of the *visitas* or periodical inspections of Chairs in 1534 the students requested more anatomical teaching, and by 1559 dissections were carried out at Alcalá, although its Chair of Surgery was not established until 1594. By the end of the eighteenth century both Spain and Portugal had undertaken considerable reforms in the university education of physicians, perhaps stimulated by the results obtained in the Schools of Surgery. In 1770 the faculty of Salamanca proposed several changes and greater emphasis on anatomical studies; in 1771 Alcalá recommended the study of mathematics and experimental physics prior to medicine, and in the same year Charles III finally established a uniform curriculum throughout the Spanish medical schools with six Chairs of Prime, Vespers, Prognosis, Method—which were known as *Institutiones Medicae*—Surgery and Anatomy.

In Portugal the *Junta da Providencia Literaria*, or educational board created by the Marquis of Pombal for the reform of higher education, gave new statutes to the University of Coimbra in 1772 and the medical studies for Bachelor of Medicine were arranged in five courses: 1st year *materia medica* and pharmacy; 2nd, anatomy and obstetrics; 3rd, medical and surgical institutions with hospital practice; 4th, aphorisms and practice, and 5th hospital practice. In 1836 Coimbra expanded and once more modernized its medical curriculum following the changes made in programs of the surgical schools at Lisbon and Oporto which in 1911 became Medical Schools in the newly established universities. In the first years of the nineteenth century several Spanish universities tried to modify their ancient curricula, and in 1804, 1807 and 1814, in spite of the political difficulties of the Peninsula during the Napoleonic wars, Salamanca tried to introduce the teaching of modern subjects with chairs of anatomy, physiology, pathology, therapeutics, internal medicine, surgery, pediatrics, gynecology, venereal diseases, and clinics. The attempt was unsuccessful and Salamanca's medical school retained the old curriculum; this led to its decline and closure in 1857. Not until 1911 did Salamanca recover its school of medicine and original status.

The medical sequence in the Spanish universities was definitively established in the Constitutions of the University of Salamanca granted by Pope Martin V in 1422. According to these, the candidates for the degree of Bachelor of Medicine, the lowest qualification for a practicing physician, had first, to be Bachelor of Arts, to follow four academic years of medical studies at the university, and to give ten public lectures. The Bachelor of Arts degree was obtained after six years of study, and no student was admitted to Salamanca unless he had a good grounding in grammar school for three years. The normal age at entrance was ten or twelve to begin the Bachelor of Arts course and at least sixteen to commence medical studies. To become a Licentiate in Medicine a minimum of four months apprenticeship after the Bachelor's degree was necessary, although Masters of Arts could obtain their medical diploma after only three academic years in the medical school; thus in Spain and Portugal it was possible for a physician to have graduated at the age of twenty-one. The academic year began the day of St. Luke, 18 October, and in some cases, ended the day of St. John the Baptist, 24 June, but there were over forty days of religious festivals and holidays and the courses often lasted until the 8 of September. The original constitutions of the University of Alcalá (1510) approved a calendar with 280 days of classes, omitting only Sundays and the usual religious feasts.

Surgical Training During the Enlightenment

Spanish and Portuguese surgeons held the lowest status in the medical hierarchy until the middle of the eighteenth century, notwithstanding that two centuries previously some of them, such as D. Daza Chacón (1510-1596), J. Fragoso (c. 1535-1597) and B. Hidalgo de Agüero (1530-1597) had been practitioners of the highest repute and authors of important surgical texts. Surgeons, however, did not receive the benefits of a university education and despite their guild of SS. Cosmas and Damian their licencing and practice were restricted by the physicians controlling the Royal Protomedicate. The Surgeons' Guild incorporated Barber surgeons, Romancist surgeons trained in vernacular texts, and Latin surgeons who took their examinations in that language. All of them were trained by apprenticeship at charity hospitals and have retained their role of sanitary assistants or *practicantes* in Spain until the present day.

Early surgical education in Portugal, as in Spain, was completely different from the academic studies of a physician. A surgeon learned his profession by apprenticeship to a master surgeon from a very early age, sometimes when fourteen years old, and worked in a hospital or *misericordia,* which was invariably run by a religious charitable order. The heart of surgical training in Portugal was the *Hospital de Todos os Santos,* or All Saints, in

Lisbon, founded in 1492. In 1556 a royal order established there an *Aula* for surgery and anatomy, subjects always linked together in the evolution of medical education, and the Chair was given to the *Físico-mor* Duarte Lopes, the Physician-in-chief. He was succeeded in 1561 by the celebrated Spanish anatomist, pupil and panegyrist of Vesalius, A. Rodríguez de Guevara, who taught for eighteen years, followed by the Catalonian A. de Monravé, the Italian B. Santucci, the Frenchman P. Dufau, and others. The teaching of surgery during the sixteenth century consisted of a daily one-hour lecture in which the professor lectured from the book of Guy de Chauliac, followed by a half-hour of questions and discussion by the students, complemented with some dissections. In 1559, during the regency of Joanna of Austria, a decree was issued making it compulsory for all apprentice surgeons to practice for two years at the Hospital of All Saints before obtaining their licence to practice.

Until the middle of the eighteenth century in Spain and the first quarter of the nineteenth century in Portugal a surgeon was trained by learning gross anatomy, the technique of phlebotomy, with details of the superficial veins, the handling of fractures and luxations, cupping, leeching, extraction of bullets, minor surgical operations, dental and ophthalmic interventions, bandaging, attention to wounds and ulcers, use of clysters and similar practices. The *Recopilacam de Cirurgia* by the surgeon of All Saints, Antonio da Cruz, published in 1601 and reprinted ten times during the seventeenth century, was the text which epitomized surgical training in that period. In 1694 the number of apprentices at All Saints was limited to ninety; in the eighteenth century surgical training was also carried out at St. Joseph's hospital and at most Portuguese charity and military hospitals. The hospitals, like the master surgeons, could issue a certificate of practice; but from 1521 the licence was given only after passing the examination with the *Cirurgião-mor* or Surgeon-in-chief, and after 1782, as in Spain, the examination of the Royal Protomedicate.

In Spain, Ferdinand VI sought better opportunities for the hospital training of surgeons by the building of the General Hospital in Madrid (1749), as new ideas in surgical education began to gain acceptance after the arrival of certain foreign practitioners such as Giuseppe Cervi (1663-1748), physician to the Queen, who became president of the Protomedicate, and a few military surgeons in the Spanish army. The renovation of surgical training was implemented by Pedro Virgili (1699-1776) who, after a humble beginning as a barber surgeon, trained at Montpellier and Paris and became a naval surgeon at Cadiz Hospital. Frustrated in Madrid by the physician's control of the Royal Protomedicate and by the vested interests of the Surgeons' Guild, Virgili was, however, able to establish in 1747 a Royal College of Surgery at Cadiz for the training of naval surgeons, where surgery was taught with particular emphasis on dissection and advanced operative techniques. Furthermore he sent some of his disciples abroad for postgraduate training: Bejar, Salvanesa and Najera went to the Schools of Leyden and Bologna for medical instruction, while Ruiz-Foribe, Navarro, Cárdenas, Manresa, Cansca and Subert went to Paris to study surgical techniques. Prior to its decline in 1796 the College of Surgery in Cádiz had produced such distinguished students as Velasco, Navas, Rodríguez del Pino, Solano and Gimbernat.

The excellent naval surgeons trained at Cádiz made it easier for Virgili to obtain permission from Charles III in 1760 to found the Royal College of Surgery at Barcelona for army surgeons, where A. Gimbernat (1734-1816) became professor of anatomy. The program of studies in 1760 was carried out in Spanish, not Latin, by five professors and covered the following subjects: anatomy with dissection of cadavers; osteology, diseases of the bones and bone surgery; physiology and pathological surgery; surgical diseases and surgical techniques; therapeutics, including bloodletting. The *Curso de Operaciones de Cirugía* (1763) by D. Velasco and F. Villaverde, professors at Cadiz and Barcelona, became a favorite text. By then surgery had become sufficiently respectable in Spain to free college graduates from the Protomedicate examination. In 1774 M. Rivas and Gimbernat left Barcelona for Leyden, Edinburgh and London to work with Percival Pott (1714-1788) and John Hunter (1728-1793); they returned to Barcelona with excellent surgical training to complete the rehabilitation of the teaching of surgery. The building of the General Hospital in Madrid was begun in 1749 and by 1754 medical teaching was offered in the wards; in 1768, rules were approved for the training of barber surgeons but still according to the old standards, though by 1775 J. Gamez was teaching anatomy in the hospital. In 1779 Gimbernat and Rivas were transferred to Madrid and the building of the General Hospital was completed in 1781. In 1783 Charles III approved the founding of the Royal College of Surgery at St. Charles, and courses in surgery commenced in 1787. In 1799 the subjects requested by Gimbernat were accepted and two new Colleges of Surgery were finally established at Burgos and Santiago.

In 1815 an anatomical theatre was built at St. Joseph's hospital in Lisbon, but in Portugal, as in Spain, the great reform in the study of surgery was due to the efforts of military surgeons. After the Surgeon-general of the army, T. Ferreira de Aguiar, drew attention to the migration of students to French and Spanish medical schools, the Royal Schools of Surgery of Lisbon and Oporto, were established in 1825, offering surgical education in a five year program: 1st anatomy and physiology; 2nd the same subjects and *materia medica* with pharmacy; 3rd hygiene and surgical pathology; 4th surgical operations and obstetrics, and 5th internal medicine. Students of surgery entered the school at fourteen, after having studied grammar, Latin, French or English. The surgeon graduates at Lisbon and Oporto, however, had a much lower professional status than the physicians of Coimbra and could only practice in localities that lacked physicians. The next step in the reform of surgical education in Portugal occurred in 1836, when the Schools of Surgery expanded their curriculum with more medical subjects and were transformed into Schools of Medicine and Surgery; but despite this the teaching and practice of medicine and surgery in Portugal remained separate for many years. Progressive legislation in 1861 and 1863 led, in 1866, to equal professional standing for the surgeons of the Lisbon and Oporto schools and the Bachelors of Medicine from Coimbra. By that time Portugal's first colonial medical school, created in 1842 at the request of Rodrigo Machado, *Físico-môr* in India, was well-established in Nova Goa.

The royal decree of Charles IV in 1799 creating the Royal College of Surgery and Medicine of San Carlos in Madrid, which had no chartered university, proposed concurrent education in surgery and medicine and intended that these should be of equal academic rank. This new educational idea failed, and in 1801 surgical education was separated from that of medicine. However, by 1804 the surgeons were free from the authority of the physicians of the Protomedicate, abolished in 1799 with control passing to a newly-created Surgical Board. In 1827 new regulations temporarily united medicine and surgery and finally, in 1843, pharmacy was included, bringing the affairs of all three bodies under the control of a joint board.

Medical Education in Spanish America and Brazil

During the colonization of America, Spain adopted two metropolitan institutions, the protomedicate and the university, which proved worthy of licencing and training surgeons and physicians. Portugal, unfortunately, did not found universities in her colonies, and Brazil was forced to import medical graduates well into the nineteenth century. Medical training by apprenticeship, particularly of surgeons, was common in America during the sixteenth century, and local examiners appointed by city councils granted licences to practice until 1542, when the new laws of the Indies placed all medical matters under the royal protomedicate. With the establishment of royal and pontifical universities in Spanish America, the European graduates began to validate their diplomas in the new universities in order to obtain the right to practice in their areas long before medical chairs were founded. The American universities adopted the constitutions of Salamanca and copied, within their own limitations, her statutes and curriculum. The conflict with the Spanish protomedicate that had long vexed Salamanca was solved in America when Philip IV decreed (1646) that the professor of Prime be appointed *ex officio* chairmen of the protomedicate in their area.

The first university in the New World was established at the Dominican College of Santo Domingo by a Bull of Paul III, 28 October 1538, confirmed by Philip II, 23 February 1558, with the same privileges as the Universities of Alcalá and Salamanca. Accordingly the University of Santo Domingo could confer degrees of bachelor, licentiate, doctor and master in all arts and sciences, and during the colonial period the University validated medical studies made elsewhere, but, contrary to general belief, the teaching at Santo Domingo was only in arts, theology, and canon law, not in medicine. The medical school was established after Santo Domingo's independence from Spain.

México

The Royal and Pontifical University of Mexico was founded by Charles V on 21 September 1551, and was confirmed by a Bull of Paul IV, 21 May 1555. México adopted the Constitutions of Salamanca, including the Chair of Medicine, which was endowed with 150 pesos in gold on 13 May 1578. The first medical chair in the American continent was then inaugurated on 21 June 1578 by the professor Juan de la Fuente (c. 1530-1595), a native of Mallorca and graduate of the Universities of Sigüenza and Seville. The Chair of Vespers was created in 1598; a Chair of Surgery and Anatomy and another of Method were established by Philip III in 1621. In 1580 the faculty approved the statutes drafted by Dr. Pedro Farfán, which included a four year course for the degree of Bachelor of Medicine: 1st year, *de elementis, de temperamentis*, some chapters from *de humoribus*, some anatomy, *de facultatibus naturalibus, de pulsibus ad tirones;* 2nd, *de differentiis febrium, ars curativa ad Glauconem, de sanguinis missione;* 3rd, Hippocrates' aphorisms, *quos et quando oportet purgari*, and book IX of Rhazes' *ad Almansorem;* 4th, *de crisibus, de diebus decretoriis* and *de methodo medendi*. In 1637 the University of México took the unusual step of creating another Chair of Astrology and Mathematics, incorporating medical geography, which was compulsory for medical students. That Chair was eventually occupied by some of the most celebrated Mexican scientists, C. Sigüenza y Góngora (1645-1700), J. Velázquez de León (1732-1786), J. I. Bartolache (1739-1790). This Medical School was outstanding in America during the colonial period. Students graduated and published theses from 1598 and during most of the seventeenth and eighteenth centuries between five and ten Bachelors of Medicine completed a training similar to that of Salamanca or Alcalá. Some local texts were published, including D. Ossorio y Peralta's *Principia Medicinae*, 1685, and M. J. Salgado's *Cursus Medicus Mexicanus*, 1727, a medical journal by J. I. Bartolache, *Mercurio Volante*, 1772, and hundreds of medical theses of Bachelors and Doctors of Medicine.

In 1645 the powerful new Rector, Archbishop J. A. de Palafox y Mendoza granted other statutes to the University of México, whereby both professors and medical students were ordered to attend the anatomical dissections to be performed every four months at the Royal Hospital for the Indians in México City, under penalty of heavy fines for non-attendance. The anatomical studies at the Royal Hospital led its Surgeon-in-chief, D. Rusi, to follow the example of Virgili in Spain and to request the establishment of a Royal School of Surgery and a Chair of Practical Anatomy in México; this was granted by Charles III in 1768. The School of Surgery was organized on the same basis as the surgical colleges of Cadiz and Barcelona, and in 1770 courses for surgeons were inaugurated with chairs of anatomy, physiology, surgical techniques, surgical clinics and forensic medicine. This training produced, as in Spain, the simultaneous but separate training of surgeons at the hospital and physicians at the university.

Medical teaching at the university declined considerably during the War of Independence (1810-1821), notwithstanding the clinical work of L. J. Montaña (1755-1820), and in 1833 the University of México with its Medical School was abolished. In its place the Vice-President of México, the physician V. Gómez Farias (1781-1858), founded the Establishment of Medical Sciences (1833), whence the present Faculty of Medicine of the National University of Mexico traces its descent. After hazardous beginnings this establishment marked a new departure in the medical education of Mexico, and the chairs created represented a fresh scientific trend in the curriculum, incorporating the French clinical programs with its texts. Anatomy was taught according to the text of J. P. Maygrier, physiology, of F. Magendie, hygiene, of E. Tourtelle, pathology, of L. C. Roche, internal medicine, of L. Martinet, external medicine, of A. Tavernier, surgery and obstetrics, of J. Coster and A. Dugès, *materica medica*, of J. B. G. Barbier, forensic medicine, of J. Briand and pharmacy of A. Chevallier. From that medical institution México produced its finest generations of clinicians by the end of the nineteenth century, C. Liceaga, M. Carpio, M. Jiménez, R. Lucio, I. Alvarado and others. At the close of the century the Medical School of México had a program of five years, with subjects similar to those of Paris and almost 400 students. At the end of the nineteenth century a School of Nursing and Obstetrics was created within the Medical School, which gained its independence in recent years. In 1925 there were already 381 nurses and 169 midwifery students.

Colonial New Spain or México had another medical school in Guadalajara, Jalisco, created by Charles IV in 1791. A Chair of Medicine under M. García de la Torre and another Chair of Surgery under I. de Brizuela y Cordero were inaugurated in 1792. The curriculum was expanded in 1826, 1834, 1847 and 1873 as a result of interesting political and religious controversies; the school still flourishes today with more than 500 students. In 1935 another medical school was created in Guadalajara at the Autonomous University which has a smaller number of students.

The city of Puebla in México created a university which has a long history. It was established in the old College of the Holy Ghost of the Society of Jesus, opened in 1578; in 1767, after the expulsion of the Jesuits, it became Caroline College and in 1825 the College of the State of Puebla. In 1647 the college graduated its first physician, B. Muñoz Parejo. The medical studies at the University of Puebla began to acquire a modern structure in 1834. Another medical school was also established in Puebla in 1907, the curriculum copied from that of Louvain; this was at the Catholic University which had an ephemeral existence.

The medical school of Morelia, Michoacán, was founded through the efforts of J. M. Gonzalez Urueña by decree of the local government, 9 November, 1829. It began with two chairs, medical and surgical pathology in the charge of Gonzalez Urueña, and anatomy under M. Ramírez. In 1847 the medical school was incorporated into the College of St. Nicholas which in 1901 adopted programs similar to those in Mexico City, and in 1917 the medical school became a part of the University of Michoacán.

Mérida in Yucatán also had an early medical school founded in 1833 by decree of the local government with two chairs, Prime under I. Vado y Lugo (1796-1853), a graduate of Guatemala University, and Vespers occuped by J. Hübbe Heyer, a graduate of Tübingen. The medical school of Mérida became a part of the University of the South-East and now gives medical education to 300 students.

The story of the medical school of Monterrey, which became part of the University of Nuevo León is one of the finest accomplishments in education by a self-trained physician. J. E. Gonzalez (1813-1888), better known as *Gonzalitos,* founded the school in 1859, after decades of teaching pharmacy and obstetrics, and became the professor of anatomy, surgery and obstetrics; he also established the chairs of chemistry, botany and pharmacy, physiology and hygiene, external pathology, internal medicine, therapeutics and forensic medicine. *Gonzalitos'* medical school grew into the second largest in Mexico and now has over 800 students.

There were other Mexican medical schools founded in the nineteenth century at San Luis Potosí (1879), Oaxaca (1826), the Military Medical School in Mexico City, first opened in 1895 and reorganized in 1917. In this century schools have been established at Chihuahua (1955), Guanajuato (1945), Pachuca (1945), Tampico (1950), and Veracruz (1952). As a new experiment in training physicians for rural areas the Polytechnic Institute in Mexico City opened another medical school which in recent years has returned to the standard curriculum. Finally, several homeopathic medical schools were started throughout Mexico at the end of the nineteenth century which are still active.

Perú

Perú had a protomedicate appointed by the city council of Lima two years after this capital was founded in 1535 and Hernando de Sepúlveda, former teacher at Salamanca and personal physician to Francisco Pizarro, the conquistador, examined and licensed practitioners for several years. The University of San Marcos in Lima was founded by Charles V on 12 May 1551 with the same privileges as Salamanca. There were several physicians in San

Marcos' faculty, which in 1571 requested the endowment of two medical chairs, Prime and Vespers; these were temporarily accepted in 1587 but not established until a half-century later. The Count of Chinchón, the celebrated Viceroy of Perú, whose name has been associated with the introduction of cinchona in therapeutics, endowed with 600 pesos the Chair of Prime on 19 November 1634 and appointed professor Juan de la Vega, his private physician, who had been in the Chair of Vespers at Seville; to the Chair of Vespers endowed with 400 pesos was appointed A. de la Rocha; medical education at San Marcos received royal confirmation from Philip IV in 1638. In 1660 the University of San Marcos further requested a Chair of Method and another of Anatomy; the Chair of Method was granted by the Viceroy in 1662 and approved by Charles II in 1690.

The evolution of anatomical teaching has particular significance in Perú. The Chair of Anatomy requested in 1660, endowed with 200 pesos in 1711, but the first professor of J. de Fontidueñas never started the course. It was P. López de los Godos who in 1723 performed weekly dissections and only in 1752 did the Chair receive royal approval. In 1780 the university built an anatomical theatre at the hospital of San Andrés, which was inaugurated in 1792 with an address by J. Hipólito Unanue (1755-1833). That event changed medical education in Perú entirely, due to weekly clinical lectures by Unanue, Valdés, J. M. Dávalos, Puente, Villalobos, C. Bueno, and others in which the surgical subjects alternated with those of internal medicine. The initial impulse allowed Unanue in 1807 to request from the Viceroy, J. F. de Abascal, the establishment of a new medical institution apart from the university, and on 13 August 1808 the Royal College of Medicine and Surgery of San Fernando was founded, in which Unanue amalgamated the curricula of Paris and Leyden. San Fernando had seven Chairs for basic natural sciences and ten for medical subjects: mathematics, physics, natural history, botany, anatomy, pathology, internal medicine, surgery, obstetrics, pharmacy, medical geography, in addition to drawing, languages and field studies. The courses began in 1812 with twelve students. Independence came to Perú in 1821, and at that date the College of San Fernando became the College of the Independence; a hospital for clinical teaching was annexed to it in 1826 but by then the College had begun to decline. Also in 1826 a maternity home was founded in Lima, directed by Mme. Cadeau de Fessel who, between 1827 and 1847, undertook the training of midwives, based on French standards. In 1850 the College again became the University Medical School, and a major reform of medical education in Perú in the middle of the nineteenth century was carried out by Cayetano Heredia (1797-1861), the professor of anatomy between 1843 and 1856. In 1876 the studies were covered in a seven years' course that included for the first time specialities such as ophthalmology, pediatrics, gynecology and even history of medicine. San Marcos has had its share of turmoil in the present century, and as a result of the university upheavals the last decade has been the creation in Lima of a private medical school named after Cayetano Heredia which seeks to promote the highest technical standard away from academic politics.

Guatemala

The University of San Carlos in Guatemala was founded on 31 January 1676 by Charles II, and by a Bull (1687) of Innocent XI received all the privileges granted to Salamanca, Lima and México. The Prime Chair of Medicine, endowed with 400 pesos, was inaugurated on 20 October 1681 by the interim professor Nicolas de Sousa, who held the position for seven years. In the eighteenth century, San Carlos enlarged the curriculum, and to the Chair of Prime were added Vespers, Anatomy with Surgery, Method and, as in México, Astrology. There were dissections every four months, which were compulsory for students and professor alike, but during the colonial period San Carlos had a very small number of students. Only seven graduated between 1681 and 1723 and none from that date to 1752; between 1753 and 1773 there were five graduates, the last of them being the most celebrated of all, José F. Flores (1751-1824). Flores became professor of anatomy and produced excellent wax models to improve teaching, besides publishing works on tetanus, balsam and cancer. Although Flores died in Madrid, he left in Guatemala a few distinguished pupils, J. A. Córdova, M. A. de Larrave and N. Esparragosa.

Narciso Esparragosa (1759-1819), well known for his obstetrical work and the spread of vaccination in Central America, was responsible for the establishment of a college of surgery at the General Hospital of Guatemala in 1804, chartered by Charles IV. Unfortunately this surgical college did not prosper due to lack of resources, and with the proclamation of Guatemala's independence in 1821 medical education practically ceased. In 1830 the medical students demanded reforms and in 1832, under a program of studies by Mariano Galvez, the old medical school of the University of San Carlos was replaced by an Academy of Studies offering a four years' course in medicine followed by two years of practice for physicians and three for surgeons. The major reform in Guatemalan medical education took place in 1840 when José Luna (1805-1888) and Quirino Flores established the Faculty of Medicine, which until 1869 had four chairs: Medicine, Surgery, Anatomy and Medical Clinics; after that date were added the Chairs of Forensic Medicine, Obstetrics and Therapeutics. Then flourished the work of J. M. Padilla (1810-1869). Luna introduced anesthesia, and great clinicians were moulded after Luna returned from training in France. Further reforms were introduced into the curriculum, which in 1920 was spread over a period of seven years.

Cuba

The University of San Gerónimo in Havana was founded by a Bull of Innocent XIII, 12 September 1721, and confirmed by Philip V (1722) with the same privileges as Santo Domingo. A graduate of the University of México, F. Gónzalez del Alamo, began medical teaching at the University in 1726, and was appointed to the Chair of Prime in 1728, only two months before his death in March of the year. Also in 1728 L. Fontayne et Culemburg, a graduate of Montpellier and Paris, began the teaching of anatomy, and another graduate of Mexico, A. Medrano y Herrera, became professor of Prime. In 1730 J. Arango Barrios was appointed to the Chair of Vespers and J. Melquiades Aparicio to that of *Methodo medendi*. The influence of the Spanish colleges of surgery stimulated anatomical teaching in Havana, and in 1797 a Chair of Practical Anatomy was established at the Military Hospital of San Ambrosio under the surgeon F. J. de Córdova y Torrebejano, a graduate of Cádiz. In 1824 surgery was separated from anatomy and a new chair was created whose professor was F. González del Valle. In 1842, San Gerónimo was secularized and became the Royal University of Havana, by which time it had produced in its 114 years of existence ninety-one physicians out of 858 graduates.

Medical education in Cuba languished between 1842 and 1899 due to the war of independence from Spain and the reforms in education of 1863, 1871 and 1892. During the occupation of Cuba by the United States of America in 1900, the Varona program was implemented, which led to a considerable increase in medical students at Havana University; during the course of 1901-1902 there were 266 out of 521 (50 per cent) studying medicine, and by 1915-1916, 1036 out of a total of 1466 (70 per cent). After its closing during the Machado regime, the University became autonomous. Currently another medical school has been created, and one more is planned; the number of professorships has been tripled since 1958, and Cuba has the only medical school in America where the education of the physician integrates historical and dialectical materialism with the history of medicine; this subject plays an aggressive role in medical training within a socialist state.

Venezuela

In Venezuela higher education developed late; the College of Santa Rosa in Caracas, founded in 1661, did not have the privilege of granting degrees; these were conferred at Santo Domingo. The Royal and Pontifical University of Caracas was established by Philip V on 22 December 1721 and was granted the same privileges as the University of Santo Domingo by a Bull of Inocent XIII (1722). The Prime Chair of Medicine was inaugurated on 21 July 1727 by Sebastián Vizena and Seixas, a Basque graduate of Toledo, but no courses were offered at that time, nor in 1738 by J. Llenes and F. Fontes, nor in 1740 with C. Alfonso y Barrios. Medical education at the University of Caracas began on 10 October 1763 when Lorenzo Campins y Ballester (1726-1785) began the course with four students. Campins was from Mallorca, graduated at Gandía. His courses in Caracas did not attract many students; his first pupil, J. F. Molina, graduated in 1775, and the second, R. Códoba Verde, in 1782. Molina succeeded Campins in the Chair in 1783, and in 1788 Felipe Tamariz, a local graduate, held it until 1814 and trained a fine group of physicians, including José M. Vargas. José Joaquín Hernandez (1776-1850) followed Tamariz in 1815 and remained medical professor until his death. During the first decade of the nineteenth century about ten students registered in the medical school and usually only three graduated each year. In 1802 Tamariz introduced Cullen's text in his teaching, and Hernandez, Bichat's anatomy and Richerand's surgery. In 1766 and 1795 the University requested the establishment of Vespers, Anatomy and Method, but not until 1824 was the Chair of Prime divided into Theory and Practice of Medicine with two professors.

The great reform in Venezuela took place in 1827 when José María Vargas (1786-1854), who was trained in Edinburgh and London, became Rector of the new University of Caracas which replaced the colonial institution; he was professor of anatomy in 1827 and surgery in 1832. The medical studies were then covered in six years and the curriculum included anatomy, physiology, pathology, surgery, internal medicine, therapeutics, obstetrics, clinics and forensic medicine. In 1883 the Medical Faculty of Caracas was abolished, and the teaching program reverted to a Faculty of Medical Sciences in the University while certain professional functions were handled by a Medical Council and a College of Physicians. The work of J. M. Vargas and J. J. Hernández paved the way for the Caracas Medical School's finest generation of physicians and educators: Luiz Razetti, professor of obstetrics; Guillermo Michelena, the great surgeon; José Gregorio Hernandez, who introduced experimental medicine; L. D. Beauperthuy, pioneer in yellow fever; F. A. Risquez, and many more. Venezuela had another medical chair during the colonial period in Mérida in which José María Vargas taught in 1805 and M. Palacio Fajardo in 1810. After the creation of the University of Mérida in 1810 several others were projected but no regular medical teaching was introduced until 1852. The first graduate was E. Fornés in 1860, and up to 1900 the medical school of Mérida, now a part of the University of Los Andes, had twenty-five graduates, and today over forty graduate annually. A second Venezuelan medical school of the Zulia University in Maracaibo was created in 1946.

Ecuador

There were in Quito from the sixteenth century on several religious colleges intent on developing medical education. The Augustinians obtained a Bull from Sixtus V (1596) and royal approval from Philip III (1621) to establish the University of San Fulgencio, with privileges which included medical degrees; it was inaugurated in 1603 but never endowed with medical chairs. The Jesuits obtained a Bull from Paul III, confirmed by several other Popes and by Philip III (1620), to establish a second university, San Gregorio Magno, which after the expulsion of the Jesuits in 1769 was joined to a third university, Santo Tomás Aquino, founded by the Dominicans by a Bull (1682) of Innocent XI, and ratified in 1688. A medical chair of Prime, insufficiently endowed, was created in 1693, and S. de Aguilar y Molina was the first professor, assisted by the protomedicate F. de Torre. The first medical graduate was D. de Herrera in 1694. Among the distinguished professors were Fr. Felipe de los Angeles, José Fisiell and Francisco Bentboll, but it was students such as F. J. E. de Santa Cruz Espejo (1747-1796) and José Mejía (1776-1813) who attained great prominence. The "solid, useful, easy and agreeable plan of studies" proposed by Bishop J. Perez Calama for the University of Santo Tomás with interesting remarks on medical texts, did not avert the decline of the medical chairs during the struggle for independence and after the death of B. Delgado, professor of Prime. Five years after the Independence of Ecuador the Faculty of Medicine of Quito was founded on 26 October 1827, becoming a part of the Central University. Major change in medical education occurred during the government of Gabriel García Moreno, when in 1864 he brought under contract from France the Professors Esteban Gayraud and Dominque Domec. They organized the medical curriculum in four chairs; anatomy, physiology with surgery, pathology with obstetrics, and therapeutics, to be studied over a period of six years. In 1874 they increased the number of chairs to seven. In 1867, the medical school in Cuenca had been established with a small student body. In 1877 the medical school of Guayaquil was opened, where the work of Julian Coronel was outstanding.

Brazil

Medical education in Brazil was much delayed by the absence of universities. Brazil's higher education was inaugurated in 1572 with the opening of the College at Bahia by the Society of Jesus, followed by a further eight in large cities, including Salvador, Olinda, Rio de Janeiro and São Paulo. After the expulsion of the Jesuits in 1759 the college had royal lecturers, but never medical teaching which, up to the time of the Napoleonic invasion of Portugal, was exclusive to Coimbra. Surgical training was offered to six students in 1803 at the *Aula de Cirurgia* in the Military Hospital of São Paulo, but the first institution for medical education in Brazil was the School of Surgery established in Bahia by John, Prince Regent, in 1808, on the advice of his chief surgeon, José Correia Picano (1745-1823). It was opened at the Military Hospital, formerly the Jesuit College, mainly for army surgeons, and had two chairs, theory and practice of surgery, under the director M. J. Estrela (1760-1840), and anatomy and surgical techniques, under J. Soares de Castro (1772-1840). Later in 1808 the Prince Regent established another School of Anatomy, Surgery and Medicine at the Military Hospital in Rio de Janeiro, also an erstwhile

Jesuit college, with chairs of anatomy, surgery, obstetrics and clinical medicine, to which were appointed J. da Rocha Mazarem (c. 1775-1849), J. J. Marqués (1765-1841), J. Lemos de Magalhães, and J. M. Bontempo (1774-1843). The School of Rio de Janeiro in 1813 and Bahia in 1815 became *Academias Médico-Cirúrgicas,* expanding their studies to a five-year program and embracing a much wider and more modern variety of subjects, anatomy, pharmacy, physiology, hygiene, pathology, therapeutics, surgery, clinical medicine and obstetrics. In spite of this expansion until 1826 graduate surgeons *aprovado* and *formado* had to take another examination with the *Cirugião-mor,* Surgeon-in-chief, before being licensed to practice, and their status was much lower than that of the physicians from Coimbra.

The major reform in Brazilian medical education took place in 1832 when the academies of Rio de Janeiro and Bahia were made *Faculdades de Medicina* and adopted the rules and program of the Medical School in Paris. The studies and diplomas of surgeons were abolished and the school provided teaching for physicians, pharmacists, and midwives. The studies for doctors of medicine extended over a period of six years and comprised: 1st year, physics and botany; 2nd, chemistry and anatomy; 3rd, anatomy and physiology; 4th, external and internal pathology, pharmacy and therapeutics; 5th, topographic anatomy, surgery, obstetrics, pediatrics; 6th, hygiene, history of medicine, forensic medicine; and before graduating candidates were required to prepare a thesis. Pharmacists were trained in a three year program and midwives in one. Dental studies were established in 1884 after the reform of Viscount of Saboia and took three years. The growth of medical schools in Brazil reflects accurately the demographic and economic explosion of the country; apart from the medical school of Porto Alegre (1898), the rest have been established during the present century: Belem in 1919; Belo Horizonte, 1951; Curitiba, 1915; Fortaleza, 1947; Juiz de Fora, 1952; Maceió, 1950; Recife, 1920, and 1950; Ribeirão Preto, 1948; Rio de Janeiro, 1912, 1920, and 1936; São Paulo, 1913 and 1933.

BOLIVIA

The University of San Francisco Javier of Chuquisaca, now Sucre in Bolivia, was established in 1623 and received from Charles III on 10 April 1789, a privilege for a chair of medicine, never implemented. After independence, Bolivia established in 1825 the Colleges of Sciences and Arts at Cochabamba, Potosí, La Paz, and Chuquisaca, with seven chairs, one of them medical, but only one student of medicine graduated at La Paz in 1833. Due to the suggestion of Jean Martin, a Paris graduate and physician to Marshal Santa Cruz, a medical school modeled on that of Paris was created at the University of San Andrés, La Paz, in 1833; Manuel Cuéllar (1810-1894) was appointed professor of anatomy and José Francisco Passamán, a native of Minorca and a graduate of Paris, was made Dean and professor of physiology, but the medical school was closed after Passamán fled from the country. During the years that followed there was no medical teaching due to civil war and the Peruvian invasion, until in 1845 medical schools were created at the Universities of Sucre, La Paz and Cochabamba, all with a five-year program of studies attended by three professors each. However, there was a continuous flow of legislation in 1853, 1863, 1866 and 1872 altering the programs, which marred any real progress in education. In 1872 La Paz was the only medical school left, but in 1877 the original three were again opened. In 1880 a fourth medical school was created at the University of Santa Cruz, which closed soon afterwards. The Instituto Médico in Sucre, with some medical teaching based on French texts, was able to survive until 1910, when the curriculum was reformed and dental studies introduced. In 1930 and 1931 a major change occurred in the universities of Bolivia, affecting the autonomy of those institutions which gave hopes of progress in medical education. In 1954 the movements of *Tupac Katari* and similarly oriented tendencies instituted the cogovernment of students and professors in the medical schools, which ended this autonomy. Medical curricula extending over seven years, and including the study of the history of medicine, have been established at the Universities of San Simón in Cochabamba, San Andrés in La Paz, and San Francisco Javier in Sucre.

Argentina

Medical education in Argentina began as a result of the efforts of the protomedicate, established at Buenos Aires in 1780, and not in the usual university chair. Michael O'Gorman (1749-1819), the Irish protomedicate of the River Plate, was appointed professor of medicine by Charles IV, 28 January 1799, and his colleague José de Capdevila, professor of surgery, but Capdevila was soon succeeded by A. E. Fabre (1734-1820). O'Gorman and Fabre began on 2 March 1801 a six-year course with fifteen students and offered the following subjects: 1st year, anatomy; 2nd, chemistry, pharmacy, botany; 3rd, institutions of medicine; 4th, wounds and diseases of the bones; 5th, surgery and obstetrics; and 6th, internal medicine. In 1802 Cosme M. Argerich (1758-1820), a Catalonian graduate and professor at Cervera, replaced O'Gorman in the chair and maintained his standard of teaching. In 1813, due to the pressing demand for surgeons during the struggle for independence, the General Assembly commissioned Argerich to organize a Military Medical Institute which accomplished the training of surgeons but closed down in 1821, shortly after Argerich's death. The second stage in medical education began after the creation of the University of Buenos Aires on 9 August 1821 with a Department of Medicine. Besides the Dean, M. de Montúfar, there was the son of Argerich, F. C. Argerich, in the Chair of institutions of surgery, J. A. Fernández in medicine, and F. Rivero in the clinics. In 1822 the young Francisco C. Argerich (1799-1849) began anatomical teaching, and with his colleagues, soon to be joined by Juan José Montes de Oca (1806-1876)—a professor when only 20 years old—and Ireneo Portela, the university's medical studies gained momentum. Unfortunately all these men went into exile during the dictatorship of General J. M. Rosas; Argerich died in Montevideo, and medical education progressively collapsed under political pressure.

Immediately after the fall of Rosas on 27 February 1852, Vicente López recommenced economic support for the University and its medical school, and in April that year appointed eight professors to chairs which gave considerable impetus to the medical curriculum: anatomy and physiology, S. Cuenca; therapeutics, L. Gómez; surgical pathology, T. Alvarez; medical pathology, M. García; surgery, J. J. Montes de Oca; internal medicine, J. A. Fernández; obstetrics and pediatrics, F. J. Muñoz; and forensic medicine and history of medicine, N. Albarellos. The importance of the 1852 reform was due to the expansion of the curriculum and particularly to the strong clinical approach of the courses, whereby the students had to attend wards from their first year. The new Regulation for higher education (1874) grouped all the Schools in the University under one Rector and gave a new structure to the medical school by creating an Academy empowered to introduce changes in the curriculum which enlarged the number of specialities. In 1880 the University of Buenos Aires became the National University, and a new Clinical Hospital was built. Between 1892 and 1906 a number of student revolts took place and in 1906 the control of the curriculum by the Academy ceased and was replaced by an assembly of professors. In 1910 the medical school began to accept the *docentes libres*, extramural professors. The medical school of Buenos Aires changed basically its system of government, selection of professors and other important aspects of medical education after the Córdoba students' revolt in 1918.

Córdoba's medical school had been established in 1877 and followed very closely the program of Buenos Aires. From 1878 it introduced clinical teaching at the old Bethlemite Hospital of San Roque, built in 1761, and in 1914 Córdoba enjoyed excellent facilities for medical training when a large Hospital of Clinics was inaugurated. The medical schools of La Plata and Rosario were created in 1919 and another in Tucuman in 1953. The important feature of medical education in Argentina which explains the decisive role of the students in the history of their institutions is their high population: Buenos Aires' medical school grew from fifteen students in 1801 to over 13,000 in 1955, and Córdoba followed suit with over 3000 in the same year.

CHILE

Philip V created the University of San Felipe in Santiago de Chile on 23 July 1738, with the same privileges and statutes as Salamanca and Lima; the chairs of theology, canons, law and mathematics were endowed with 600 pesos, medicine with 400 and anatomy was honorary. The University was inaugurated in 1747 but the professor to the medical Chair of Prime, Domingue Nevin, a graduate of Rheims, was not appointed until the 4 August 1756. Nevin died in 1770, but left some graduates, Verdugo, Núñez Delgado and P. M. Chaparro. I. J. Zambrano succeeded Nevin, and in 1776 the chair was occupied after contest by the Chilean José Antonio Rios against the opposition of P. Manuel Chaparro, who objected to the appointment. In the independent period between 1811 and 1814 a National Institute was established with two chairs of medicine and one of anatomy which remained vacant. The National Institute was re-opened in 1818 but no medical education took place in Chile for the next 25 years, excluding certain examinations of practitioners by the protomedicate. A new stage in the medical education of Chile started when the National Institute inaugurated a six year course of Medical Sciences on 17 April 1833. The Chairs were in the charge of William C. Blest (c. 1800-1884), medicine; Pedro Morán (1771-1840), anatomy and physiology; and J. Vicente Bustillo (1800-1873), pharmacy; to whom the government added professors brought under special contract from France, including Lorenzo Sazie (1807-1865), professor of obstetrics and surgery; François J. Lafargue, successor to Moran in anatomy; and other foreign graduates such as A. N. Cox, Thevenot and many more. The texts used were mostly French: Bichat, Velpeau, Chaussier, Maygrier, although Blest favored Cullen's book. At the Hospital of San Juan de Dios a dissecting theatre was built in 1833, and the first four graduates of these studies, L. Ballester, J. Mackenna, F. Rodriguez and J. Tocornal, received their degrees in 1842. The medical school expanded the curriculum in 1863 and took over the courses of phlebotomy and dentistry that since 1854 had been offered at the Hospital of San Juan de Dios. In 1864 a new strong clinical tendency was given to medical training when hospital internship was made compulsory from the second year of study. Meanwhile, the growth of student population was considerable and the four students of 1833 became thirty-seven in 1860, 108 in 1870 and 338 in 1876. During those years specialties began to be added to the curriculum; in 1872 microscopic anatomy and private teaching outside the medical school were introduced to complement the work in the congested classrooms at the Practical School of Medicine. In 1890 a new medical school was built, and the graduates, who had been sent to Europe since 1874, began to return with specialized medical training. A second medical school in Chile was created in 1924 at Concepción, a Catholic school in Santiago in 1930, and recently in 1957 another has been established in Valparaiso.

Colombia

The early history of medical education in Columbia is complex, due to the competitive privileges received by the first religious institutions in colonial New Granada. In 1563 the Dominicans established the college of Santo Domingo at Bogotá which in 1619 received a Bull of Paul V to grant degrees; medical teaching was offered at Santo Domingo from 1639 by the protomedicate Diego Henriquez, but it did not prosper. The college of San Bartolomé, established in 1592, was granted university privileges under the Jesuits by Philip III (1602), confirmed by a Bull of Clement VIII (1604), and years later offered medical teaching. It was, however, to the college of

Nuestra Señora del Rosario, founded in 1653 at Bogotá by the Archbishop Cristóbal de Torres, that Philip IV granted a privilege for studies, and chairs were endowed for philosophy, law and medicine. José de la Cruz was appointed to the chair of medicine in 1715 but did not take possession of it; in 1733 Francisco Fontes was appointed but no students registered, and it was not until 1753 that medical teaching was inaugurated with the appointment of the protomedicate Vicente Román Cancino as professor of Prime at the Rosario College. Before Cancino's death in 1765 two distinguished men had graduated from the college, Alejandro de Gastelbondo and Juan Bautista de Vargas. Vargas was in turn appointed professor of Prime on 7 January 1767, and gave lectures from that date including some on the circulation of the blood, but in 1770 he left Bogotá to practice at Popayán. No medical teaching took place between 1770 and 1799 until P. Miguel de Isla (c. 1740-1807), an arts graduate who became director of San Juan de Dios Hospital, was appointed professor of Prime on the advice of the celebrated José Celestino Mutis (1732-1808). Charles IV approved Isla's appointment in 1801 and Mutis, a medical graduate of Seville, organized the first complete medical curriculum in 1802, reformed in 1805. Medical studies lasted eight years, five of theory and three of practice: 1st, anatomy, surgery, obstetrics; 2nd, physiology; 3rd, institutions of medicine; 4th, the work of Hippocrates; and 5th, clinics. Among the first seven students was Vicente Gil de Tejada; the first graduate in 1805 was Joaquín Cajiao. Gil de Tejada graduated in 1806 and after Isla's death became the professor and was responsible for the training of famous physicians, J. F. Merizalde, M. Ibáñez, J. C. Zapata, B. Osorio and F. Quijano. Benito Osorio became professor at the Rosary College and José Felix Merizalde (1787-1868) inaugurated in turn a medical school at the old college of San Bartolomé. After the independence of Colombia the medical schools at San Bartolomé and the Rosario were opened again and courses of anatomy, surgery, pathology, physiology, obstetrics and therapeutics were offered. In 1823 Pierre Paul Broc and Bernard Daste arrived under contract for the teaching of basic sciences, particularly anatomy.

The great step forward in medical education in Colombia occurred when a new Faculty of Medicine at Bogotá was inaugurated on 3 February 1827 with Juan Ma. Pardo y Pardo as Dean, with a modern curriculum that followed the French outline. Further improvements in medical studies took place after the creation of the National University of Colombia on 22 September 1867; from that school came Antonio Vargas Reyes, founder of the Private Medical School and afterwards Dean of the National Medical School, J. M. Buendía, A. M. Pardo, A. Ospina and many more. Cartagena was the second city of Colombia to have a medical school in 1828; Medellín established another in 1872; the Jesuits at Bogotá opened the Javeriana in 1942 which became the second largest; Popayán established one in 1949, and Manizales another in 1952. Cali medical school, opened in 1950, has reached the highest standards thanks to philanthropic support from abroad.

Uruguay

During the colonial period Uruguay depended on the Buenos Aires protomedicate, and no medical education was planned until the Larrañaga law on 11 June 1833 established studies of law, medicine and economics at the House of General Studies in Montevideo. The medical department with two chairs was never inaugurated and there was no medical education even after the establishment of the University in 1849. Pedro Visca was one of the students sent to France on a medical scholarship aimed at training professors, but on his return in 1873 the three chairs of medicine proposed by the Rector of the University, Plácido Ellauri, were not endowed. The Faculty of Medicine of Montevideo was at last established by decree of 15 December 1875. A chair of anatomy with professor Julio Jurkowski, another of physiology under Francisco Suñer y Capdevila, (1842-1916), and that of chemistry under J. J. González Vizcaíno were inaugurated in 1876; new chairs were established in 1877: Antonio Serratosa in general pathology; J. Crispo Brandis, internal medicine; Joaquín Miraljeix, surgical pathology; Eduardo Hemmerich, therapeutics; Guillermo Leopold, clinical medicine and surgery. Other subjects and specialities were added in 1882, including forensic medicine and obstetrics, and this school which started in 1876 with only one student in 1955 had over 1200 students.

Paraguay

In Paraguay the development of medical education was late, and like its political history, had a military touch. The director F. Solano López appointed William Stewart, a British army surgeon during the Crimean War, Surgeon-general in charge of the medical corps between 1865 and 1870, during the war against Brazil. With the help of another British surgeon, Frederick Skinner, Stewart organized not only an early school for army surgeons in 1857 but also the general sanitation of the country, vaccination and registration of births and deaths; he also became Dean of the first medical school. In 1892, following the pattern of French clinical teaching, the new Medical School was founded. It had such a limited number of students that in 1933 during the Chaco war against Bolivia there were only ten physicians available to the army, and 100 medical students had to be made into military surgeons; at present their number has increased to over 500.

Puerto Rico

Puerto Rico has had anatomical teaching since 1814, improved in 1818 when José Espaillat, a surgeon at San Juan Military Hospital, founded a Department of Medical Instruction with the help of the distinguished Venezuelan educator J. M. Vargas. In 1845 the Department became the School of Medicine and Pharmacy with four chairs and a curriculum aimed at the training of surgeons; the first graduate was Emigdio Antique, but due to the lack of a charter the graduates had to take the licencing examination with the Delegation of Medicine and Surgery that controlled the powers of the old protomedicate during the Spanish dominion. In 1942 fresh efforts were made by O. Costa Mandry, commissioned by the medical association to create a new medical school, but this has not yet materialized.

CENTRAL AMERICA

The medical schools of Central America appeared rather late. Nicaragua, Honduras, El Salvador and Costa Rica depended on the University in Guatemala during the colonial period. Panamá was part of Colombia until the opening of the Canal and its medical students—Sebastián López Ruiz the most distinguished and also the most unhappy among them—went to study medicine at Bogotá, and not until 1951 did Panama City have a medical school. Nicaragua established a medical school at León in 1814 which still maintains a high student population, El Salvador at San Salvador in 1847, Honduras at Tegucigalpa in 1882, and Costa Rica at San José in 1957 after an early trial in 1877. Old historical bonds remained, and as a result of high policies, in spite of the distances involved, one in ten of the students at the medical schools of mother Salamanca and old Alcalá, now the University of Madrid, speak with the accent of Spanish America.

Medical Students and University Autonomy

The medical student was made a man of privilege by grace of Ferdinand III in 1252, when Salamanca students were freed from taxation. In 1267 Alfonso X confirmed their class privileges and gave them the right of travel without control or payment of tolls. In 1387 John I exempted them from sharing board with troops or being deprived of personal effects, and in 1388 he allowed the students to bring wine and food into Salamanca without passing through customs. In 1390 Henry III exempted the students from patrol and army duties, and from 1391 they were granted immunity from the normal process of justice, save in cases of serious offences or homicide. Moreover, colleges were created to protect their education, requiring in return just two things: not to bear arms in the university nor to keep concubines in their rooms, and there is abundant picaresque literature to prove that these two demands were not usually complied with. Indeed, the backbone of the Spanish universities were the *Colegios Mayores* or senior colleges, and Lérida University appears to have had the earliest foundation, the Asunta (1386); Salamanca had four, San Bartolomé (1401) (which has always been considered the oldest in Spain), Cuenca (1500), Oviedo (1517), and Fonseca (1521); Valladolid two, the Santa Cruz (1479) and San Gregorio (1488); Sigüenza, San Antonio Portaceli (1477); Toledo, Santa Catalina (1485); Alcalá three, San Ildefonso (1499), Madre de Dios (1513) and Trilingual (1528); there were others at Sevilla, Huesca, and Valencia. Their main role was to provide residences and tutorial teaching for the students. In some the founders determined the proportion of medical scholarships, for instance in Valladolid the Santa Cruz had three medical students out of twenty. In fact, the first of the Spanish colleges was not founded in Spain itself but in Italy, when in 1364 the exiled D. Gil de Albornoz established the Spanish College of San Clemente at the University of Bologna, with provision for twenty-four scholarships, some for medical students for a period of up to eight years, the recipients being already Bachelors working towards the Doctor's degree. Some years later another Spanish college was founded in Bologna, the Vives, named after Dr. Andrés Vives, who had been physician to the Grand Turk, and in the University of Montpellier the College of Gerona founded by the Catalonian Juan Bruguera. Medical students at the colleges without scholarships had rooms in special residences or in boarding houses run by widows; they were subject to academic discipline and usually grouped by faculties to avoid the frequent quarrels between different disciplines.

After the introduction of printing in Spain and Portugal, the medical students, even those in colleges with good medical libraries, continued to depend on class notes prepared by other students or by the *stationarii* who sold or rented manuscripts, in preference to the expensive books. Language marked the basic difference between physican and surgeon. The medical student was a university man reared in Latin. The apprentice surgeons, on the other hand, learned their craft in the vernacular and rose to superior status when able to take examinations in Latin and receive the diploma of Latin surgeons. The strong theoretical background was not only reflected in the method of teaching but in the examinations and graduation procedures. The test for the Bachelor's degree in medicine after the five years of study was a lecture; for a Licentiate the lecture was followed by debate, and the Doctorate was preceded by the traditional vexation. In the background of these tests and the academic discussions pertaining to graduation can be found the heavy ballast of scholastic education with its logical argumentation and the show of *desembargada lengua*, unbridled tongue, which are still a nightmare on the road to sound scientific education in Spain.

The status of the medical student in respect to his colleagues reflected the social hierarchy of the physician during those centuries when nations were run by theologians, lawyers and philosophers; the physician was merely a highly qualified craftsman. The growth of medical chairs in the universities is a reliable index pointing to the changing social interest from the humanities to the medical sciences, but this can be better analyzed in the trend of student population in Spain from the fifteenth to the twentieth century. The most accurate sources for the early period are the records of Salamanca and Alcalá, but it must be pointed out that by the seventeenth century medical education began to decline there and medical students flocked to Valencia, Zaragoza and elsewhere.

The important fact is that in Spain around 1500 medical students made up less than 3 per cent of the student population; by 1600 the figure reached 4.6 per cent; by 1700 almost 5 per cent; in 1800 over 5 per cent, and after

TABLE 8

MEDICAL EDUCATION IN SPAIN, 1785
Population 10,200,000

University	Period of Education		Chairs		Students	
	General	Medical	Total	Medicine	Total	Medicine
Alcalá	1508–1833	1508–1824	31	5	450	25
Avila	1550–1807	0	11	0	0	0
Almagro	1553–1807	0	10	0	0	0
Cervera	1714–1842	1714–1842	36	6	891	50
Granada	1531–fl.	1537–fl.	28	4	150	10
Oñate	1540–1807	0	12	0	892	0
Orihuela	1552–1807	1569–1790	24	0	283	0
Osma	1555–1777	0	16	0	100	0
Osuna	1548–1820	1549–1807	17	4	226	42
Oviedo	1608–fl.	0	20	0	385	0
Salamanca	1215–fl.	1252–fl.	52	8	1851	43
Santiago	1504–fl.	1642–fl.	34	5	1036	10
Seville	1502–fl.	1508–fl.	22	4	518	62
Sigüenza	1476–1807	1552–1807	9	0	0	0
Toledo	1520–1807	1529–1807	24	3	416	15
Valencia	1411–fl.	1500–fl.	44	8	1174	209
Valladolid	1260–fl.	1346–fl.	44	6	1299	20
Zaragoza	1476–fl.	1618–fl.	24	6	1171	230
Total			458	59	10,842	716

fl.: flourishing.

1900 over 30 per cent of the total number of students. The strong upsurge in the number of medical students in Spain that followed the spread of positivism and the emphasis on the teaching of natural sciences recommended by A. Comte was not exclusive to Spain but extended to Portugal, Spanish America and Brazil. The medical student was never a social force during the colonial period, not owing to lack of ideas on social reform but by reason of numbers: México had five to ten medical graduates a year, Guatemala one, Perú up to three. But the seed was there and the course of the Cuban war of independence changed when the Spanish governor shot a handful of medical students. Mexico, from ten medical students at the beginning of the independent period in 1821 reached 400 to 1900 and over 1200 in 1965; then things began to happen.

The scarcity of physicians in Brazil by the end of the eighteenth century was such that in 1800 the city of Rio de Janeiro was ordered by the king to nominate annually two students to attend medical and surgical studies in Portugal. In 1810, after the opening of the surgical schools at Bahia and Rio de Janeiro, three students were sent to Edinburgh to learn surgery. After Brazil's secession from Portugal in 1821 the only Brazilians allowed to remain in Portugal were the medical students in Coimbra. The movement in student population in Brazil closely followed that of other Spanish American

TABLE 9

MEDICAL EDUCATION IN SPAIN IN 1920
Population 21,340,000

University	Students			Medicine			Women		Graduates	
	Total	Intra-mural	Extra-mural	Total	Intra-mural	Extra-mural	Total	Medicine	Total	Medicine
Barcelona	3354	1843	1511	1199	659	540	292	15	315	101
Cadíz	583	329	254	482	245	137	19	1	80	45
Canarias	122	38	84	0	0	0	3	0	1	0
Granada	1775	924	851	371	318	53	59	1	163	41
Madrid	8958	4393	4565	2722	1743	979	344	21	705	230
Murcia	848	121	727	0	0	0	10	0	70	0
Oviedo	636	103	473	0	0	0	4	0	54	0
Salamanca	881	456	425	386	240	166	12	0	73	32
Santiago	1019	675	344	377	333	44	14	2	153	77
Sevilla	973	505	468	289	200	89	35	2	103	39
Valencia	1330	947	383	597	466	131	60	11	113	54
Valladolid	1767	976	791	696	585	111	32	10	212	93
Zaragoza	1262	872	390	528	420	108	36	1	157	60
TOTAL	23,508	12,182	11,266	7647	5209	2358	920	64	2199	772

areas. In order to stimulate registration, in 1820 John VI established twelve scholarships for students of limited resources at the Medical Academy of Rio de Janeiro and an equal number at Bahia, and by 1832, the last year of the Academy's existence, 145 students were enrolled in Rio de Janeiro. After becoming a *Faculdade* in 1833 the medical school of Rio de Janeiro produced five doctors in 1835 and ten in 1836. By 1871 the school had 466 students, of whom fifty-six graduated in medicine, thirty-three in pharmacy and two in midwifery. In 1862 Bahia had 116 medical and thirty pharmacy students and, by 1871, 215 medical and seventy pharmacy students and fifty-three graduate physicians. In 1837 Bahia's medical school played a leading role during the *Sabinada*, the republican and secessionist movement led by Dr. F. Sabino Vieira, lecturer in anatomy, and two other liberal professors, V. F. de Magalhaes and J. F. de Almeida. The students became directly involved in the struggle against the Regency, Vieira was put into prison and the medical schools closed. In Rio de Janeiro there were serious incidents in 1848 because certain medical students refused to kiss the hand of the Emperor on receiving their degrees. Brazil declared war on Paraguay in 1864, and a number of professors and students rushed to the front and the hospitals. Others condemned the war, declaring themselves pacifists or republicans, and refused to register for military service. In 1871 there was a students' revolt against the new examining regulation and the students took over the school building; it was recovered by the police who freed the two professors held as hostages.

Much of the greatness and tragedy of the Latin American university stem from the autonomy Salamanca possessed at its birth. Study of the original constitutions of Salamanca, Coimbra or Alcalá offers perfect examples of democratic government, and they seem the utopias dreamed of by republics of scholars. Due to their properties, donations and rents, as well as the income from registrations and examination fees, the Spanish and Portuguese universities were economically self-supporting. The students had the most decisive role in university government; at Lérida they elected the Rector directly, at Salamanca the Rector was elected by the eight councilors nominated by the students. Alcalá was perhaps less democratic. The students also had the decisive vote in selecting the professors. The power of the students at Coimbra since the Constitutions of 1309 was very much the same; they elected the Rector, teachers and wardens. Medical chairs in most universities became vacant after four years, although at Alcalá after 1666 the tenure was six years, and candidates for professorships were required to register for the contest. The public exercises consisted of an hour's lecture and commentary from three different sections in a text opened at random, with argument and discussion by other contestants. The chair was given by the Rector and his three councilors to the candidate who received the largest number of votes from the registered students. Although this system was subject to bribery or the students' fancy, the universities were able to resist royal appointments of medical professors, and the student body retained legally in Spain until 1634, the right to appoint its teachers.

The turning point in academic representation through out the Spanish American universities occurred as a result of the students' strike at the University of Córdoba, Argentina, on 21 June 1918. In their manifesto the students protested against the "divine right" of the university professors and demanded a democratic government on the grounds that in the academic *demos* the sovereignty and the right of self-government should rest with the students. They were rebelling against the control of a class of professors with vested interests, their mediocrity, amoral groups backed by the Jesuits, the awkward system of administration and its teaching methods, the untouchable concept of authority in a university divorced from science, out of contact with current ideas, and dedicated to the unending repetition of old texts. The student movement conceived of the university as an instition integrated by three states, each of them with equal representation in its government: the students, the graduates and the professors. They requested in particular the cooperation of all graduates towards that aim, and as a first step the Federation of University Students took over the University of Córdoba on 9 September 1918, and appointed the presidents of student unions of law, medicine and engineering to direct their schools and select the professorial staff. The movement had been to a great extent instigated in July 1918 by medical students who, at the First National Congress of University Students held in Córdoba, adopted the following resolutions: to establish a hostel for students, internship in the hospitals, a relief fund for students, compulsory courses in philosophy, a closer relationship between the schools of the university, the spread of university culture among the working classes, together with the improvement of social hygiene, the teaching of social medicine and medical ethics, the setting up of a social committee in the university, and an improvement in the relations of the university with the press in an effort to spread university culture to its social environment.

The ideas expressed by the Córdoba students were adopted not only by other Argentinian universities but by the students of Perú and Chile in 1920, and thereafter spread to every nation in South and Central America. Between 1925 and 1930 the original movement of the medical students had gone beyond its initial tenets and incorporated social ideas which had been fermenting for many decades in Spanish America. In some cases the movement espoused the cause of the wretched Indians in Spanish America, in others it violently opposed the so-called Yankee Imperialism which stemmed from the increasing political and economic dependence of Spanish American countries upon the United States of America. The most important outcome of these events, particularly after the IX National Congress of Students in México in 1932, was that most national universities had become autonomous, México in 1929, with much greater student, and in certain cases graduate student, participation in the government.

Medical Educators and Political History

The movement throughout Europe of Spanish and Portuguese scholars was of considerable importance in medical education ever since medieval times. The movement of physicians was in some cases in search of further education, but in many instances the migration of large numbers of professionals was a result of racial segregation or ideological conflicts, which unfortunately have received little attention from political historians. A noble tradition was broken in 1492 when great numbers of Jewish physicians,who for generations had been among the finest scholars, teachers and practitioners, left Spanish soil rather than be converted to the Christian faith. Portugal was only a temporary asylum because in 1496 they were forced to leave and settle in the Low Countries, England and elsewhere, thus depriving the Peninsula of almost one quarter of a million people, including not only their wealthiest inhabitants but some of the best physicians.

During the first half of the sixteenth century the Italian and French universities were stimulated by the arrival of Spanish medical scholars who moved from Castile and Aragon to the dissecting rooms of Bologna, Padua, Montpellier or Paris, and so greatly advanced the teaching of anatomy in Spain and Portugal. Some Spaniards never returned, among them anatomists whose teaching and publishing was done abroad such as L. Vasse and J. de Valverde. Others, on the contrary, gave Spain and Portugal their golden century in medicine, and the anatomical teaching of A. Laguna and R. Reinoso in Alcalá, C. de Medina in Salamanca, P. Ximeno, L. Collado and J.

Calvo in Valencia, and A. Rodríguez de Guevara at Valladolid, Coimbra and Lisbon, set the high professional standards of that period. In 1550 Philip II issued a notorious decree forbidding natives of the Spanish realms to attend foreign universities on the grounds that Spanish teaching was of the highest quality and that the departure of students was leaving the Spanish universities depopulated, which was due to their rapid multiplication to the point where, at the beginning of the seventeenth century, there were thirty-two universities in Spain. The law affected not only those who left to learn but also those who went to teach. Behind the official reason lay the increasing interest of Spanish scholars in the Reformation, the corruption of morals abroad, and the export of currency. The decrees excluded those students and teachers who went to Bologna because of the Spanish colleges there, Naples where the physician Miguel Vilar became outstanding among educators, and Coimbra, where John III had installed a selected group of teachers from Salamanca, including Antonio Reinoso, the professor of medicine, and André de Gouveia from Bordeaux. In 1563 Philip II appended to his 1550 decree on the migration of scholars another regulation pertaining to medicine; this forbade the transfer of students from one university to another in order to avoid physicians obtaining their diplomas on false records of studies. Similar to this law was another passed by Charles III in 1770. At the same time as Catholic Spain closed its frontiers to prevent the departure of native scholars, colleges for foreign Catholics were created on Spanish soil and their arrival led later to the easy promotion of Irish physicians to high positions. The first of the colleges "for martyrs" under the care of the Jesuits was established in 1589 at the University of Valladolid for English nationals, and another for the English was opened in 1593 at Seville. An Irish college was established in 1592 at the University of Salamanca and a second in 1650 at Alcalá, where scholarships also existed for the Dutch and Flemish.

Contrary to general belief, medical profesors in the universities of Spain and Portugal were at the bottom of the faculty hierarchy. In 1309 Coimbra endowed the Chair of Law with 600 pounds, and Canons received 500, while Medicine received only 200. The position of Medicine in respect of Surgery and Anatomy was very similar, and at Salamanca or Alcalá in 1566 the professor of Prime received 200 ducats a year, and in 1665 the professor appointed for Surgery received 100, and that one for Anatomy only sixty ducats. In México during the seventeenth century Prime was endowed with 500 pesos, Vespers with 300 pesos and Surgery and Anatomy with 100 pesos per year. As late as 1804 Charles VI still consulted the universities as to whether medical doctors could be elected university rectors, against the ruling of the constitutions, although San Marcos in Lima had several physicians as rectors of the university.

Medical education in Spain during the seventeenth century reflected the intellectual collapse of the nation, but for a few isolated individuals. In the middle of the eighteenth century the initiative of foreign doctors in the royal household and certain surgeons educated abroad made it possible for young Spanish graduates to travel to Paris, Leyden, and Edinburgh and later become professors at the colleges of Surgery in Cádiz, Barcelona and Madrid. Their efforts, however, were short-lived; the Napoleonic wars in the Peninsula placed many of the following generation in the dilemma of choosing between patriotic fervor and the acceptance of modern scientific ideas introduced by the foreign invader. The aftermath crippled medical education in Spain for over a century. After the defeat of Napoleon and the failure of the Spanish 1812 liberal constitution, Ferdinand VII returned to the Spanish throne. His return forced the migration of large numbers of physicians and medical educators with liberal ideas such as M. Seoane, B. Ordás, D. Argumosa, J. Ardevol, M. Lagasca, A. Cibat, T. García, P. P. Montesino to London, and M. Batllés and others to Edinburgh; while men like P. Catelló, J. B. Foix, J. F. Vendrell, R. Trujillo, J. Alíx, J. M. Brull, R. Lopez Mateo and J. Mosacula languished in Spanish prisons or were "purified" and displaced. A century was to elapse before Spain began to plan anew, and in 1907 a programme of medical education was drawn up by the *Junta para Ampliación de Estudios e Investigaciones Científicas*, a Board of Postgraduate Studies created on the advice of S. Ramón y Cajal (1852-1934). During the first thirty years of the present century the Junta's scholarships enabled the most promising medical graduates in Spain to train in Europe's leading laboratories and clinics and return to occupy the Chairs of Spanish medical schools. But it is significant that the Spanish laureate who succeeded Ramón y Cajal received his Nobel award in exile. The Spanish Civil War (1936-1939) ended with another mass migration, in which physicians once again proved most sensitive to social injustice and foremost in the defense of intellectual freedoms. As a result—their names are legion—the greater part of the faculty in Spanish medical schools went into exile and the rank and file of Spanish intellectual life was "purified" as it had been a century before. But, what the mother lost the daughters received, and much of the effort Spain had put into incorporating modern scientific ideas into medical education fell on fertile ground in Spanish America. Distinguished physicians and educators left concentration camps to find generous chairs and laboratories which have stimulated medical knowledge and contributed to a notable improvement of medical education in Spanish America since 1939.

Influence of Philanthropy, Social Security and New Humanism

With a background of colonial education, the clinical reforms of the past century and the great social changes now taking place in Spanish America, there are several factors which are currently shaping medical education in that area: the role of private philanthropy in the training of professors and the technical improvement of institutions, the spiritual guidance of a handful of medical humanists of every nation, and the social demands of a fast growing, progressively industrialized population, which for centuries had low standards of health, food and sanitation.

Philanthropic efforts have frequently been devoted to medical education, but in South America, as a result of religious indoctrination, they have been invariably channeled into Catholic charities. Philanthropy in that area came from outsiders, most notably, both in scope and results, the work of the Rockefeller Foundation. From its incorporation in 1913 the Rockefeller Foundation became interested in public health programs, such as hookworm infestation which had considerable medical and economic importance for Central and South America. Influenced by the facts revealed in the Flexner Report (1910) and backed by the experience obtained between 1913 and 1920 in the Peking Union Medical College, the Rockefeller Foundation created a Division of Medical Education from which Spanish America was to receive decisive leadership and support. Initially in the São Paulo Medical School in 1913 the help took the form of building facilities, but with the passage of years the Rockefeller Foundation focussed its interest on the improvement of teaching standards and medical research in Spanish America. Medical education in Spanish America received accrued benefits from the Rockefeller Foundation due to its programs in public health and fellowships. In the first instance, large rural areas became beneficiaries of the attack against the most frequent causes of disease in tropical and temperate climates, and in the process the sanitary ranks for field work were trained. In the second instance, the program of fellows trained in the United States of America selected groups of medical graduates who returned to teaching positions in their country of provenance, thus raising the standards of medical education, in particular those pertaining to pre-clinical subjects. In recent years the Rockefeller Foundation has committed its efforts of wholehearted support of the full aspects of medical training of an entire area. Such is the example of the Universidad del Valle, Cali, Colombia, where a medical school, teaching hospital, nursing school and a rural health center have been created.

The exchange and discussion of teaching programs during the first Pan-American Congress of Medical Education at Lima in 1951 leveled out most of the disparities in curricula in Spanish American medical schools. The problem, however, could not be entirely solved by standard courses because social and human factors were involved as much as pedagogic or technical facilities. The philosophy behind medical education in Spanish America has been analyzed by Chávez, its foremost medical educator, who considered that medical schools in Spanish America should aim at the training of general practitioners with great limitation in the extent of their programs: concentration on objective teaching, avoiding theoretical or memorizing procedures; predominance of clinical teaching but grounding in true scientific training; to make the prevention of disease as important as its cure; to place equal emphasis on the somatic as on the psychic problem of the patient. Besides defining the technical problems of the curriculum, Chávez made repeated pleas for the spiritual needs in medical education, in order to make a more humane and a more cultured physician.

The creation of national health insurance and the trend towards socialized medicine in Spanish America has led the medical schools to reexamine their programs in the light of national needs. A fact clearly shown by the historical growth of the medical schools is that they have been unable to keep up with the most rapidly growing population in the world, although the number of medical students has increased considerably since 1821. There are areas in Spanish America where agrarian reform and labor unions have produced social changes with stable systems of medical care and social benefits, in which the physician plays the leading role. However, there are other Spanish American areas where the medical school is operating against a background of squalor, medieval agrarian systems or backward industrial compounds; there the medical student and the medical educator have traditionally led social reform because they know that no medical education can flourish in such conditions.

BIBLIOGRAPHY

The establishment of medical schools in Spain and Portugal has been discussed in detail by Fuente (1884), and the evolution of medical and surgical training summarized by Sánchez Granjel (1962). The information on chairs, curricula and students of Alcalá medical school by Alonso Munoyerro (1945) provides accurate information on Spain from 1500 to 1820. Teaching at Coimbra, the only medical school in Portugal and its dominions for many centuries, occupies special chapters in the work of Ferreira (1948). There is an outline of medical teaching in Spanish America during the colonial period and a guide to selected literature by Guerra (1953), and the evolution of medical education in Brazil, with many interesting biographical and political sidelights, has been presented by Santos Filho (1947). Several books and monographs have been published in recent years on the history of medical education in Latin America, besides the usual chapters in the national histories of medicine, but the most comprehensive and reliable survey is still the work by Moll (1944).

ABASCAL, H., Los primeros estudios médicos en Cuba; fundación de la Universidad. *Rev. Argent. Hist. Med.*, 1942, 1: 7-24.

ALBARELLOS, N., Apuntes históricos sobre la enseñanza de la medicina en Buenos Aires. *Rev. Farm. B. Aires*, 1857, 3: 151-154.

ALGERIA, C., La educación superior en el interior de Venezuela—escuela de medicina. *Rev. Soc. venez. Hist. Med.*, 1960, 8: 63-76.

ALONSO MUNOYERRO, L., Provisión de cátedras de medicina en Alcalá de Henares (1509-1641), in *Act. X Congr. Int. Hist. Med.*, Madrid, 1935, 1 (2): 71-200.

———, *La Facultad de Medicina en la Universidad de Alcalá de Henares.* Madrid, 1943.

Alvarez Blazquez, D., Una protesta histórica contra el método de enseñanza de la medicina en Galicia en el siglo XVIII, in *Act. I Congr. Esp. Hist. Med.*, Madrid, 1963, pp. 63-69.

Alvarez-Ude, F., Escuela de medicina del estudio general de Navarra. *Medicamenta, Madr.*, 1961, **35**: 32-35, 102-104.

———, La Facultad de Medicina de Barcelona. *Medicamenta, Madr.*, 1960, **33**: 41-45, 96-166, 225-230.

———, La Facultad de Medicina de Granada. *Medicamenta, Madr.*, 1960, **33**: 355-359; **34**: 39-43, 100-104, 291-294.

Ariza Martin, R., *Real Colegio de Cirugía, Medicina y Farmacia de Pamplona*, Barcelona 1962. Thesis.

Arreguín, V. A., Preparación técnica del médico en la Escuela Superior de Medicina Rural dependiente del Instituto Politécnico. *Medicina, Mex.*, 1946, **26**: 109-112.

Baroja, P., Desde la última vuelta del camino; II. Familia, infancia y juventud, in *Obras Completas*, Madrid, 1949, Vol. 8, pp. 569-611.

Beltran, J., *Historia del Protomedicato de Buenos Aires. Estado de los conocimientos sobre Medicina en el Río de la Plata, Durante la Época Colonial*, Buenos Aires, 1937.

Blest, G. C., Alocución . . . en la apertura [de los cursos médicos del Instituto Nacional en 17 de abril de 1833 . . . Santiago], *An. Chil. Hist. Med.*, 1959, **1**: 299-305.

Botella, J., La enseñanza universitaria médica, vista desne la cátedra, *Clinica Lab.*, 1951, **51**: 392.

Brito, M., *Description de l'école de médecine de México*. Paris, 1862.

Bruni Celli, B., Historia de la facultad médica de Caracas. *Rev. Soc. Venez. Hist. Med.*, 1958, **6**(16-17): 1-415.

Camacho, O., La selección de estudiantes en la Facultad de Ciencias Médicas de la Universidad Central de Venezuela, in *Trab. y Act. Final I Congr. Panamer. Ed. Méd.*, Lima, 1951, pp. 169-171.

Campos, E. de S., O problema do ensino médico. *Med. Cirurg, Farm.*, 1946, pp. 419-423.

———, Problemas de la enseñanza médica, in *Trab. y Act. Final I Congr. Panamer. Ed. Méd.*, Lima, 1951, pp. 58-70.

Campos Fillol, R., *Crónica de la Facultad de Medicina de Valencia (de 1866 a 1946). Ochenta años de la vida de una Facultad*. Valencia, 1955.

Canton, E., *Historia de la Universidad de Buenos Aires y de su influencia en la cultura argentina. La Facultad de Medicina y sus Escuelas*. Buenos Aires, 1921, 4 vols.

Carrasquel de Vazquez, M., Apuntes para una historia de la enfermería en Venezuela. *Tec. Hosp.*, 1960, **7**(2): 11-22.

Carter, A., Medical training in Mexico: some highlights in its history. *Bull. Pan. Am. Un.*, 1933, **67**: 850-856.

Castro Villagrán, J., L'éducation médicale d'aujourd'hui. *Sem. Hop. Paris*, 1951, **27**: 3163-3164.

———, La educación médica en la época actual, in *Trab. y Act. Final I Congr. Panamer. Ed Méd.*, Lima, 1951, pp. 79-84.

Chávez, I., *La evolución de la medicina y la formación profesional de los médicos*. Monterrey, 1951.

———, *Reflexiones en torno a la educación médica y a la elevación del nivel de nuestras escuelas de medicina.* Mexico, 1963.

CID DOS SANTOS, J., O problema do ensino nas faculdades de medicina, os escolhos da carreira médica e as deficiencias da reorganização hospitalar. *Jorn. Méd.*, 1953, **22**: 936-940.

COLINAS, V., La facultad de medicina de Madrid en los años 1877-1883. *Medicamenta, Madr.*, 1952, **18**(221): 77-81.

CORTEZO, C. M., La medicina madrileña en el siglo XIX. *Siglo Méd.*, 1924, **73** (3680): 620-623; (3681): 648-652.

COSTA, A. C. DA, Posição da medicina na universidade de hoje. *Jorn. Méd.*, 1951, **18**: 829-839.

COSTA, C., Los estudios médicos en Chile durante la colonia. *An. Chil. Hist. Med.*, 1960, **2**(2): 37-102.

COSTA, D. G. DE, Ensino médico nos Estados Unidos: sugestões para sua aplicação em nosso meio. *Med. Cirurg. Farm.*, 1946, pp. 610-618.

DIEZ TORTOSA, J., La Facultad de Medicina de Granada veinticinco años antes de aparecer Actualidad Médica *Actualidad Méd.*, 1950, **26**(30): 268-270.

DUQUE HERNANDEZ, O., La facultad de medicina de la Universidad de Antioquia y los problemas de la educación médica en Colombia. *Antioquia Méd.*, 1954, **4**: 61-69.

ENRIQUEZ DE SALAMANCA, F., Deficiencias que se observan en la enseñanza de la medicina, y propuesta general para su reforma. *Clin. Lab.*, 1951, **51**: 392.

ESCRIBANO GARCÍA, V., Notas y recuerdos de la época de Cajal. *Bol. Univ. Granada*, 1952, **24**: 20-40.

FEIJOO, B. J., De lo quesobra y falta en la enseñanza de la medicina. *An. Chil. Hist. Med.*, 1960, **2**(2): 193-198.

FERNANDEZ DEL CASTILLO, F., *La Facultad de Medicina según el Archivo de la Real y Pontificia Universidad de México.* Mexico, 1953.

FERRER, D., *Historia abreviada del Real Colegia de Cirugía de Cádiz, 1748-1834.* Cádiz, 1960.

———, Esbozo histórico del 'Real Colegio de Cirujanos de la Armada' hasta la fundación de la Facultad de Medicina (Cádiz, 1748-1844). *An. Univ. Hispalense*, 1959, **19**(4): 1-11.

———, De la unión del estudio de la medicina y la cirugía. *Med. Hist.*, Barcelona, 1966, **24**: 14.

FIGUEROA, M. H., *Historia de la fisiología en Guatemala.* Guatemala, 1958.

FISHBEIN, M., Medical education in Latin America. *J. Am. Med. Ass.*, 1948, **137**: 8-16.

FUENTE, V. DE LA, *Historia de las universidades . . . y demás establecimientos de enseñanza en españa.* Madrid, 1884-1889. 4 vols.

GARCÍA CHUECOS, H., Los estudios de medicina en Mérida. *Rev. Soc. Venez. Hist. Med.*, 1953, **1**: 756-761.

———, Documentos históricos; los estudios de medicina y cirugía en Caracas, en los primeros años del siglo XIX; informe rendido a este respecto por el doctor Felipe Tamariz. *Rev. Soc. Venez. Hist. Med.*, 1956, **4**(10): 61-67.

GARCIA DEL CARRIZO, M. G., *Historia de la Facultad de Medicina de Madrid, 1843-1931.* Madrid, 1963. Thesis.

García Maldonado, L., *Nueva insistencia sobre reforma de estudios médicos en Venezuela.* Caracas, 1950.

García Rosell, O., Tendencia actual de la educación médica, in *Trab. y Act. Final I Congr. Panamer. Ed. Méd.,* Lima, 1951, pp. 29-57.

Garretón Silva, A., Los estudios médicos en Chile; la reforma de 1943. *Día Méd.,* 1946, **18**:407-410.

Garzón Maceda, F., *Historia de la Facultad de Ciencias Médicas de la Universidad de Córdoba.* Córdoba, 1927. 2 vols.

Gicklhorn, R., Ein historisch interessantes Fakultätsgutachten aus Peru (1792). *Arch. Gesch. Med.,* 1960, **44**: 180-182.

Gonzalez-Galván, J. M., La escuela de medicina de Guadalupe. *Medicamenta, Madr.,* 1945, **3**(80): 381-382.

Cornall, J. G., A Spanish royal decree of 1617 concerning the examination of physicians and surgeons. *Med Hist.,* 1961, **5**: 290-292.

Gracián-Casado, M., *Necesidad de la investigación en la enseñanza médica* (Primer Seminario de Educación Médica en Colombia ... Comisión de Medicina Preventiva). Cali, 1955.

Granier-Doyeux, M., Bosquejo histórico de los estudios médicos en Venezuela. *An. Univ. Cent. Venez.,* 1955, **40**: 149-156.

Guerra, F. *Historiofrafía de la medicina colonial Hispano-Americana.* México, 1953.

———, Medical colonization of the New World. *Med. Hist.,* 1967, **7**: 147-154.

——— The medical student in Spanish America. *Br. Med. Stud. J.,* 1964, **19**(2): 44-47.

Guijarro Oliveras, J., La facultad de medicina de Granada en el siglo XVIII, in *Act. I Congr. Esp. Hist. Med.,* Madrid, 1963, pp. 129-133.

Guirao Gea, M., La facultad de medicina de Granada a través de los tiempos. *Actualidad Méd.,* 1950, **26**(301): 5-60.

Gutierrez Galdo, J., Los planes de estudio de la facultad de medicina de Granada en los siglos XVI, XVII y XVIII. *Actualidad Méd.,* 1965, **41**(488): 643-655.

Hernando, T., *La enseñanza de la medicina en España.* Madrid, 1934.

Horwitz, A., Evolución de la educación médica en la América Latina. *Bol. Of. Sanit. Pan-Am.,* 1962, **52**: 281-286.

Ibáñez, J., *Manifestación hecha al Consejo de Salubridad de Puebla, con el objeto de introducir algunas reformas necessarias en el ejercicio de las ciencias médicas.* Puebla, 1866.

Ibarra, J. A., La situación del médico en México. *Médico Méx.,* 1952, **2**(4): 14-15.

Indelicato, J., *Respuesta al libelo infamatorio publicado por el Lic. Antonio Escoto bajo el título de contestación del jalisciense a las nuevas Reflecsiones que sobre el reglamento de la enseñanza médica publicó el autor del aviso.* Guadalajara, 1841.

———, *Contestación del Jalisciense a las Nuevas Reflecsiones que sobre el Reglamento de la Enseñanza Médica Publica el Autor del Aviso.* Guadalajara, 1841.

Izquierdo, J. J., *Balance cuatricentenario de fisiología en México.* México, 1934.

———, Orígenes y culminación de nuestro primer movimiento renovador de la enseñanza médica. *Gac. Méd. Méx.,* 1958, **88**: 521-532.

Jimenez Catalan, M., *Memorias para la historia de la Universidad Literaria de*

Zaragoza. *Roseña bio-bibliográfica de todos sus grados mayores en las cinco facultades, desde 1583 a 1845.* Zaragoza, 1926.

Krumdieck, C. F., *Bases de la Educación Médica. Ponencia Peruana al Segundo Congreso Médico-Social Panamericano.* Lima, 1949.

Lapham, M. E., Goss, C. M., and Berson, R. C., *A survey of medical education in Colombia, 1953.* Washington D.C. 1953.

Lapuente Mateos, A., Breve semblanza histórica de la Facultad de Medicina de Madrid. *Medicina, Madr.*, 1948, 16(1): 60-66.

Larregla Nogueras, S., *Aulas médicas en Navarra. Crónica de un movimiento cultural.* Pamplona, 1952.

Larrey, D. J., *Commission scientifique du Mexico. Programme d'instructions sommaires sur la médecine.* Paris, 1864.

Lastres, J. B., *Historia de la medicina peruana.* Lima, 1951.

[Laval, M. E.], El plan de estudios de medicina del Padre Chaparro. *An. Chil. Hist. Med.*, 1960, 2(2): 175-180.

Lazarte, C., and Solari, E., Selección de alumnos en las facultades de medicina, in *Trab. y Act. Final I Congr. Panamer. Ed. Med.*, Lima, 1951, pp. 176-181.

Leal, I., La cátedra de medicina en la Universidad de Caracas. *Rev. Hist., Caracas*, 1962, 3(10): 13-60.

León, N., Apuntes para la historia de la enseñanza y ejercicio de la medicina en México. *Gac. Méd. Méx.*, 1915, 10 (3rd ser.): 466-489; 1919, 11 (3rd ser.): 210-286.

———, Apuntes para la historia de la enseñanza y ejercicio de la medicina en México; 2a. parte, 1582-1600. *Gac. Méd. Méx.*, 1915, 10 (3rd. ser.): 210-286.

———, Apuntes para la historia de la enseñanza y ejercicio de al medicina en México, desde la conquista hispana hasta el año 1833. (3a. parte). *Gac. Méd. Méx.*, 1921, 55: 3-48.

Llopis, J. M., Campins y Ballester, fundador de los estudios médicos en Venezuela. *Médico, Mex.* 1962, 12(1): 55-58.

Long, E. C., *The development to 1880 of medical education in Guatemala.* Durham, N.C., 1966.

Lopes, R. S., Aspectos do ensino médico. *Folha Méd.*, 1951, 32: 49-51.

López Laguarda, J. J., *Formación del médico y su ejercicio profesional en la Valencia del siglo XVIII.* Valencia, 1948.

López Sanchez, J., The teaching of medicine in Cuba, its past and present state, prospects of its future development. *Finlay*, 1965, 5: 63-70.

Lorenzo-Velázquez, B., La medicina española como contributiva en la formación de los estudiantes hispano-americanos. *Arch. Fac. Med., Madrid*, 1967, 12: 371-384.

Marcos del Río, P. F., Enseñanza de las ciencias médicas en la Universidad de Salamanca durante los cursos 1674-1678. *Act. X Congr. Int. Hist. Med.*, Madrid, 1935, 1(1): 186.

Mariscal García, N., *Don Alfonso X el Sabio y su influencia en el desarrollo de las ciencias médicas en España.* Madrid, 1922.

Marques, A., Da eficiencia no ensino médico. *Impr. Med., Rio de J.*, 1952, 28(461): 25-35.

Martínez Durán, C., La fundación de la Cátedra de Prima de Medicina en la

Universidad de San Carlos de Guatemala. *Reforma Méd.*, 1940, **25**: 62-64, 196-198.

Martínez Pérez, F., and Aznar López, J., Los estudios de medicina, cirugía y farmacia en la Universidad de Guatemala durante los siglos XVII y XVIII. *Medicamenta, Madr.*, 1958, **29**(319): 127-128.

———, El doctorado en medicina en las Universidades hispano-americanas durante el siglo XVII. *Medicamenta, Madr.*, 1959, **31**(388): 223, 224.

Mazo, G. del (Ed.), *La reforma universitaria. El movimiento argentino. Propagación americana. Ensayos críticos (1918-1940)*. La Plata, 1941.

Medina, J. T., *La medicina y los médicos en la Real Universidad de San Felipe (capítulo de un libro inédito)*. Santiago, Chile, 1928.

Menéndez de la Puente, L., La Facultad de Medicina en la Universidad de Huesca en los siglos XVI y XVII. *Clinica Lab.*, 1964, **78**(461): 149-160.

Mexico, Real y Pontificia Universidad. *Constituciones de la Real y Pontificia Universidad de México*. Mexico, 1775, 2nd ed.

Migliaro, E. F., Reflexiones sobre la formación de docentes universitarios en América Latina. *Act. Cient. Venez.*, 1962, **13**: 83-86.

Molinari, J. L., Antecedentes históricos de la enseñanza médica en Buenos Aires. *Sem. Méd. B. Aires*, 1964, **78**: 127-150.

Molinari, J. L., and Hernández, H. H., Los estudios médicos en el Virreinato del Río de la Plata hacia la época de la Revolución de Mayo en 1810. *An. Inst. Invest. Hist., Rosario, Argentina*, 1960, **6**(4): 597-648.

Moll, A. A. *Aesculapius in Latin America*. Philadelphia, 1948.

Moncayo de Monge, G., La Universidad de Quito; su trayectoria en tres siglos (1551-1930). *An. Univ. Cent. Ecuad.*, 1943, **71**: 191-374.

Muñoz, M. E., *Recopilación de las leyes pragmáticas reales, decretos y acuerdos del real proto-medicato. Hecho por encargo y dirección del mismo Real Tribunal*. Valencia, 1751.

Nicaise, M. E., La enseñanza de la medicina en la Edad Media. *Bol. Inst. Libre Enseñanza, Madr.*, 1891, **15**(450): 264-266; (351): 277-280.

Oliver, F., L'enseignement de la médecine en Espagne dans le Moyen Age, in *IX Int. Congr. Hist. Med., Bucharest*, 1932, pp. 638-640.

———, Die medizinische Fakultät zu Zaragoza; historische Vorgänge um das Medizinstudium in Europa. *Grunenthal Waage*, 1961, **2**: 110-115.

Palafox Marques, S., Las ideas médico-pedagógicas del Doctor Letamendi. *Arch. Ibero-Amer. Hist. Med.*, 1960, **12**: 201-248.

Palanca, J. A., La facultad de medicina de Granada a principios del siglo XX. *Actualid. Méd. Granada*, 1950, **26**(301): 89-91.

Pardal, R., La pompa en la toma de grados en las universidades coloniales de América. *Revta Méd. Cub.*, 1960, **71**: 57-65.

Paredes Borja, V., La facultad de medicina de Quito (1693-1956). *Arch. Ibero-Amer. Hist. Med.*, 1957, **9**: 407-413.

Paul, J. D., Birthplace of American medical education. [Mexico]. *Penn. Med. J.*, 1965, **68**: 64.

Paz Soldán, C. E., La organización de la enseñanza clínica en Lima. *Reforma Méd.*, 1939, **25**: 863-868.

———, Isaac Newton y los albores de la escuela médica peruana. *An. Soc. Peru. Hist. Med.*, 1942, **4**: 63-68.

———, El cuatricentenario de la 'Fábrica' de Andreas Vesalius y las bases anatómicas de la Escuela Médica de Lima. *An. Soc. Peru. Hist. Med.,* 1944, **6**: 1-136.

Perez de Petinto, M., Mis cincuenta años en la Facultad de Medicina de la Universidad de Madrid. *Bol. Inform. As. Nac. Méd. For.,* Madrid, 1962, **29-30**: 395-435.

[Peset Llorca, V.], Informe del Claustro de Medicina de Valencia sobre renovación de estudios (1721). *Arch. Ibero-Amer. Hist. Med.,* 1961, **13**: 143-155.

Peset, Llorca, V., and Faus Sevilla, P., Los médicos en el Libro de Oposiciones a Cátedras de 1720 a 1751 de la Universidad de Valencia, in *Act. I Congr. Esp. Hist. Med.,* Madrid, 1963, pp. 165-170.

Peset y Cervera, V., *Noticia histórica del catedrática Valenciano de materia médica, Dr. Juan Plaza. Solemne sesión apologética celebrada por la Facultad de Medicina de Valencia para honrar la memoria de sus antiguos catedráticos los Doctores Plaza, Collado y Piquer.* Valencia 1895.

Pessõa, S. B., Formação de pesquidores e o papel das faculdades de medicina. *Revta Paul. Med.,* 1952, **41**: 37-43.

Pina, L. de., Humanisme et éducation médicale au Portugal. *Arch. Ibero-Amer. Hist. Med.,* 1956, **8**: 59-86.

Pou Orfila, J., Reflexiones sobre la educacion médica. *Gac. Méd. Méx.,* 1925, **56** (Supl. al No. 3): 39 pp.

Prieto Carrasco, C., *La Medicina en la Universidad de Salamanca. Lo que se sabe y lo que se puede suponer de sus orígenes y periodo floreciente y de su decadencia. Cuestiones que plantea este interesante problema histórico.* Salamanca, 1936.

———, *La Enseñanza de la anatomía en la Universidad de Salamanca.* Salamanca, 1936, 4°, 16 pp.

Puebla, G., *Reglamento para el estudio y ejercicio de las ciencias médicas en el Departamento de Puebla.* Puebla, 1842; 2 ed., 1856.

Puschmann, T., *A History of medical education from the most remote to the most recent times.* London, 1891.

Rio de la Loza, M., Algunos apuntes históricos sobre la enseñanza médica de la capital. *Gac. Méd. Méx.,* 1892, **27**:48-60.

Rocha, J. M. da, Rumos do ensino médico e socialização da medicina. *Rev. Brasil. Med.,* 1954, **11**: 334-337.

Rodríguez Tejerina, J. M., La Escuela Mallorquina de Anatomía y Cirugía, in *Act I Congr. Esp. Hist. Med.,* Madrid, 1963, pp. 171-175.

Romero, H. Organización y enseñanza de salubridad. *Arch. Soc. Méd. Valdivia,* 1947, **1**: 12-30.

Romero Flores, J., *Historia de la Escuela de Medicina en Michoacán.* Mexico, 1937.

Rubio Mane, I., Historia de la Escuela de Medicina y Cirugía de Yucatán. *Rev. Med. Yucatán,* 1934, **17**: 503-536.

[Sánchez] Granjel, L., Pragmaticas y leyes sobre la ordenación de la enseñanza ejercicio de la medicina en España en los siglos XVI y XVII. *Medicamenta, Madr.,* 1949, **12**(168): 114-116.

———, *Discurso sobre el pasado de la enseñanza del saber y el arte médico en la Universidad de Salamanca.* Salamanca, 1953.

Sánchez, G., and Revuelta Ramírez, J., Historia del Monasterio de Guadalupe y de su escuela de medicina. *Trab. Cáted. Hist. Crít. Med.*, Madrid, 1934, **3**: 213-234.

SANSON, R. D. DE, Ensino médico. *Folha Med.*, 1948, **29**: 102-104.

———, Comentarios a margem do ensino. *Rev. Brasil. Med.*, 1953, **10**: 292-299.

SANTOS FILHO, L., *Historia da medicina no Brasil. (Do Século XVI ao Século XIX)*. São Paulo, 1947, 2 vols.

SARRIA, E., Historia de una cátedra fundada en el siglo XVI, en la primera facultad médica establecida en la Américas. *Trab. Cáted. Hist. Crít. Med.*, Madrid, 1936, **8**: 347-366.

SERRE DE MIRABEAU, B. A., *Memoria histórica e commemorativa de Faculdade de Medicina ... desde ... 1772 até o presente*. Coimbra, 1872.

SILVA LEAL, M. DA, A reforma das faculdades de medicina. *Jorn. Méd.*, 1949, **13**: 541.

SOLARES ECHEVERRÍA, J., *Contribución al progreso de la enseñanza de las ciencias médicas en Guatemala*. Guatemala, 1950.

TEMPEL, F. J. P., 150 Jahre Medizinstudium in Brasilien. *Münch. Med. Wschr.*, 1958, **100**: 1184-1185.

TORRE, J. M., enseñanza de la medicina en México. *Gac. Méd. Méx.*, 1963, **93**: 259-269.

TORRES PERENA, M., and MIRÓ CABEZTANY, F., La Universidad de Lérida y la medicina catalana de su tiempo. *Trab. Cáted. Hist. Crít. Med.*, Madrid, 1936, **7**: 225-246.

VALDIZÁN, H., *La Facultad de Medicina de Lima*, 2 ed., Lima, 1927, 3 vols.

VAN LIERE, E. J., Some observation on medical education and practice in Peru. *W. Virginia Med. J.*, 1950, **46**: 68-70.

VARGAS, F. L., Enseñanza médica preclínica; selección de alumnos, in *Trab. y Act. Final I Congr. Panamer. Ed. Méd.*, Lima, 1951, pp. 159-168.

VARGAS BASURTO, F. R., La enseñanza médica. *Bol. Sanid. Milit.*, 1952, **5**: 168-169.

VASCONCELLOS, I. DE, O conselheiro Dr. José Correa Picanço, fundador do ensino médico no Brasil. *Rev. Brasil. Hist. Med.*, 1957, **8**: 223-228.

VÁZQUEZ DOMÍNGUEZ, A., La formación del médico en la Universidad de Cervera (1717-1842). *Arch. ibero-amer. Hist. Med.*, 1956, **5**: 177-206.

VILLOSLADA ACOSTA, J., La facultad de medicina de Granada; su origen, sus maestros. *Actualid. méd., Granada*, 1945, **31**: 451-462.

WORLD HEALTH ORGANIZATION, *World Health Statistics Annual, 1962. III. Health personnel and hospital establishments*. Geneva, 1966.

ZUBIRÁN, S. Consideraciones sobre algunos problemas de la educación médica en México, in *Trab. y Act. Final I Congr. Panamer. Ed. Med.*, Lima, 1951, pp. 85-95.

ZUCKERMANN, C., *La Enseñanza médica en México. Proyecto de reformas*. México, 1943.

———, Necesaria reforma en la enseñanza médica; nuevo plan de estudios. *Bol. Ofic. Sanit. Panamer.*, 1952, **32**: 26-34.

ZÚÑIGA CISNEROS, M., Educación médica: evolución y tendencias. *Rev. Soc. Venez. Hist. Med.*, 1961, **9**(23): 371-391.

MEDICAL EDUCATION IN THE UNITED STATES BEFORE 1900

WM. FREDERICK NORWOOD
School of Medicine, Loma Linda University
Loma Linda, California

History is composed of three major elements: geography, people, and time. People become social forces and if circumstances permit a people or even peoples become a society or a nation.

Time and geography seem to be the most stable and predictable elements in this triad of history. They set the stage for the dynamics of change which are indissolubly linked with the life of man and other forms of life (1).

Health, disease, and medical care are vital factors in the becoming and continuing processes of history. The historian, at least the medical historian with some anthropological awareness, looks at his problem somewhat like a physician who examines the history as well as the physical condition of the patient, including his total environment.

Before 1751

The historic English and British experiment of planting elements of their culture on the Atlantic shores of North America were episodes in the history of established cultures. But, given time, the third element, they became the roots of new cultures, the birth of peoples who eventually united to form a nation.

After the founding of the English colony at Jamestown in 1607, 158 years passed before the establishment of the first collegiate curriculum in medicine in the English colonies. The early colonists were too much concerned with the grim business of providing enough food and shelter and adequate protection from hostile natives to be concerned with the esoteric problems of a mature society. They were, nevertheless, fully aware of their great need for medical assistance. The devastating impact of malnutrition, respiratory diseases and gastroenteritis during the initial "starving time" at Jamestown (1607) and the bleak first winter (1620-1621) at Plymouth fill imperishable pages in an epic of physical and spiritual endurance (2).

Furthermore, early settlers in the colonies were doomed to be decimated by repeated epidemics of communicable diseases, if not imported by themselves, introduced earlier to the indigines by explorers, conquistadores, or African slaves. Possessing little or no immunity on the first visitation of such scourges as smallpox and diphtheria, Indian tribes suffered high mortality rates. According to a likely legend the Indians, in what Shryock aptly termed a "biologic warfare", evened the score by introducing the early invaders to syphilis which, of course, regardless of its origin, was pandemic in western Europe before the later planting of the colonies. In spite of the frequency of epidemics there were intervals when colonial diarists and correspondents recorded joyful comments regarding their salubrious climate and freedom from distempers.

Colonials were, nevertheless, subject to a host of diseases and disorders due in part to ignorance of personal and public hygiene and poor understanding of dietary and therapeutic principles. In the long run, endemic diseases caused more deaths than the dreaded epidemics. Even cancer, stroke, and heart disease were apparent among those who lived to middle or old age but such ailments failed to impress practitioners or laity as major problems in a society which offered at birth an average life expectancy of perhaps no more than thirty years (3).

Among pious colonials there was general agreement that in the religious economy disease and death were the will of God to be borne with humble submission. It was also assumed at times that the Almighty sent disease and disaster as retributive judgments for wrongdoing or in severity to the heathen as a special favor to the elect. The Reverend Cotton Mather, a third-generation colonial clergyman in Massachusetts, who also practised medicine, noted with ecclestiastical pride the providence of God in bringing the Mayflower to rest at a point "wonderfully prepared for their Entertainment by a sweeping Mortality that had lately been among the natives . . . which carried away not a tenth but *Nine Parts of Ten* (yea 'tis said *Nineteen of Twenty*) among them; So that the woods were almost cleared of those pernicious creatures to make room for a better growth" (4).

Although Reverend Mather counselled profound submission to the will of God at times of death he faltered in the face of the appalling rate of infant mortality. Perhaps he recalled the loss of most of his own little ones when he declared feelingly: "O how unsearchable the judgments of God and His Ways past finding out! The lamps but just litt up, and blown out again." (5).

Stimulated by an article in the *Transactions of the Royal Society of London*, which introduced inoculation as a preventive measure against smallpox, Mather influenced Dr. Zabdiel Boylston of Boston to employ the modality during the 1721 epidemic. Boylston's inoculations proved successful, but he was bitterly opposed by the outspoken and sometimes correct William Douglass, the only M.D. practitioner in Boston at that time. Douglass withdrew from his untenable position within a decade, which added historical luster to Mather's medical reputation. Beall and Shryock, after a careful study of Mather's life and medical philosophy have labeled him the "first significant figure in American medicine" (6).

In addition to the Puritan divines who added medicine to their professional duties, some clergymen in other colonies for the same reason added physic to their duties. So also did courageous and willing colonials in other callings and trades. Adventurous spirits who formed the first settlements usually had with them or imported barber-surgeons, apothecaries, or lay practitioners. Their professional knowledge usually included a smattering of regular medicine, well seasoned with folk remedies, to which they ultimately added some of the therapeutic wisdom of the aborigines, who were esteemed for their closeness to nature.

In early colonial times few degreed practitioners chose to abandon the ease of an established practice in some British or continental city where graduate physicians tended to establish themselves, and to emigrate to a pioneer society. Barber-surgeons, apothecaries, and lay practitioners were more venturesome. Furthermore, in Britain there was a cold war between the Royal College of Physicians and the Worshipful Society of Apothecaries throughout the seventeenth and early eighteenth century. The circumstances tended not only to control rigidly the output of physicians but also to increase the number of apothecaries and to expand the scope of their professional coverage. Hence, England and Scotland exported barber-surgeons and apothecaries much more freely than physicians (7). Likewise numerous lay practitioners with whom the British Isles were then well stocked found their way to the New World. Midwives were prominent because male obstetricians were not in demand or even tolerated until later (8).

Perhaps the most fortunate of the colonial ventures was William Penn's "holy experiment", which began with the founding of Philadelphia and Pennsylvania in 1682. With Penn came three Welsh physicians who were unable to stay a deadly epidemic of smallpox which claimed one third of the emigrants before they reached the promised land. Among the three was one referred to by George Norris as the colony's first eminent physician, "tender Griffith Owen, who both sees and feels" (9). It may very well be that Quakerism's utilitarian ideology which marked Pennsylvania's development encouraged and cultivated the best in medicine and set the stage for Philadelphia to become the first medical center in the British colonies.

The planting of British and continental culture in the New World highlighted the necessity of perpetuating various arts and skills vital to the growth and permanency of the colonies. Sufficient practitioners of the healing arts could be provided through one or more of three methods: (*a*) increasing the number of medical immigrants, (*b*) sending young men abroad to study, (*c*) adapting the well-known apprenticeship system to meet the needs and limitations of a callow, struggling society. The stream of medical immigrants was not impressive and did not contribute much in quality until the eighteenth century. Because economic affluence was long in coming to the New World there was no substantial flow of young colonials to British and continental medical schools until after the middle of the eighteenth century. Malloch made an extensive though incomplete study of the medical interchange between the British Isles and America before 1801. He identified about 600 practitioners involved in this two-way passage of medical culture, without attempting to assess the total impact of foreign trained military surgeons stationed with British, mercenary and French troops during the Revolutionary War (10).

Meanwhile, apprenticeship flourished as a necessity and without effective restrictions on the who and how of preceptorship. As early as 1629 the Massachusetts Bay Company contracted with surgeon Lambert Wilson to serve the settlers and nearby Indians and to give medical training to one or more young men. Nothing is known of Wilson's teaching but a few years later (1647) the missionary to the New England Indians, John Eliot, wrote to a Cambridge clergyman:

> Our young students in Physick may be trained up better than yet they bee, who have onely theoreticall knowledge, and are forced to fall to practise before ever they saw an Anatomy made, or duely trained up in making experiments, for we never had but one Anatomy in the Countrey, which Mr. Giles Firmin (now in England) did make and read upon very well, but no more of that now (11).

Recognizing this deficiency the General Court on 27 October 1647 decided that it was "very necessary to such as studies physick, or chirurgery may have liberty to reade anatomy and to anatomize once in four years some malefactor in case there be such as the court shall alow of" (12).

Anatomy was central to the study of medicine but autopsies were rarely requested or approved and bodies for dissection were very difficult to secure. Hence young men might continue as apprentices for three to seven years without a practical knowledge of anatomy. Indeed it was possible for a preceptor to attempt instruction while being himself limited to textbook anatomy. In addition to many poorly trained preceptors, there were numerous quacks, armed with nothing but their ignorance and willingness to pose as healers, even educators. The need for first-hand anatomical experience was great throughout the seventeenth and eighteenth centuries. When there was a legal dissection of a hanged criminal preceptors and their students were sometimes permitted to participate or observe. Krumbhaar recounted a number of such recorded events (13). Samuel Sewall recorded in his diary on 22 September 1676 that he spent the day with Dr. Brakenbury and five other observers "dissecting the middlemost of the Indian executed the day before", and that one of the group, taking the heart in hand, "affirmed it to be the stomack" (14), a gesture which dramatized the ignorance of the time. Packard reported six different autopsies in New England from 1674 to 1678, which perhaps have more significance for medical education than for criminology. Hartwell, writing in 1881, cited an Order of Council of Lord Baltimore of 20 July 1670 in which "Chyrurgeons" John Stansley and John Pearce were ordered to view the head of one Benjamin Price, supposed to have been killed by Indians. There is no indication that this autopsy was of educational significance except for the two surgeons. The celebrated autopsy on Governor Henry Slaughter of New York (1691) was the result of his sudden and unexpected death soon after the questionable conviction and execution of two of his political enemies for high treason. Dr. Johannes Kerfbyle, Leyden graduate, and five associates made the post-mortem examination for which they received eight pounds eight shillings. The governor's death has been variously ascribed to poisoning, delirium tremens and to bitter remorse for the execution of guiltless men. The abstract of the report from Dr. Kerfbyle and associates suggested to Walsh's perceptive mind that pulmonary embolism was the immediate cause of death: ". . . That the late Governour dyed of a defect in his blood and lungs occasioned by some glutinous tough humor in the blood which stopped the passage thereof and occasioned its settling in the lungs, which by other accidents increased until it caryed him off a sudden . . . ." (15).

The extent to which medical apprentices may have benefitted from the Slaughter autopsy is not clear, but some of the participants were no doubt preceptors. About sixty years later there was another New York incident, that of the executed criminal, Hermannus Carroll. In this case not an autopsy but an injection and dissection were conducted by Drs. John Bard and Peter Middleton specifically for the instruction of the young men then engaged in the study of medicine (16).

According to Krumbhaar a more formal and significant effort to instruct with the use of a cadaver took place in Philadelphia almost two decades earlier. Thomas Cadwalader, descendant of two early Philadelphia families, had served an apprenticeship before going to the University of Rheims and to England to further his studies. Presumably under the celebrated William Cheselden in England, he developed skill and reputation as a prosector, an anatomical skill rare in Europe and virtually unknown in the colonies. On his return to Philadelphia about 1730 and at the insistence of physicians and students, he consented to give a course of public lectures on the cadaver. Among his practitioner pupils was William Shippen, Sr., whose son William later received an Edinburgh M.D. degree and was associated with John Morgan in founding the College of Philadelphia's first medical curriculum (17).

The controversial William Douglass, M.D. of Boston was a physician to be reckoned with throughout the first half of the eighteenth century. He was ultimately distinguished for his definitive paper on the scarlet fever epidemic which invaded Boston in 1735-1736 (18). His preface addressed to the medical society in Boston had distinct educational overtones. He wrote because he "thought it might prove a piece of humanity and benevolence" after diligent observation of many cases which he endeavored "to reduce . . . to some easy distinct Historical and Practical Method". Secondly, he hoped to influence physicians in other colonies to record their observations of "Epidemical Distempers" to the "considerable advantage in Physick" (19).

Public spirited and frequently referred to as a learned physician, Douglass continued to be disputatious throughout life. He appears to have been respected more than loved. Writing a century and a half after the death of Douglass, James Mumford in 1903 named Douglass with Zabdiel Boylson and Cadwallader Colden as the three conspicuous physicians of their generation (20). Douglass' contemporaries were not always so charitable. Dr. Alexander Hamilton of Annapolis, Maryland, a 1737 Edinburgh graduate, visited New England in the summer of 1744. As a voracious tourist he recorded in his *Itinerarium* candid, charming, and even severe criticisms of whatever and whomsoever he observed. He rated William Douglass as "a man of good learning but mischievously given to criticism and the most compleat snarler ever I knew". He disparaged Douglass for his dependence on "empirycism or bare experience as the only firm basis upon which practice ought to be founded" (21). Hamilton's remarks may have been influenced by the fact that Douglass was a Utrecht graduate (1712) and not a fellow alumnus of Edinburgh. The disparity could have been due in whole or in part to the mere fact that Hamilton was of a quarter of a century later vintage. Some scholars have suggested that colonial medicine was as much as fifty years behind that of Britain and the continent, a possible handicap for Douglass even though originally educated abroad.

Hamilton had more to say about Boston's distinguished fifty-three year old bachelor physician and educator:

He has got here about him a set of disciples who greedily draw in his doctrines and, being but half learned themselves, have not wit enough to discover the foibles and mistakes of their preceptor. This man I esteem a notorious physicall heretick, capable to corrupt and vitate the practice of the place by spreading his erroneous doctrines among his shallow brethren (22).

After a week of visiting nearby towns Hamilton returned to Boston and one evening found himself in the company of "Dr. Clerk, a gentleman of fine naturall genius, who, had his education been equivalent, would have outshone all the other physitians in Boston." But Dr. Douglass was present and gave direction to the conversation with a tirade against Dr. Herman Boerhaave, noted Dutch medical educator and godfather of the University of Edinburgh medical school. Hamilton defended Boerhaave with an appropriate salvo directed at Douglass. The battle apparently continued throughout the evening. Hamilton later confided to his diary:

I could not learn his reasons for so vilifying this great man, and most of the physitians here (the young ones I mean) seem to be awkward imitators of him in this railing faculty. They are all mighty nice and mighty hard to please, and yet are mighty raw and uninstructed (excepting D. himself and Clerk) in even the very elements of physick. I must say it raised my spleen to hear the character of such a man as Boerhaave picked att by a parcell of pigmies (23).

Douglass can be forgiven much of his egocentric ostentation simply because he did assume the role of educator at a time when educated educators were in critical shortage. Apart from certain theories there was probably little difference between the opinions of Hamilton and Douglass regarding the quality of prevailing medical practice and what was happening under the apprenticeship system in all the colonies. Near the close of his career Douglass assumed to speak for all the colonies (1749) when he asserted

... if we deduct persons who die of old age, of mala stamina vitae or original bad constitutions, of intemperance, and accidents, there are more die of the practitioner than of natural course of the distemper under proper regimen. The practitioners generally without any considerable thought fall into some routine method, and medicines, such as repeated blood-lettings, opiates, emetics, cathartics, mercurials, Peruvian bark (24).

"In our various colonies", he reasoned, "to prevent a notorious depopulation from mal practice in medicine or care of diseases", there should be regulation of the profession, "which at present is left quite loose". With specific reference to the preceptorial system he added:

A young man without any liberal education, by living a year or two in any quality with a practitioner of any sort, apothecary, cancer doctor, cutter for the stone, bone-setters, tooth-drawer, etc. with essential fundamental of ignorance and impudence, is esteemed to qualify himself for all branches of the medical art, as much or more than gentlemen in Europe well born, liberally educated (and therefore modest likewise) have travelled much, attended medical professors of many denominations, frequented city hospitals, and camp infirmaries, etc. for many years (25).

Douglass further insisted that in practice "a dull application" was no answer. "There must be a suitable genius, and sometimes a particular paroxysm of imagination, as is remarkable in poets and painters, and as I have observed in myself, in the diagnostick part of our profession." Knowledge, that is observation and sagacity, he held to be the two prime requisites in a physician. Many of America's mid-eighteenth century practitioners he held to be "an impudent delusion and a fraud" (26).

Douglass at least could have footnoted his dark picture with the observation that a very few young colonials began going to Leyden and elsewhere to study medicine early in the century (27). He perhaps did not know that of those who began going to Edinburgh late in the second quarter of the century, John Moultrie of South Carolina was the first (1749) to graduate. It is not known precisely how many Britishers or continentals, recent medical graduates from Edinburgh or some other school in the British Isles or on the continent, were going out to enjoy the professional emoluments to be found in established new world communities by or before mid-century, but there was such a trend. Bridenbaugh in his "Introduction" to Dr. Alexander Hamilton's *Itinerarium* through New England and the middle colonies (1744) notes that "well-educated young Scots, particularly physicians, were going out to America at this time in search of a better market", to which Hamilton's diary bears witness (28). One reason for this trend toward the New World was answered in part by Benjamin Franklin in a letter to David Hume of Edinburgh after Franklin announced that soon he would be returning to Pennsylvania from London where he had been serving as agent of the colony. Hume had lamented Franklin's forthcoming departure as a substantial loss to the intellectual life of British society. Noting the practical law of supply and demand Franklin replied as one speaking for all professional and intellectual Britishers who might migrate to the New World: "You have here at present just such a plenty of wisdom. Your people are, therefore, not to be censured for desiring no more among them than they have; and if I have any, I should certainly carry it where, from its scarcity, it may probably come to a better market" (29).

Certainly the trend by 1750 had long range possibilities for enriching and upgrading the various provincial cultures which land, people, and time were formulating in the Western Hemisphere. There were symptoms of social, professional, and educational awareness among perceptive medical practitioners who collectively began to sense some of their responsibilities as medical communities, especially in the larger cities.

The Revolutionary Period, 1751-1800

John Adams who came to maturity during this period remarked that the American Revolution took place "in the minds and hearts of the people" before the first shots were fired at Lexington in 1775. His compatriot, Dr. Benjamin Rush of Philadelphia, declared in 1783, two years after Yorktown, "The American war is over but this is far from being the case with the American Revolution". It was apparent to both men that the military hostilities were central to but not all inclusive to the Revolution. Adams knew the Revolution was long in the making and Rush merely meant that, the war being over, the stage was set for the further shaping of a new social, political and cultural order in an emerging nation, based upon the principles for which the colonials had fought (30).

The vicissitudes of colonial life and commerce had wrought changes in thought and action that ultimately culminated in the Revolution. From the accession of William and Mary (1689) to the fall of Quebec (1763), New France exerted continuous pressure on English America. Hit and run engagements between English and French forces in America reflected the greater build-up of hostilities then in progress in Europe. The English colonies, more or less geographically isolated and founded under a variety of legal instruments, had drifted into the virtues and vices of provincialism before the "Gallic peril" strongly suggested intercolonial cooperation. The ensuing war (French and Indian, 1754-1763) convinced discerning colonials of the value of cooperation. Provincial legislatures with pride appropriated sums for military purposes, granted supplies to British forces, supervised their own colonial militia, and contended that their own men under colonial officers were equal to British regulars.

Many of the American regimental surgeons and assistants were not well trained or qualified but the experience was often educational for men of lesser experience when teamed with older, better prepared surgeons. The educational opportunity was even better when young American practitioners were assistant surgeons in British regiments whose surgeons were usually well trained and equipped. In military medicine the Seven Years' War, as the conflict was known in Europe, proved to be for American surgeons a curtain raiser for the more enervating experience of the Revolution.

The most significant medical event in the decade of the 1750's was the chartering (1751) and opening (1752) of the Pennsylvania Hospital, the first such institution in the British Colonies planned and designed to be a general hospital (31). The petition to the Assembly of Pennsylvania for a charter was written by Benjamin Franklin, who was the principal fund raiser and promoter, but in his *Autobiography* he gives full credit to Dr. Thomas Bond for conceiving the idea (32).

When the Pennsylvania Hospital Board of Managers in 1755 asked Benjamin Franklin to provide the inscription for the cornerstone of the institution's first permanent home, Philadelphia's bellwether of science and culture scribed as follows:

> In the year of Christ
> MDCCLV
> George the Second happily reigning
> (For he sought the happiness of his people)
> Philadelphia flourishing
> (For its inhabitants were public spirited)
> This building
> By the bounty of the government
> And of many private persons,
> Was piously founded
> For the relief of the sick and miserable:
> May the God of mercies
> Bless this undertaking (33).

Physicians, managers, donors, and a select concourse of citizens on 27 May 1755, marched from the temporary quarters (first occupied in 1752) to the new site where the white marble bearing Franklin's deathless prose was made fast with mortar.

Although the new building was not occupied until 1765, the ceremony was symbolic of a germinating community and provincial awareness of the social and cultural obligations of a society which aspired to greatness. Dr. Thomas Bond and Franklin built firmly on a clear definition of the proposed institution's charitable role and a fiscal plan whereby the Provincial Assembly would match funds with private donors (34). Thus a pattern of public and private cooperation marked the beginning of the hospital movement in the English colonies. Although the original articles of incorporation of the Pennsylvania Hospital did not spell out educational objectives, they readily fell into place within the scope of the broad charter when circumstances decreed the need (35).

As a matter of fact, the founding of the Pennsylvania Hospital was a significant preamble to the founding of the colonies' first medical school at the College of Philadelphia in 1765, or to change the metaphor, the Pennsylvania Hospital was a fountainhead of the mainstream of formal medical education in the United States. Early in the history of the hospital, members of the attending staff utilized hospital patients in the training of their apprentices who, to some extent, served as house officers and hospital pharmacists.

Apart from the establishment of the Pennsylvania Hospital and scattered efforts to establish local or provincial societies within the profession, medical men had not yet begun to manifest the measure of gregariousness and sense of obligation which later characterized the profession. Cooperative planning of institutional medical education did not receive the collective attention of medical men themselves before 1765. Consequently American colonial physicians were a heterogeneous, although not an inconsequential lot. Of an estimated 3500 practicing physicians at the outbreak of the Revolutionary War, no more than 400 are believed to have held a *bona fide* medical degree (36).

Certain significant educational processes were in motion before and after the Revolution, but the apprenticeship continued to be for the majority the conventional access to the practice of physick and chirurgery. By the end of the eighteenth century, thirty-two American colleges possessed charters, authorized by royal decree or legislative act, which were sufficiently broad in their wording to justify the conferring of medical degrees, but only ten instituted any kind of medical instruction before 1 January 1801.

Before the Revolutionary War two educational institutions began offering curriculums leading to a medical degree, the College of Philadelphia in 1765, and King's College in New York City in 1767. Due to the exigencies of the war both of these programs were interrupted. During its brief span of pre-war existence, the College of Philadelphia conferred twenty-nine Bachelor of Medicine degrees and five Doctor of Medicine degrees. Within a shorter period, King's College conferred twelve of the former and two of the latter, in addition to four *ad eundem* M. D. degrees (37).

Before the turn of the century, medical instruction was started under ten different collegiate charters as follows: College of Philadelphia (1765), King's College (1767), William and Mary College (1779), Harvard College (1783), University of the State of Pennsylvania (1783), University of Pennsylvania (1792), Columbia College (1792), Queen's College (1793), Dartmouth College (1797), and Transylvania University (1799). During the

Revolutionary War (1779) William and Mary College established a chair in anatomy and medicine, but only one medical degree was ever conferred, an honorary M.D. in 1782 (38). Two other institutions, Yale College and Washington College of Chestertown, Maryland, offered no instruction during this period but gave both *ad eundem* and honorary M.D. degrees. Transylvania University conferred no degrees in the eighteenth century. The charter of the Connecticut State Medical Society empowered it to confer the doctor of medicine degree. Since this society offered no instruction, the fourteen M.D. degrees which it handed out classify as *ad eundem* or honorary.

Frederick Waite has logically classified the medical degrees of this late eighteenth century period approximately as follows:

M.B.—the first medical degree, offered by America's first medical schools at the College of Philadelphia and King's College, and patterned after the British system. All schools in time abandoned it for the initial M.D.

Advanced M.D.—available to holders of the M.B. degree, after an interval of time and evidence by examination or presentation of a thesis, of satisfactory professional progress.

Initial M.D.—the conventional Doctor of Medicine degree which all but Harvard and Dartmouth had substituted for the M.B. and the advanced M.D. by the turn of the century.

Honorary M.B. or M.D.—degrees given as a mark of distinction to apprentice-trained or self-made physicians who were deemed competent and worthy of the honor.

Ad eundem M.B. or M.D.—degrees conferred on physicians who had previously received medical degrees. At times, the conferring of an *ad eundem* degree was not distinguished from an honorary degree.

According to Waite's classifications and findings, each institution listed in Table 10 (including the pre-war activities of College of Philadelphia and King's College) conferred its first degree in the year indicated and conferred the types and number of degrees as enumerated, before 1 January 1801 (39).

Including all types of medical degrees, a total of 335 were conferred before the end of the century. Waite's researches reveal that twenty-three of the total were duplicates. This means that only 312 individuals received degrees. Almost two-thirds (199) of the total number of degrees were conferred in Philadelphia, eighty-five were in New England, and forty-nine in New York.

The history of most of the twentieth century medical schools which trace their lineage from the eighteenth century is characterized by legal tangles and professional quarrels in their early years. The medical schools of the University of Pennsylvania, Harvard College, and Dartmouth College only entered the nineteenth century with a good lease on life.

The budding of American institutional medical education during the last

TABLE 10
AMERICAN MEDICAL SCHOOL STATISTICS BEFORE 1 JANUARY 1801

First Med. Degree	Institution	M.B.	Adv. M.D.	Initial M.D.	Hon. M.B.	Ad Eundem M.B.	Hon. M.D.	Ad Eundem M.D.	Total
1768	College of Philadelphia	29	8	8					45
1780	University of the State of Pennsylvania	67	7					1	75
1792	University of Pennsylvania		1	78					79
1768	King's College	12	2					4	18
1793	Columbia College			16					16
1792	Queen's College	3		6			3	3	15
1783	Harvard College	29	1		1		16	4	51
1782	Dartmouth	9				1	3		13
	Yale College						5	2	7
1782	William and Mary College						1		1
	Washington College						1		1
1793	Connecticut State Medical Society						10	4	14
	Totals	149	19	108	1	1	39	18	335
	Duplications								23
	Individuals who received degrees								312

three and one-half decades of the eighteenth century was extensively influenced by old-world medical education, especially Scottish, English and Dutch. The medical apprenticeship in England was seven years in length, patterned somewhat after the guild system of the crafts. A youth wishing to study medicine was indentured to his master who entered into a contract with the parents or guardian. The medical apprenticeship in America was rarely more than three years, and legal indenture was not common. In the eighteenth century an increasing number of Britishers began entering medicine by way of university study which did not prescribe an apprenticeship. The University of Edinburgh, in particular, became the mecca of American youth who sought institutional instruction after an apprenticeship and were financially able to study abroad. Clinical medicine was taught in a more challenging and effective fashion at Edinburgh than at Oxford or Cambridge. There was also a very practical reason why Americans, until early in the nineteenth century, favored Scottish medical education. The English schools required a B.A. degree for admission and after three years of attendance on lectures in Latin, a Bachelor of Medicine degree was granted. The M.D. degree came only to holders of the M.B. degree after seven years in practice, presentation of a satisfactory thesis, and the passing of additional examinations. At Edinburgh, although no preliminary degree was required, some intellectual competence and a knowledge of Latin, natural history, and related areas were expected. Lectures were in English. Attendance on three terms, submission of an acceptable thesis, and passing of examinations were the requirements for a Doctor of Medicine degree. John Morgan and William Shippen, who pioneered the College of Philadelphia curriculum in medicine in 1765, were Edinburgh graduates, as were also 112 other Americans by the end of the century. An equivalent number studied in Edinburgh without receiving the degree. Morgan, Shippen and other Americans studied Anatomy under John and William Hunter and received clinical instruction in London Hospitals which was more extensive than bedside instruction in Edinburgh. Lesser numbers frequented continental schools (40).

The founders of the College of Philadelphia medical school in 1765 conceived of their two-term curriculum as a supplement to and not a substitute for the preceptorial system. As a matter of fact, the requirements for the medical degree in this first school included evidence of a satisfactory apprenticeship and a knowledge of pharmacy. Students entering upon a course in medicine were required to be college graduates, or in the opinion of the trustees and professors to have acquired a satisfactory knowledge of Latin, mathematics, and natural and experimental philosophy. Each student was to attend one course of lectures in anatomy, *materia medica,* chemistry, and theory and practice of physic. The specified requirements also included one course of clinical lectures and one year of attending the practice of the Pennsylvania Hospital. The proximity and adequacy of this hospital was a vital factor in the initial success of America's first medical school.

On completion of this curriculum, candidates recommended by the faculty were permitted to appear for a public examination leading to the M.B. degree. The M.D. degree was reserved for candidates who, three years or more after receiving the M.B. degree, were at least twenty-four years of age and had prepared and publicly defended a thesis approved by the faculty (41).

In 1789 the reorganized medical faculty of the College of Philadelphia abolished the M.B. degree. When in 1792 the medical department of the University of Pennsylvania superseded the two previous medical institutions in Philadelphia it offered no M.B. degree but conferred the M.D at the close of a three-year preceptorship and two terms of residence in course, which activities could run concurrently. The premedical requirements were eliminated and a required course in natural and experimental philosophy was offered as a part of the medical curriculum. The thesis requirement was retained but made optional in Latin or English (42).

Early in the history of the University of Pennsylvania and other eighteenth century medical schools, serious functional problems arose from a shortage of textbooks and teaching materials and at times from a dearth of teachers. The faculty compensated by requiring each student to attend two complete courses of lectures, the second year being a repetition of the first, rather than two terms of graded curriculum. In this way, it was believed the student would more thoroughly cover the subjects. King's College probably set the pattern for this policy of doubtful value by requiring their advanced M.D. candidates to spend one term repeating the regular lectures. This obvious pedagogic foible, at first justified as necessary, became by 1820 the universal practice in American medical schools and was naively justified because it was the custom in Philadelphia (43).

With these early modifications in a soundly devised plan of medical education, a depreciated and diluted scheme—including attendance at all lectures each year—began to characterize medical education in the United States. It is a somber fact that the great majority of American medical schools, rapidly increasing in numbers during later decades contributed actively or by default to the debasement of standards for admission to the profession.

The Reluctant Years, 1801-1860

Before evaluating medical education in the United States during the nineteenth century it is well to review the American scene. For the young republic it was a time of vast geographic expansion, domestic turmoil, social upheaval, and the beginning of the industrial revolution. The Louisiana Purchase (1803) and the Monroe Doctrine (1823) were unmistakable signs of national consciousness and responsibility. Watered by the well springs of personal freedom, ideologies of many hues and casts appeared to flourish in the shallow intellectual soil of a rising civilization. Some movements rooted and became a part of America; others withered but enriched the seedbed of a new culture.

Alexis de Tocqueville, who visited extensively in the United States during the middle of this period, was impressed with American emphasis on the practical while the exploration of the common sources of knowledge and the development of basic scholars in the sciences was grossly neglected. He attributed this exceptional outlook principally to America's Puritan origin, its exclusively commercial habits, and to the very nature of the country being occupied. Only the proximity of Europe with its cultivation of the arts and sciences, he observed, permitted such neglect in the United States, without a relapse into barbarism (44).

There were actually some signs of relapse. By 1830 practically all states had some sort of regulations designed to protect the public and to restrict, to some extent, admission to the practice of medicine. But the groundswell of Jacksonian democracy, 1830-1850, swept away these protective acts and left the profession and public almost without definitive and protective laws in their respective commonwealths. Thus for the space of several decades, while American legislators burned incense to the rights of the "common man", the portals of medical practice were open to thousands of poorly qualified Americans, not to mention hordes of irregulars who rode the crest of succeeding waves of immigration from Europe. Not until late in the century were intelligent forces within and without the profession able to bring about a restoration of public understanding and responsibility.

Close behind the cutting edge of the American frontier followed men of ideas and special competence—merchants, artisans, lawyers, doctors, teachers, clergymen. Where the social order was elementary, industrious pioneer physicians often combined one or two other callings with medicine in order

TABLE 11
Estimates* of the Number of Medical Students and Medical School Graduates at Intervals

Year	Medical School Attendance	Medical School Graduates
1810	650	100
1840	2500	800
1860	5000	1700
1870	6500	2000
1880	12,000	3200
1890	16,500	5000
1900	25,000	5200

* Made by the author or drawn in whole or in part from the following sources: N. S. Davis, *Contributions to the History of Medical Education and Medical Institutions*, 1877; Samuel W. Butler, *The Medical Register and Directory of the United States*, 1878; *Polks Medical Directory*, 1890, 1910; *Report of the Commissioner of Education*, 1872, 1876, 1881, 1890–1891, 1899–1900; A. D. Bevan, Report of the Council on Medical Education, *Journal of the American Medical Association*.

to survive. Others, especially the herbalists and leeches, became roving practitioners. Medicine in some form kept up with a shifting frontier (45).

Historian Dixon Ryan Fox once pointed out that four stages are discerned in the transit of culture (46). As applied to the passage of the body of medical knowledge from Europe to America, these four steps are: first, when the Colonial pioneer communities received their medical succor from foreign trained practitioners who accompanied or soon joined them; second, when native youth returned to the old country for professional indoctrination in medicine; third, when schools were instituted in the new land; fourth, when there were sufficient native institutions to make the profession self-sustaining (47). Within the borders of the American scene, with its repeated waves of westward migration, professional competence was approached in much the same order, from provincial to metropolitan status. The transit of medical education through these years of "manifest destiny", although lighted occasionally by flashes of individual brilliance, were for the most part dull decades of educational stagnation and trade school competition.

The American system of medical education, a combination of apprenticeship and public lectures developed into a characteristic pattern with unique features during the early decades of the nineteenth century. It was an era marked by mushroom growth of medical schools. The trend continued unabated, only to be modified in character by certain social forces and some pressure from within the profession in the decade of the fifties. Subsequently the exigencies of the Civil War closed practically all of the southern schools, weakened many northern schools through the withdrawal of southern students and the departure of some northern students and faculty members for military service.

No significant uniformity prevailed in the manner by which physicians associated themselves together in the capacity of a medical faculty of some university, college, or independent school. There were both apparent and hidden motives. Usually a desire to elevate the standards of the profession was declared to be the principal motive. The provincial urge to make or at least to advertise the facilities of an institution or community as equal or superior to that of its competitors, played an important role. All too often petty differences, selfishness, pride, and a lust for publicity and a large income were controlling factors with the individual.

Although many methods were employed in formulating plans for and the establishment of medical schools, they group rather conveniently into five classifications with some allowance for variation. Typical illustrations will serve to emphasize differences and to highlight the panorama which spans these years.

The medical school conceived and developed as an integral part of a well-established collegiate institution had the best claim to legitimacy and longevity. By "integral part" is meant a medical school or department subject to general administration and to the trusteeship of a board competent to interpret the needs of medical education. It was a sign of a broad educational understanding when the initiative to add such a department emanated largely from the trustees and not from an outside group. The medical department of the College of Philadelphia, previously noted, falls under this classification. The trustees were definitely committed to instituting the school before John Morgan delivered his famous two-session commencement address in which he advocated such an establishment (48). The University of Michigan medical school, similarly founded in 1849, paid its medical professors salaries. Paying of regular salaries to teachers in medical schools was all but an unknown custom in the middle of the past century (49). The policy evidenced strong administrative ties between the regents and the medical school.

A more common scheme by which teaching faculties sprang into being as departments of established institutions was "grafting". Self-appointed and self-organized groups of practitioners often knocked at the door of a college, asking to be accepted as an adult medical department. Almost as often trustees replied favorably but they seldom accepted any financial responsibility. They saw in a medical department, though it might be located in another state, an opportunity to boast of expansion and enhance the school's name. The arrangement was simple since the conventional college charter of that time authorized the giving of any and all degrees. The medical professors continued to carry on as they saw fit with their added prestige. Such faculties gave little attention to mother institutions except at the commencement season, when medical graduates sallied forth with their parchments bearing the seal of a college which they may never have seen and which had scant knowledge of its medical alumni. When Queen's College of New Brunswick, New Jersey, in 1792 and 1793 conferred fifteen medical degrees, it was through the agency of a medical faculty in New

York City. Another alliance of this sort was the liaison in 1820 between Castleton Medical Academy of Castleton, Vermont, and Middlebury College in the same state. The academy had previously received a charter to grant medical degrees, but the Castleton professors were meeting obstacles which they felt the new relationship would correct. By 1827 the ties were severed, and Castleton continued as an independent proprietary college (50). The Clinical School of Medicine (1827) of Woodstock, Vermont, allied itself with Waterville (Colby) College, Waterville, Maine, in 1828, with Middlebury College in 1833, and received an independent charter as Vermont Medical College in 1835. Schools following such a history did not enjoy a healthy existence. Professorial feuds or quarrels between the trustees and the adopted faculties usually brought about dissolution but not always without a struggle. In 1854 the trustees of Hampden-Sidney College in the south of Virginia sought through the Virginia legislature to retain control of its seceding medical department in Richmond, to which it had never contributed a dime. The professors won and were rechartered as the Virginia Medical College (51).

A medical school planned and executed as one of the departments in the initial structure of a university or college is a third classification. Perhaps the best illustration of this type is the medical department of the University of Virginia, established in 1825. A medical faculty was incorporated in Thomas Jefferson's original plan for the institution. Although only one chair was established at first and clinical instruction was minimized for a time, the university relationship was sound and broadly based for that time. The trustees not only acknowledged medicine as a profession but incorporated it in the curriculum as a discipline worthy of the attention of an educated gentleman (52).

The fourth grouping comprises schools sponsored by professional societies. The College of Physicians and Surgeons of New York City was established in 1807, by an enabling act of the legislature which named the newly organized Medical Society of the City and County of New York as the corporate body of the proposed school (53). A local professional society with all its politics and professional rancor did not prove to be a suitable governing body. Somewhat less trouble and conflict were experienced by the College of Medicine at Maryland in Baltimore, founded in 1807 as a child of the Medical and Chirurgical Faculty—the state society (54). Neither institution, however, continued long under professional society control or influence.

The fifth and largest group includes the many independent medical schools, proprietary in nature, which multiplied rapidly during this period. The first of such institutions was the medical department of the University of Maryland. The faculty of the College of Medicine of Maryland, impressed by its initial success, envisioned itself as the nucleus of a university to be composed of several schools. In 1812 it secured the passage of a legislative act authorizing it to "constitute, appoint, and annex to itself" faculties in law, divinity, and the arts and sciences. The act endowed each faculty with the power to fill vacancies in its teaching staff and made the provost and the professors of the several departments or schools the regents of the university. The professors apparently failed to sense that universities traditionally had developed as a superstructure on the well laid foundation of a school of liberal arts and that medical schools thrived best when planned and developed as an adjunct to a sturdy central structure. The charter which they secured created a sort of mythical university, since all departments, medicine excepted, failed to materialize with any degree of permanency for some years (55). The professors and regents were one and the same body, endowed with the power of self-perpetuation. This vesting of the powers of trusteeship in the faculty was a departure from sound practice, marking the beginning of evil days for American medical education.

In the decades following, other state legislatures chartered independent groups of self-appointed professors over whom there was exerted little or no restraint. Not all proprietary schools held uniformly low standards, but the pressure of competition and the absence of detached and independent governing bodies without personal pecuniary interests forced standards downward.

When the west opened up, the wilderness was soon spotted with medical faculties determined to participate in the race for local or provincial priority in the training of physicians for the migrating frontier. Individuals and faculties laid great emphasis on their inalienable constitutional right to freedom of action. Doctor Samuel Annan at the opening of the Kentucky school of Medicine, in 1850, publicly declared:

> Thought and action, within the limits prescribed by regard for the public weal, are free as the air which goeth where it lists; and it would, therefore, be no more consistent with truth and fact to assert that individuals or associations have not the right to engage in the instruction of Students of Medicine, than to say they have no permission to organize schools, academies, and colleges for general education (56).

Wars between schools and faculty feuds over questions of establishing or moving schools inspired some of the most bitter enmity imaginable. The public weal, as Annan labeled the general welfare, was often not a deciding factor in faculty and trustee decisions. The presence of two or more rugged individualists on the same faculty or on neighboring faculties held the possibilities of factional strife with shocking emotional outbursts.

The success of the Transylvania medical department in the early thirties inspired some Kentucky physicians who were not enjoying the emoluments of lecture fees to think and talk of another medical school. Charles Caldwell, Transylvania's ablest dealer in caustic repartee, delivered an address entitled: *Thought on the Impolicy of Multiplying Schools of Medicine,* which was published in 1834. The unnamed object of Caldwell's diatribe was Transylvania's alumnus, James Conquest Cross, distinguished more for brilliancy than for emotional stability. There ensued a two-way abusive harrangue. When news leaked out that Cross had secured permission to establish a medical department in Louisville, under the charter of Centre College of Danville, the friends of Transylvania attacked the scheme with telling vigor. In a pamphlet acknowledging defeat, Cross matched Caldwell in the art of unbridling the pen.

> We have stifled the barking of their curs, but now upon our trail, we hear the deafening and petrifying cry of their unkennelled bloodhound, who is to hunt us down, and achieve our destruction. But to speak less figuratively, Dr. Caldwell, the Champion of the Transylvania Medical School, has come out in a pamphlet of between thirty and forty pages, filled with abuse, which in malice, hatred, virulence and vindictiveness, has never been surpassed (57).

Doctors Caldwell and Cross achieved early distinction as educators but their vitriolic natures made them increasingly unacceptable to their colleagues. Apparently having failed to develop the arts of tolerance and unselfishness in their professional relationships, both men gravitated from the center of the stage and died in relative obscurity.

The Madison *Wisconsin Argus* announced on 26 September 1848, the forthcoming opening, in Rock Island, Illinois, of a branch of Madison Medical College of Wisconsin. The fact was that plans for a school in Madison never matured, so the branch had no trunk. Dean M. L. Knapp's rhetorical skill must have impressed the aggregation of plainsmen who, on the first day (7 November), crowded the lecture hall on the Illinois side of the Mississippi.

> No honor could be more congenial to my feelings for since enduring some fifteen years of toil in the profession in Illinois, to find myself at last in this El Dorado of the flow'ry West, on the banks of a lovlier than the "Blue Moselle," presiding as accoucheur at the birth of a new institution of Medical Learning, pure, promising, and undefiled by perfidy, comely in every feature, and limb, matchless, indeed, at her birth, is, to me, a source of more unalloyed happiness than I could enjoy were I elected to the Chief Magistracy of a State (58).

After a term of three months, Dr. Knapp's matchless infant was moved to Davenport, Iowa, where it lived briefly under a new charter as the College of Physicians and Surgeons of the Upper Mississippi. In the spring of 1850 the peripatetic dean again packed his valise and relocated at Keokuk, Iowa, where his child was adopted as the medical department of the State University of Iowa in 1851 (59).

The ephemeral nature of many medical schools, especially the country or frontier institutions, is well illustrated by La Porte University medical department founded by a New Yorker, Daniel Meeker, who took his first course of medical lectures at the College of Physicians and Surgeons of the Western District of the State of New York, in a village named Fairfield.

Continuing his trek westward Meeker paused long enough to attend a second term at Willoughby Medical College (named for one of Fairfield's favorite professors) in the Western Reserve. Settling at La Porte, Indiana, Dr. Meeker began practising and training office pupils. An attorney friend who also had students joined with him in projecting a university composed of two professional schools—law and medicine. Formal instruction in medicine began in 1842 (60). There was a chamber-of-commerce tinge to an announcement in the university *Circular* for 1844-1845: "A Female Seminary is about being organized and the town of La Porte promises to become the Athens, as it is now Eden, of the Northwest" (61).

Attendance reached a peak three years later but female pulchritude, Grecian culture, and Edenic landscape were not enough to sustain the university. It died for lack of support in 1850, mainly because of a shortage of anatomical material and clinical facilities. According to Helen Clapesattle, Dr. William V. Mayo, an 1850 graduate, (later a pioneer physician in Rochester, Minnesota) and two colleagues tried to revive the school in 1852. Destruction of the buildings by fire during their first term disheartened the promoters who abandoned the project (62).

The major cities of the United States, as well as rural communities, witnessed the struggle to advance medical education and to benefit personally by participation. Some events in Philadelphia are sufficient to illustrate. The idea of a second medical school in Philadelphia after the merging of the two schools in 1792 remained dormant for some years. After George McClellan, a Yale arts graduate, finished in medicine at the University of Pennsylvania in 1819, he made it easier for ambitious or disgruntled practitioners to begin thinking of a second school. Within two years McClellan had so many private pupils he had to open a lecture room on Walnut Street. Before long the young preceptor was carrying the torch for a new school which had both zealous support and ardent opposition. The movement culminated in the establishment of Jefferson Medical College of Philadelphia under the charter of Jefferson College of Canonsburg, Pennsylvania. The new school opened in November 1825, with its 107 students having no significant effect on medical attendance (440) at the University of Pennsylvania (63). The spirit of proprietary independence which characterized Jefferson from the first became very real in 1838 when legal ties were severed by mutual consent, and the legislature launched Jefferson on an independent existence with the same powers and restrictions as at the University of Pennsylvania.

Complete independence may have precipitated the faculty quarrel which resulted in the vacating of all faculty chairs in 1839. George McClellan was not among the old faculty re-appointed. At forty-three years of age he set out with characteristic vigor to organize another medical school, the third for Philadelphia. He consummated a rather loose agreement with Pennsylvania College of Gettysburg, but with full corporate privileges as the medical department. In 1840 he secured from the legislature a charter amendment which confirmed the right of Pennsylvania College to confer medical degrees in Philadelphia (64). The 1840 *Announcement* noted that the clinical facilities of Blockley and Pennsylvania Hospitals were accessible to its students. An 1849 charter change made the medical department of Pennsylvania College little short of independent. The faculty was actually given the right to fill vacancies in the board of trustees. Although attended by some success this school eventually suffered fatally from internal feuds and financial difficulties. In 1859 its faculty resigned and its assets were merged with the Philadelphia College of Medicine, another proprietary school, which was struggling to exist under the leadership of Doctor James McClintock but succumbed in 1861. This, however, was not the end of proprietary medical education in Philadelphia.

A survey of all medical schools in the country during this period reveals that the proprietary factor in some form and in varying degrees characterized the organic structure of practically every school. Most of the medical faculties adopted by collegiate institutions continued virtually as autonomous institutions. Rarely did the foster parent take any financial responsibility. The lack of adequate clinical facilities and sufficient anatomical material was an insoluble problem for rural faculties and for some urban schools. The realization that they were of necessity the masters of their fate bred a spirit of independence among many medical faculties, which frequently manifested itself in quarrels, dissension, and even disreputable conduct.

A student enrolling in medical school paid a matriculation fee of three to five dollars. He then paid a lecture fee to each teacher whose course he expected to take. A printed card of admission signed by the professor was the student's access to the course. Teachers without formally printed cards scribbled and signed the equivalent on the backs of playing cards. Separate tickets were issued for practical anatomy and hospital attendance. With a set of such cards the student could go to the dean's office and enroll in the school register or "album". The cost of tickets varied throughout the country, fifteen dollars being the average for lecture tickets. Dissecting tickets were five to ten dollars, and the hospital or clinic charge was never more than ten. A graduation fee ranged from ten to forty dollars, but was usually fifteen or twenty dollars. Schools required a statement certifying that the candidate had served acceptably a term of three years of apprenticeship. The apprenticeship fee which ranged from fifty to 100 dollars per year must be added to the cost of medical education. If the location of the school was convenient the student could attend lectures during the course of his preceptorship. With five or six lecture fees to pay each term, plus all accessory fees including the apprenticeship, the cost of medical education could easily amount to 350 to 650 dollars and three years of time (65).

Although some schools attempted to operate with no income other than students' fees, funds for capital investments—of necessity—had to come from outside sources. At times sacrificial professors dedicated part or all of their fees to developing a physical plant. More often they clung to every lecture fee offered them and joined in soliciting or luring students from whatever source available. There are many examples of legislative loans and gifts, even to privately owned institutions. Public subscription, often including the gifts of prospective professors, frequently made up the initial sum for establishing a new school. Universities and colleges were very slow to recognize or to accept any financial responsibility for such capital expenditures. Even lotteries, by state authorizations, were indulged in to raise money for buildings and to pay off mortgages (66).

A variety of terminology was employed in the classification of subject matter and the meaning of courses. Anatomy, physiology, and chemistry were regarded as the basic pre-clinical subjects before the Civil War. Modern bacteriology and pharmacology had not yet been born. Pathology as a separate discipline was in process of formulation (67). Botany and natural history were also regarded as important until late in the period, but all schools did not offer them. The principal clinical subjects were: surgery, medicine, therapeutics, *materia medica* or pharmacy, and obstetrics and diseases of women and children. Medical jurisprudence was added in many schools before 1840. Two or more subjects were commonly combined in one chair, such as anatomy and surgery.

Before the nineteenth century American medical professors were dependent on foreign texts. English authors and American or English translations of European writers were popularly received throughout this period, but after 1800 there was a growing number of American authors of textbooks. These sources, along with writers in indigenous medical journals, exercised more and more influence in American medical education.

Every medical school had a collection of books called a library. At times the school library was little more than the assembled collections of the professors. Three hundred to 1500 volumes was average for an institutional library. The library of the Pennsylvania Hospital was a valuable adjunct to medical education in Philadelphia. These decades were a period of gestation for some of America's great medical libraries.

A prime problem in all but a few city medical schools was the quiet, if not legitimate, securing of an adequate supply of bodies for dissection. After the Massachusetts anatomy act of 1831 which gave unclaimed bodies to recognized medical institutions, other states slowly followed suit. The situation relaxed somewhat, but public sentiment still made grave robbing an apparent necessity in some quarters for many years.

Of thirty-three schools reporting to the newly organized American Medical Association in 1849, only eight reported hospital attendance as obligatory. Twelve reported that no such requirement was made. The remaining thirteen were significantly silent. The schools of Philadelphia, New York, Boston, Baltimore, and other cities, which had access to relatively good hospitals and clinics, in their announcements laid great emphasis on their wealth of clinical facilities. It is known that the ward walks, the conventional method of clinical instruction, during the lecture terms were frequently so crowded that only a few students close to the clinician actually benefitted from these excursions through the hospitals. According to the historical evidence available to this author, the New Orleans School of Medicine was the first to institute a teaching method akin to the clinical clerkship. In 1857 this school began assigning each student to an individual patient. The student then did a complete history and physical which was recorded and reported. The plan included regular conferences between student and teacher which were conducted in a constructively critical manner. Students followed their cases carefully until patients recovered or died. Whenever possible, autopsies were secured and performed after the teacher and student discussed the anticipated findings (68). Clinical instruction was a major problem of nearly all medical schools of this time and a serious concern of the committee on medical education of the American Medical Association.

The movement to open the medical profession to women achieved a measure of success before the Civil War. Elizabeth Blackwell had enrolled at Geneva Medical College in the fall of 1847 and graduated with honors in 1849. By 1865, four medical schools for women had been established and several other schools had permitted the enrollment of women. Two of the women's schools were devoted to sectarian medicine not acceptable to the regulars. Certain liberals supported medical education for women because they believed in equal rights for the sexes. Curiously, some conservatives found themselves in the same camp because they believed that male physicians should not care for women (69).

Sectarianism in medical education during this period is a story in itself. The schools of the isms were often ephemeral and soon passed away, with the exception of the eclectic, homeopathic and physio-medical colleges, a few of which maintained themselves into the twentieth century. The sectarian schools existing before the Civil War contributed to the continuation of low standards of medical education.

The glaring defects in the system of medical education in the United States were long noted by competent and intelligent educators and practitioners (70). As early as 1827 delegates from medical societies in New York and the New England states met in Northampton and passed a remarkable set of resolutions designed to elevate the standards of medical education, but no school had the moral courage to embark on the recommended program (71). The Northampton plan came to naught as did many other well planned suggestions during succeeding decades because reluctance to act was more prevalent than determination. But the agitation for reform gradually grew and expanded, and became a major factor in the founding of the American Medical Association in 1847. A survey of the *Transactions* of the Association from 1847 to 1861 reveals some isolated efforts to elevate the standards of medical education but no nationwide movement was implemented by all the sincere and meaningful resolutions passed at national conventions. With no administrative authority over the practice of medicine and medical education, and with virtually no effective state regulatory acts to invoke, the American Medical Association did well in its first fourteen years to check the downward trend, by pointing the profession and its educational institutions in the direction of reform (72).

No American medical schools prior to the Civil War were comparable to the best foreign schools, but not all were short lived or flagrantly afflicted with the weaknesses of the prevailing pattern. In spite of various hindrances several university medical departments enjoyed consistent patronage at home and in some respects abroad. Among the better schools of the country when the Civil War disrupted the routine of American life were the University of Pennsylvania School of Medicine, Jefferson Medical College, College of Physicians and Surgeons of Columbia University, University of the City of New York Medical Department, Chicago Medical College (later a school of Northwestern University) and New Orleans School of Medicine.

The profession, in spite of many weaknesses, produced during this period some brilliant educators, men whose names live on in the annals of medical education—John Morgan, William Shippen, David Hosack, the Warrens, Daniel Drake, Nathan Smith, to name a few. More like them and a multitude of lesser known teachers and physicians spawned a breed of practitioners not unlike themselves, who in a time of alert provincialism and migrating frontiers succored a race of men who were dedicated to a defense of the rights of the individual and man's quest for wealth.

Years of Strife and Reconstruction, 1861-1875

Before the Southern states seceded physicians in the south were giving thoughtful attention to the contentions of Dr. Samuel A. Cartwright whose major thesis was in support of what John Duffy has termed "States Rights Medicine" (73). Cartwright believed that in addition to obvious differences of pigmentation in the whites and blacks there were other striking differences in skeletal structure and shape, size and weight of the brain, and volume of red blood in the pulmonary and arterial systems. Other writers expanded Cartwright's observations, but many Southern physicians were only willing to concede that there were some differences between a white and a slave practice. A more convincing argument developed in support of slavery was based on the assumption that the climate and diseases of the South required a different form of therapeutics. General acceptance of this and other arguments used to buttress a slave society led to the logical assumptions that young men expecting to practice in Southern states should study in Southern schools. Thus it was possible to throw the weight of sectional patriotism behind the establishment and promotion of Southern medical schools. By the time secession was a reality a number of new medical schools had been founded in Southern states, and the enrollment in previously existing schools had increased noticeably. Patronage of Northern schools by Southern students, nevertheless, had continued to be very substantial.

The Civil War had a tremendous impact on American culture. Apart from an emphasis on military medicine and surgery, the War had little effect on the curriculum of medical schools. When war seemed imminent hundreds of Southern students in northern schools transferred to schools in the slave states. But before long practically all southern schools closed for want of teachers and students who had become early and enthusiastic volunteers. It was short-sighted of the Confederacy to permit this drying up of the source of medical personnel. In both the North and South students who had completed the first year of medicine were given credit for the second year if they served as surgical assistants in the field or in hospitals for one year. Before the War was over much was learned empirically about hospital construction and administration, and the transportation of patients and hospital supplies. With ether and chloroform generally available to allay pain during operations, many surgeons were alleged to have performed amputations beyond the dictates of wisdom. With drugs treated as contraband of war by the North, the South had to substitute whiskey for ether and chloroform in the last part of the War. The years of hostility provided an abundant opportunity to study infection of wounds, both battlefield and surgical. Unfortunately Lister's antisepsis was still unknown and the high mortality rate bears witness to the inability to control wound infection. Skill in amputation technique was no doubt improved in spite of the high mortality rate. Public health benefitted from experience in hygiene and isolation. Apart from the observations of a few keen minds like S. Weir Mitchell, W. W. Keen, William Hammond, Hunter McGuire, and Joseph Jones medical science did not gain substantially from the vast wartime experience.

The tyranny of utilitarianism which shaped the course of American medicine prior to the Civil War continued to be the dominant force in medical education during hostilities and for decades thereafter. The growth of urban centers and the reactivation of westward expansion during reconstruction years stimulated the demand for medical service. Hence numerous medical schools were founded during the reconstruction decade, most of them proprietary in character, established in urban centers. Previous promoters of country schools had already learned the hard way that without teaching beds and bodies for dissection they could not compete or long survive.

Shortly before the War (1859) Lind University in Chicago took the lead in establishing a two-year graded curriculum in medicine. There had been previous but unsuccessful efforts to establish graded curriculums. When General John Eaton, United States Commissioner of Education, made his Report in 1870 only two colleges had a graded course of lectures occupying three winter sessions, Chicago Medical College (formerly Lind University Medical School, soon to become the medical school of Northwestern University) and the Woman's Medical College of the New York Infirmary. Northwestern did not make the three-year curriculum compulsory until some years later (74).

By 1875 the United States Office of Education reported seventy-eight medical schools in operation, twenty-six of which were established after 1865. That two of the latter were eclectic and six homeopathic in character suggest that sectarian medicine was still in vogue and that licensing laws were still loose or non-existent.

In the last years of the third quarter of the century a few events of significance occurred. The Harvard medical faculty came to an agreement and announced for 1871-1872 a three-year, graded course of instruction with thirty-seven weeks in each term (75). The University of Pennsylvania medical faculty, facing strong local competition, was a little slower in matching this reform but compensated in 1874 by opening a university hospital, the first of such institutions planned and constructed as a medical teaching facility (76).

THE QUEST FOR MATURITY, 1875-1900

In spite of indications that reformation was in the making the last quarter of the century was marked by a steady increase in the number of medical schools, the great majority of which were sub-standard according to the goals endorsed by the few better schools. By 1881 the Commissioner of Education reported the current existence of ninety-six schools, thirteen of which were still offering a two-year curriculum. By 1890 the Commissioner reported 120 schools, including nine eclectic, fourteen homeopathic and two physiomedical. Approximately fifteen per cent of the matriculates were reported as holding a degree in letters or science. Ten years later the Commissioner's *Report* (1899-1900) listed 151 active medical schools, of which eight were eclectic and physiomedical and twenty-two homeopathic. Of the reported 25,000 matriculates it would appear from the data that after ten years there was only a slight increase in the number holding baccalaureate degrees. Actually many matriculates entered upon the study of medicine without a high school diploma, and the qualifying examinations given to such applicants were usually designed to pass all but the grossly illiterate (77).

In addition to witnessing a rapidly increasing and thriving number of low grade allopathic, eclectic and homeopathic city schools, the American public in the post bellum decades was introduced to osteopathy which was destined to occupy its share of the expanding vacuum created by popular aversion to the vagaries of regular medicine. To broaden the choice for the medically illiterate public the chiropractic art was added a decade after the introduction of osteopathy. All these branches of the healing art developed their own educational institutions for perpetuating their health ideologies. Because eclectic and homeopathic schools conferred the M.D. degree they were usually licensed along with allopathic (regular) graduates. The *laissez-faire* system permitted this heterogeneous state, and the American public tolerated, even defended it.

In spite of flourishing marginal cults and isms there were increasing signs of self-analysis and therapeutic action within the regular profession and among its leading educators. Scholars have referred to the time 1870-1900 as the era of German influence. American students and young doctors began going to Germany after the Civil War. They were impressed with full-time professorships, teaching departments, institutes, clinics, and hospital beds, all within the state-supported university complex. Within this academic and professional structure they also witnessed the role of graduate education and research. They saw advances in both physical and chemical aspects of physiology which tended to re-emphasize the whole organism in contradistinction to the French emphasis on pathology of organs and structures (78).

Some Americans returned home to pursue full-time practice, but some became identified with medical education. A few schools, principally those associated with universities, began much needed reforms in departmental organization, curriculum, grading, and admission and graduation requirements. Eventually university hospitals were founded and research laboratories were established. The many-faceted roles of the public health movement gradually emerged. The founding of specialty societies and journals and libraries all contributed to an American renaissance in medical education. The private and church-related hospital movement hastened the rapprochement between hospitals and medical education.

In 1877, a group of medical school delegates met in Chicago and formed the American Medical College Association which agreed to minimal standards of medical education. Six years later (1883) the organization foundered over the proposal that three instead of two years of graded curriculum be required. Low grade medical education exercised a slowly declining dominance throughout the rest of the century. Fortunately, however, fundamental and influential changes were in the making.

The 1890's were the watershed of American economy and culture. Roads, canals, river transportation, railroads, and commerce had completed the phasing out of the frontier, so long regarded as a deciding factor in determining the character of American culture. In spite of the crucial importance of agriculture, the industrial revolution had decreed the dominance of industry. The population, between sixty and seventy million, was increasingly concentrated in urban areas with critical public health problems. Scientific movements and professionalism were besieging higher education. The public was beginning to grasp the importance and significance of both to society. The intelligent laity, especially industrialists and other custodians of wealth, were beginning to see a practical connection between their merchandising and industrial pursuits and scientific advances in medicine. Further, by this time research was identified with hospitals, a traditional object of American humanitarianism. There were signs of maturation on numerous fronts of the American scene. The nation was entering adulthood.

The time was ripe for medical education to come into its own as a mature university discipline. In 1890 representatives of sixty-six medical schools revived their moribund organization and relabeled it the Association of American Medical Colleges. Demonstrating capacity for self discipline, they adopted a code which inaugurated institutional habits of continuous self-improvement.

The most impressive and significant event of the 1890's was the opening of The Johns Hopkins University School of Medicine in 1893. Johns Hopkins, a Baltimore merchant, provided funds for establishing the University in 1876. At this early date the founder included in his master plan the development of a hospital and medical school to be patterned after the best in medical education. The result was a blend of the best as observed by John Shaw Billings in England, France, and Germany, with emphasis on the latter.

The Johns Hopkins Hospital, although under a separate board, was planned to operate as an integral part of the Hopkins' medical complex. It was no happenstance that the teaching hospital was staffed and operating in 1889 four years before the School of Medicine opened. The four-year graded curriculum, a planned educational program for interns and residents, full time faculty members in the basic sciences, and clinical and laboratory research were characteristic of the plan, and set the pattern for the future of American medical education.

In spite of brilliant leadership and sound planning at the Johns Hopkins, concerned medical educators in various places were fully aware of regrettable conditions which still prevailed in many quarters. In an erudite discussion of "The Best Preliminary Education for the Study of Medicine", Flavel S. Thomas, in 1886, noted that the time was passed in some schools when all that a man had to do to earn an M.D. was "to serve as chore-boy to a physician for three years, attend two courses of lectures of four months each and pass four out of the seven examinations" (79). He was pleased that seven schools were offering a four-year course, twenty-three required three years, and ninety required a matriculation examination. The object of Thomas' principal thrust was "the cheap, low-standard, short-course schools, which fight against us, not for the love of the cause but for money".

Addressing the newly organized Association of American Medical Colleges, in 1892, H. D. Allyn made a plea for the utility of a liberal education for students wishing to study medicine, because the experience, he believed, gave to the student comprehensiveness of vision, rid him of pretensions, prejudices and vanity, supplied him with a valuable acquaintance, inspired him with laudable ambition, spurred him to praiseworthy effort, and taught him how to study—qualities not always detectable in baccalaureate candidates of later generations. Doctor Helen F. Warner, who discussed Allyn's paper, thought that the lack of preliminary education put American graduates at great disadvantage in European schools and that it was "perhaps responsible for the discourteous refusal of the Prussian authorities to accept American diplomas" (80).

The committee of the American Academy of Medicine on the requirements for preliminary education in the various medical colleges reported in 1887 and again in 1893. Frederick Henry Gerrish who reported for the committee, being skeptical of data provided by some cheap school to the Illinois State Board of Health, employed detective means to clarify or confirm his doubts. A child who could barely write was engaged to direct written inquiries to certain schools. Each inquiry expressed ignorance of some branch of learning declared essential by the school and revealed the general illiteracy of the "applicant". Half of the schools showed some disposition to abide by the regulations for preliminary education. The remainder "fairly tumbled over each other in their indecent scramble to secure this prospective student who frankly proclaimed his unfitness, even according to their miserable standards." Further, the schools hastened to give assurance that they never intended to exclude a student so well prepared as this particular applicant.

A Southern school in response to a different inquiry in 1893 replied:

Yes, we advise a man to know Latin and Natural Philosophy if he can—! but if a practical man, who can write a good letter (like yours)—and is well balanced and educated in all the practical walks of life, and desires to enter our college, I have, as the proper examiner, always given him the advantage of his practical knowledge of the world! as I have found these men better balanced in Judgment, and generally make better practitioners (81).

Obviously no school rose above the level imposed by the moral character limitations imposed by the men who administered them.

In his presidential address before the Association of American Medical Colleges in 1901, Albert R. Baker called attention to the medical preceptor as an antiquated institution even though it had rendered a commendable service when no other means were available. Admitting an excessive number of medical schools, many with low standards, Baker suggested that classes be subdivided into small groups and that more teachers be employed, so that some of the advantages of preceptorial training might be retained. In his opinion there were two to three times as many schools as necessary and already enough graduates to satisfy the national demand for five to ten years. He urged that poorly educated graduates return to a good school and repeat their education, noting that of 170 graduates of the Cleveland College of Physicians and Surgeons from 1896 to 1901, thirty-three had qualified for a second M.D. (82).

Baker believed that the highest possible standards that could then be demanded were incorporated in the 5 June 1899 resolutions proposed by the joint committee of the AAMC and the Confederation of State Medical Examining and Licensing Boards which called for: (a) Preliminary education equivalent to that of a graduate from a good high school; (b) four courses of graded instruction not less than six months in length, given in four separate years (83). Baker praised the work and uplifting influence of the Association during its first decade but looked to the Confederation as the suitable organization to establish and regulate a uniform standard for medical education in the future.

Meanwhile, the Johns Hopkins Medical School more than any other American institution was blending medical education, service, and research in a manner reminiscent of John Morgan's ideology as it had never before been accomplished in America. In 1900 William Welch, the principal builder of the Hopkins medical complex spoke humbly but with conviction: "Our own contribution to this process may now be small, but America is destined to take a place in this forward movement commensurate with her size and importance" (84).

NOTES

In the preparation of this study the author has made use of several of his former published papers, usually without specific documentation, such as: American medical education from the Revolutionary War to the Civil War, *J. Med. Educ.*, 1957, **32**: 433-448; Medical education and the rise of hospitals, *J. Am. Med. Ass.*, 1963, **186**: 1008-1012; Critical incidents in the shaping of medical education in the United States, *J. Am. Med. Assn.*, 1965, **194**: 715-718; and Medicine in the era of the American Revolution, *Int. Rec. Med.*, 1958, **171**: 391-407.

1. As used here, geography must ultimately include outer space; however, we are not concerned here with an expansion of geography. The animal kingdom is involved in the course of history but only man has attempted with purpose to influence or record it.
2. Percy's discourse, in ALEXANDER BROWN, ed., *The genesis of the United States*, Vol. 1, Boston, 1890, pp. 153-169; COTTON MATHER, *Magnolia Christi Americana*, Vol. 1, Hartford, 1855, pp. 46-56.
3. A thorough study in depth is that of St. JULIEN RAVENEL CHILDS, *Malaria and colonization in the Carolina low country 1526-1696*. Baltimore, 1940. Useful secondary sources with valuable bibliographies are: P. M. ASHBURN, *The ranks of death*, New York, 1947, *passim;* JOHN R. BLAKE, Diseases and medical practice in colonial America, in FÉLIX MARTÍ-IBAÑEZ, ed., *History of American medicine*, New York, 1958, *passim;* JOHN DUFFY, *Epidemics in colonial America*, Baton Rouge, 1953, *passim;* RICHARD H. SHRYOCK, *Medicine in America*, Baltimore, 1966, pp. 1-17.
4. COTTON MATHER, *Magnolia Christi Americana*, Vol. 1. Hartford, 1855, p. 55.
5. Quoted in O. T. BEALL, JR., and R. H. SHRYOCK, *Cotton Mather, first significant figure in American medicine*. Baltimore, 1954, p. 76.
6. *Ibid., passim.* In WILLIAM DOUGLASS, *A summary, historical and political of the . . . present state of the British settlement in North America*, Vol. 2, Boston, 1755, p. 409, the author claims that he lent the *Transactions of*

the Royal Society to Reverend Mather in order "that he might have the imaginary honor of a new fangled notion" and that Mather, without consulting him (Douglass) let "a rash undaunted operator" employ the method. Since Mather was elected to the Royal Society in 1713 one wonders what reason there was to lend him the *Transactions*.

7. LESTER S. KING, *The medical world of the eighteenth century*, Chicago, 1958, pp. 1-29; Sir GEORGE CLARK, *A history of the Royal College of Physicians of London*, Vol. 1, London, 1964, *passim;* H. CHARLES CAMERON ed., (from manuscript notes of Cecil Wall, revised, annotated and edited by E. ASHWORTH UNDERWOOD) *A history of the Worshipful Society of Apothecaries of London*, Vol. 1, 1617-1815, London, 1963, pp. 1-57 *passim.* According to the latter source, many apothecaries in the seventeenth century had a much better practical knowledge of medicine than did the theoretically trained university graduates with the M.D. degree, *ibid.*, pp. 76-90.

8. Among the earliest known practitioners were chirurgeons Will Wilkerson and Thomas Wooton, and doctor of physicke Walter Russell at Jamestown; barber Jan Peterson with the Swedes on the Delaware; deacon Samuel Fuller, at Plymouth; Giles Firmin, apothecary, Boston; Thomas Lord, Connecticut; Dr. John Clark, Rhode Island; surgeons Gerrit Schult and Hans Kierstedt, New Amsterdam. See FRANCIS R. PACKARD, *History of medicine in the United States*, Vol. 1, New York, 1932, pp. 3-43, and various histories of medicine by colonies or states.

9. GEORGE W. NORRIS, *The early history of medicine in Philadelphia*, Philadelphia, 1866, p. 11. There were a few settlers in the Pennsylvania area before Penn's colony. Among them was Chirurgeon John Goodson of the Society of Free Traders, who later settled in Philadelphia after Penn arrived. PACKARD, *op. cit.* pp. 7 f.

10. ARCHIBALD MALLOCH, *Medical interchange between the British Isles and America before 1801*, London, 1939, *passim.* Other useful sources of names are: R. W. INNES SMITH, *English-speaking students of medicine at the University of Leyden*, Edinburgh, 1932, *passim;* University of Edinburgh, *List of the graduates in medicine in the University of Edinburgh from 1705 to 1866*, Edinburgh, 1867, *passim.*

11. Eliot's letter was dated 24 September 1674. *Massachusetts Historical Society Collections*. Vol. 4 (3rd ser.), Cambridge, 1854, p. 57. Reference to Lambert Wilson is in PACKARD, *op. cit.*, p. 25.

12. Cited by EDWARD B. KRUMBHAAR, The early history of anatomy in the United States, *Ann. Med. Hist.*, 1922, 4: 271; quoted from *General Court Records*, Vol. 2, 1647, p. 175.

13. KRUMBHAAR, *loc. cit.*, pp. 271 ff.

14. Diary of Samuel Sewall. *Massachusetts Historical Collection*, Vol. 5 (5th ser.), p. 21.

15. Quoted by JAMES J. WALSH, *History of medicine in New York*, Vol. 1, New York, 1919, p. 36, from *Minutes of New York Council*, Vol. 6, (30 July 1691) p. 42.

16. PACKARD, *op. cit.*, p. 297.

17. CASPAR WISTAR, in his 1809 *Eulogium on Dr. William Shippen,* cited by KRUMBHAAR, *op. cit.,* pp. 272-273, makes reference to Dr. Cadwalader's important contribution to medicine and medical education through his dissections and relatively frequent autopsies.
18. WILLIAM DOUGLASS, *The practical history of a new epidemical eruptive military fever . . . which prevailed in Boston New England in 1735 and 1736,* Boston, 1736, *passim.* PACKARD, *op. cit.,* Vol. 1, p. 98, believed, as did Samuel A. Green before him, that the epidemic was "undoubtedly diphtheria" but GEORGE H. WEAVER, in his Life and writings of William Douglass, M.D. (1691-1752), *Bull. Soc. Med. Hist., Chicago,* 1917-1921, 2: 229-259, holds strictly to the scarlet fever interpretation. Duffy offers the partially clarifying statement that the scarlet fever outbreak occurred in conjunction with diphtheria attacks in New England in 1735-1736; DUFFY, *op. cit.,* p. 131.
19. DOUGLASS, *op. cit.,* pp. I-II.
20. JAMES G. MUMFORD, *A narrative of medicine in America,* Philadelphia, 1903, p. 44.
21. CARL BRIDENBAUGH, ed., *Gentleman's progress, the itinerarium of Dr. Alexander Hamilton, 1744,* Chapel Hill, 1948, p. 116.
22. *Ibid.,* pp. 116-117.
23. *Ibid.,* p. 132. By way of explanation, Hamilton learned that Douglass had been a disciple of Archibald Pitcairne, physician and poet, who once held a chair at Leyden and had spirited exchanges with Boerhaave, *ibid.,* and note 305, p. 242.
24. WILLIAM DOUGLASS, *A summary, historical and political,* etc. Vol. 1, p. 383.
25. *Ibid., loc. cit.*
26. *Ibid.,* pp. 383-384. Douglass rightly held for his time that mathematical precision could not be applied to the solution of medical problems as in astronomy. He contended "that in our microcosm or animal oeconomy, there are so many inequalities as not to admit of any fixed rules, but must be left to the sagacity of some practitioners, and to the rashness of others."
27. See the Introduction by JOHN D. COMRIE to R. W. INNES SMITH, *English-Speaking students of medicine at the University of Leyden,* Edinburgh, 1932, pp. I-XIX, for a discussion of English-speaking students at Leyden and other continental schools at this time. Smith lists a dozen or more Anglo-Americans who registered at Leyden before 1750. Some of these were residents from the British West Indies.
28. BRIDENBAUGH, *op. cit.,* p. XII f.
29. JARED SPARKS, *The works of Benjamin Franklin,* Vol. 7, Boston, 1838, p. 238. David Hume's letter to Franklin is in Vol. 6, pp. 243-245.
30. The quotations from John Adams and Benjamin Rush are from the Editors' Foreword to EVARTS BOUTELL GREENE, *The revolutionary generation, 1763-1790,* New York, 1943, p. XV. The Foreword, presumably by Arthur M. Schlesinger and Dixon Ryan Fox, does not document the quotations.
31. It is well known that a few hospitals in Mexico and Canada antedated the Pennsylvania Hospital. There have also been wordy arguments on behalf

of Philadelphia General Hospital, which sprang from the Philadelphia Almshouse built in 1732. Packard argues that the Philadelphia General Hospital, the New Orleans Charity Hospital and a few other hospitals, the outgrowth of institutions established for the care of the indigent poor or aged ("in other words almshouses") had not achieved the status or character of hospitals by 1752 when the Pennsylvania Hospital accepted its first patient. Packard's argument is logical, but the debate is productive only in that it has uncovered considerable information of value. PACKARD, *op. cit.*, Vol. I, pp. 181-270.

32. BENJAMIN FRANKLIN, *The autobiography*, Boston, 1906, p. 129. See also BENJAMIN FRANKLIN, *Some account of the Pennsylvania Hospital*, Philadelphia, 1754, *passim*. T. G. MORTON and F. WOODBURY, *The history of Pennsylvania Hospital*, Philadelphia, 1895, trace the origin of Pennsylvania Hospital back to agitation for a school and hospital in a Friends' monthly meeting on 25 September, 1709, pp. 4 f., but nothing happened until 1751.

33. BENJAMIN FRANKLIN, *Some account of the Pennsylvania Hospital*, Philadelphia, 1754, p. 15.

34. FRANCIS R. PACKARD, *Some account of the Pennsylvania Hospital*, Philadelphia, 1938, p. 2.

35. T. G. MORTON, *The history of the Pennsylvania Hospital*, Philadelphia, 1897, pp. 29-32.

36. J. M. TONER, *Contributions to the annals of the medical progress and medical education in the United States, before and during the War of Independence*, Washington, D.C., 1874, p. 106; N. S. DAVIS, *history of medical education and institutions in the United States of America*, Chicago, 1851, pp. 9-10.

37. The five M.D. degrees at the College of Philadelphia and the two at King's College were conferred on previous M.B. candidates who had completed specified additional work. King's College M.D.'s were the first in American History.

38. The vital statistics of medical education during the colonial and early national period are principally the product of careful research by FREDERICK C. WAITE, Medical degrees conferred in the American colonies and in the United States in the eighteenth century, *Ann. Med. Hist.*, 1937, 9: 314-320. The one William and Mary honorary M.D. degree was conferred on John F. Coste, first physician of the French Army operating in the Virginia area; see *The history of the William and Mary College . . . 1693-1870*, Baltimore, 1870, p. 145; FREDERICK C. WAITE, The age of Harvard Medical School in relation to that of other existing medical schools in the United States, *New Engl. J. Med.*, 1937, 216: 418. The University of the State of Pennsylvania and the University of Pennsylvania were actually lineal descendants of the College of Philadelphia.

39. See WAITE, *op. cit.*, pp. 317-319 for individual tables on each school and composite tables for the eighteenth century, which I have reproduced.

40. WHITFIELD J. BELL, Medical students and their examiners in eighteenth century America, *Trans. Stud. Coll. Physns. Philad.*, 21 (4th ser.): 14-24. Bell

quotes Dr. Samuel Johnson as having told a young American, Arthur Lee of Virginia, that Oxford and Cambridge did not permit students to enter physic until seven years of literary study had qualified them for the M.A. degree. On the European influence see also NORWOOD, *Medical education in the United States before the Civil War*, Philadelphia, 1944, p. 27.

41. JOSEPH CARSON, *A history of the medical department of the University of Pennsylvania*, Philadelphia, 1869, pp. 77-85.
42. *Ibid.*, pp. 43 f. N. S. DAVIS, *op. cit.*, p. 54, comments on this early event in the decline of medical education standards in the United States. Apparently attendance on the Pennsylvania Hospital wards ran concurrently with the didactic lecture schedule, with this down-grading of the regulations.
43. HENRY BURNELL SHAFER, *The American medical profession, 1783-1850*, New York, 1936, p. 35. Shafer cites Minutes of the Medical Faculty, University of Pennsylvania 1767-1814 (manuscript in 4 volumes), p. 102.
44. ALEXIS DE TOCQUEVILLE, *Democracy in America*, Newark, 1945, Vol. 2, pp. 36-38. See also HENRY STEELE COMMAGER, The *American mind*, New Haven, 1950, pp. 3-40.
45. For a few examples see GEORGE H. WEAVER, Beginnings of medical education in and near Chicago, *Bull. Soc. Med. Hist., Chicago*, 1925, 3: 342.
46. DIXON RYAN FOX, Civilization in transit, *Am. Hist. Rev.*, 1927, 32: 754-755.
47. It is refreshing to note that in the case of dentistry, the profession in America antedated its European colleagues by shedding the traditions of charlatanism and establishing indigenous techniques superior to those of European countries. The first American dental school (Baltimore College of Dental Surgery, 1840) was established nineteen years before the first European school (English) which dated from 1859. Thus the transit of culture was reversed. See RICHARD H. SHRYOCK, *The development of modern medicine*, New York, 1947, p. 338-339.
48. JOHN MORGAN, *A discourse upon the institution of medical schools in America*, Philadelphia, 1765.
49. NORWOOD, *op. cit.*, pp. 350-352. In the paying of salaries Michigan was antedated by the University of Virginia which, in 1825, imported Robley Dunglison, a young Englishman who had been editor of the *London Medical Repository*, to fill a general chair of medicine at an annual salary of $1000 with a guarantee of additional emoluments of $500. A full medical faculty developed as a result of this beginning.
50. FREDERICK C. WAITE, *The first medical college in Vermont, Castleton, 1818-1862*, Montpelier, 1949, pp. 49-78.
51. WYNDHAM B. BLANTON, *Medicine in Virginia in the nineteenth century*, Richmond, 1933, pp. 42-50.
52. BLANTON, *op.cit.*, pp. 19 ff; HERBERT M. ADAMS, Thomas Jefferson and the University of Virginia, in *United States Bureau of Education Circular*, No. 1, Washington, D.C., 1888, pp. 86-98.
53. DAVID HOSACK, *Observations on the establishment of the College of Physicians*

and Surgeons in the City of New York, New York, 1811, pp. 2 ff.
54. Eugene Fauntleroy Cordell, *University of Maryland, 1807-1907,* New York, 1907, pp. 6-15; and *The medical annals of Maryland, 1799-1899,* Baltimore, 1903, pp. 54-59.
55. *Ibid.* pp. 36-43. The act of 1812 was eventually declared unconstitutional by reason of contravening the rights and privileges of the charter of 1807.
56. Samuel Annan, *An introductory lecture delivered at the opening of the Kentucky School of Medicine,* 1850, p. 3. In defense of rural medical schools see T. Romeynn Beck, *On the utility of country medical institutions,* Albany, 1825, delivered as an introductory lecture at the College of Physicians and Surgeons of the Western District of the State of New York, December 13, 1824.
57. James Conquest Cross, *Thoughts on the policy of establishing a school of medicine in Louisville, together with a sketch of the present condition and future prospects of the Medical Department of Transylvania University,* Lexington, 1834, p. 5.
58. M. L. Knapp, *An address delivered at the opening of the Rock Island Medical School,* Chicago, 1848, p. 5.
59. The original intention of Knapp and his colleague, George W. Richards, (both recently withdrawn from LaPorte Medical College in Indiana) was to open a school in Illinois. When they ran into legal difficulties they turned to the callow Wisconsin legislature which authorized the Madison Medical College, "with power to create a branch of the same." Their Madison associates apparently expected the school to be in or near Madison, with authority to create a branch. The "branch" provision actually became a betrayal of the original intent. See Norwood, *op. cit.,* pp. 347-350.
60. W. H. Kemper, *A medical history of the state of Indiana,* Chicago, 1911, pp. 48-56.
61. *Annual Circular and Catalogue of LaPorte University,* 1844-1845, p. 3.
62. Helen B. Clapesattle, *The Doctors Mayo,* Minneapolis, 1941, p. 30.
63. Friends of the University became increasingly apprehensive in 1825-1826, when the Jefferson agreement with the Canonsburg institution was altered to provide a local board in Philadelphia. The university trustees memorialized the legislature and were successful in blocking Jefferson's right to confer medical degrees in Philadelphia. Doctor McClellan made a marathon drive with horse and sulky to Harrisburg and returned with an unhampered charter before the first graduation on 8 April 1826.
64. The second section of the act secured by McClellan declared that "hereafter it shall not be lawful for any college incorporated by the laws of the State, to establish any Faculty for the purpose of conferring degrees, either in Medicine or the Arts, in the City or County of the Commonwealth, other than that in which said college is or may be located." See Samuel Gring Hefelbower, *The history of Gettysburg College.* 1932.
65. For further information on the cost of medical education see Norwood, *op. cit.,* pp. 392 ff.
66. *Ibid.,* pp. 387-391.

67. The introduction of the microscope as a practical aid to medical science came late in the prewar period, not soon enough to establish pathology and bacteriology on a rational basis.
68. ALBERT E. FOSSIER, History of medical education in New Orleans from its birth to the Civil War, *Ann. Med. Hist.*, 1934, **6**: 432-435.
69. FREDERICK C. WAITE, American sectarian medical colleges before the Civil War, *Bull. Hist. Med.*, 1946, **13**: 148-166. NORWOOD, *op. cit.*, pp. 416-421. See also ALEX BERMAN, "Neo-Thomsonianism in the United States, *J. Hist. Med.*, 1956, **11**: 133-155.
70. Even medical students came in for a share of the criticism leveled at medical education. See DANIEL DRAKE, *Strictures on some of the defects and infirmities of intellectual and moral character, in students of medicine* . . . 1847; DAVID HOSACK, *An introductory lecture on medical education* . . . 1801. Beginning in 1832, Elisha Bartlett edited a short-lived journal beamed specifically to medical students entitled *The Monthly Journal or Medical Students Gazette;* IRVING A. BECK, An early American journal keyed to medical students: A pioneer contribution of Elisha Bartlett, *Bull. Hist. Med.*, 1966, **40**: 124-134.
71. *Proceedings of a Convention of Medical Delegates held at Northampton in the State of Massachusetts,* 1827, pp. 3-10, cited by SHAFER, *op. cit.*, pp. 91-92.
72. The American Medical Association, from its inception, had advocated better pre-professional education, longer lecture terms (13-16 weeks was conventional), more extensive clinical instruction, and more rigid examinations. By 1860 few schools had terms of less than eighteen weeks, and a few were as long as five or six months. The decline of country schools and the increase in the number of city schools meant that more students were receiving better anatomical and clinical instruction. Americans continued to go to Europe for basic and advanced medical education and observation. This practice provided a small but steady stream of new blood for the profession. London and Paris were the more popular places of study in the first half of the nineteenth century. Nevertheless, large numbers of young men were still entering practice with apprentice training only or with the precepteeship and less than two terms of lectures, and hence degreeless. See N. S. DAVIS, Address of N. S. Davis, *Trans. Am. Med. Ass.*, 1866, **16**: 71-75. CHARLES ROSENBURG, The American medical profession: mid-nineteenth century, *Mid-America*, 1962, **44**: 163-171, traces the decline in public esteem experienced by the regular American medical profession, which he suggests, reached its nadir in 1851 when the Georgia legislature appropriated $5000 for a Botanic Medical College (p. 164).
73. Illinois State Board of Health, *Conspectus*, Springfield, 1884, p. XXIII.
74. *Ibid.* pp. XXII f.
75. *Ibid.*
76. W. F. NORWOOD, Critical incidents in the shaping of medical education in the U.S., *J. Am. Med. Ass.*, 1965, **194**: 715-718; GEORGE W. CORNER, *Two centuries of medicine. A history of the School of Medicine, University of Pennsylvania,* Philadelphia, 1965, p. 137 ff.

77. For these statistics see *Report of the Commissioner of Education* . . . 1872, 1876, 1881, 1890-1891, 1899-1900; JOHN H. RAUCH, *Reports on medical education, medical colleges and the regulation of the practice of medicine in the United States and Canada,* Vol. 2, Chicago, n.d., bound volumes of periodic reports of the Illinois State Board of Health. It should be noted that under the enthusiastic leadership of local newspapers, especially the *Philadelphia Record,* the low-grade medical schools and diploma mills of Philadelphia were exposed to the public and forced to close by 1880. See HAROLD J. ABRAHAMS, *Extinct medical schools of nineteenth century Philadelphia,* Philadelphia, 1966, pp. 418-564 *passim.*
78. Medical men contemplating a visit to or a period of study in European medical schools after 1883 found practical assistance in HENRY HUN, *A guide to American medical students in Europe,* New York, 1883. A definitive study on the German influence on American medicine is in THOMAS N. BONNER, *American doctors and German universities* . . . *1870-1914,* Lincoln, 1963.
79. FLAVEL S. THOMAS, The best preliminary education for the study of medicine, *New Engl. Med. Mon.,* 1886, **5**: 343.
80. H. D. ALLYN, Value of academic training preparatory to the study of medicine, *Bull. Am. Acad. Med.,* 1892, **1**: 184-185.
81. FREDERIC H. GERRISH, Report of the committee on the requirements for preliminary education in the various medical colleges in the United States, *Bull. Am. Acad. Med.,* 1894, **1**:435, 439.
82. ALBERT R. BAKER, Evolution of the American Medical College, *Bull. Am. Acad. Med.,* 1901, **7**: 493.
83. *Ibid.,* p. 496.
84. Welch's statement quoted by RICHARD H. SHRYOCK, *American medical research,* New York, 1947, p. 73.

MEDICAL EDUCATION IN THE UNITED STATES: LATE NINETEENTH AND TWENTIETH CENTURIES

JOHN FIELD
UCLA School of Medicine
Los Angeles, California

> We cannot contemplate with any great satisfaction the early history of medical education in America. Probably medical education had nowhere, at any time, fallen to such a low estate as it did during a large part of the last century in our country. *William H. Welch (1)*

The stage was set for reformation of American medical education—then in the sad state described by Welch (1), Flexner (2), Norwood (3), Shryock (4-7) and others—by the broad surge of social and economic changes in this nation following the Civil War. These included rapid increase in population and wealth accompanied by burgeoning development of transportation, communication systems, industrialization and urbanization. The industrial revolution, which so rapidly increased national affluence, also created large individual fortunes "which could be used to assist medical science—if and when inclination arose" (5). Thus by the 1870's the time was ripe for reformation in medical education. Moreover, increasing pressure for such reform was exerted by the ever-enlarging group of American physicians trained in Germany (8), and by farsighted educators such as Charles W. Eliot and Daniel C. Gilman.

CRUCIAL FACTORS IN REFORMATION OF AMERICAN MEDICAL EDUCATION, 1870-1920

In addition to those permissive factors in the cultural environment listed above, there were several factors relating to the "internal logic" of medical science and education which were especially important in the initiation and accomplishment of reformation of American medical education that marked the era 1870-1920. Four such factors will be examined in this section. These were not separate and distinct but rather interrelated and interacting and as such reinforced the others in achieving reform. Moreover, all were mushrooming continuations of earlier trends.

European Progress in Medical Science and Education, 1850-1900

The second half of the nineteenth century was a period of revolutionary and massive improvement in medical research (creation of new knowledge in medicine) and medical education. While France and England contributed importantly to these changes, as shown by the work of Pasteur, Claude Bernard, Lister and many others, major developments occurred in Germany and other German-speaking nations. Some idea of the imposing scope and depth of German progress in the medical sciences can be obtained by scanning the great series of German *Handbücher* (in just one field, physiology, these included R. Wagner's *Handwörterbuch der Physiologie mit Ruchsicht auf Physiologisches Pathologie* (1842-1853); L. Hermann's *Handbuch der Physiologie* (1879-1883); and W. Nagel's *Handbuch der Physiologie des Menschen* (1905-1910). An early "*Handbuch*" in English was the *Text-Book of Physiology*, in two volumes, edited by E. A. Schaefer (1898, 1900). Evidence of German dominance in research is illustrated by the fact that most of the references in Schaefer (later Sir Edward A. Sharpey-Schaefer) are to the German literature, although all the contributors to this handbook were British.

Reinforcing the impression of the dominance of German medical science in the period 1850-1900—indeed until 1914—is the panorama of famed names set forth in William H. Welch's *Introduction* to the English translation of Billroth's account which covers the explosive rise of German medicine (9). Billroth's book is of unique value because he himself was a major actor in this drama of science and he was personally familiar with many of the participants and events. Welch's colorful vignette tells the story well:

> Billroth's book . . . traces the historical development of medical education, especially in the German universities . . . and sets forth the underlying ideas and conditions of medical teaching, study and research in these universities, at a time when they had far outstripped those of other lands in their plan of organization, scientific spirit, facilities for imparting knowledge, and productivity. They offered advantages for scientific and practical study never before attained and to this day [1924] hardly equalled, certainly not surpassed, in other countries throughout the whole field of medicine. The amazing rise of the German universities to a position where they had become the Mecca of aspiring students from other countries had been the work of scarcely more than three decades. These were the days . . . of Henle, Hyrtl, Kölliker, His, Waldeyer in anatomy; of Du Bois-Reymond, Ludwig, Pflüger, Brücke, Heidenhain in physiology; of Virchow, Von Recklinghausen, Klebs, Cohnheim, Weigert in pathology; of Frerichs, Traube, Wunderlich, Kussmaul, Von Ziemssen, Wagner, Leyden in internal medicine; of B. von Langenback, Voltmann, Billroth, Thiersch, Esmarch in surgery; and in various specialties of Von Graefe, Arlt, Hebra, Kaposi, Politzer, Sigmund, Credé, Meynert and many others of distinction.

Rudolf Virchow, who founded the great journal, *Archiv für pathologische Anatomie und Physiologie und für klinische Medizin* (1847 ff.), and whose book *Cellularpathologie* (1858) "was without doubt the most influential work on disease to appear in the nineteenth century" (10), epitomized the contributions of German medical science by 1877 as follows:

It is no longer necessary today to write that scientific medicine is also the best foundation for medical practice. It is sufficient to point out how completely even the external character of medical practice has changed in the last thirty years. Scientific methods have everywhere been introduced into practice. The diagnosis and prognosis of the physician are based on the experience of the pathological anatomist and the physiologist. Therapeutic doctrine has become biological and thereby experimental science (10, p. 10).

Not listed in Welch's *Introduction* to Billroth, perhaps because he was considered a physicist in Welch's day, was one of the greatest medical scientists and teachers of all time, Hermann von Helmholtz, M.D. (1842), successively professor of physiology at Königsberg, Bonn and Heidelberg, and finally professor of physics in Berlin. His independent discovery of the law of conservation of energy and his accomplishments in optics, acoustics, electrodynamics, mathematics, instrumentation, and other fields comprised no small part of the achievements of German biomedical science in the nineteenth century. His great treatises, *Die Lehre von den Tonemfindungen, als physiologische Grundlage für die Theorie der Musik* and *Physiologische Optik* were models of scientific method and logic. Although supplemental and to some extent modified by later work, they are still major references. Koenigsberger's biography of Helmholtz (11) not only recounts his achievements but also provides a picture of the interactions of German scientists and their vigorous creativity that aids in the understanding of the "information explosion" in Central Europe at that time.

As may be inferred from Welch's long list of famous men (9), the thrust and breadth of progress in German biomedical science in that remarkable period was due in part to the sheer number as well as the talent and industry of German teacher-investigators. For example, in the 1890's there were some 300 biomedical scientists in Germany. In striking contrast, there were less than 100 in Britain and very few in the United States (6). Thus the German "explosion" was due, in part to a "critical mass" of brains (Walter Cannon's "fecundity of aggregation)" (12).

The impact of German achievements on American medical training in research and education was profound and lasting—far greater than earlier English and French influences. This was due not only to the demonstrated excellence of German university schools of medicine (13, 14) but also to a degree of mass communication unparalleled in professional history to that time. Thus an estimated 15,000 Americans were trained in German medical schools in the period 1870-1914 (8). Most of these were physicians seeking advanced training in a clinical specialty—training such as the United States could not then provide. A relatively small number sought advanced training in basic medical sciences and academic medicine, which likewise could not be obtained in America then. From this little group, whose subsequent achievement, influence and reputation was out of all proportion to their numbers, the faculties of Hopkins, Harvard, Michigan, Yale, as well as other institutions drew the men "who later created modern medical schools in the United States" (8, cf. also 15).

The history of the founding of the American Physiological Society (1887) is germane to the point just made. At least sixteen of twenty-eight original members had been or later were trained in Germany; four more had studied in England or France (16). This small group included many of the leaders in reformation of American medical education, such as H. P. Bowditch (Harvard), W. H. Howell (Hopkins), H. N. Martin (Hopkins), C. S. Minot (Harvard), William Osler (then at Pennsylvania, later Hopkins and Oxford) and William H. Welch (Hopkins). Subsequently Welch became the most influential leader in the reform and advancement of medical education in the United States (17, 18, 19).

Thus European medical science provided and European medical schools imparted to Americans the knowledge and procedures which led to reform of American medical education and later to the massive rise of American medical research. Moreover, the practical achievements of men such as Pasteur, Lister and Koch—who showed that medicine had become more than an intellectual and professional tradition and that it really worked (5)—served both to win public confidence and to inspirit medical teachers and practitioners. While this reaction probably exceeded the degree warranted by the facts, it was important in developing a favorable climate of opinion for the improvement of medical education. Its intensity can be appraised by comparing Helmholtz's overview of medicine in 1842 with Welch's assessment of medical progress in the decade 1880-1890.

Describing his feelings of dismay and frustration while a medical student and a young physician in the 1840's, Helmholtz wrote:

My education fell within a period of the development of medicine when among thinking and conscientious minds there reigned perfect despair. It was not difficult to understand that the older and mostly theorizing methods of treating medical subjects had become absolutely useless ... How the science must be built up the example of other natural sciences had made clear, but yet the new task stood of giant height before us. A beginning was hardly made, and the first beginning was often very crude. We cannot wonder if many honest, serious, thinking men then turned away in dissatisfaction from medicine or if they from principle embraced an extreme empiricism (20).

In contrast, William H. Welch, influenced by his own experience in Germany during the 1870's (19) and by the impressive record of achievement of medical sciences then and later (21), in a very different vein wrote of the picture some forty years after Helmholtz's student days:

At the end of that wonderful decade, 1880-1890, perhaps the most wonderful decade in the history of medicine, there had been a revolution in medical thought through the discovery of agents causing infectious diseases—discoveries [such] as the bacillus of tuberculosis, of Asiatic cholera, of diphtheria, of typhoid fever and other infectious diseases. Those living today [1914] can hardly realize the enthusiasm and youthful spirit which was stirred not only among medical men but in the general public by these discoveries (22).

Thus the period 1840-1890 marked a transition from frustration and despair to achievement and hope in the minds of many significant medical scientists and educators. Reformation of American medical education and sequentially of American medical practice was undertaken by Americans directly acquainted with the real triumphs of European medical science from 1870 to 1914. The catalytic effect of these triumphs is clear.

Development of Graduate Education in the United States, 1876-1900

In the whole history of education it is hardly possible to find another institution whose influence was so far-reaching and revolutionary as Gilman's Johns Hopkins University and Medical School (18).

The "university revolution" in the United States—which began with the establishment of Johns Hopkins University in 1876 and, as a "revolution", ended with establishment of the Association of American Universities in 1900 (23)—was an event that presaged and conditioned reformation in American medical education (24).

From the start, Hopkins was a University in the nineteenth century German pattern—an institution which fostered and emphasized creation of new knowledge (research) rather than the process of imparting the knowledge and skills already known (teaching). There is impressive documentation on this point (17-19, 23, 24). In this sense Hopkins' priority and its exemplary significance were soon appreciated as the addresses of Woodrow Wilson, Charles W. Eliot and William R. Harper at the Hopkins quarter century celebration made very clear (18).

The impact of the university revolution on higher education in America was very pervasive. "The University revolution did far more than add universities to the nation's higher educational structure, build new universities around old colleges, and multiply professional, technical and graduate schools. It invaded the college itself and altered profoundly the content of a college education. Interest in vocational training was growing. During the last quarter of the nineteenth century the number of students going to college more than doubled—which was both a consequence and a cause of changes in curricula. The growth of scientific knowledge went on with such rapidity that it confounded the old idea of a fixed body of study; there was so much that could be known, there were careers in so many specialties that the idea of some satisfactory over-all educational 'coverage' became increasingly untenable. It is obviously impossible," Huxley observed to his audience at Johns Hopkins [in 1876] "that any student should pass through the whole of the series of courses offered by a university" (25).

Establishment of a Model School of Medicine at Hopkins

This was the first medical school in America of genuine university type, with something approaching adequate endowment, well equipped laboratories conducted by modern teachers, devoting themselves unreservedly to medical investigation and instruction, and with its own hospital in which the training of physicians and the healing of the sick harmoniously combine to the infinite advantage of both. The influence of this new foundation can hardly be overstated. It has finally cleared up the problem of standards and ideals; and its graduates have gone forth in small bands to found new establishments or to reconstruct old ones. (2, A. Flexner, p. 12).

Not much can be added to Flexner's insightful if somewhat overstated appraisal of the significance of Hopkins Medical School in reformation of American medical education. The School, which opened in 1893, was developed by men who had helped to create Hopkins University and who shared its aspirations and its goals (24). Moreover most of these men had experience in and much knowledge about the great German university medical schools of the time. While German influence had been reflected in reforms at a few of the better American medical schools (e.g. Harvard, Michigan, Pennsylvania, Syracuse) in the 1870's and 1880's, substantial change in long-established schools proved difficult and slow (19, 26, 27). The fresh start at Hopkins was not hampered by traditions and commitments.

Notable innovations in administrative policy and operation and in educational planning and procedure which became guide lines for reform in other American medical schools were summarized by William H. Welch in his celebrated lecture to the Harvey Society (1916) on *Medical Education in the United States* (1). These included:

(*a*) Recruitment of faculty based upon a nation-wide search for outstanding men, guided by consultation with experts in Europe and in other U.S. medical schools. The first four appointees—Welch, Osler, Halsted and Kelly—were selected this way. This procedure was revolutionary. Other schools, even the better ones, usually appointed prominent local men (often their own graduates), to vacancies or new posts.

(*b*) Higher admission requirements with specified prerequisites. Hopkins was the first medical school in the United States to require for admission graduation from an approved four-year college or scientific school and to demand that the candidate's undergraduate training include biology, chemistry, physics, French, German and Latin. The Latin requirement was later dropped.

(*c*) A four-year undergraduate curriculum which included a logical sequence of training in basic medical sciences for two years, with emphasis on laboratory work. In the last two years patients were studied in the outpatient clinics and at the bedside. This was not the first four-year medical teaching program, but it set a standard of excellence without peer at that time.

(*d*) Advanced training: Internship and residency appointments were available to approved candidates who had won the M.D. degree.

(*e*) Students and house staff (interns and residents) were encouraged to take part in research endeavors.

(*f*) Close academic ties were maintained with the parent university.

(g) Medical school and hospital were interlocked and interacting from the start. Thus the professor of medicine was physician-in-chief to the hospital; the professor of surgery was chief of the surgical service; the professor of pathology was hospital pathologist, etc. All medical appointments to the hospital staff were made by the school of medicine.

Hopkins Medical School provided a standard of excellence adapted to the American scene and a source of talented faculty for the medical schools of the United States. These were its major contributions to the reformation of medical education in this country—and great contributions they were. Not only did Hopkins provide "small bands of graduates to found new establishments or to reconstruct old ones" (2), but also junior members of its faculty were sought for appointment elsewhere (e.g. Councilman to Harvard, Simon Flexner to Pennsylvania). As might be expected, Hopkins men were attracted to the stronger schools or to promising new ones—an example of the Matthew effect (28). Thus their influence became especially significant at California, Duke, Harvard, Minnesota, Rochester, Stanford, Vanderbilt, Virginia, Washington (St. Louis), Wisconsin and Yale (24). So successful has been the work of her eminent sons and their colleagues that today Hopkins is but one of a number of outstanding medical schools in the United States—an outcome that would surely have delighted President Gilman and Doctors Welch, Osler, Halsted and Kelly.

Efforts of Organized Medicine and of Private Foundations to Improve Medical Education in the United States

From the time of its inception the American Medical Association (AMA) has been formally concerned with improvement of medical education in the United States. The call of the Medical Society of the State of New York, which led to a preliminary national convention in New York (1846) and to organization of the AMA in Philadelphia (1847), began with N. S. Davis' statement "... it is believed that a National Convention would be conducive to the elevation of the standard of medical education in the United States..." (29, pp. 7, 887). However, the efforts of the successive Committees on medical education to achieve that goal were not fruitful. This was due in part to continued opposition from members of the AMA who had vested interests in weak medical schools, (7) and in part to sectarian medical groups outside the AMA (30, p. 2). Reorganization of the structure of the AMA in 1902 and creation of a permanent Council on Medical Education in 1904 (cf. 29, pp. 1195-1197 for membership 1904-1947) changed the picture—partly because the new organizational structure facilitated action but chiefly because the members of the new Council were chosen wisely. These members were A. D. Bevan (Rush—Surgery), Chairman, W. T. Councilman (Harvard—Pathology)—C. H. Frazier (Pennsylvania—Surgery), V. C. Vaughan (Michigan—Dean), and J. A. Witherspoon (Vanderbilt—Medicine).

These distinguished medical educators acted promptly and effectively. Under their direction the quality of medical education in the United States was assessed during 1904-1905. The results showed that, in an over-all sense, American medical education was far inferior to concurrent medical training in Germany, France and England. Notable deficiencies were: (a) low admission requirements (and those poorly enforced); (b) poorly trained teachers; (c) inadequate facilities; and (d) insufficient financial support (cf. 31). The general picture was later revealed in the first Flexner report (2). On the basis of the Committee's surveys, Chairman Bevan reported to the AMA House of Delegates in 1905 that, "It is evident from a study of the medical schools of this country and their work that there are five especially rotten spots which are responsible for most of the bad medical instruction. They are: Illinois with fifteen schools, Missouri with fourteen . . . Maryland with eight . . . Kentucky with seven . . . and Tennessee with ten . . . That is, fifty-four medical schools in these five states and not more than six of these can be considered acceptable" (29, p. 894).

In 1905 the Council set forth an "ideal standard" medical curriculum, based on the curricula of the better schools of England, Germany and France, and this they recommended for eventual general adoption. The admission requirement would be an education sufficient to qualify for entrance to recognized universities. The medical curriculum would be, "A five year medical course; the first year devoted to physics, chemistry and biology; the next two years to laboratory sciences of anatomy, physiology, pathology, and pharmacology, and two years to the clinical branches, with close contact with patients in both dispensary and hospital. A sixth year as an intern in the hospital" (29, p. 894).

Modest though it now appears, this "ideal standard" was revolutionary at that time. The Council, determined to achieve the "ideal" ultimately, realized that it was not practical as an immediate goal. Accordingly an interim standard was recommended for immediate adoption. This comprised, ". . . four years of high school for admission, a four-year medical course and satisfactory performance in a state licensing examination" (29, p. 895).

Carrot-and-stick tactics were used to obtain compliance with the interim standard. The council devised a rating system for grading medical schools. On this basis existing schools were classified in three categories: A (acceptable), B (doubtful), and C (unacceptable). Results of the first ratings were presented by Chairman Bevan to the American Medical Association in 1907. Although the grading was "very lenient", forty-six of 160 schools were in the B category and thirty-two in C.

Reaction to this rating system was considerable. Between 1906 and 1910 (the year of the Flexner report) the number of medical schools in the United States decreased by twenty-nine (cf. 32, p. 208). This decrease involved several mergers in cities having two or more schools and closure of some Class C schools because State Boards of Medical Examiners refused to permit their graduates to take examinations (29).

About this time (1908) the Carnegie Foundation for the Advancement of Teaching became interested in appraisal of the quality of professional education in the United States. Learning of this interest, the Council sought the cooperation of the Foundation in its continuing efforts to raise the standards of medical education. A momentous decision was made in December 1908, when the Foundation agreed to undertake an assessment of American medical education. Abraham Flexner began this task a few days later (17).

The result of this endeavor, *Medical Education in the United States and Canada,* Bulletin No. 4, Carnegie Foundation for the Advancement of Teaching, was published in 1910. This was the famous "Flexner Report" described by William H. Welch as ". . . one of the most remarkable and influential publications in educational literature. It had not only a large influence upon professional opinion, but especially a large influence on universities and upon public opinion" (1).

President Pritchett's *Introduction* to Bulletin No. 4, in its own right a major document in the history of American medical education, reviewed the report insightfully and succinctly, as the following excerpt shows:

> The report which follows is divided into two parts. In the first half the history of medical education in this country and its present status are set forth. The story is there told of the gradual development of the commercial medical school, distinctly an American product, of the modern movement for transfer of medical education to university surroundings, and of the effort to procure stricter scrutiny of those seeking to enter the profession. The present status of medical education is then fully described and a forecast of possible progress in the future is attempted. The second part of the report gives in detail a description of the schools in existence in each state of the Union and in each province of Canada (2)

The pervasive influence of Hopkins on the Flexner report is very evident. Thus in President Pritchett's *Introduction* the name of William H. Welch heads the list of those thanked for "constant and generous assistance"—a list which includes Simon Flexner, Arthur D. Bevan, N. P. Colwell (Council secretary) and F. C. Zapffe (secretary of the Association of American Medical Colleges). Flexner's autobiography tells the story quite fully in the following passage:

> Having finished my preliminary reading, I went to Baltimore—how fortunate for me that I was a Hopkins graduate!—where I talked at length with Drs. Welch, Halsted, Mall, Abel and Howell, and with a few others who know what a medical school ought to be, for they had created one. . . . I became thus intimately acquainted with a small but ideal medical school, embodying in a novel way, adapted to American conditions, the best features of medical education in England, France and Germany. Without this pattern in the back of my mind, I could have accomplished little. (17, p. 115).

Part II of the Flexner Report disclosed disgraceful conditions in many medical schools and unsatisfactory standards in all but a few (cf. 2 p. 28). While these facts were known by the Council and by many medical educators and physicians, they were not widely known by the public or even in the university world. The publicity provided by the *Report* itself and by discussion of Flexner's findings in the news media ". . . largely closed the gap between private knowledge of departures from accepted standards and public acceptance of those standards" (33 p. 18).

The achievements and limitations of the organizations and educational leaders involved in the reform of American medical education have recently been reviewed by Shryock (7): "By 1930 nearly all medical schools required an arts degree for admission and provided a three—or four—year graded curriculum, improved hospital facilities and clinical instruction. In addition, boards in some thirty states insisted—in cooperation with the American Hospital Association (founded in 1898)—that candidates take a year's internship as well as a recognized degree. . . . As a result of these trends the ratings of medical schools rose rapidly. In 1913, for example, twenty-four were graded A+, thirty-nine as A and the rest as B. By 1918 out of the nearly sixty schools which were members of the Association of American Medical Colleges fifty-six were graded A and only three listed as B. Although a number of low grade schools still existed in that year, such colleges continued to disappear . . . By 1925 a B grade was rare and during the 1930's and 1940's such a rating became a near catastrophe" (7, pp. 63-64). These grades and numbers show that there was a very great improvement in the selection and education of doctors as a result of the reform movement.

But unfortunately this is not the whole story. To some extent the gain won by elimination of inferior and commercial medical schools was offset by the development and legal recognition of substandard practitioners of the healing arts—men who do not qualify for the M.D. degree—and of substandard institutions in which they are trained.

Growth and Diversification: 1920-1945

It is impossible to impart the entire content of medical and surgical science to the student. One cannot even impart the content of a single subject of the curriculum. The utmost to be expected is to give the student a fair knowledge of the principles of the fundamental subjects of medicine, and a power to use the instruments and methods of his profession; to give him the right attitude toward his patients and his fellow members in the profession; and above all to put him in a position to carry on the education which he has only begun in medical school. *William H. Welch (34)*

The Undergraduate Medical Curriculum

Welch put the case very well. The undergraduate medical curriculum of 1920-1950, fairly well standardized in the pattern developed at Hopkins, was designed to provide a "fair knowledge of the principles of the fundamental subjects of medicine"—the common core of training necessary for all physicians.

The gross structure of the undergraduate medical curriculum, in terms of major subjects and their sequential order, was much the same in the approved medical schools of the United States from 1920 to 1950. In fact the overall strategy for curriculum planning set forth by Welch in his address to the Association of American Medical Colleges in 1910 (34), and the descriptions of the undergraduate medical curriculum provided by the Commission on Medical Education (AAMC) in 1932 (35), by Weiskotten et al., in 1940 (32), and by Deitrick and Berson in 1953 (36) are, in a general way, similar in terms of courses included and their sequences. There was, however, one significant progressive change (with perturbations) throughout the period 1910-1960. This was the decrease in hours scheduled for anatomy. The time gained in this way was distributed chiefly to physiology and biochemistry. The changing role of anatomy, reflected in shortening of the courses, has been well described by Fuhrman in his recent monograph on the multidisciplinary laboratory (37).

Many schools made one notable change in the sequence of training in the clinical clerkship years (third and fourth). The Hopkins pattern of assigning third year student clerks to outpatient clinics and fourth year clerks to hospital wards was reversed. There are advantages in this shift, because patients in the wards are usually sicker, the signs and symptoms of disease are more fully developed, and follow-up is more easily scheduled. Thus the clerk is better prepared for outpatient study by this sequence.

Thus over the first half of the twentieth century, the organization of the undergraduate medical curriculum changed little in respect to major topics. For example, the curricular structures set forth in the Hopkins *Catalogue* of 1905-1906 (38) and in the Stanford *Announcement* of 1916-1917 (39) resemble the general patterns described by the Commission on Medical Education in 1932 (35), Weiskotten et al., in 1940 (32), and Deitrick and Berson in 1953 (36). Modifications of these patterns have developed at different schools and at various times.

Although the gross structure of the curriculum changed little from 1900 to 1950, the period was one of marked growth, development and diversification of medical knowledge. The discoveries of Einthoven; Ehrlich; Landsteiner; Eijkman; Hopkins; Banting, Best and Macleod; Sherrington; Adrian; Whipple and Domagk among others (cf. 40-42), and the vigorous exploitation of these discoveries led to pervasive changes in the microstructure of the curriculum—the weighting of subtopics, the content of lectures and laboratory periods, the foci of conferences and seminars and the like. Constancy of gross curricular structure masked dynamic change in course content. An informal picture of much of the scientific progress of the period is set forth in a volume of selected prefaces of the Annual Reviews of Biochemistry, Physiology, Pharmacology, and Physical Chemistry (43). These prefaces were initiated by V. E. Hall, then and now editor of the *Annual Review of Physiology,* when he wrote an invitation to senior contributors. Dr. Hall's letter read in part: "This is an invitation, the motive behind which is our desire to make the *Review* something more than a consideration in detail of the current advances in our science. Physiology is a form of human activity as well as an accumulation of knowledge. As such it has a history of hopes, ambitions, enthusiasms, fashions and phobias. Older members of the profession are retiring, newer ones rising to prominence. New institutions are being founded, older ones changing their form and function. New patterns of financial support, of teaching and research are emerging with various influence on their conduct . . ." (43). The responses provide a panorama of medical science "in living color" from the 1940's to the 1960's (43)—an "inside" story unique in the literature. Of particular interest in the context of this paper is Erlanger's *A Physiologist Reminisces.* This is the autobiography of an immigrant's son who graduated from the University of California in 1895, from Hopkins Medical School in 1899, interned at Hopkins, was invited to a faculty post at Hopkins by W. H. Howell, and was one of the "small band" from Hopkins who, in Flexner's words, "founded new institutions or reconstructed old ones". Erlanger's outstanding achievements in neurophysiology were recognized by award of the Nobel Prize in 1944 shared with H. S. Gasser (40). Long before that award Erlanger's work had been incorporated in the courses in physiology forming part of the "standard" curriculum of American medical schools—a prime example of change in the microstructure and of its tone, flavor and atmosphere.

Internship and Residency Training

An internship year—a year which followed the M.D. degree and which rounded out and extended the clinical training of the undergraduate medical curriculum by additional hospital experience under supervision—became an almost universal part of American medical education by the 1930's (cf. 35, p. 141). The history of the development of internship training in the United States, a description of its structure, and an assessment of its value are set forth in the reports of the Commission on Graduate Medical Education in 1940 (44), and of Deitrick and Berson in 1953 (36).

The most significant change in medical education in the United States during the decades following 1920 was the increase in number, in duration, and expansion in size of residency training programs (cf. 44, 36). Whereas the internship had been designed to produce a "safe" general practitioner, the purpose of the residency was to provide specialists—experts in special fields. Residency training in the modern sense, denoting several years of supervised, progressive specialized training, was first developed at Hopkins (45, Vol. I, p. 161). At the start, teaching hospitals and a few non-teaching hospitals established their own patterns and standards. The length of a resident's training was often determined by his own wishes. "In a teaching hospital the resident was responsible for professional care of patients under the supervision of the chief of his service. He usually participated in the instruction of the medical students, helped to supervise the work of the interns, and frequently he carried out some research" (36).

The basic cause of the growth of residency programs was recognized by William H. Welch when he wrote in 1910, "It is impossible to impart the entire content of medical and surgical science to the student" (34). The information explosion in science was forcing specialization in medicine as it was in the other fields. At first specialty training was obtained by Americans in European clinics (8), or by securing one of the few good residencies available in the United States, or by acting as an assistant to a competent specialist, or by combining these procedures. "No standard, or examinations, existed for men wishing to claim that they had special qualifications. A physician could become a specialist simply by declaring that he was one" (36).

The history of the development of specialty boards which established standards of quality in the several specialties and required examination for certification is presented in the reports of the Commission on Graduate Medical Education (44), Deitrick and Berson (36) and Johnson and Weiskotten (31) as well as in the History of the A.M.A. edited by Fishbein (30). Exemplary case histories of development of residency programs at Harvard Medical School (Massachusetts General Hospital) and at the University of Pennsylvania have been written respectively by Cope (12, 46) and by Corner (27). Current information on intern and residency programs is provided in the *Annual Reports* of the Council on Medical Education and Hospitals in the annual *Education Number* of the *Journal of the American Medical Association* and in the Association's yearly *Directory of Approved Internships and Residencies*.

Postgraduate Medical Education

The "information explosion" in medical science has often been noted in this paper. A major aspect of that phenomenon, exponential increase in the rate of discovery of significant remedies as a function of time, was recently described by Paton (47). The vigor and thrust of current medical investigation is revealed more completely by the burgeoning number of journals of original publication, review journals, *Recent Advances*, thematic monographs and abstracts.

While a physician is an intern or resident in a teaching hospital the medical faculty and the medical library keep him informed of significant new developments in medical science. In this situation the research of yesterday is the teaching of today and the practice of tomorrow (cf. 48, p. 1650). When the resident leaves the medical school these resources are no longer directly available. Postgraduate medical education is designed to lessen the consequent communications gap. Descriptions of programs in this field are included in references 32, 35, 36 and 44, and the subject is reviewed each year in the *Annual Reports* of the Council on Medical Education and Hospitals.

Medical Education in Transition: 1940-1966

There ought to be in the Harvard Medical School an extended instruction far beyond the limits of any one student's capacity. This involves, of course, some optional or elective system within the school itself, whereby the individual student should take what is, for him, the best four years' worth, the faculty supplying teaching which it might take a single student eight, twelve or twenty years to pursue. *Charles W. Eliot, 1895 (49)*

In a very general way the history of American medical education from 1920-1945 resembles that of European medical education (particularly German) from 1870 to 1914. Both eras were marked by considerable stability in the gross structure of the curriculum, by continuous improvement in course content, advances in research and increasing interaction between research and teaching. In both periods the general level of education was high, although there was a wide range in excellence among the several medical schools. *Minerva Medica,* a resident in Germany from 1870 to 1914, was believed to be an American in the 1920's—as Osler had predicted in 1908 (cf. 50). There was something of the "Proud Tower" atmosphere (51) in American medicine during the long armistice.

War and Peace: 1940-1947

The year 1940 marked the beginning of a new era in the relations of the federal government and science. So far as a line can be drawn across the continuous path of history, this date separates the first century and a half of American experience in the field from what has come after. As the scale of operations changed completely, science moved dramatically to the center of the stage. By the time the bombs fell on Hiroshima and Nagasaki, the entire country was aware that science was a political, economic and social force of the first magnitude. *A. Hunter Dupree (52)*

The story of the establishment of the National Research Defense Council (NRDC) in 1940, of its inclusion in the broader structure and mission of the Office of Scientific Research and Development (OSRD), and of the crucial administrative leadership of Vannevar Bush, J. B. Conant, K. T. Compton and F. B. Jowett in providing the link between the universities (and their medical schools) and government research for national security has been set forth in two major reports (52, 53). OSRD's mission was "to serve as a center for mobilization of the scientific personnel and resources of the nation in order to assure maximum utilization of such personnel and resources in developing and applying the results of scientific research to defense purposes." OSRD was also empowered to coordinate, aid and supplement the scientific research activities relating to national defense of the Department of War, Navy and other departments and agencies of the Federal Government (52).

Thus OSRD became a central office for the entire government. Bush, as director, was responsible directly to the President. The magnitude of war research effort, which rose to a climax in 1945, far exceeded both the World War I program and the level of activity in 1940. Thus expenditures in 1945 reached $1.6 billion in contrast to $100 million in 1940. For practical purposes, money was no longer a limiting factor. As Dupree put it, "the scale of operations changed completely" (52).

Political interest in the relation of government to science rose during the war as the budget of OSRD skyrocketed. Before closing down the OSRD, in line with Bush's intentions, the President asked Bush for his recommendations regarding future relationships of Government and Science. The President's letter to Bush (17 November 1944) posed four questions (53). These were:

First: What can be done, consistent with military authorities, to make known to the world as soon as possible the contributions which have been made during our war effort to scientific knowledge? . . .

Second: With particular reference to the war of science against disease, what can be done now to organize a program for continuing in the future the work which has been done in medicine and related sciences? . . .

Third: What can the Government do now and in the future to aid research activities by public and private organizations? The proper roles of private research, and their interrelation, should be carefully considered.

Fourth: Can an effective program be proposed for discovering and developing scientific talent in American youth so that the continuing future of scientific research in this country may be assured on a level comparable to what has been done during the war? . . .

Bush appointed committees of distinguished scientists and educators to study each of the four questions. Question two was referred to a committee representing medical research (53) "Bush's report, *Science—the Endless Frontier*, attempted a profile of American science and a prescription for the future. The basic principle of the interrelated system appears in the body of the report" (53, 54).

The Government should accept new responsibilities for promoting the flow of new scientific knowledge and the development of scientific talent in our youth. These responsibilities are the proper concern of Government, for they vitally affect our health, our jobs and our national security. It is in keeping with basic United States policy that the Government should foster the opening of new frontiers and this is the modern way to do it. For many years the Government has wisely supported research in the agricultural colleges and the benefits have been great. The time has come when such support should be extended to other fields.

The effective discharge of these new responsibilities will require the full attention of some over-all agency devoted to that purpose. There is not now in the permanent Government structure receiving its funds from Congress an agency adapted to supplementing the support of basic research in the colleges, universities and research institutes, both in medicine and the natural sciences, adapted to supporting research on new weapons for both Services, or adapted to administering a program of science scholarships and fellowships.

Therefore I recommend that a new agency for these purposes be established. Such an agency should be composed of persons of broad interest and experience, having an understanding of scientific research and scientific education. It should have stability of funds so that long-range programs may be undertaken. It should be recognized that freedom of inquiry must be preserved and should leave internal control of policy, personnel and the method and scope of research to the institutions in which it is carried on. It should be fully responsible to the President and through him to the Congress for its program (54).

Bush envisaged a National Research Foundation of comprehensive scope. It would include Divisions of Natural Sciences, Medical Research and National Defense. Embodying many features and characteristics of OSRD, it was to be administered under the control of a part-time Board composed of people not otherwise in government (53, 54). This Board would appoint the Director, formulate policy and would be empowered to make appropriate grants and contracts. The powers and political independence proposed for the Board seemed too broad to some, and legislation including the features outlined was vetoed by President Truman.

Thus when OSRD terminated a National Research Foundation had not been established. This had been foreseen by the Committee on Medical Research of OSRD, and arrangements had been made to transfer a number of ongoing research programs to other Federal agencies for support. These included the National Institutes of Health (NIH), the Office of Naval Research (ONR), and other appropriate agencies. The agencies concerned and the history of this transition have been described in several monographs and papers. (53, 55-61).

In 1950, the National Science Foundation (NSF) was authorized with a more limited scope and more conventional administration than Bush had proposed for a "National Research Foundation" (cf., National Science Foundation Act of 1950). NSF was established in 1951—with a miniscule budget. Science now had pluralistic support among Government agencies, much like the system of pluralistic support from private foundations developed in the period 1900 to 1940.

Let us now turn from the institutional setting for fostering creative activity in science to the record of achievement in medicine during the war. Here the term medicine is used in a broad sense to include medical research, practice and institutions as well as like aspects of public health. The account which follows is merely a vignette, touching a few high points.

By far the greatest achievement in biomedical science during the war was the development of Fleming's discovery (1928) of penicillin and exploitation of its chemotherapeutic effects. This was one of the great scientific breakthroughs of all time. The vast collaborative effort required to prepare penicillin in pure form, to explore its curative effects in infectious diseases in animals and then in man, and to determine its chemical structure—an undertaking which laid the foundations of antibiotic therapy—was briefly and incisively described by Goodman and Gilman (62) and set forth in detail by major participants (63-65). A general review placing this classic undertaking in context, accompanied award in 1945 of the Nobel Prize for *Physiology or Medicine* to Fleming, Chain and Florey (40, pp. 77-145).

As exemplified in this classic case and as Bush pointed out (54), most wartime research involved application of existing scientific knowledge to the problems in hand rather than creation of new knowledge.

Other medical achievements in World War II were reflected in reduction of the death rate in the United States Army from 14.1 per thousand (World War I) to 0.6 per thousand (World War II). Such diseases as yellow fever, typhus, tetanus, pneumonia and meningitis were "all but conquered" by penicillin, the sulfa drugs, better vaccines, and improved hygienic measures (including use of the insecticide DDT). These overwhelming demonstrations that medicine "really works" created a climate of opinion favorable for the support of medical research after the war.

American medical schools participated in the war effort not only in augmented research endeavors germane to military needs but also by speeding up the process of medical education to provide more physicians. This was done by reducing required premedical training from four years of college work to two or less, and by compression of the four year medical curriculum into three calendar years.

Impact of Increasing Federal Research Support on Medical Education

The crucial decision to transfer OSRD commitments to NIH initiated a new era of Federal aid to medical research. During the next two decades the Federal government rapidly increased its contribution to medical research and research training (66-68). Private foundations and health agencies continued to render significant aid, but the percentage of the total cost of research provided from these sources decreased (cf. 69).

Quantitative measures of these changes were set forth recently by J. A. Shannon, Director of NIH, in his Alan Gregg Memorial Lecture (68). On that occasion Shannon pointed out that over-all expenditure for medical research rose from $87 million (Federal share 31 per cent) in 1947 to $2.05 billion (Federal share 68 per cent) in 1966. In 1947, NIH provided (in money terms) ten per cent of the national effort; in 1966 the NIH contribution was 40 per cent. In those two decades the total dollar support for medical research increased 24-fold. This enormous augmentation resulted in, ". . . expansion of research manpower and facilities, in the size of our graduate enrollments; in the extent and diversity of conferences, meetings and publications; in the growing pace of scientific progress; and, unfortunately, in the burden of administrative mechanisms, forms and details" (68). The relative growth of the several NIH programs (such as direct research, research grants and contracts, from 1945 to 1962, was shown in an informative chart recently published by Kidd (67, Fig. 4.3, p. 99).

The impact of NIH programs on American medical education was appraised by Turner in 1967 (70), who wrote that ". . . this vast program has enormously strengthened the medical schools of the country and greatly enhanced the quality of medical education, although, until the Health Professions Educational Assistance Act of 1965, no funds were voted by Congress directly for medical education. The great university medical centers of America—all eighty-eight of them, not just a few—have become the vital links in the health program of the country in at least two major respects. First, the university medical centers are the principal sources of health and medical manpower of the nation; second, for the communities in which they are located, they constitute focal points of excellence in medical practice, points from which improvements in preventive and curative medicine are most likely to stem." These considerations alone justify all the money the Congress has voted and the people of the country have provided.

The large NIH program of research support paid an unexpected dividend as Turner observed, "For almost solely as a result of this program, during the past fifteen years, it has been possible to bring about a substantial increase in the yearly number of medical graduates in the United States (from 6135 in 1951 to 7677 in 1966, a 25 per cent increase) and further increments are projected. We now have the medical and scientific manpower to staff the enlarged schools and the new schools that the nation needs so badly. Had this program not been operative, significant expansion of education facilities could have been accomplished only at the risk of a decline in quality of medical education" (70).

Evidence in support of Turner's insightful analysis was provided by an NIH staff study of twenty medical schools (eleven state-owned and nine private) focused on the decade 1949-1959 (71). This study showed that the number of faculty members (full- and part-time) employed by these schools doubled during the decade and that, with the aid of Federal construction grants, facilities were expanded. These circumstances permitted significant increase in the number of medical students, interns and residents, graduate students in basic medical sciences, clinical fellows and trainees.

CHANGING PATTERNS OF MEDICAL EDUCATION

There is a tendency everywhere, and perhaps especially in America, to equate what is new with what is good, and to confuse the means with the ends, especially when the means are striking and dramatic. The present trend is to concentrate on the teaching and to neglect the teacher, on the assumption that if only the right methods can be found, all will be well. This seems to me to put the cart squarely before the horse. Given good teachers and good students, it does not matter much what passes between them; the product will also be good. *D. C. Sinclair (72)*

The undergraduate curriculum of American medical schools, in terms of courses and their sequence, changed little from 1920 to the 1950's. While revision and development were continuous within each course, the information explosion could not be wholly contained by such methods. Little by little the curriculum became overloaded and beset with makeshift modifications. For example, at Stanford some 900 hours were added to the curriculum between 1940 and 1955. This virtually eliminated options and free time—and what free time was left was poorly distributed (73). Other schools had the same problem. By the early 1950's it was clear that, "the apparatus that had worked so well up through World War II was beginning to creak" (74).

In the atmosphere which prevailed in 1953 Dean Berry of Harvard proposed a critical reexamination of the structure of medical education in an attempt to make it more serviceable (75). To achieve this end, he proposed a series of Teaching Institutes to be sponsored by the Association of American Medical Colleges (AAMC). These Institutes would be attended, on invitation, by medical faculty from all medical schools. This plan was favorably received and the first Institute was held in the fall of 1953.

Berry's interest in restructure of the curriculum stemmed chiefly from his concern with "comprehensive medicine" (cf. 75) and his hope to make room for training in this field "without diminishing scientific medicine." This was to be done by "elimination of what is archaic", and addition of significant new material. Thus the curriculum would be redesigned around "an essential scientific core" in such a way that curricular time could be provided for the social behavioral sciences germane to comprehensive medicine.

As might be anticipated, Berry's strategic plan was supported by some medical educators and opposed by others. For example, in the view of R. F. Loeb, the curricular time for essential basic science training was already inadequate. Moreover, incorporation of social sciences was of doubtful value because "those disciplines have not yet reached a definitive maturity which can find tangible application to medicine" (76, cf. also R. K. Merton, 77, pp. 6-7).

The two views coexist today. The Institutes, nevertheless, did prove valuable in many ways. They provided a forum for reexamination of the totality of medical and premedical training as a continuum, for appraising the strength and weakness of NIH policy and training and for review of medical school administration. In brief summary, the first three Institutes dealt with basic medical sciences, the next two with medical school applicants and medical students. *Six* and *Seven* were concerned with the strategy and tactics of clinical teaching. *Eight* was devoted to an examination of the interactions of medical education and medical care. *Nine*, a highlight in the series, comprised a review in depth of the relations and interactions of teaching and research. In the tenth report the interrelations of medical education and medical practice were scrutinized. The last three *Reports* focussed on administrative matters. A complete list of the titles of the Reports of the Institutes, with bibliographic citations, is included in Littlemeyer's paper (78).

Turning now to the changing patterns of medical education as they evolved in medical schools, it is clear that they were influenced by the proceedings of the several Institutes, and, in turn, influenced those proceedings.

The first substantial curricular change in the postwar era occurred at Western Reserve. Here administrative developments facilitated a new look at the medical teaching program. Appointment of a new dean, interested in curricular reform, coincided with retirement of a number of key department chairman (79). Thus reappraisal was natural and relatively easy. As noted in a Stanford committee report of the early 1950's ". . . the task of introducing change into and from within a given academic culture, of reaching common consent among several hundred highly individualistic faculty members to new programs and new curricular patterns is difficult, complex and frequently less than an orderly affair" (73, p. 1068).

The nature and extent of restructure of the curriculum at Western Reserve has been described by the participants (80, 81,) and also in Lee's review (79). Because the time allocated to the common core of courses required of all students was reduced, and opportunity for elective courses was correspondingly increased, the Western Reserve curriculum may be considered an early application of the "core and elective principle" (cf. Stowe, 73).

The several phases of the new Western Reserve program were designed and directed by "subject committees" rather than departments. Integration of subject matter was stressed throughout. A notable operational innovation was provision of "multi-discipline laboratories" for student laboratory work, research and study (cf. 37).

Comparison of the content of the new curriculum at Western Reserve with that of the old one shows that the change was not very radical (79). The subject committees endeavored to preserve the informational content germane to the present, to eliminate that which seemed less relevant, and to provide for diversification in the elective component. These had been aims of the departments too, but they were easier to achieve in the new operational pattern.

Lee's thoughtful evaluation of the changeover at Western Reserve may have broad application. "The most significant change at Western Reserve was one of attitude, both toward the content of medical education and toward the student. The most obvious results can be seen by superficial observers in the enthusiasm of the well-informed faculty and the relaxed yet mature behavior of the medical students" (79). No small part of the benefit of change in a teaching program, when the new pattern is welcomed by faculty and students, may be a "placebo effect". Like some other useful placebos a new curriculum is expensive, presented with enthusiasm by the doctor and deemed effective by the patient (cf. 82, 83).

Eight other case histories of "experimentation in medical education" were reported in Lee's monograph (79). These described a broad spectrum of change, including extension of the aegis of the medical school to encompass pre-medical college training at Hopkins, Northwestern and Boston; introduction of "comprehensive care" teaching programs into the undergraduate medical curriculum at Cornell, Colorado, Temple and North Carolina ("definitions of comprehensive care will vary a good deal depending on the point of view of the institutions professing to practice or teach it", 79, p. 30), and development of a curriculum at the (then) new medical school of the University of Florida.

Change was indeed in the air. In 1964, the Council on Medical Education undertook a survey to appraise the number and extent of significant curricular changes in American medical schools to 1964, and the number of major changes anticipated. To that end the *Annual Questionnaire* of the American Medical Association for 1964, sent to all eighty-eight medical schools in the United States at that time, included two questions on the curriculum. These were: (*a*) Has there been a major change in the curriculum at your school during the past few years? and (*b*) Have definite plans been made for major changes in the curriculum in the future? In reply to the first question, fifty-four schools reported no changes in the past few years, seven described minor changes, and eighteen major changes. To the second question, forty-three answered that no changes were anticipated in the immediate future; the remainder described a variety of plans for future curricular change or development.

The pattern of change described by schools which had altered or planned to alter their educational programs was fairly uniform, ". . . the emphasis invariably being placed on greater use of elective time, further reductions in the amount of scheduled didactic lectures and conferences, and the introduction of courses synthesizing basic and clinical sciences." Three plans were described in some detail. These were the new curricula at the University of Missouri, the Medical College of Virginia and the University of Rochester, used for the entering classes of 1964-1965 (84, pp. 751-752).

All three of these new programs comprised a common core of basic science and clinical training required of all students and substantial blocks of elective time—they were designed in line with the core-elective principle described by Stowe (73). The elective program and the free time included in these curricula provide for flexibility and diversity. These features permit the student to mold his medical education, in part at least, to his abilities, interests and goals. They permit earlier specialization.

In all schools having core-elective curricula the required core includes basic science and clinical components. The structure of the required core is fairly constant from school to school. There appears to be considerable range in the number and variety of elective course offerings.

Thus there are two curricular formats in use in American medical schools today, one old and one new. The older type is the standard 1920-1960 pattern developed at Hopkins and described by Flexner (2), which became general after the reforms early in this century. The newer is the core-elective program, of which early examples were the Western Reserve (79) and Stanford (73) plans, and the still more recent curriculum at UCLA (87). The major difference in structure between the two categories is the increased use of elective time in the newer group. The trend, as shown by reports of the plans of new schools and of curricular modifications in older ones, is strongly toward the core-elective type (84-88).

The similarity in structure of the core-elective programs suggests that the forces leading to change were general. What were these forces and why was the 1920-1960 format no longer satisfactory? D. H. Solomon, one of the chief architects of the revised curriculum at UCLA (87), provided the following answers in a statement remarkable for incisiveness and lucidity:

> The largest factor by far was the postwar explosion of information relevant to medicine, both in the scientific disciplines fundamental to medicine and in the clinical fields themselves. Specialties and subspecialties sprang forth and flowered. The violent expansion of knowledge and proliferation of fields of expertise led inevitably to overlap, duplication and uncertain responsibility in the academic arena. The new subject matter tended either to be by-passed or taught in unduly fragmented segments by several departments, each recognizing the importance of the new material and attempting to work it into the existing structure. A mechanism was needed to allow increased coordination of teaching efforts between departments and simultaneously provide the flexibility to introduce entirely new courses into the medical curriculum. Just as the handling of new information presented a problem, so did the rapid obsolescence of previously accepted "facts" and principles. Finally, the volume of new subject matter simply strained the limits of the existing curricula. Was the medical school experience to be lengthened, or was greater selectivity to be exercised in choice of subject matter, or both?
>
> Another major force for change was the rapid elevation of standards of scientific information in high school and college [cf. 89]. Students approached their medical school career better prepared, with a more sophisticated knowledge of career alternatives and a more mature understanding of their own goals. And meanwhile the practice of medicine was changing radically as a career. Only in its very core is medicine a single profession today. Outside the core lies a host of different careers, varying from public health and hospital administration to biochemical research, from gynecologic endocrinology to cardiac surgery, from academic medicine to family practice (74).

In addition to the information explosion and better preparation of premedical students, forces which pertain to the internal logic of medical education, there was a third force, contextual in nature. This was social need for an increased medical capability (cf. 90-97). Clearly the flexibility of the core-elective format permits appropriate response to this factor—which may include elective programs in "comprehensive medicine", a term with several connotations (cf. 79, 90-95).

Perhaps this section on "Changing Patterns of Medical Education" can be put in perspective by the following quotation from one of the newer curricular plans:

The prime requisites for excellence in medical education remain, of course, the maintenance of the highest possible quality of faculty, students, facilities and intellectual environment. However an optimal curricular framework is required for full development of these resources (87).

Possibly Sinclair would agree that this statement of principle places the horse where he belongs (cf. 72).

New Era of Growth and Expansion

Three major phenomena have combined to create a growing need for physicians for the nation. The first is the rapid growth of population, with a more than proportionate increase in the younger and older age groups which need the most medical service. The second is the increase in the individual use of medical services accompanying improvements in living standards, increased urbanization, more education, widespread use of health insurance and advances in medical knowledge. And the third is the increase in the number of physicians required for specialized services such as research and teaching. *Bane Report (98, p. xiii)*

By 1959 it was clear that an immediate strenuous program of action would be necessary to meet the country's need for physicians—a situation not foreseen a decade earlier (30, pp. 355-396). Three sequential appraisals of current and projected medical manpower concurred in recommending expansion of existing medical schools and creation of enough new schools to increase the number of medical graduates from an annual figure of 7000 in 1959 to 11,000 by 1975 (98-101). The consultants who wrote the Bane *Report* (98) concluded that the number of new schools needed might range from twenty to twenty-four, depending on the rate and extent of expansion of existing schools as well as other factors (98, pp. 57, 75). The Jones Committee concurred (100, p. xvi).

The impact of these reports, augmented by that of an impressive position paper issued by the Director of the AAMC and by the endeavors of other groups within and without the Government, can be appraised by comparing the rate of development of new medical schools from 1925 to 1962 with the subsequent rate, as shown by the data in Smythe's recent report (102).

From 1925 to 1962, twenty-one schools initiated four-year teaching programs. This number represented thirteen new schools, one school of osteopathy converted to medicine, and seven two-year schools expanded to four year programs (102, Table 1). In contrast, since 1964, sixteen American medical schools have been established or planned and have either admitted, are planning to admit, or have been authorized to admit students (102,

Table 3). "These schools will eventually enroll 1340 first-year students and can be expected to graduate at least 1200 physicians per year. At the same time seventy-four of the eighty-seven schools which are fully operating have reported projected increases which are expected to total at least 885 first-year students per year by 1972" (102).

The architectural and operational planning, photographs of existing buildings or architect's models of proposed buildings and in many cases the proposed curricular plans, with brief reviews of present status, of eleven developing medical schools (Arizona, Brown, University of California at San Diego, University of Connecticut, Hawaii, Pennsylvania-Hershey, Michigan State, Mt. Sinai, New Mexico, Rutgers and South Texas) are included in the *Report* of the Council on Medical Education for 1964-1965 (84). The 1965-1966 *Report* (85) provided similar information for the remaining five developing schools (University of California at Davis, University of Massachusetts, Toledo State, SUNY-Stonybrook and Louisiana State-Shreveport). In 1966-1967 an illustrated report (86) on expansion of facilities at older schools (Colorado, Hahneman, Bowman-Gray, Western Reserve, Georgetown, Loma Linda, Miami, SUNY-Downstate and the University of Pennsylvania) furnished much information on the rate and extent of the medical manpower increase anticipated from these programs.

While the sixteen developing schools and the planned expansions at older ones will serve to maintain the long-term physician—population ratio of 132 ± 2 per 100,000 (99, p. 34; 100, p. 31) and probably to increase it slightly, it appears unlikely that the increase now envisaged will meet even the domestic demand for more physicians (90, 98-103). To achieve this the rate of creation of new schools and expansion of old ones must be accelerated further. Moreover, if American medical manpower is to play a significant part on the international scene, the increase must be greater still (cf. 104).

SIGNIFICANT OMISSIONS

This review had focussed chiefly on the genesis and spread of the Hopkins curricular format, on its continuous adaptation to the creation of new knowledge in medicine and cognate fields, and its transition to the core-elective pattern in response to the post-war information explosion stemming from the vast increase in research support. Many important topics have been touched tangentially if at all. Some of these will be listed, with references, so that the interested reader may have a toehold for further progress. The list includes:

1. Growing concern with medical education as a continuum, from the beginning of college (or even high school) education, through the four-year medical curriculum and the average (today) of four to five years of house staff training. The M.D. degree is awarded at about the mid-point of this continuing education experience (cf. 84-86, 90-92).
2. The lifelong process of continuing medical education (cf. 35, 36, 44, 84-86, 90-92).

3. Major trends related to health care and their implications (90-92, 97).
4. American participation in world-wide medical education and health care (104).

Finally, significant omissions to this point include the names of a few scholars whose influence has been pervasive rather than apposite to the particular problem in hand. These are Claude Bernard (105), A. L. Bloomfield (106), R. J. Dubos (107), T. Puschmann (108), and A. N. Whitehead (109).

Trends

In the space of 176 years the Lower Mississippi has shortened itself 242 miles. That is an average of a trifle over one mile and a third per year. Therefore any calm person, who is not blind or idiotic, can see . . . that in 742 years from now the Lower Mississippi will be only a mile and three quarters long, and Cairo and New Orleans will have joined their streets together. . . . *Mark Twain* (110)

Although, as noted by Mark Twain (and others), extrapolation is a risky procedure, several trends which became increasingly distinct in the recent past seem likely to burgeon in the decade to come. These trends will now be listed and characterized briefly:

1. Growing tendency to plan medical education as a continuum, with its sequential components (premedical and residency) developed and coordinated under the guidance of medical schools and their parent universities.
2. Spreading adoption of the core-elective curriculum with modifications in line with local interests and capabilities. This format, with built-in potential for adaptation to advances in medical knowledge, to the student's talents and goals, and to new areas of emphasis (e.g. comprehensive medicine, community medicine) has the flexibility and versatility needed in an era of change.
3. Expansion in the scope and depth of continuing medical education under the aegis of medical schools. This trend is now necessary for provision of good medical care. The information explosion in sciences germane to medicine not only provides new knowledge and techniques but also compels re-evaluation and modification of older concepts and practices.
4. Heightened rate of establishment of new medical schools and expansion of old ones. While the number of physicians needed in the years ahead cannot be predicted accurately and while there is yet no adequate measure of the "need for health care", it seems likely that this affluent society will demand more doctors than existing, expanded and planned medical schools can provide. In addition, under foreseeable circumstances, the emerging nations would be greatly benefited by American physicians' participation in their programs for medical education. A plan to provide "doctors for health" might complement the "atoms for peace" endeavor.
5. Proliferating spiral in the numbers of persons training for the professions and occupations related to health care. "The spectrum of skills now in use is broad and it seems certain that it will become progressively broader. It seems equally certain that more persons with specialized skills will need to be trained to buttress the physician-led team and compensate for the fact that fewer physician will be available than might be desirable" (90).

6. Continuing increase and reappraisal of support for biomedical research and education is so vital to the health needs of the nation, including education of physicians, that it now appears as "manifest destiny". Leaf (111) recently published a cogent and compelling statement of the need for general support of medical schools together with their affiliated teaching hospitals. He emphasized that, for best results, support should not distinguish between their educational and research endeavors.

Hopefully, in the decades to come, those responsible for working out changes in the strategy and tactics of medical education will bear in mind Flexner's judicious admonition: "Too much planning, too conscious planning, too much articulation, too conscious articulation will destroy the freedom of the individual upon which in the last resort progress depends" (112, p. 119).

REFERENCES

1. WELCH, W. H., Medical education in the United States. *Harvey Lect.*, 1915-1916, 11: 366-382.
2. FLEXNER, A., *Medical education in the United States and Canada*. New York, 1910. (Bull. No. 4, Carnegie Foundation).
3. NORWOOD, W. F., *Medical Education in the United States before the Civil War*. Philadelphia, 1944.
4. SHRYOCK, R. H., *Medicine in America: historical essays*. Baltimore, 1966.
5. ———, *Development of modern medicine*. New York, 1947.
6. ———, *American medical research; past and present*. New York, 1947.
7. ———, *Medical licensing in America*. Baltimore, 1967.
8. BONNER, T. N., *American doctors and German universities*. Lincoln, Nebraska, 1963.
9. BILLROTH, T., *The medical sciences in German universities*, introd. by W. H. Welch. New York, 1924.
10. RATHER, L. J., *Disease, life and man*. Stanford, California, 1958.
11. KOENIGSBERGER, L., *Hermann von Helmholtz*, transl. by F. A. Welby. New York, 1965.
12. CANNON, W. B., quoted, p. 209, *The Masachusetts General Hospital, 1935-1955*, ed. N. W. FAXON, Cambridge, Mass., 1959.
13. FLEXNER, A., *Medical education in Europe*. New York, 1912. (Bull. No. 6, Carnegie Foundation).
14. FLEXNER, A., *Medical education: a comparative study*. New York, 1925.
15. GERARD, R. W., The United States, pp. 160-168 in *Perspectives in physiology*, ed. I. VIETH, Washington, D.C., 1954.
16. MEEK, W. J., HOWELL, W. H., and GREENE, C. W., *History of the American Physiological Society semicentennial 1887-1937*. Baltimore, 1938.
17. FLEXNER, A., *I remember: an autobiography*. New York, 1940.
18. ———*Daniel Coit Gilman: creator of the American type of university*. New York, 1946.
19. FLEXNER, S., and FLEXNER, J. T., *William Henry Welch and the heroic age of American medicine*. New York, 1941.
20. HELMHOLTZ, H. VON, quoted in MERTZ, J. T., *European thought in the nineteenth century*, 4th ed., Vol. 1. Edinburgh, 1923: 209-210.

21. Rosen, G., Critical levels in historical process, *J. Hist. Med.*, 1958, **13**: 179-185.
22. Welch, W. H., Twenty-fifth anniversary of the Johns Hopkins Hospital, 1889-1914. *Bull. Hopkins Hosp.*, 1914, **25**: 363-366.
23. Berelson, B., *Graduate education in the United States.* New York, 1960.
24. Shryock, R. H., *The unique influence of the Johns Hopkins University on American medicine.* Copenhagen, 1953.
25. Hofstadter, R., Hardy, C. and DeWitt, *The development and scope of higher education in the United States.* New York, 1952.
26. Morison, S. E., *The development of Harvard University.* Cambridge, Mass., 1930.
27. Corner, G. W., *Two centuries of medicine.* Philadelphia, 1965.
28. Merton, R. K., The Matthew effect in science. *Science*, 1968, **159**: 56-63.
29. Fishbein, M., ed., *A history of the American Medical Association: 1847-1947.* Philadelphia, 1947.
30. Burrow, J. G., *AMA: voice of American medicine.* Baltimore, 1963.
31. Johnson, V., and Weiskotten, H. G., *A history of the Council on Medical Education and Hospitals of the American Medical Association, 1904-1959.* Chicago, 1960.
32. Weiskotten, H. G., et al., *Medical education in the United States.* Chicago, 1940.
33. Merton, R. K., Some preliminaries to a sociology of medical education, pp. 3-79 in *The student physician,* ed. R. K. Merton, G. Reader, and P. L. Kendall, Cambridge, Mass., 1957.
34. Welch, W. H., The medical curriculum. *Bull. Am. Acad. Med.*, 1910, **11**: 720-726.
35. *Final report of the Commission on Medical Education.* New York, 1930.
36. Deitrick, J. E., and Berson, R. C., *Medical schools in the United States at mid-century.* New York, 1953.
37. Fuhrman, F. A., *Multidiscipline laboratories for teaching the medical sciences.* Stanford University, 1968.
38. *The Johns Hopkins University circular: catalog and announcement for 1905-1906, of the Medical Department.* New Series, 1905, No. 8, October, 1905.
39. *School of Medicine. Annual announcement, 1916-17, Leland Stanford Junior University,* April 1916. Second Series, No. 90.
40. *Nobel lectures in physiology or medicine.* 1901-1962. Amsterdam, 1964-1967. 3 vols.
41. Schuck, H., et al.,*Nobel. The man and his prizes.* Stockholm, 1950.
42. *Les Prix Nobel.* Stockholm, 1901-1967. 66 vols.
43. *The excitement and fascination of science.* Annual Reviews, Inc., Palo Alto, California, 1965.
44. *Graduate medical education. Report of the Commission on Graduate Medical Education.* Chicago, 1940.
45. Chesney, A. M., *The Johns Hopkins Hospital and the Johns Hopkins University School of Medicine,* Baltimore, 1943.
46. Cope, O., pp. 158-186 in N. W. Faxon, ed., *The hospital in contemporary life.* Cambridge, Mass. 1949.

47. Paton, W. D. M., The impact of pharmacology on medicine. *Med. Press*, 1956, **236**: 213-217.
48. Medical education in the United States and Canada. *J. Am Med. Assoc.*, 1956, **161**: 1637-1681.
49. Eliot, C. W., quoted in: The Harvard Medical School, 1906-1956, *New Eng. J. Med.*, 1956, **255**: 1035-1041.
50. Osler, W., Vienna after thirty-four years. *J. Am. Med. Ass.*, 1908, **50**: 1523-1525.
51. Tuchman, B. S., *The proud tower*. New York, 1966.
52. Dupree, A. H., *Science in the Federal Government: a history of policies and activities to 1940*. Cambridge, Mass., 1957.
53. Committee on Science and Public Policy, National Academy of Sciences. *Federal support of basic research in institutions of higher learning*. Washington, D. C., 1964.
54. Bush, V., *Science—the endless frontier*. Washington, D.C., 1945.
55. Stewart, I., *Organizing scientific research for war: the administrative history of the Office of Scientific Research and Development*. Boston, 1948.
56. Wehl, F. J., ed., *Research in the service of national purpose*. Proceedings of the Office of Naval Research Biennial Convocation. Washington, D. C., 1966.
57. Swain, D. C., The rise of a research empire: NIH, 1930 to 1950, *Science*, **138**: 1233-1237.
58. Rosen, G., Patterns of health research in the United States, 1900-1960. *Bull. Hist. Med.*, 1965, **39**: 201-221.
59. Brand, J. L., The National Mental Health Act of 1946: a retrospect. *Bull. Hist. Med.*, 1965, **39**: 231-245.
60. Dupree, A. H., The structure of government-university partnership after World War II. *Bull. Hist. Med.*, 1965, **39**:245-251.
61. Shannon, J. A., et al., Discussion; symposium on the federal government and health research, 1900-1960. *Bull. Hist. Med.*, 1965, **39**: 252-260.
62. Goodman, L. S., and Gilman, A., *The pharmacological basis of therapeutics*. New York, 1955.
63. Fleming, A., History and development of penicillin, in *Penicillin: Its Practical Applications*. Philadelphia, 1946.
64. Florey, H. W., Historical introduction, in *Antibiotics*, Vol. 1. New York, 1949.
65. Chain, E. B., The development of bacterial chemotherapy. *Antibiot. Chemother.*, 1954, **4**: 215-241.
66. Kidd, C. V., *American universities and federal research*. Cambridge, Mass., 1959.
67. ———, The expansion of research support. *J. Med. Educ.*, 1962, **37**: 95-100.
68. Shannon, J. A., Advancement of medical research: a twenty-year view of the role of the National Institutes of Health. *J. Med. Educ.*, 1967, **42**: 97-108.
69. Weaver, W., *U.S. philanthropic foundations. Their history, structure, management and record*. New York, 1967.
70. Turner, T. B., The medical schools twenty years afterwards: impact of the

research programs of the National Institutes of Health. *J. Med. Educ.*, 1967, **42**: 109-118.
71. *A study of twenty medical schools. Report to the Director, National Institutes of Health.* Washington, D. C., 1959.
72. SINCLAIR, D. C., Basic science: some observations in the United States. *Lancet*, 1953, **1**: 463-467.
73. STOWE, L. M., The Stanford plan: an educational continuum for medicine. *J. Med. Educ.*, 1959, **34**: 1059-1069.
74. SOLOMON, D. H., Cutting some chains: a new medical curriculum at UCLA. in Press, 1968.
75. BERRY, G. P., Medical education in transition. *J. Med. Educ.*, 1953, **28**: 17-42.
76. LOEB, R. F., Values in undergraduate medical education. *Trans. Ass. Am. Physicians*, 1955, **68**: 1-5.
77. MERTON, R. K., *Social theory and social structure.* Glencoe, Illinois, 1957.
78. LITTLEMEYER, M. H., Annotated bibliography on current changes in medical education. *J. Med. Educ.*, 1968, **43**: 14-28.
79. LEE, P. V., *Medical schools and the changing times. Nine case reports on experimentation in medical education.* Evanston, Illinois, 1962.
80. WEARN, J. T., et al., Reports on experiments in medical education. *J. Med. Educ.*, 1956, **31**: 515-565.
81. HAM, T. H., Medical education at Western Reserve University. *New Eng. J. Med.*, 1962, **267**: 868-874.
82. ROUECHÉ, B., Placebo. *New Yorker*, 15 October 1960.
83. LASAGNA, L., Placebos. *Sci. Amer.*, 1955, **193**: 68-71.
84. Medical education in the United States: 1964-1965. *J. Am. Med. Ass.*, 1965, **194**: 731-823.
85. *Report of the Educational Policy and Curriculum Committee UCLA School of Medicine, 1964-1965.* Unpublished.
86. Medical education in the United States: 1965-1966. *J. Am. Med. Ass.*, 1966, **198**: 847-944.
87. Medical education in the United States: 1966-1967. *J. Am. Med. Ass.*, 1967, **202**: 725-832.
88. POPPER, H., ed., *Trends in new medical schools.* New York, 1967.
89. FUNKENSTEIN, D. H., The changing pool of medical school applicants. *Bull. Am. Coll. Physicians*, 1967, **8**: 376-382.
90. COGGESHALL, L. T., *Planning for medical progress through education.* Evanston, Illinois, 1965.
91. ATCHLEY, D., Changing patterns of medical care and support. *J. Med. Educ.*, 1966, **41**: 325-331.
92. *The graduate education of physicians.* (Report of the Citizens' Commission on Graduate Medical Education, J. S. Millis, Chairman). Evanston, Illinois, 1966.
93. *Meeting the challenge of family practice.* (Report of the Ad-Hoc Committee on Education for Family Practice of the Council on Medical Education, W. R. Willard, Chairman). Chicago, 1966.
94. JACOBSON, E. D., Revolution in the medical curriculum. *J. Med. Educ.*, 1967, **42**: 1081-1086.
95. FUNKENSTEIN, D. H., Implications of the rapid social changes in universities

and medical schools for the education of future physicians. *J. Med. Educ.*, 1968, **43**: 433-454.
96. MELLINKOFF, S. M., Miracles in medicine and problems in pedagogy. *J. Am. Med. Ass.*, 1966, **198**: 629-636.
97. BRESLOW, L., Changing patterns of medical care and support. *J. Med. Educ.*, 1966, **41**: 318-324.
98. *Physicians for a growing America* (Report of the Surgeon-General's Consultant Group on Medical Education, F. Bane, Chairman). Public Health Service Pub. No. 709, October, 1959.
99. *The advancement of medical research and education through the Department of Health, Education and Welfare* (Final Report of the Secretary's Consultants on Medical Research and Education, S. Bayne-Jones, Chairman). U. S. Dept. H.E.W., June, 1958.
100. *Federal support of medical research.* (Report of the Committee of Consultants to the Subcommittee on Departments of Labor and Health, Education and Welfare of the Committee on Appropriations, U. S. Senate, 86th Congress, Second Session, B. Jones, Chairman). May, 1960.
101. Proposals for the support of medical education by the federal government. (Communication from the Executive Director of the Association of American Medical Colleges). *J. Med. Educ.*, 1961, **36**: 730-736.
102. SMYTHE, C. M., Developing medical schools: an interim report. *J. Med. Educ.*, 1967, **42**: 991-1004.
103. FEIN, R., *The doctor shortage.* Washington, D.C., 1967.
104. H. VAN A. HYDE, ed., Manpower for the world's health, *J. Med. Educ.*, 1966, **41**: 1-344.
105. BERNARD, C., *An introduction to the study of experimental medicine*, transl. by H. C. GREENE. New York, 1927.
106. BLOOMFIELD, A. L., *A bibliography of international medicine*, Chicago, 1958.
107. DUBOS, R. J., *Louis Pasteur—free lance of science.* Boston, 1950.
108. PUSCHMANN, T., *A history of medical education from the most remote to the most recent times*, transl. by E. H. HARE. London, 1891.
109. WHITEHEAD, A. N., *The aims of education.* New York, 1929.
110. MARK TWAIN (Samuel Clemens), *Life on the Mississippi.* New York, 1874.
111. LEAF, A., Government, medical research and education. *Science*, 1968, **159**: 604-607.
112. FLEXNER, A., *Universities, American, English, German.* New York, 1930.

INDEX

A

ʿAbd Allāh al-Shaykd al-Sadīd, 56
ʿAbd al-Raḥīm al-Dakwār, 56
ʿAbd al-Raḥmān b. Muh. Futays, 42
Abū al-ʿAlāʾibn Zuhr (Avenzoar), 59-60, 76, 83
Abū ʿAli al Husayn Ibn Sīnā (see Ibn Sīnā)
Abū Bakr Muh. b. Zarrarīyā al Rāzī (see al-Rāzī)
Abū al-Faraz ibn al-Quff, 57
Abū al-Ḥasan Ḥubaysh b. al ʿAsam, 45
Abū al-Ḥasan Saʿīd al-Nīlī, 48
Abū Māhir Mūsā b. Sayyār, 57
Abū Naṣr Saʿid b. Abī al-Khayr, 48
Abū al-Rayḥan al-Bīrūnī, 54
Abū Naṣr Saʿid b. Abī al-Khayr, 48
Abū al-Qāsim al-Zahrāwī (Abulcasis) see al-Zahrāwī
Abū Zayd Ḥunayn b. Isḥāq al-ʿIbādī (Ḥunayn, Johannitius) see Ḥunayn
Achrelius, Eric, 290
Acland, Henry, 242
Acrel, Olof af, 268, 270, 271, 274
Adams, John, 470
Anderson, Elizabeth Garret, 247
ʿAdud al-Dawlah, King, 40
Agüero, B. Hilgado de, 428
Aguiar, T. Ferreira de, 430
Aguilar y Molina, S. de, 439
Aḥmad al Harrānī, 57
Aḥmad ibn al-Jazzār, 55
Aḥmad, son of Yūnis al-Ḥarrānī, 57
ʿAlāʾ al-Dīn ibn-Nafīs (see Ibn al-Nafīs)
Albertini, Ippolito Francesco, 114
Albornoz, D. Gil de, 447
Alcmeon of Crotona, 4-5, 11
Alcuin, 73
Alderotti, Taddeo, 81, 83
Alexander II of Russia, 315-316, 405
Alexander III of Russia, 320
Alexander V, Pope, 125
Alexander VI, Pope, 419, 420, 421

Alfonso II of Aragon, 420
Alfonso V of Aragon, 420
Alfonso IX of Leon, 419
Alfonso X of Leon, 419
ʿAli b. ʿAbbas al-Majūsī (Haly Abbas), 46
ʿAli b. ʿIsā, 45
ʿAli b. Riḍwān al-Miṣrī, 47
ʿAli b. Sahl Rabbān al-Ṭabarī (al-Ṭabarī), 45
ʿAli b. Yaḥyā al-Munajjim, 42
All-India Institute of Medical Sciences, 360-361
Allyn, H. D., 490-491
Almeida, J. F. de, 449
Almeida, Luis d', 394
Altenstein, Karl Freiherr Stein zum, 182, 184
American Medical Association (AMA), 485-486, 507, 521
AMA Council for Medical Education, 507-509
American Medical Association of Vienna, 228
American Medical College Association (later Association of American Medical Colleges, AAMC), 489, 490, 491, 511, 519
American Physiological Society, 504
Anatomical teaching in classical antiquity, 11-13, 14, 21; Denmark, 285-287, 288; England, 236, 238, 239, 241; France, 131, 136-137, 140, 141, 144, 146, 147, 158-160; India, Vedic period, 344-345; Japan, 397-399, 410; medieval, 80, 84; microscopic, 106-107; Netherlands, 201, 202-203; Pavia, 108-112; Portugal, 429; Russia, 307; Scotland, 256; sixteenth century, 94, 96-101; Spain, 427; Sweden, 266, 269; United States, 466-467, 484
Anatomical wax models, 115-117
Angeles, Fr. Felipe de los, 439
Annan, Samuel, 480
Aoki Kenzo, 407
Aoki Konyō, 397-398

Aparicio, J. Melquiades, 437
Apollo, 3
Apollonides, 10
Apollonios of Cition, 13, 15
Apothecaries, England, 236, 237, 240; Act of 1815, 240, 241, 259
Archibios, 15, 25-26
Arculano of Verona, Giovanni d', 85
Aretaeus, 48
Argentina, Medical education in, 441-442
Argerich, Cosme M., 441
Argerich, Francisco C., 441-442
Arib b. Sa<īd al-Qurtubī, 55
Aristogenes of Cnidos, 13
Aristophanes, 6
Aristotle, 5, 6, 10, 12, 14, 23, 25, 27, 91, 285
Arnisaeus, Henning, 286
Articella, 77-79, 93
Arzānī, Akbar, 355
Asclepiads, 3-4
Asclepius, 3, 24
Aselli, Gaspare, 106
Association of American Medical Colleges, *see* American Medical College Association
Athenaios of Attaleia, 20
Ātreya, 336
Auenbrugger, Leopold, 221, 225
Augustine, 73
Augustus, Emperor, 18
Aurelius-Aesculapius, Corpus of, 75
Avenzoar, *see* Abū al<Alā> ibn Zuhr
Avicenna, *see* Ibn Sīnā

B

Bäck, Abraham, 268, 269
Badr al-Dīn b. al-Qāḍī <Abd al-Raḥmān, 41
Baer, K. E. von, 313
Baker, Albert R., 491-492
Bakhtīshu<, son of Jūrjis b. Bakhtīshu, 43
Balfour, Andrew, 246
Bamberger, Heinrich von, 227
Bang, Oluf Lundt, 288
Barber-Surgeons' Company, London, 100, 239
Barber-Surgeons, Sweden, 263, 265-266
Bard, John, 467

Barrios, J. Arrango, 437
Barssuk-Moysseiev, Foma J., 311
Bartholin, Caspar, the elder, 285, 286
Bartholin, Caspar, the younger, 286
Bartholin, Thomas, 286
Bartolache, J. I., 432
Barziza of Padua, Cristoforo, 85
Basel, printing at, 92
Bauduin, Antonius F., 411
Baveriis of Bologna, Baverio de, 85
Beaton, John, 251
Beauperthuy, L. D., 438
Benedict XIII, Pope, 419
Benedict XIV, Pope (Prospero Lambertini), 115-116
Benedetti, Alessandro, 99
Benivieni, Antonio, 84
Bentham, Jeremy, 243
Bentinck, Lord, 357
Benzi, Ugo, 83
Berengar of Tours, 76
Berengario da Carpi, Giacomo, 97, 98, 101
Bergius, Peter Jonas, 269
Berlin, University of, 178-179
Bernard, Claude, 157, 159, 288
Bernoulli, Daniel, 306
Berry, G. P., 519
Berzelius, Jons Jacob, 273
Bevan, A. D., 507, 509
Bharadwāja, 335-336
Bhela, 335
Bhela, Sāleh bin, 352
Bichat, Marie François Xavier, 146, 148, 157, 166
Bidloo, Govard, 207, 209-210, 305
Bidloo, Nicolaus, 305-306
Billroth, Theodor, 173, 175-176, 182, 192, 218, 227, 502
Biochemistry, Netherlands, 211
Biology, medical, France, 158-159
Al-Bīrūnī, 60
Bischoff, Theodor Ludwig Wilhelm, 182
Black, Joseph, 255
Blaes, Ger, 397
Blignière, Barbier de, 136
Bloch, Oscar Thorvald, 289
Blumentrost, Laurentius, 308
Boë, Franciscus de le (Sylvius), 96, 206
Boeck, Carl Wilhelm, 296
Boeck, Christian Peter Bianco, 295

INDEX

Boerhaave, Herman, 96, 107, 203, 206-211, 223, 238, 257, 468
Bolivia, Medical education in, 440-441
Bologna, University of, 116
Bonafide, Francesco, 96
Bond, Thomas, 471
Bondt, Gerardt de, 202, 204
Bonghi, Ruggiero, 113
Boniface VIII, Pope, 124, 419
Bonner, Thomas Neville, 228
Bontempo, J. M., 440
Bordeu, Théophile de, 145
Borja, St. Francis of, 423
Boswell, James, 251
Botanical gardens, 96, 138, 202, 269, 291
Botany, France, 138, 141, 145; Netherlands, 202; Pavia, 109, 110, 111-112; Sweden, 269
Botkin, Sergei P., 318, 319
Bowditch, H. P., 504
Boxer, Charles R., 399
Boy, Nils, 267
Boylston, Zabdiel, 464, 468
Brahmā, 335
Bramley, M. J., 357-358
Brazil, Medical education in, 439-440, 448-449
Brest, School of naval medicine, 138
Bretton, Surgeon, 356-357
British Medical Association, BMA, 244
British Postgraduate Medical Federation, 247
Brizuela y Cordero, I. de, 433
Broc, Pierre Paul, 444
Bromelius, Magnus, 266
Brosse, Gui de la, 141
Bruguera, Juan, 447
Buchheim, Liselotte, 177
Buchheim, Rudolph, 313
Buchwald, Johannes de, 287, 289
Buddhism, 331-332
Budé, Guillaume, 141
Burton, William, 211
Bush, J. F., 314
Bush, Vannevar, 514-516

C

Cabanis, George, 147, 148
Cadwalader, Thomas, 467
Caelius Aurelianus, 74, 75, 78
Caius, John, 93, 101
Calama, Bishop J. Perez, 439
Caldwell, Charles, 480-481
Caldwell, William, 258
Calvo, J., 452
Campins y Ballester, Lorenzo, 437-438
Cancino, Vicente Román, 444
Cardano, Gerolamo, 106
Cartwright, Samuel A., 486
Casserio, Giulio, 98-99
Castro, J. Soares de, 439
Cavazzoni, Angelo Michele, 117
Celsus, 5, 14, 25, 26, 132
Central America, Medical education in, 446
Cervi, Giuseppe, 429
Chaparro, P. Manuel, 442-443
Chaptal, Jean, 151, 165
Charaka, 334-353, *passim*
Charlatans, England, 240; Greece, 9; India, Vedic period, 343
Charles II of Spain, 435, 436
Charles III of Spain, 427, 433
Charles IV of Spain, 431, 433, 436, 441
Charles V, Emperor (Charles I of Spain), 420, 422, 432
Charles IX of France, 137
Chartres, Theological school of, 76
Charukovski, Prohor, 314
Chauliac, Guy de, 97, 132, 429
Chemistry, France, 141, 145-146, 158-160; Pavia, 109, 111-112
Cheselden, William, 239, 467
Chesnecopherus, Johannes, 263
Chevreul, Eugène, 141
Chicoyneau, François, 143
Chile, Medical education in, 442-443
China, Academy of Medical Sciences, 374
Chinchón, Count de, 435
Chiron, son of Kronos, 3
Christensen, Anders, 285
Christian III of Denmark, 285
Christian V of Norway and Denmark, 293-294
Christina of Sweden, 264, 267
Chrysippus, 13
Church, influence of in France, 121-122
Cisneros, Cardinal F. Ximenez de, 421, 425
Classical Chinese medicine in modern times, 375

Classical Vietnamese medicine in modern times, 380-381
Clayton, Thomas, 238
Clearchus of Soloi, 12
Clement V, Pope, 420
Clement VI, Pope, 420
Clement VII, Pope, 422
Clement VIII, Pope, 443
Clement X, Pope, 423
Clinical teaching, England, 236, 237, 239, 245; Finland, 290; France, 122, 137-138, 158-162, 163-165; Germany, 186-189; Pavia, 109-110, 111-112; Russia, 305, 314, 320-321; sixteenth century, 95-96; Spain, 426; Sweden, 266-267, 268, 271-276; Vienna, 217-228. See also, Hospital teaching
Clot, Antoine B., 63
Coiter, Volcher, 99
Colden, Cadwallader, 468
Collado, L., 451
Collège de France, 141, See also France, Collège Royal
Colombia, Medical education in, 443-444
Colombo, Realdo, 98
Colwell, N. P., 509
Compton, K. T., 514
Comte, Auguste, 448
Congenis, William de, 80-81
Conant, J. B., 514
Conradi, Andreas Christian, 295
Constantine the African, 76, 91
Cop, Nicholas, 94
Copernicus, Nicolaus, 90
Corbeil, Gilles de, 78
Córdova y Torrebejano, F. J. de, 437
Coronel, Julian, 439
Corvisart, Jean-Nicholas, 147, 225
Coudray, Angélique le Boursier du, 138
Councilman, W. T., 507
Cross, James Conquest, 481
Cruikshank, William C., 239
Cruveilhier, J. B., 166
Cruz, Antonio da, 429
Cuba, Medical education in, 436-437
Cuéllar, Manuel, 440
Cullen, William, 255
Cuneo, Gabriele, 106

D

Dahrendorf, Ralf, 193

Daste, Bernard, 444
Dāwūd al-Antākī, 63
Daza Chacón, D., 428
Dekkers, Frederik, 210
Democedes of Crotona, 4
Democritus, 11
Denmark, Collegium Medicum, 288; Royal Academy of Surgery, 287, 288, 289, 294
Desault, Pierre-Joseph, 137, 144, 146, 147
Desnoues, Guillaume, 117
Dexippos, 10
Dhanwantari, 336
Días, Mendo, 419
al-Dīnawarī, 48
Diocles of Carystus, 12
Dionis, Pierre, 141
Dioscorides, 20, 48, 53, 61, 138
Disputations, medieval, 76-77, 82
Dissection, Classical antiquity, 11, 21; India, modern period, 358-359; India, Vedic period, 344-345; medieval, 84; Japan, 397-399, 410. See also Anatomical teaching
Divōdāsa, King, 336
Dodonaeus, Rembert, 202
Domec, Dominque, 439
Domitian, Emperor, 19
Donatus, 73
Douglass, William, 464, 467-469
Duchanoy, Claude-François, 137
Dubois, Antoine, 144, 146, 147
Dufau, P., 429
Duffy, John, 486
Dupuytren, Guillaume, 166
Durand, Maurice, 369
Du Verney, Guichard-Joseph, 141

E

East India Company, effect on Indian medicine, 356
Ecuador, Medical education in, 438-439
Eder, Georg, 220
Edict of Marly, 1707, 123, 125, 126, 137
Edinburgh, Foreign students at, 258
Egeberg, Christian August, 296
Egerton, C. C., 358
Ekströmer, Carl Johan, 273-274
Eliot, Charles W., 501, 505
Eliot, John, 466
Emerich, Franz, 220, 222

Embryology, India, Vedic period, 345
Empedocles, 4
Empiricists, 14-15, 17, 21
Engel, Arthur, 280
England, College of Physicians, London, 237, 243; Dissenting academies, 243; Royal College of Surgeons, 240, 241-242, 243, 277 (*See also* Barber-Surgeons' Company, London); Royal Society, 238
Erasistrateans, 14, 17
Erasistratus, 16
Erasmus, Johann Friedrich, 310
Eric XIV of Sweden, 267
Erlanger, J., 512
Espaillat, José, 445
Esparragosa, Narciso, 436
Estienne, Charles, 100
Estrela, M. J., 439
Euler, L., 306
l'Externat, 165-166, 170

F

Fabre, A. E., 441
Fabrizi d'Acquapendente, Girolamo, 98, 99, 106
Faber, Knud Helge, 289
Fabricius, Jacob, 286
Falloppia, Gabrielle, 98, 99
Farfán, Pedro, 432
al-Fazārī, 352
Feldshers, Russian, 316
Fenger, Christian, 289
Ferdinand of Aragon, 421, 422
Ferdinand I of Austria, 220-222
Ferdinand VI of Spain, 429
Fernández, J. A., 441
Fernel, Jean, 101, 204
Ferrari of Pavia, Gianmatteo, 85
Fessel, Mme. Cadeau de, 435
Fichte, Johann Gottlieb, 178
Filomafitski, Alexei M., 314
Fincke, Thomas, 285
Finland, Clinical Institute, Helsinki, 292-293; Collegium Medicum, 291
Finlayson, George, 370
Fischer, Johann Bernhard, 307
Fleming, Alexander, 517
Flexner, Abraham, 173, 177, 188-189, 192, 245, 501, 506, 509, 522
Flexner, Simon, 509

Flexner Report, 245, 362, 508, 509, 510
Flores, José F., 436
Flores, Quirino, 436
Folsom, George, 406
Fontana, Felix, 117
Fontayne et Culemburg, L., 437
Forensic medicine, Pavia, 112
Forlì, Jacopo da, 83, 85
Folk medicine, Scotland, 251
Foster, Michael, 242
Fourcroy, Antoine de, 141, 145, 146, 147, 148, 150, 225
Fox, Dixon Ryan, 477
Fragoso, J., 428
France, Academy of Surgery, 143-144, 147, 149, 166, 167 (*See also* France, St. Côme, Fraternity of; France, College of Surgery); College of Hospital Physicians, 164; Colleges of Medicine, 141-145; Collège Royal, 122, 141 (*See also* Collège de France); College of Surgery, 143 (*See also* France, Academy of Surgery; France, St. Côme, Fraternity of); Jardin du Roi, 122, 138, 141; Royal Society of Medicine, 145, 165, 166; St. Côme, Fraternity of, 100, 138, 142-143
Francis I of France, 141, 142
Francis II of Austria, 226
Franck, Johannes (Franckenius), 263-264
Frandsen, Hans (Johannes Franciscus Ripensis), 285
Frank, Johann-Peter, 110, 225-227, 313
Frank, Joseph, 313
Franklin, Benjamin, 470-471
Frazier, C. H., 507
Frederick I of Norway and Denmark, 293
Frederick II, Emperor, 95
Frederick V of Denmark, 288
Freer, Robert, 255
French Revolution, changes following, 148-151, 164
Frerichs, Friedrich Theodor, 188
Frese, H., 309
Fries, Lorenz, 98
Frigimeliga, Francesco, 95
Frisius, Gemma, 90
Frois, Father Luis, 393-394
Frugardi, Roger, 80

Fuchs, Leonhart, 94, 98, 101
Fuente, Juan de la, 432
Fukuzawa Yukichi, 408

G

Gadolin, Johan, 291
Galen, 3, 5, 7, 8, 9, 17, 21-29, 44, 45, 46, 48, 49, 58, 63, 74, 77, 79, 85, 91-95, 97, 132, 138, 236, 264, 285, 426
Galenism, 91, 92, 94, 97, 100-101, 124
Galilei, Galileo, 105, 106
Galli, Giovanni Antonio, 115-116
Galvani, Luigi, 113, 116
Gamez, J., 430
Garbo, Bono del, 81
Gardie, Magnus Gabriel de la, 265
Gariopontus, 76
Gattinaria of Milan, Marco, 85
Gayraud, Esteban, 439
General Medical Council, England, 243, 244, 361
Gerard of Cremona, 62, 91
Gerrish, Frederick Henry, 491
al-Ghāfiqī, 61
Ghini, Luca, 96
Gilbertus, Anglicus, 78, 83
Gilman, Daniel C., 501
Gimbernat, A., 430
Gladebach, Marcus, 303
Godos, P. López de los, 435
Goercke, Johann, 181
Gondoin, Jacques, 143
Gónzalez del Alamo, F., 437
Gónzalez, J. E., 434
Goodenough Committee, England, 247, 362
Goodeve, H. H., 358
Gourmelen, E., 132
Gouveia, Andre de, 452
Grazioli, Gerolamo, 106
Grundman, Herbert, 219
Guatemala, Medical education in, 436
Guevara, A. Rodríguez de, 429, 452
Guillotin, Joseph-Ignace, 148
Guinter of Andernach, Joannes, 93-94, 97, 98
Guldberg, Gustav Adolph, 294
Gupta, Pandit Madhusūdan, 357, 358
Gustav Adolph II of Sweden, 264, 267

H

Haartman, Gabriel Erik, 291
Haartman, Johan Johansson, 291
Hall, V. E., 512
Haen, Anton de, 223-224, 227, 228
al-Ḥakam II, Caliph, 57
al-Ḥākim, Fatimid, 55
Haldane, Lord, 245
Haldane Commission, 245
Hales, Stephen, 238
Haller, Albrecht von, 211
Hamilton, Alexander, 468-469
Hammond, William, 487
Hare, David, 358
Harper, William R., 505
Harvey, William, 99
Haviland, John, 242
Heberden, William, 238
Hedin, Sven Anders, 271
Heiberg, Christen, 296
Heiberg, Jacob Munch, 294
Heidenhain, Rudolph, 190
Heister, Lorenz, 398
Helmholtz, Hermann von, 503, 504
Henry IV of France, 121, 138
Heracleides of Taras, 18
Heredia, Cayetano, 435-436
Hermann, L., 502
Hernandez, José Gregorio, 438
Hernandez, José Joaquín, 438
Herodotus, 10
Herophileans, 14, 17
Herophilus of Chalcedon, 13, 14, 16
Heuermann, Georg, 287
Heurne, Jan van, 96, 204
Heurne, Otto van, 96, 204-205
Hewson, William, 239
Hiärne, Urban, 269
Hibat Allāh Ismaʿīl ibn Jumayʿ, 47
Hibat Allāh ibn al-Tilmīdh, 56
Hindu civilization, 329-331, 333-334; education, 330-331, 333-334
Hinduism, 329-331
Hippocrates, 3, 5, 7, 10, 22, 27, 46, 48, 56, 58, 77, 79, 85, 91-95, 132, 133, 138, 204, 236
Hippocratic Corpus, 5, 7, 8, 19, 48, 50
Hippocratic Oath, 7, 51, 52, 338
Hjelt, Otto Edvard August, 293
Hoang Cao Khai, 371
Hoffman, Theodore, 413
Hoffvenius, Petrus, 264, 267, 269, 290
Holmes, Andrew F., 258
Holst, Axel, 295
Holst, Frederick, 294
Homer, 3, 4

Hoorn, John von, 269
Hospitallers, 222
Hospitals, Åbo, Country Hospital, 291; Aix, Hospital of St. Jacques, 137; Angers, Hospital of St. Jean, 137; Avignon, Hospital of Sainte Marthe, 137; Baghdad, <Adudī, 40, 55; Muqtadirī, 56; Baltimore, Johns Hopkins, 490; Barcelona, Hospital of the Holy Cross, 422; Bologna, Santa Maria della Morte, 115; Santa Maria della Vita, 115; Cairo, Mansuri, 41, 55; Cambridge, England, Addenbrooke's Hospital, 239; Chicago, Cook County Hospital, 289; Copenhagen, King Frederick Hospital, 288, 289; Damascus, Nūrī, 40-41, 55; Edinburgh, Royal Infirmary, 259; Glasgow, Royal Infirmary, 255; Gothenburg, Sahlgren Hospital, 280; Guatemala, General Hospital, 436; Kronstadt, 306; Leyden, Caecilia Hospital, 205, 212; Lisbon, All Saints Hospital, 394, 428-429; St. Joseph's Hospital, 430; London, Bethlem, 237; Guy's Hospital, 239; Hammersmith Hospital, 246-247; The London, 239; Middlesex Hospital, 239; Royal Free Hospital, 247; St. Bartholomew's Hospital, 237; St. George's Hospital, 239; St. Thomas's Hospital, 237, 238; University College Hospital, 243; Westminster Hospital, 239; Lund, Munck House, 279; Madras, General Hospital, 356; Madrid, General Hospital, 429, 430; Malmö, 279-280; México City, Royal Hospital for the Indians, 432; Montpellier, Hospital of the Holy Spirit, 80-81; Hospital of St. Eloi, 137; Moscow, 304; Nagasaki, Yōjōsho Hospital, 410-411; Nantes, Hôspital de la Charité, 138; Norway, National Hospital (Rikshospitalet), 295, 296; Oxford, Radcliffe Infirmary, 239; Padua, Hospital of San Francesco, 96, 220; Paris, Bicêtre, 144, 162; Charité, 132, 137, 144, 147, 149, 161; Hospice de la Vieillesse, 144; Hôtel-Dieu, 132, 137, 138, 144, 146-147, 149, 161, 164, 165; Necker, 162; Notre Dame de la Pitié, 144, 161; Sainte Anne, 162; St. Louis, 162; Salpêtrière, 161, 162; Pavia, Hospital of San Matteo, 109-110, 111; Petersburg, 306, Philadelphia, Pennsylvania Hospital, 471-472; Rome, San Spirito, 107; Salamanca, St. Mary the White, 426; Santiago de Chile, San Juan de Dios, 443; Stockholm, Danvik Hospital, 267; Garrison Hospital, 272; Johannis Cottage Hospital, 267; Maternity Hospital, 269, 270; Orphan Asylum, 267, 273; Princess Louisa Children's Hospital, 282; Serafimer, 267-271, 273, 274, 276, 277, 283; Vienna, Civic Hospital, 220, 223, 224; General Hospital, 222, 225, 226; Holy Trinity Hospital, 221, 224, 225

Hospital teaching, France, 122, 146-147, 161-162, 163-165; Netherlands, 204-206, 209; Spain, 420-421; United States, 484-485

Hotton, Petrus, 209
Howell, W. H., 504
Ḥubaysh b. al-Hasan, 46
Ḥubaysh b. Ibr. al-Tiflīsī, 47, 48
Hufeland, Christian Wilhelm, 408
Hufeland, Friedrich, 175, 191
Hugh of Lucca, 81
Humboldt, Wilhelm von, 174, 176, 178-181, 184, 185, 189, 191
Hume, David, 470
Humphry, George, 242
Hunayn, 43, 44, 45-48, 51, 55, 56, 57, 59, 62, 77, 91
Hundt, Magnus, 98
Hunter, John, 239, 430, 475
Hunter, William, 239, 255, 256, 347, 475
Huss, Magnus, 274-275, 288
Hwasser, Israel, 273, 292
"Hypertai", 8-9

I

Iatromathematicians, 22-23
Iberia, Medical education in, 419-431
Ibn Abd Rabbih, 61
Ibn al-Baytar, 48, 61
Ibn Dhan, 352
Ibn Jubayr, 40-41
Ibn Juljul, 48
Ibn Jumay, 59
Ibn Masawayh, 62
Ibn al-Nafis, 41, 56, 62-63
Ibn al-Quff, 62
Ibn Ridwan, 54-55
Ibn Rushd (Averroes), 60

Ibn Sa id, 39
Ibn Sina (Avicenna), 42, 47, 54, 56, 58-60, 79, 80, 85, 91, 94, 95, 133, 218, 236, 264, 285, 424, 426, 427
Ibn al-Sūrī, 48
Ibn al-Tilmīdh, 48
Ibn Usaybi<ah, 41
Ibn Wāṣif, 57
Iceland, Medical education in, 297
al-Idrīsī, 48
Ilmoni, Immanuel, 292
India, Age of Dharmashāstrās, 332; Ayurvedic medicine, 335-352, *passim*, 356; Puranic age, 332; Upanishadic age, 332; Vedic medicine, 332, 334-335
Indian Association for the Advancement of Medical Education, 363
Indian medicine under Arab rule, 352-355
Innes, John, 257
Innocent XI, Pope, 436
Innocent XIII, Pope 437
Inozemtsev, Fedor J., 315
l'Internat, 165-166, 170
Isaac Judaeus, 91
Isabella of Castille, 421, 422
Isḥāq, son of Ibn al-Tilmīdh, 48
Isḥāq b. Sulaymān, 55
Isḥāq al-Ruhāwī, 52
Islam, decline of medical practice, 57-60; factors influencing acquisition of medical knowledge, 39-40; growth of hospitals, 40-41; growth of libraries, 41-42; medical education, tenth to twelfth centuries, 49-55; modern medical education, 63-64; translations, 42-49; types of medical education, 55-57
Istifān b. Basīl, 48
Italy, Accademia degli Inquieti, 113-114; Accademia dell'Istituto, 114-116; Istituto delle Scienze, 114
Itō Genboku, 405, 407
Itsing, 333
Iwasaki Kanyen, 405
<Izz, son of Yūnis al-Harrānī, 57

J

Jacobus, James, 203
Jamāl al-Dīn al Qifti, 42
Jamieson, James, 356

James II of Aragon, 420
Japan, American medical system in, 414; Chinese medicine in, 392, 396; Dutch influence in, 395-401, 403, 405-412; Franciscan influence in, 394; German medical system in, 412-414; Institute for Western Medicine (East College), 412; Jesuit influence in, 391-394; modern medicine in, 413-414; national reactions against Western teaching, 394-395, 404, 405, 406; Western medicine in, 391-392, 395-396; 397-414
Jauregg, Wagner von, 227
Jefferson, Thomas, 479
Jenner, Edward, 254
Jibrā> il, grandson of Jūrjīs b. Bakhtīshu<, 43
Jīvaka, 341-342
Johannitius, *see* Ḥunayn
Johan III of Sweden, 263, 267
John III, Pope, 420
Johns Hopkins University, 490, 505 (*See also* under Hospitals, Medical Schools)
Johnson, Samuel, 251
Jones, Joseph, 487
Joseph II of Austria, 224-225
Jowett, F. B., 514
Julius II, Pope, 421, 422
Julius III, Pope, 422, 423
Jumelin, Jean-Baptiste, 137
Jundī-Shapūr, 40, 55-56
Jūrjīs b. Bakhtīshu<, 43
Jurkowski, Julio, 445

K

Kaempfer, Engelbert, 396, 400
Kaldah, Hārit Ben, 352
Kalm, Matthias, 292
Karl IX of Sweden, 267
Karolinska Institute, Stockholm, 270-276, 277-279, 281, 283, 284
Katsuragawa Hoshū, 400-401
Kaufmann, Franz Joseph, 224
Keen, W. W., 487
Kerfbyle, Johnnes, 467
Kerstens, Johann Christian, 310
al-Kindī, 45, 49
Kirstenius, Petrus, 264
Knapp, M. L., 481
Knox, John, 254

INDEX

al-Kohen b. al-ʿAṭṭār, 61
Kosuki Genteki, 398
Kratzenstein, Christian Gottlieb, 287
Krumbhaar, Edward B., 466, 467
Kulmus, Johann, 398
Kwan A-to, 368
Kwan Roku, 392

L

Laennec, R. T. H., 137, 157, 288
Laguna, A., 451
Lambert of St. Omer, 74
Lancisi, Giovanni Maria, 107
Lanfrank of Milan, 81
Langerhans, Paul, 193
Lan-ông, 370
Laon, Theological school of, 76
Lapeyronie, François de, 115, 143
Larrey, Dominique-Jean, 139, 157
Lassone, François de, 145
Latin in medical education, Germany, 174-177; Ibero-America, 447; Russia, 307, 319, 322; Sweden, 276
Latin America, Jesuit influence in, 438, 439, 443; medical education in, 431-455
Lay medicine in Classical antiquity, 20
Lebedev, N. D., 314
Lee, P. V., 520-521
Leegaard, Christopher Blom, 295
Lelli, Ercole, 115-117
Lémery, Nicolas, 141, 146
Lemnius, Vilhelmus, 267
Lenz, Max, 179, 191
Leo X, Pope, 421
Leoniceno, Niccolò, 92-93
Leyden, Ernst von, 176
Leyden, University of, 201, 202 (See also Medical schools)
Linacre, Thomas, 93, 94
Linné, Carl von, (Linnaeus), 266, 399
Littlemeyer, M. H., 520
Locke, John, 169
Loeb, R. F., 519
Lomonosov, Michael V., 310
London Hospital Medical College, 240
London School of Hygiene and Tropical Medicine, 246
London, University of, 243 (See also Medical schools)
Longoburgo, Bruno, 81
Lopes, Duarte, 429

López, Solano F., 445
López, Vicente, 442
Louis II of Anjou, 125
Louis XIII of France, 143
Louis XV of France, 115, 126, 143, 144
Lowe, Peter, 252
Ludwig, Karl, 190
Lull, Raymond, 422
Luna, José, 436
Luther, Martin, 179
Lyons, printing at, 92

M

Machado, Rodrigo, 431
Machaon, 4
Mādhava, (Mādhavakara), 351
Maeno Ryotaku, 398-399
Magalhães, J. Lemos de, 440
Magalhaes, V. F. de, 449
Magnenus, Johannes Chrisostomus, 106
Maison Médicale of the Czar, 304
Major, Herman Wedel, 295
al-Majūsī (Haly Abbas), 51, 53, 55, 57, 62
Malmstein, Per Henrik, 275
Malthe, Alexander Ludvig Normann, 296
al-Maʾmūn, Caliph, 42
Mandry, O. Costa, 446
Manka (Mānikya), 352
al-Manṣūr, Caliph, 352
al-Manṣūr Qalāwūn, King, 41
Manton, Jo, 247
Manutius, Aldus, 93
Manzolini, Anna Morandi, 115, 117
Manzolini, Giovanni, 117
Malpighi, Marcello, 106, 107, 109, 113, 114
Manfredi, Eustachio, 113, 114
Maqrīzī, 42
Marat, Jean Paul, 254
Mareschal, Georges, 143
Marqués, J. J., 440
Maria Theresa of Austria, 220, 222
Marinus, 21
Marsilio de Sancta Sophia, 83, 85
Marsili, Antonio Felice, 113, 114
Marsili, Luigi Fernando, 114
Martin V, Pope, 419, 428
Martin, John, 440
Mather, Cotton, 464
Martin, H. N., 504
Martin, Martin, 251-252

Maslamah al-Majrītī, 56
Materia medica, France, 131 (*See also* Pharmacology); India, Vedic period, 347; Pavia, 109, 111-112
Mathematics in Classical medicine, 22-23
Matsumoto Ryōjun, 409-410, 411
Maty, Matthieu, 209
Mayne, Robert, 255
Mayo, William V., 482
Mazarem, J. da Rocha, 440
McClellan, George, 482-483
McClintock, James, 483
McCosh, J. 358
McGuire, Hunter, 487
Mead, Richard, 238, 239
Mechanism in medicine, Netherlands, 207-209
Meckel, Johann Friedrich, 181
Medical Act of 1858, Britain, 236, 243-244, 253, 254, 259
Medical apprentices, England, 237, 240; Greece, 7-9, 19; Spain, 420-421, 428; United States, 474
Medical botany, sixteenth century, 96
Medical Council, India, 361-362
Medical curriculum, Germany, 180-183; India, Vedic period, 343-344; Mexico, 432; Netherlands, 203, 204, 210-211; Pavia, 108-110, 111-112; Portugal, 426, 427; Russia, 304, 305, 309, 314, 320; Scotland, 253, 256; sixteenth century, 90-91, 93-95, 100; South Vietnam, 384; Spain, 424-425, 426, 427; Sweden, 264, 276-279; United States, 475, 484, 510-512
Medical ethics, Classical antiquity, 19; medieval, 75
Medical examinations, Denmark, 290; England, 240, 241, 242, 244, 245; France, 132-135, 160-161; Germany, 184-185, 191-193; India, modern period, 359-360; Vedic period, 341; medieval, 82-83; Netherlands, 203; Portugal, 421; Russia, 306, 308, 321; sixteenth century, 92; Spain, 421, 447; Sweden, 270, 272, 273, 276-278
"Medical Institutions", Pavia, 109
Medical journals, France, 167
Medical libraries, France, 140, 167-168
Medical qualifications, France, 127, 153; Germany, 191-193; Russia, 321, 324; Scotland, 253, 254; Spain, 428; Sweden, 265, 277; United States, 473
Medical research, France, 158-159; Germany, 189-191; Vienna, 225
Medical schools, Aarhus, 289; Aberdeen, 254-255; Åbo (Turku), 290-292, 293; Aix-en-Provence, 125, 128, 137, 142; Aix-Marseilles, 152; Alcalá de Henares, 421, 423, 425-426, 427, 428, 447, 450; Alexander, 316-317; Alexandria, 12, 13, 24; Amiens, 142, 152; Amsterdam, 201, 206; Angers, 122, 123, 125, 127, 130, 133, 134, 135, 152; Astrakhan, 322; Avignon, 122, 124, 127, 128, 130, 131, 135, 138; Baghdad, 55; Bahia, 439, 440, 449; Barcelona, 420, 422, 423; Baroda, 362; Beirut, 63; Belem, 440; Belo Horizonte, 440; Bergen, 296; Berlin, 178-179, 189, 190, 191; Besançon, 125, 126, 128, 130, 131, 133, 134, 152; Bogotá, 443-444; Bologna, 79, 81, 90, 91, 96, 97, 106, 113-116, 238; Bombay, 360; Bordeaux, 125, 128, 142, 152, 155, 156, 157, 163; Bourges, 125, 128, 134; Brest, 152; Buenos Aires, 441-442; Caen, 122, 123, 125, 127, 130, 132, 133, 134, 136, 138, 152; Cahors, 124, 127, 128, 130, 135, 136; Cairo, 63; Calcutta, 356-359, 362; Cali, 444; Cambridge, 81, 83, 89, 90, 236-237, 238, 242, 243, 245, 248; Canton, 368, Caracas, 437; Cartagena, 444; Castleton Medical Academy, 479; Cervera, 422, 423; Châlons, 142; Chicago Medical College, 486, 487 (*See also* Lind University); Chihuahua, 434; Christiana (Oslo), 294-297; Clermond-Ferrand, 142, 152; Cnidos, 5, 10; Cochabama, 441; Coimbra, 419, 426, 427, 430, 450; College of Medicine, Baltimore, 479-480; College of Philadelphia, 472, 473, 474, 475, 478; College of Physicians & Surgeons of Columbia University, 486; College of Physicians & Surgeons of New York City, 479; Colorado, 521; Columbia College, 472; Concepción, 443; Copenhagen, 285-290; Córdoba, 442, 450-451; Cornell, 521; Cos, 5, 10; Cuenca, 439; Curitiba, 440; Damascus, 56; Dartmouth College, 472, 473; Delhi, 355;

INDEX

Dijon, 142, 152; Dniepropetrovski, 321; Dôle, 125, 126, 130, 131, 133, 134; Dorpat, 311, 313, 319; Douai, 122, 126, 127, 128, 129, 130, 131, 133, 134, 135; Edinburgh, 137, 239, 241, 255, 257-259, 363, 474-475; Edo, 401, 405, 407; Erivan, 322; Fortaleza, 440; Franeker, 202; Freiburg im Breisgau, 94, 98, 101, 173, 174, 177, 180, 182, 225; Gandía, 423; Glasgow, 253, 255-257; Glasgow, Anderson's College, 255; Gothenburg, 280; Göttingen, 189; Granada, 422; Grenoble, 125, 142, 152; Groningen, 202, 207; Guadalajara, 433; Guanajuato, 434; Guatemala, 436; Guayaquil, 439; Hanoï, 371, 378-379; Harderwijk, 202, 207; Havana, 436-437; Harvard, 472, 473, 488, 513, 514; Heidelberg, 94, 96, 98, 173, 174, 182; Helsinki (formerly Helsingfors), 292-293, 317, 319; Hue, 381; Huesca, 420; Ingolstadt, 94; Jefferson Medical College, Philadelphia, 482, 486; Johns Hopkins, 363, 490, 505-507, 510-511, 512, 521, 522; Juiz de Fora, 440; Kazan, 312, 316; Kharkov, 312, 316; Kiev, 316; King's College, New York, 472, 473, 475; Krasnodar, 322; Kronstadt, 306-309; Kyoto, 407, 413; La Paz, 440-441; La Plata, 442; La Rochelle, 142; Leipzig, 96, 177; Leningrad, 323; León, 446 Lerida, 420; Leyden, 96, 137, 202, 203, 204-211, 220, 223, 238, 239, 257; Lille, 142, 152, 156, 157, 162; Lima, 434-436; Limoges, 142, 152; Lind University, Chicago (See also Chicago Medical College), 487; Lisbon, 420, 427; London, 239-240, 243, 246, 247; Louisville, 481; Louvain, 94, 126, 203; Lucknow, 355; Lund, 265, 276, 279, 280; Lyons, 142, 152, 155, 156, 157, 162, 163; Maceió, 440; Madras, 359; Madrid, 423; Manchester, 241; Maracaibo, 438; Manizales, 444; Marburg, 263; Marseilles, 152, 156, 162, 163; Medellín, 444; Medical College of Virginia, 521; Michigan, 506; Mérida, 434; Mexico City, 432, 433, 434; Middlebury College, Vermont, 479; Milan, 108; Minsk, 322; Monterey, 434; Montevideo, 445; Montpellier, 80, 81, 83, 89, 90, 91, 94, 95, 96, 98, 99, 122, 123, 124, 128, 129, 130, 132, 133, 134, 135, 137, 138, 143, 148-149, 151, 152, 155, 156, 157, 218; Montreal, 258; Morelia, 434; Moscow, 304-311, 314, 315, 318-319; Moulins, 142; Munich, 190; Nagasaki, 403, 409; Nancy, 122, 126, 127, 128, 138, 142, 152, 156, 157, 163; Nantes, 122, 123, 125, 128, 130, 132, 133, 136, 138, 152; Neelamegnan, 356; Newcastle, 241, 242; New Orleans School of Medicine, 486; Nice, 152; Nîmes, 142; North Carolina, 521; Northwestern University (See also Chicago Medical College, Lind University), 487-488, 521; Oaxaca, 434; Odessa, 316; Omsk, 322; Oñate, 422; Oporto, 427; Orange, 122, 125, 127, 128; Orihuela, 423; Orléans, 122, 142; Osaka, 401, 407, 408; Oslo (See also Christiana), 296; Osuna, 422; Oulu, 293; Oxford, 81, 83, 89, 90, 99, 236-237, 238, 242-243, 245, 248; Pachuca, 434; Padua, 89, 90, 91, 94, 95-96, 97, 99, 106, 107, 204, 219, 220; Palma, 423; Pamplona, 422, Panama City, 446; Paris, 81, 89, 90, 92, 93, 94, 96, 98, 101, 122, 123-124, 127, 128, 130, 131, 132, 133, 134, 135, 136, 137, 138, 140, 143, 148-149, 151, 152, 155, 157, 162, 163, 166, 218; Pavia, 106, 108-112, 225; Peking, 368, 374; Pennsylvania College, 482-483; Perpignan, 125, 128, 134, 135; Pest, 225; Petersburg, 306-309, 313-314, 317-318, 320, 321; Philadelphia College of Medicine, 483; Phom-Penh, 387, 388; Pisa, 96; Poitiers, 125, 127, 128, 130, 131, 133, 134, 152; Pondichéry, 371; Pont-à-Mousson, 126, 127, 128, 130, 132, 133, 142; Popayán, 444; Prague, 225; Puebla, 433; Queen's College, New Brunswick, 472, 478; Quito, 438-439; Recife, 440; Rennes, 152; Reykjavik, 297; Rheims, 122, 126, 127, 128, 130, 133, 134, 136; Ribeirâo Preto, 440; Rio de Janeiro, 439-440, 449; Rome, 107; Rosario, 442; Rostov, 317; Rouen, 142, 152; Saīgon, 381-385; St. Andrews, 240, 254; Salerno, 77, 78; Salamanca, 419, 420, 421, 424, 427-428, 447, 450; San José, 446; San Luis

Potosí, 434; San Salvador, 446; Santiago de Chile, 442-443; Santiago de Compostela, 421; Santo Domingo, 431-432; São Paulo, 440; Seville, 421-422; Sigüenza, 422; Smolensk, 322; Stanford, 511, 522; Strasburg, 122, 126, 127, 128, 130, 133, 134, 137, 138, 148-149, 151, 152, 163; Sucre, 440-441; Syracuse, 506; Tampico, 434; Tashkent, 322; Tbilisi, 322; Tegucigalpa, 446; Temple, 521; Tien-Tsin, 368; Toledo, 422; Tomsk, 316; Toulouse, 122, 124, 125, 129, 130, 132, 133, 134, 138, 152, 162; Tours, 142, 152; Transylvania University, 472, 473, 480-481; Troyes, 142; Tucuman, 442; Tübingen, 94, 192-193; Turku (formerly Åbo), 293; U.C.L.A., 522; Ulm, 195; Umea, 280; University of the City of New York, 486; University of Iowa, 481; University of Maryland, 479-480; University of Missouri, 521; University of Pennsylvania, 472, 473, 475, 482, 486, 488, 506, 513; University of Rochester, 521; University of the State of Pennsylvania, 472; University of Virginia, 479; Uppsala, 263-265, 266, 276, 279, 280, 290; Utrecht, 202, 204; Valencia, 420, 422, 423; Valence, 125, 127, 134, 135, 142; Valladolid, 420, 421, 427; Valparaiso, 443; Veracruz, 434; Vermont Medical College, 479; Vienna, 94, 95, 98, 110, 137, 217-228; Voronezh, 322; Warsaw, 317; Washington College, Chestertown, 473; Western Reserve, 520, 522; William & Mary College, 472, 473; Willoughby Medical College, 482; Wilno (now Vilnius), 311-312; 313; Würzburg, 402; Yale College, 473; Zaragoza, 420, 423

Medical students, France, 129-130, 135-136, 156-158; Germany, 178, 180-183; Ibero-America, 446-451; India, Vedic period, 337-341; medieval, 75, 82-83; Russia, 319-320; sixteenth century, 90, 92, 98; South Vietnam, 382-383, 385; Sweden, 275-276, 277; Vienna, 218-219, 221, 227

Medical teachers, France, 128-129, 153-155, 158-159; Germany, 183-189; Greece, 6-7; medieval, 75; South Vietnam, 382-383

Medina, C. de, 451
Medrano y Herrera, 437
Meeker, Daniel, 481-482
Meerdervoort, Pompe van, 409-412
Mejía, José, 439
Merizalde, José Felix, 444
Merton, R. K., 519
Méry, Jean de, 147
Mesmer, 146, 313
Methodists, 17, 18, 21, 78
Mettrie, Julien de la, 126
Mexico, Medical education in, 432-434, 448, 451
Mianowski, J., 317
Michelena, Guillermo, 438
Middleton, Peter, 467
Midwives, France, 138, 163; Peru, 435; Russia, 308; South Vietnam, 386; Sweden, 269
Milan, University of, 105
Military medicine, France, 139, 163; Germany, 181; Russia, 317-318, 320; Spain, 430
Milne-Edwards, Henri, 141
Minot, C. S., 504
Mitchell, S. Weir, 487
Mohanjo-Daro, excavations at, 334
Mohr, Otto Louis, 294
Molina, J. F., 438
Molinelli, Pier Paolo, 115
Mondeville, Henri de, 75, 81
Mondino da Luzzi, 84, 92, 97, 106
Monravé, A. de, 429
Monte, Giambattista da, 95-96, 106, 204, 219, 220, 225
Monteaux, Nicolas Chambon, 137
Montúfar, M. de, 441
Moreno, Gabriel García, 439
Morgan, John, 258, 467, 474-475, 478
Morgagni, Giovanni Battista, 107, 113-114, 116-117
Morsing, Christern Thorckelsen, 285
Moultrie, John, 469
Mudrov, Matvei, Y., 314
Mueller, Leopold, 413
Muhadhdhab al-Dīn <Abd al-Rahīm b. <Alī (al-Dukhwār), 41
Muh. <Alī, 63
Muh. b. Ibrāhīm al-Ansārī (Ibn al-Akfānī), 45

INDEX

Muh. b. Jarīr al-Ṭabarī, 48
Mukhin, Efrem O., 315
al-Muqtadir, Caliph, 53, 56
al-Mukhtār b. Buṭlān (Ibn Buṭlān), 47, 54-55
Müller, Friedrich von, 188
Müller, Johannes, 190, 295
Mumford, James, 468
Munro, Alexander, I, 257
Museums, Copenhagen, 286; France, 141, 168; Italy, 110, 116, 117; Vienna, 117
al-Mustanṣir, Caliph, 55
al-Mu<taṣim, 52
Mutis, José Celestino, 444
Muwāffaq, Abū Mansūr, 353
Myrepsius, 138

N

Nakagawa Junan, 398, 400-401
al-Nadīm, 43, 49, 56
Nagel, W., 502
Najīb al-Dīn al-Samarqandi, 50
Nassau, William of, 125
Nauck, Ernst Theodor, 173, 177, 182, 218
Naval medicine, France, 138; Spain, 429-430
Nevin, Domingue, 442
Newman, Charles, 236, 259
Newton, Isaac, 238
Nguyen Huu, 382
Nguyen van Nha, Mrs., 382
Niccolò da Reggio, 92
Nicholas II of Russia, 292
Nicholas IV, Pope, 420
Nicolaysen, Julius, 296
Nizam al-Mulk, 55
Noro Genjō, 397
Numata Jiro, 411
Nūr al-Dīn Zinkī, King, 40, 55

O

Obraztsov, V. P., 319
Obstetric models, 115
Obstetrics, Denmark, 288-289; eighteenth century, 115; France, 131, 138, 146, 159-161; Pavia, 108, 111-112; Russia, 308; Sweden, 269, 270
Obstetrics and Gynecology, India, Vedic period, 348-349
Oca, Juan José Montes de, 441

Oddi, Oddi degli, 204
Ogata Kōan, 407, 408-409, 410
O'Gorman, Michael, 441
Oishi Ryohei, 407
Oppolzer, Johann, 227, 228
Orabasius, 48, 132
O'Shaughnessy, William B., 358
Osler, William, 166, 169-170, 504
Ostroumov, A. A., 319
Ōtsuki Gentaku, 401
Ovsyannikov, Filip V., 313
Owen, Griffith, 465

P

Padilla, J. M., 436
Paget, George, 242
Palafox y Mendoza, J. A. de, 432
Panum, Peter Ludwig, 288
Paraguay, Medical education in, 445
Pardo y Pardo, Juan M., 444
Paré, Ambroise, 139, 146
Parejo, B. Muñoz, 433
Paris, College of Pharmacy, Corporation of Apothecaries, 139
Pashutin, V. V., 319
Passamán, José Francisco, 440-441
Passavant, Jean de, 81
Pasteur, Louis, 159
Pathology, France, 131, 159-160; India, 345-346; Pavia, 111-112; Russia, 319
Paul of Aegina, 48, 132, 204
Paul III, Pope, 422, 438
Paul IV, Pope, 423, 432
Paul V, Pope, 443
Paulli, Simon, 286
Pauw, Pieter, 202
Pavia, University of, 105, 107-108, 112, 116
Pavlov, Ivan P., 318
Pearce, John, 466
Pediatrics, India, Vedic period, 349
Penn, William, 465
Peralta, D. Ossorio, 432
Perry, Matthew, 391, 406
Perú, Medical education in, 434-436
Peter of Abano, 83
Peter IV of Aragon, 420
Peter the Great of Russia, 304, 308
Petit, Marie-Antoine, 164
Pham Biêu Tâm, 382
Pharmacology, France, 145, 159-160
Pharmacy, France, 131, 139, 141; Islam,

60-61; Montpellier, 139
Philip the Good, Duke of Burgundy, 125
Philip II of Spain, 126, 201, 423, 426
Philip III of Spain, 426, 432, 438, 439, 443
Philip IV of Spain, 422, 431, 435, 444
Philip V of Spain, 422, 423, 437, 442
Philosophers, Greek, 4-5, 10
Philosophic medicine, Classical antiquity, 5, 16, 24-25; medieval period, 76-77, 83
Phoebus, Philip, 187, 190
Physician, Greek type of, 3-4, 8-9, 11, 13, 14, 18-19, 23-24, 27
Physick, Philip Syng, 258
Physiology, Denmark, 288; France, 131, 140, 146, 159-160; India, Vedic period, 345-346; Pavia, 111-112
Picano, José Correia, 439
Piermarini, Giuseppe, 110
Pinel, Philippe, 148
Pirogov, Nikolai, I., 313
Pitcairne, Archibald, 257
Platter, Felix, 91
Platter, Thomas, 91, 95
Plato, 4, 7, 8, 10, 14, 23, 27
Plummer, Andrew, 257
Pneumatists, 16, 17
Podaleirios 4
Political intervention in medical education, Ibero-America, 451-453; Russia, 312, 315, 316, 319, 322
Politzer, Adam, 228
Pollack, Leopold, 110
Polunin, A. J., 319
Pombal, marquis of, 427
Portal, Antoine, 167
Portela, Ireneo, 441
Posidonius, 16
Pott, Percival, 430
Pozzo, Francesco dal, 106
Practical medicine, medieval, 74, 83-85, 95
Premedical education, Germany, 174-178; Russia, 323
Private teaching, England, 239, 243; France, 145-146; Scotland, 255
Protospatharius, Theophilus (Philaretus), 85, 91
Ptolemaeus, 23

Ptolemy I, 12
Ptolemy VIII, 13
Puerto Rico, Medical education in, 445-446
Puschmann, Theodor, 169, 173, 176, 181, 192, 218
Pythagoras, 4

Q

Qantoorī, Ghulām Hussain, 355
Qiftī, 43
Quarin, Joseph, 226
Quarin, Peter, 222

R

Rabban-al-Tabarī, Abū Sahl Alī, 353
Rabelais, 94
Raicus, Johannes, 264
Ramée, Pierre de la, 137
Ramón y Cajal, S., 453
Ranzaburō, Ōtori, 401
Rashīd al-Dīn <Alī b. Khalīfah, 41
al-Rashīd, Hārūn, 352
Rayer, Dean, 162
Razetti, Luiz, 438
al-Rāzī, 40, 43, 44, 49-51, 55, 56, 57, 59, 76, 91, 285, 426
Récamier, Joseph, 164
Regelius, Olof, 290
Reil, Johann Christian, 179, 190, 191
Reinoso, Antonio, 452
Reinoso, R., 451
Reisch, Gregory, 98
Requirements for medical school, Germany, 174, 178; Russia, 305, 319-320, 322; sixteenth century, 90-91
Retzius, Anders, 273
Ricardus Anglicus, 78
Ricci, Matteo, 395
Rietz, Grégoire François du, 264
Rinder, Y., 309
Rios, José Antonio, 443
Rivas, M., 430
Robertson, William, 258
Rochefort, Desbois de, 137, 147
Roberg, Lars, 266
Rocha, A. de la, 435
Rockefeller Foundation, 454
Rondelet, Guillaume, 100
Rosas, Anton von, 221
Rosenstein, Nils Rosén von, 267, 282

INDEX

Rudbeck, Johannes Johannis, the elder, 263
Rudbeck, Olof, 264-265, 269
Rudbeck, Peter Johannis, the elder, 263
Rudnev, M. M., 319
Rudolphi, Karl Asmund, 191
Ruel, Jean, 94
Rufus of Ephesus, 20, 21
Ruiz, Sebastián López, 446
Runeberg, Johan Wilhelm, 293
Rush, Benjamin, 258, 470
Rusi, D., 432
Russia, Academy of Sciences, 306, 308; Chancellery of Apothecaries, 304; Hospital schools, 304-310
Rutherford, John, 257

S

Sābūr, 61
Sahlgren, Anton Niclas, 280
St. Clair, Alexander, 257
Salerno, 77, 78
Salgado, M. J., 432
Sālih ibn Sallūm 63
Salmawayh b. Binān, 52
Samoilovich, D., 308
Sandberg, Ole Rømer Aagaard, 295
Sandri, Giacomo, 114
Santa Cruz Espejo, F. J. E., 439
Santesson, Carl Gustav the elder, 275
Santucci, B., 429
Sareshal, Alfred, 83
Savonarola, Michael, 76, 85
Saxtorph, Mathias, 288
Sbaraglia, Giovanni Gerolamo, 106, 113
Scaliger, Joseph, 202
Scarpa, Antonio, 110
Schacht, Herman Oosterdijk, 210, 211
Schacht, Lucas, 206, 210
Schaefer, E. A., 502
Schein, Martin, 307
Schelling, Friedrich Wilhelm Joseph, 178, 313
Schelsky, Helmut, 178, 179, 180, 194
Schiller, Friedrich von, 178
Schleiermacher, Friedrich, 178
Schmidt, Joseph Hermann, 181, 182, 184
Schmiedeberg, Oswald, 313
Schoenlein, Johann Lucas, 187-188
Schools of Health, France, 148-150
Schreiner, Kristian Emil, 294
Schreiber, J. F., 307
Schulzenheim, David Schulz von, 269
Schützer, Salomon, 266
Scientific method in medicine, Germany, 96, 106-108, 113-115, 140, 179-180, 182, 184, 187-191, 194
Scotland Faculty of Physicians & Surgeons of Glasgow (now Royal College of Physicians & Surgeons of Glasgow), 252-253; Royal College of Physicians of Edinburgh, 257; Royal College of Surgeons of Edinburgh, 252
Screvenius, Ewald, 205
Scribonius Largus, 19
Seal, Motilal, 360
Sechenov, Ivan M., 318, 319
Semashko, Nikolai, 322
Senfelder, Leopold, 220
Sepúlveda, Hernando de, 434
Sewall, Samuel, 466
Shannon, J. A., 517-518
al-Sharīf, 48
Shchepin, Konstantin J., 308
Shige, Dennoshin, 404
Shippen, William, Sr., 467
Shippen, William, Jr., 258, 467, 474-475
Shryock, R. H., 501, 510
Sibbald, Robert, 257
Siddiqi, 45
Siebold, Philipp Franz Balthasar, 402-407
Sinān b. Thābit, 40, 53, 56
Sinclair, D. C., 518-519, 523
Sixtus IV, Pope, 420, 422
Sixtus V, Pope, 438
Sjöman, Sten Edvard, 292
Skeyne, Gilbert, 254
Skinner, Frederick, 445
Skjelderup, Michael, 294, 295
Sklifovsky, Nicolai V., 319
Slaughter, Henry, 467
Slaves as physicians in Classical antiquity, 4, 8-9, 18-19
Society of History of Medicine, France, 167
Société Royale de Médecine, France, 167
Solomon, D. H., 522
Soranus, 19
pseudo-Soranus, 74, 75, 78

Sørenssen, Nils Berner, 294, 295
Spain, Dominican influence in, 423; Jesuit influence in, 423, Moslem influence in, 424; the Protomedicato, 421, 428, 429, 430, 431; Royal College of Surgeons, 423, 429, 430
Spallanzani, Lazzaro, 107-108
Specialization, Classical antiquity, 6, 25-26; Finland, 293; France, 162-163; India, modern period, 362; Russia, 323-324; Sweden, 280-284; Vienna, 227-228
Spöring, Herman Diedrich, 290-291
Spranger, Eduard, 178, 181
Stainpeis, Martin, 219
Stansley, John, 466
Stefan, F., 309
Stensen, Niels (Nicolaus Steno), 116, 286-287
Stephenson, John, 258
Stewart, Ferdinand Campbell, 169
Stewart, William, 445
Stifft, Joseph Andreas, 226
Stoll, Maximilian, 221, 224, 225, 226
Stopius, Martin, 220
Straeten, Willem van der, 204
Strumpell, Adolph von, 183, 190
Stübler, Eberhard, 173, 182
Student revolts, Ibero-America, 450-451
Sue, Jean-Joseph, 144, 146
Sugita Gempaku, 398-399
Sulaymān b. Juljul, 53, 55
Suñer y Capdevila, Francisco, 445
Surgeons, Denmark, 287; England, 236, 237, 239; medieval, 80-85; Russia, 315, 319; sixteenth century, 97, 100; Sweden, 265, 269, 270, 271, 276
Surgery, Denmark, 287-288; France, 131, 132, 139, 141, 142-145, 146-147, 159-161; India, Vedic period, 344, 347-348; Islam, 61-63; Pavia, 108, 111-112; Portugal, 428-431; Russia, 307, 314, 319; Scotland, 252, 253; Spain, 428-431
Surgical teaching, Bologna, 81; Greece, 15; medieval, 80-83; Montpellier, 80; sixteenth century, 94, 97, 99-100
Susini, Clemente, 117
Sushruta, 335-353, *passim*
Sweden, Collegium Medicum, 264, 265, 266, 269, 270, 271; Royal Health Collegium, 271, 272, 273; Surgical Society, Stockholm, 265-266, 268, 269, 270
Swieten, Gerard van, 210, 217, 220, 222-223
Sylvius, Jacobus, 90, 93
Symcotts, John, 237

T

Tagault, Jean, 132
Takano Chōei, 405
Tamariz, Felipe, 438
Thābit b. Sinān, 57
Thauer, Rudolf, 183
Theatres, anatomical, 110, 124, 136, 203, 265, 286, 291, 430
Theodoric of Cervia, 81
Theophilus, 85
Theoretical medicine, medieval, 73-80, 83, 85; Pavia, 109, 111-112
Thessalus, 17, 22, 24
Thinh, Nguyên van, 372
Thomas, Flavel S., 490
Thompson, William Gilman, 194
Thouret, Augustin, 149
Thulstrup, Magnus Andreas, 294, 295, 296
Thunberg, Carl Pieter, 399-400
Tillandz, Elias, 290
Tissot, Simon-André, 110
Titsingh, Isaac, 401
Tocqueville, Alexis de, 476
Tokugawa Ieyasu, 394-395
Torre, M. García de la, 433
Torres, Archbishop Cristóbal de, 444
Totsukā Seikai, 405
Trafvenfelt, Erik Carl, 271
Traube, Ludwig, 190
Trier, Seligmann Meyer, 288
Trousseau, Armand, 158
Tsāng Hiuen (Yūan Chwāng), 333
Tuê, Ting, 370
Turner, T. B., 518
Tyler, John, 356-357

U

Uchermann, Vilhelm Krisian, 296
Umayyah b. Abī al Salt
U.S.A., Association of American Universities, 505
U.S.A., Bane Report, 523
U.S.A., Carnegie Foundation for the

Advancement of Teaching, 509
U.S.A., Commission on Medical Education, 511; Council on Medical Education, 521, 524
U.S.A., Edinburgh medical graduates in, 474-475
U.S.A., German influence in medicine (1875-1900), 489, 503-505
U.S.A., Internship and residency training, 512-513
U.S.A., Jones Committee, 523
U.S.A., Medical education before 1751, 463-470; 1751-1860, 470-486; 1861-1900, 486-492; 1901-1945, 506-514; 1945-1959, 514-523; 1960-1968, 523-526
U.S.A., National Institutes of Health, 516, 517-518; National Research Defense Council, 514; National Science Foundation, 516; Office of Scientific Research & Development, 514-516
Uranue, J. Hipólito, 435
Urban VIII, Pope, 422
Uruguay, Medical education in, 445
Usaybi<ah, 41, 51, 52-53, 57

V

Valignano, Alessandro, 394
Valle, F. Gonzáles del, 437
Valsalva, Antonio Maria, 114, 116-117
Valverde, J. de, 451
Vargas, José María, 438, 445
Vargas, Juan Bautista de, 444
Vasa, Gustav of Sweden, 263
Vasse, L., 451
Vaughan, V. C., 507
Vedās, 334-335
Vega, Juan de la, 435
Velásquez de Leon, J., 432
Vellanski, D. M., 313-314
Venezuela, Medical education in, 437-438
Veniaminov, P. D., 311
Venice, printing at, 92
Verbeck, Guido Fridolin, 412
Verde, R. Códoba, 438
Verle, Giambattista, 116-117
Vesalius, Andreas, 91, 95, 97-99, 101, 136, 399, 425
Vesling, Johann, 397
Vetter, Rudolph Aloys, 226
Vicq d'Azyr, Félix, 141, 145, 146, 148

Vieira, F. Sabino, 449
Vigevano, Guido de, 75
Vilar, Miguel, 452
Virchow, Rudolph, 176, 177, 178, 182, 190, 288, 502-503
Virgili, Pedro, 429-430
Vives, Andrés, 447
Vizcaíno, J. J. González, 445
Volta, Alessandro, 107-108
Voss, Joachim Andreas, 294

W

Wagner, R., 502
Waite, Frederick, 473
Wallich, N., 358
Walsh, James J., 467
Warner, Helen F., 490-491
Weber, Karl Otto, 184
Weitbreicht, J., 306
Welch, William H., 492, 501, 502-507, 509, 510, 513
Wickersheimer, Ernest, 218
Wilhelm II of Germany, 177
Willebrand, Knut Felix von, 293
William of Saliceto, 81, 82
William the Silent, 202
Wilson, Lambert, 466
Wilson, Woodrow, 505
Winge, Emanuel Fredrik Hagbarth, 295
Winslow, Jacques-Bénigne, 136, 141
Witherspoon, J. A., 507
Women medical students, China, 374; England, 247; India, 359, 360; Islam, 55; Russia, 320; South Vietnam, 381; Sweden, 277; United States, 485
Wonk Cheuk-hing, 368
Worm, Oluf, 286
Worm-Müller, Jacob, 295
Wyllie, James, 314

X

Xavier, St. Francis, 391-392
Ximeno, P., 451

Y

Yamawaki Tōyō, 397-398
al-Ya<qūbi, 48
Yersin, A., 371, 372, 378
Yoshimune Tokugawa, 396-397
Yūhanna b. Māsawayh (Ibn Māsawayh), 40, 43-44, 45

Yūsuf ibn-al-Maghribī, 56

Z

Zabolotny, D. K., 321
Zagorski, Peter A., 313
al-Zahrāwī (Abulcasis), 43, 48, 53-54, 55, 56, 59, 60, 61, 62, 132
Zakharin, Gregorii A., 318-319
Zapffe, F. C., 509
Zemstvos, 315-316
Ziemssen, H. W. von, 190
Zimmermann, Johann Georg, 212
Zumbo, Giulio Gaetano, 117
Zybelin, Semion G., 310-311